Student's Solutions Manual

to accompany

College Algebra

Second Edition

John W. Coburn

St. Louis Community College at Florissant Valley

Written by
Rosemary M. Karr, Ph.D.
Collin County Community College

 **Higher Education**

Boston   Burr Ridge, IL   Dubuque, IA   New York   San Francisco   St. Louis
Bangkok   Bogotá   Caracas   Kuala Lumpur   Lisbon   London   Madrid   Mexico City
Milan   Montreal   New Delhi   Santiago   Seoul   Singapore   Sydney   Taipei   Toronto

Higher Education

Student's Solutions Manual to accompany
COLLEGE ALGEBRA, SECOND EDITION
JOHN COBURN

Published by McGraw-Hill Higher Education, an imprint of The McGraw-Hill Companies, Inc., 1221 Avenue of the Americas, New York, NY 10020. Copyright © 2010 and 2007 by The McGraw-Hill Companies, Inc. All rights reserved.

This book is printed on recycled, acid-free paper containing 10% post consumer waste.

2 3 4 5 6 7 8 9 0 QWD/QWD 0 9

ISBN: 978-0-07-334907-7
MHID: 0-07-334907-0

www.mhhe.com

College Algebra Second Edition

College Algebra Second Edition

R.1 Exercises

1. Proper subset of; element of

3. Positive; negative; $7; -7$; principal

5. Order of operations requires multiplication before addition.

7. a. $\{1, 2, 3, 4, 5\}$
   b. $\{\ \}$

9. True

11. True

13. True

15. $\dfrac{4}{3} = 1.\overline{3}$

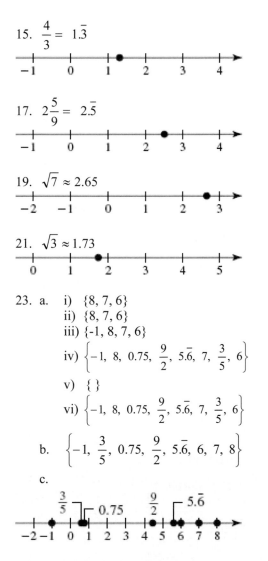

17. $2\dfrac{5}{9} = 2.\overline{5}$

19. $\sqrt{7} \approx 2.65$

21. $\sqrt{3} \approx 1.73$

23. a. i) $\{8, 7, 6\}$
   ii) $\{8, 7, 6\}$
   iii) $\{-1, 8, 7, 6\}$
   iv) $\left\{-1,\ 8,\ 0.75,\ \dfrac{9}{2},\ 5.\overline{6},\ 7,\ \dfrac{3}{5},\ 6\right\}$
   v) $\{\ \}$
   vi) $\left\{-1,\ 8,\ 0.75,\ \dfrac{9}{2},\ 5.\overline{6},\ 7,\ \dfrac{3}{5},\ 6\right\}$

   b. $\left\{-1,\ \dfrac{3}{5},\ 0.75,\ \dfrac{9}{2},\ 5.\overline{6},\ 6,\ 7,\ 8\right\}$

   c.

25. a. i) $\left\{\sqrt{49},\ 2,\ 6,\ 4\right\}$
   ii) $\left\{\sqrt{49},\ 2,\ 6,\ 0,\ 4\right\}$
   iii) $\left\{-5,\ \sqrt{49},\ 2,\ -3,\ 6,\ -1,\ 0,\ 4\right\}$
   iv) $\left\{-5,\ \sqrt{49},\ 2,\ -3,\ 6,\ -1,\ 0,\ 4\right\}$
   v) $\left\{\sqrt{3},\ \pi\right\}$
   vi) $\left\{-5,\ \sqrt{49},\ 2,\ -3,\ 6,\ -1,\ \sqrt{3},\ 0,\ 4,\ \pi\right\}$

   b. $\left\{-5,\ -3,\ -1,\ 0,\ \sqrt{3},\ 2,\ \pi,\ 4,\ 6,\ \sqrt{49}\right\}$

   c.

27. False; not all real numbers are irrational

29. False; not all rational numbers are integers.

31. False; $\sqrt{25} = 5$ is not irrational.

33. c; IV

35. a; VI

37. d; III

39. Let a represent Kylie's age: $a \geq 6$ years.

41. Let n represent the number of incorrect words: $n \leq 2$ incorrect words.

43. $|-2.75| = 2.75$

45. $-|-4| = -4$

47. $\left|\dfrac{1}{2}\right| = \dfrac{1}{2}$

49. $\left|-\dfrac{3}{4}\right| = \dfrac{3}{4}$

51. $|-7.5 - 2.5| = |-10| = 10$;
    $|2.5 - (-7.5)| = |10| = 10$

R.1 Exercises

53. -8 and 2

55. Negative

57. $-n$

59. Undefined; since $12 \div 0 = k$, implies $k \cdot 0 = 12$.

61. Undefined, since $7 \div 0 = k$, implies $k \cdot 0 = 7$.

63. a. Positive
    b. Negative
    c. Negative
    d. Negative

65. $-\sqrt{\dfrac{121}{36}} = -\dfrac{11}{6}$

67. $\sqrt[3]{-8} = -2$

69. $9^2 = 81$ is closest.

71. $-24 - (-31) = -24 + 31 = 7$

73. $7.045 - 9.23 = -2.185$

75. $4\dfrac{5}{6} + \left(-\dfrac{1}{2}\right) = 4\dfrac{5}{6} + \left(-\dfrac{3}{6}\right) = 4\dfrac{2}{6} = 4\dfrac{1}{3}$

77. $\left(-\dfrac{2}{3}\right)\left(3\dfrac{5}{8}\right) = \left(-\dfrac{2}{3}\right)\left(\dfrac{29}{8}\right) = -\dfrac{58}{24} = -\dfrac{29}{12}$
    or $-2\dfrac{5}{12}$

79. $(12)(-3)(0) = 0$

81. $-60 \div 12 = -5$

83. $\dfrac{4}{5} \div (-8) = \dfrac{4}{5} \cdot \left(-\dfrac{1}{8}\right) = -\dfrac{4}{40} = -\dfrac{1}{10}$

85. $-\dfrac{2}{3} \div \dfrac{16}{21} = -\dfrac{2}{3} \cdot \dfrac{21}{16} = -\dfrac{42}{48} = -\dfrac{7}{8}$

87. $12 - 10 \div 2 \times 5 + (-3)^2$
    $= 12 - 10 \div 2 \times 5 + 9$
    $= 12 - 5 \times 5 + 9$
    $= 12 - 25 + 9$
    $= -4$

89. $\sqrt{\dfrac{9}{16}} - \dfrac{3}{5} \cdot \left(\dfrac{5}{3}\right)^2$
    $= \dfrac{3}{4} - \dfrac{3}{5} \cdot \dfrac{25}{9}$
    $= \dfrac{3}{4} - \dfrac{75}{45}$
    $= \dfrac{3}{4} - \dfrac{5}{3}$
    $= \dfrac{9}{12} - \dfrac{20}{12}$
    $= -\dfrac{11}{12}$

91. $\dfrac{4(-7) - 6^2}{6 - \sqrt{49}} = \dfrac{-28 - 36}{6 - 7} = \dfrac{-64}{-1} = 64$

93. $2475\left(1 + \dfrac{0.06}{4}\right)^{4 \cdot 10} = 4489.70$

95. $D = \dfrac{d \cdot n}{n + 2} = \dfrac{5 \cdot 12}{12 + 2} = \dfrac{60}{14} \approx 4.3 \text{ cm}$

97. $50 + (-3)(6) = 50 - 18 = 32^\circ$ F

99. $134 - (-45) = 134 + 45 = 179^\circ$ F

101. $3\dfrac{1}{7} = \dfrac{22}{7} \approx 3.14286$;

  $\dfrac{355}{113} \approx 3.14159$;

  $\dfrac{62,832}{20,000} = 3.1416$;

  $\sqrt{10} \approx 3.1623$;

  $\pi \approx 3.141592654$;

  Tsu-Ch'ung-chih: $\dfrac{355}{113}$

103. Negative

R.2 Exercises

1. Constant

3. Coefficient

5. $-5 + 5 = 0$; $-5 \cdot \left(-\dfrac{1}{5}\right) = 1$

7. Two; 3 and -5

9. Two; 2 and $\dfrac{1}{4}$

11. Three; -2, 1 and -5

13. One; -1

15. $n - 7$

17. $n + 4$

19. $(n - 5)^2$

21. $2n - 13$

23. $n^2 + 2n$

25. $\dfrac{2}{3}n - 5$

27. $3(n + 5) - 7$

29. Let w represent the width. Then $2w$ represents twice the width and $2w - 3$ represents three meters less than twice the width.

31. Let b represent the speed of the bus. Then $b + 15$ represents 15 mph more than the speed of the bus.

33. Let h represent the altitude of the helicopter. Let b represent the building's height.
$h = b + 150$

35. Let L represent the length. Let W represent the width.
$L = 2W + 20$

37. Let N represent the cost of a gallon of milk in 1990. Let M represent the cost of a gallon of milk in 2008.
$M = 2.5N$

39. Let g represent the number of gallons of insecticide. Let T represent the total charge.
$T = 12.50g + 50$

41. $4x - 2y$; $4(2) - 2(-3) = 8 + 6 = 14$

43. $-2x^2 + 3y^2$; $-2(2)^2 + 3(-3)^2$
$= -2(4) + 3(9) = -8 + 27 = 19$

45. $2y^2 + 5y - 3$; $2(-3)^2 + 5(-3) - 3$
$= 2(9) - 15 - 3 = 18 - 15 - 3 = 0$

47. $-2(3y + 1)$;
$-2(3(-3) + 1) = -2(-9 + 1) = -2(-8) = 16$

49. $3x^2y$; $3(2)^2(-3) = 3(4)(-3) = -36$

51. $(-3x)^2 - 4xy - y^2$;
$(-3 \cdot 2)^2 - 4(2)(-3) - (-3)^2$
$= (-6)^2 - 8(-3) - 9 = 36 + 24 - 9 = 51$

53. $\dfrac{1}{2}x - \dfrac{1}{3}y$; $\dfrac{1}{2}(2) - \dfrac{1}{3}(-3) = 1 + 1 = 2$

55. $(3x - 2y)^2$;
$(3 \cdot 2 - 2(-3))^2 = (6 + 6)^2 = 12^2 = 144$

57. $\dfrac{-12y + 5}{-3x + 1}$; $\dfrac{-12(-3) + 5}{-3(2) + 1} = \dfrac{36 + 5}{-6 + 1} = \dfrac{-41}{5}$

59. $\sqrt{-12y} \cdot 4$;
$\sqrt{-12(-3)} \cdot 4 = \sqrt{36} \cdot 4 = 6 \cdot 4 = 24$

61. $x^2 - 3x - 4$

x	Output
-3	$(-3)^2 - 3(-3) - 4 = 14$
-2	$(-2)^2 - 3(-2) - 4 = 6$
-1	$(-1)^2 - 3(-1) - 4 = 0$
0	$(0)^2 - 3(0) - 4 = -4$
1	$(1)^2 - 3(1) - 4 = -6$
2	$(2)^2 - 3(2) - 4 = -6$
3	$(3)^2 - 3(3) - 4 = -4$

-1 has an output of 0.

63. $-3(1 - x) - 6$

x	Output
-3	$-3(1 - (-3)) - 6 = -18$
-2	$-3(1 - (-2)) - 6 = -15$
-1	$-3(1 - (-1)) - 6 = -12$
0	$-3(1 - (0)) - 6 = -9$
1	$-3(1 - (1)) - 6 = -6$
2	$-3(1 - (2)) - 6 = -3$
3	$-3(1 - (3)) - 6 = 0$

3 has an output of 0.

65. $x^3 - 6x + 4$

x	Output
-3	$(-3)^3 - 6(-3) + 4 = -5$
-2	$(-2)^3 - 6(-2) + 4 = 8$
-1	$(-1)^3 - 6(-1) + 4 = 9$
0	$(0)^3 - 6(0) + 4 = 4$
1	$(1)^3 - 6(1) + 4 = -1$
2	$(2)^3 - 6(2) + 4 = 0$
3	$(3)^3 - 6(3) + 4 = 13$

2 has an output of 0.

67. a. $-5 + 7 = 7 + (-5) = 2$

b. $-2 + n = n + (-2)$

c. $-4.2 + a + 13.6 = a + (-4.2) + 13.6$
$= a + 9.4$

d. $7 + x - 7 = x + 7 - 7 = x$

69. a. $x + (-3.2) + \underline{3.2} = x$

b. $n - \dfrac{5}{6} + \dfrac{5}{\underline{6}} = n$

71. $-5(x - 2.6) = -5x + 13$

73. $\dfrac{2}{3}\left(-\dfrac{1}{5}p + 9\right) = -\dfrac{2}{15}p + 6$

75. $3a + (-5a) = 3a - 5a = -2a$

77. $\dfrac{2}{3}x + \dfrac{3}{4}x = \dfrac{8}{12}x + \dfrac{9}{12}x = \dfrac{17}{12}x$

79. $3\left(a^2 + 3a\right) - \left(5a^2 + 7a\right)$
$= 3a^2 + 9a - 5a^2 - 7a$
$= -2a^2 + 2a$

81. $x^2 - \left(3x - 5x^2\right) = x^2 - 3x + 5x^2 = 6x^2 - 3x$

83. $(3a + 2b - 5c) - (a - b - 7c)$
$= 3a + 2b - 5c - a + b + 7c = 2a + 3b + 2c$

85. $\dfrac{3}{5}(5n - 4) + \dfrac{5}{8}(n + 16)$

$= 3n - \dfrac{12}{5} + \dfrac{5}{8}n + 10$

$= \dfrac{24}{8}n - \dfrac{12}{5} + \dfrac{5}{8}n + \dfrac{50}{5}$

$= \dfrac{29}{8}n + \dfrac{38}{5}$

87. $\left(3a^2 - 5a + 7\right) + 2\left(2a^2 - 4a - 6\right)$
$= 3a^2 - 5a + 7 + 4a^2 - 8a - 12$
$= 7a^2 - 13a - 5$

89. $R = \dfrac{kL}{d^2}$;

$R = \dfrac{(0.000025)(90)}{(0.015)^2} = \dfrac{0.00225}{0.000225} = 10$ ohms

91. Let j represent the speed of the jet. Let t represent the speed of the turbo-prop.

(a) $t = \dfrac{1}{2}j$

(b) $t = \dfrac{1}{2}(550) = 225$ mph

93. Let W represent the width. Let L represent the length.
(a) $L = 2W + 3$
(b) $L = 2(52) + 3 = 107$ ft

95. Let c represent the cost of the 1978 stamp. Let t represent the cost of the 2004 stamp.
$t = c + 22$; $t = 15 + 22 = 37¢$

97. Let t represent the number of hours of labor. Let c represent the total cost.
$c = 25t + 43.50$; $c = 25(1.5) + 43.50 = 81.00

99. a. positive odd integer

R.3 Exercises

1. Power

3. $20x$; 0

5. a. cannot be simplified, unlike terms.
   b. can be simplified, like bases.

7. $\dfrac{2}{3}n^2 \cdot 21n^5$

$= \dfrac{2}{3}\left(\dfrac{21}{1}\right) \cdot n^2 \cdot n^5 = 14n^7$

9. $\left(-6p^2 q\right)\left(2p^3 q^3\right)$
$= -6 \cdot 2 \cdot p^2 \cdot p^3 \cdot q \cdot q^3 = -12p^5 q^4$

11. $(a^2)^4 \cdot (a^3)^2 \cdot b^2 \cdot b^5$
$a^8 \cdot a^6 \cdot b^2 \cdot b^5 = a^{14}b^7$

13. $\left(6pq^2\right)^3 = 6^3 (p)^3 \left(q^2\right)^3 = 216p^3 q^6$

15. $\left(3.2hk^2\right)^3 = (3.2)^3 (h)^3 \left(k^2\right)^3 = 32.768h^3 k^6$

17. $\left(\dfrac{p}{2q}\right)^2 = \dfrac{(p)^2}{(2)^2 (q)^2} = \dfrac{p^2}{4q^2}$

19. $\left(-0.7c^4\right)^2 \left(10c^3 d^2\right)^2$
$= (-0.7)^2 \left(c^4\right)^2 (10)^2 \left(c^3\right)^2 \left(d^2\right)^2$
$= 0.49c^8 \cdot 100c^6 d^4$
$= 49c^{14}d^4$

21. $\left(\dfrac{3}{4}x^3 y\right)^2 = \left(\dfrac{3}{4}\right)^2 \left(x^3\right)^2 (y)^2 = \dfrac{9}{16}x^6 y^2$

23. $\left(-\dfrac{3}{8}x\right)^2 \left(16xy^2\right)$
$= \left(-\dfrac{3}{8}\right)^2 (x)^2 \left(16xy^2\right)$
$= \dfrac{9}{64}x^2 \cdot 16xy^2 = \dfrac{9}{4}x^3 y^2$

25. a. $V = S^3$;
$V = \left(3x^2\right)^3$
$V = 3^3 \left(x^2\right)^3$
$V = 27x^6$
   b. $V = 27x^6$;
$V = 27(2)^6$
$V = 27(64)$
$V = 1728$ units^3

27. $\dfrac{-6w^5}{-2w^2} = 3w^3$

29. $\dfrac{-12a^3 b^5}{4a^2 b^4} = -3ab$

31. $\left(\dfrac{2}{3}\right)^{-3} = \dfrac{2^{-3}}{3^{-3}} = \dfrac{3^3}{2^3} = \dfrac{27}{8}$

R.3 Exercises

33. $\dfrac{2}{h^{-3}} = 2h^3$

35. $(-2)^{-3} = \left(\dfrac{-1}{2}\right)^3 = \dfrac{-1}{8}$

37. $\left(\dfrac{-1}{2}\right)^{-3} = (-2)^3 = -8$

39. $\left(\dfrac{2p^4}{q^3}\right)^2 = \dfrac{2^2\left(p^4\right)^2}{\left(q^3\right)^2} = \dfrac{4p^8}{q^6}$

41. $\left(\dfrac{0.2x^2}{0.3y^3}\right)^3 = \dfrac{(0.2)^3\left(x^2\right)^3}{(0.3)^3\left(y^3\right)^3}$

$= \dfrac{0.008x^6}{0.027y^9} = \dfrac{8x^6}{27y^9}$

43. $\left(\dfrac{5m^2n^3}{2r^4}\right)^2 = \dfrac{(5)^2\left(m^2\right)^2\left(n^3\right)^2}{(2)^2\left(r^4\right)^2} = \dfrac{25m^4n^6}{4r^8}$

45. $\left(\dfrac{5p^2q^3r^4}{-2pq^2r^4}\right)^2 - \left(\dfrac{5pq}{-2}\right)^2$

$= \dfrac{5^2\,p^2q^2}{(-2)^2} = \dfrac{25p^2q^2}{4}$

47. $\dfrac{9p^6q^4}{-12p^4q^6} = \dfrac{3p^2}{-4q^2}$

49. $\dfrac{20h^{-2}}{12h^5} = \dfrac{5}{3}h^{-2-5} = \dfrac{5}{3}h^{-7} = \dfrac{5}{3h^7}$

51. $\dfrac{\left(a^2\right)^3}{a^4\cdot a^5} = \dfrac{a^6}{a^9} = \dfrac{1}{a^3}$

53. $\left(\dfrac{a^{-3}b}{c^{-2}}\right)^{-4} = \dfrac{a^{12}b^{-4}}{c^8} = \dfrac{a^{12}}{b^4c^8}$

55. $\dfrac{-6\left(2x^{-3}\right)^2}{10x^{-2}} = \dfrac{-6\left(4x^{-6}\right)}{10x^{-2}} = \dfrac{-24x^{-6}}{10x^{-2}}$

$= \dfrac{-12x^{-6-(-2)}}{5} = \dfrac{-12x^{-6+2}}{5} = \dfrac{-12x^{-4}}{5} = \dfrac{-12}{5x^4}$

57. $\dfrac{14a^{-3}bc^0}{-7\left(3a^2b^{-2}c\right)^3} = \dfrac{14a^{-3}bc^0}{-7\left(27a^6b^{-6}c^3\right)}$

$= \dfrac{14a^{-3}bc^0}{-189a^6b^{-6}c^3} = -\dfrac{2a^{-3-6}b^{1-(-6)}c^{0-3}}{27}$

$= \dfrac{-2a^{-9}b^{1+6}c^{-3}}{27} = \dfrac{-2b^7}{27a^9c^3}$

59. $4^0 + 5^0 = 1 + 1 = 2$

61. $2^{-1} + 5^{-1} = \dfrac{1}{2} + \dfrac{1}{5} = \dfrac{5}{10} + \dfrac{2}{10} = \dfrac{7}{10}$

63. $3^0 + 3^{-1} + 3^{-2} = 1 + \dfrac{1}{3} + \dfrac{1}{3^2}$

$= \dfrac{9}{9} + \dfrac{3}{9} + \dfrac{1}{9} = \dfrac{13}{9}$

65. $-5x^0 + (-5x)^0 = -5(1) + 1 = -5 + 1 = -4$

67. $6{,}600{,}000{,}000 = 6.6 \times 10^9$

69. $6.5 \times 10^{-9} = 0.0000000065$ m

71. $\dfrac{465{,}000{,}000}{17{,}500} = \dfrac{4.65 \times 10^8}{1.75 \times 10^4} = \dfrac{4.65}{1.75} \times \dfrac{10^8}{10^4}$

$= 2.657142857 \times 10^4 = 26571.42857$

≈ 26571 hours

$\dfrac{26571}{24} = 1107.125 \approx 1107$ days

73. $-35w^3 + 2w^2 + (-12w) + 14$
    Polynomial; None of these; Degree 3

75. $5n^{-2} + 4n + \sqrt{17}$
    Nonpolynomial because exponents are not whole numbers; NA; NA

77. $p^3 - \dfrac{2}{5}$

Polynomial; Binomial; Degree 3

79. $7w + 8.2 - w^3 - 3w^2$

$= -w^3 - 3w^2 + 7w + 8.2$

Lead coefficient: -1

81. $c^3 + 6 + 2c^2 - 3c = c^3 + 2c^2 - 3c + 6$

Lead coefficient: 1

83. $12 - \dfrac{2}{3}x^2 = -\dfrac{2}{3}x^2 + 12$

Lead coefficient: $-\dfrac{2}{3}$

85. $\left(3p^3 - 4p^2 + 2p - 7\right) + \left(p^2 - 2p - 5\right)$

$= 3p^3 - 4p^2 + 2p - 7 + p^2 - 2p - 5$

$= 3p^3 - 3p^2 - 12$

87. $\left(5.75b^2 + 2.6b - 1.9\right) + \left(2.1b^2 - 3.2b\right)$

$= 5.75b^2 + 2.6b - 1.9 + 2.1b^2 - 3.2b$

$= 7.85b^2 - 0.6b - 1.9$

89. $\left(\dfrac{3}{4}x^2 - 5x + 2\right) - \left(\dfrac{1}{2}x^2 + 3x - 4\right)$

$= \dfrac{3}{4}x^2 - 5x + 2 - \dfrac{1}{2}x^2 - 3x + 4$

$= \dfrac{3}{4}x^2 - 5x + 2 - \dfrac{2}{4}x^2 - 3x + 4$

$= \dfrac{1}{4}x^2 - 8x + 6$

91.
$$
\begin{array}{r}
q^6 + 2q^5 + q^4 + 2q^3 \\
- \left(\quad q^5 + 2q^4 + \quad q^2 + 2q \right) \\
\hline
q^6 + q^5 - q^4 + 2q^3 - q^2 - 2q
\end{array}
$$

93. $-3x\left(x^2 - x - 6\right) = -3x^3 + 3x^2 + 18x$

95. $(3r - 5)(r - 2)$

$= 3r^2 - 6r - 5r + 10 = 3r^2 - 11r + 10$

97. $(x - 3)\left(x^2 + 3x + 9\right)$

$= x^3 + 3x^2 + 9x - 3x^2 - 9x - 27 = x^3 - 27$

99. $\left(b^2 - 3b - 28\right)(b + 2)$

$= b^3 + 2b^2 - 3b^2 - 6b - 28b - 56$

$= b^3 - b^2 - 34b - 56$

101. $(7v - 4)(3v - 5)$

$= 21v^2 - 35v - 12v + 20$

$= 21v^2 - 47v + 20$

103. $(3 - m)(3 + m)$

$= 9 + 3m - 3m - m^2$

$= 9 - m^2$

105. $(p - 2.5)(p + 3.6)$

$= p^2 + 3.6p - 2.5p - 9$

$= p^2 + 1.1p - 9$

107. $\left(x + \dfrac{1}{2}\right)\left(x + \dfrac{1}{4}\right)$

$= x^2 + \dfrac{1}{4}x + \dfrac{1}{2}x + \dfrac{1}{8}$

$= x^2 + \dfrac{1}{4}x + \dfrac{2}{4}x + \dfrac{1}{8}$

$= x^2 + \dfrac{3}{4}x + \dfrac{1}{8}$

109. $\left(m + \dfrac{3}{4}\right)\left(m - \dfrac{3}{4}\right)$

$= m^2 - \dfrac{3}{4}m + \dfrac{3}{4}m - \dfrac{9}{16} = m^2 - \dfrac{9}{16}$

111. $(3x - 2y)(2x + 5y)$

$= 6x^2 + 15xy - 4xy - 10y^2$

$= 6x^2 + 11xy - 10y^2$

113. $(4c + d)(3c + 5d)$

$= 12c^2 + 20cd + 3cd + 5d^2$

$= 12c^2 + 23cd + 5d^2$

R.3 Exercises

115. $\left(2x^2+5\right)\left(x^2-3\right)$
$= 2x^4 - 6x^2 + 5x^2 - 15$
$= 2x^4 - x^2 - 15$

117. $4m-3$; Conjugate: $4m+3$
$(4m-3)(4m+3)$
$= 16m^2 + 12m - 12m - 9 = 16m^2 - 9$

119. $7x-10$; Conjugate: $7x+10$
$(7x-10)(7x+10)$
$= 49x^2 + 70x - 70x - 100 = 49x^2 - 100$

121. $6+5k$; Conjugate: $6-5k$
$(6+5k)(6-5k)$
$= 36 - 30k + 30k - 25k^2 = 36 - 25k^2$

123. $x+\sqrt{6}$; Conjugate: $x-\sqrt{6}$
$\left(x+\sqrt{6}\right)\left(x-\sqrt{6}\right)$
$= x^2 - \sqrt{6}x + \sqrt{6}x - 6 = x^2 - 6$

125. $(x+4)^2 = x^2 + 2(4\cdot x) + 16 = x^2 + 8x + 16$

127. $(4g+3)^2$
$= 16g^2 + 2(4g\cdot 3) + 9 = 16g^2 + 24g + 9$

129. $(4p-3q)^2$
$= 16p^2 - 2(4p\cdot 3q) + 9q^2$
$= 16p^2 - 24pq + 9q^2$

131. $\left(4-\sqrt{x}\right)^2$
$= 16 - 2\left(4\cdot\sqrt{x}\right) + \left(\sqrt{x}\right)^2$
$= 16 - 8\sqrt{x} + x$

133. $(x-3)(y+2) = xy + 2x - 3y - 6$

135. $(k-5)(k+6)(k+2)$
$= (k-5)\left(k^2 + 2k + 6k + 12\right)$
$= (k-5)\left(k^2 + 8k + 12\right)$
$= k^3 + 8k^2 + 12k - 5k^2 - 40k - 60$
$= k^3 + 3k^2 - 28k - 60$

137. $M = 0.5t^4 + 3t^3 - 97t^2 + 348t$

t	M
1	$M = 0.5(1)^4 + 3(1)^3 - 97(1)^2 + 348(1)$ $= 254.5$
2	$M = 0.5(2)^4 + 3(2)^3 - 97(2)^2 + 348(2)$ $= 340$
3	$M = 0.5(3)^4 + 3(3)^3 - 97(3)^2 + 348(3)$ $= 292.5$
4	$M = 0.5(4)^4 + 3(4)^3 - 97(4)^2 + 348(4)$ $= 160$
5	$M = 0.5(5)^4 + 3(5)^3 - 97(5)^2 + 348(5)$ $= 2.5$

a. $M = 0.5(2)^4 + 3(2)^3 - 97(2)^2 + 348(2)$
$M = 0.5(16) + 3(8) - 97(4) + 696$
$M = 8 + 24 - 388 + 696$
$M = 340$ mg;
$M = 0.5(3)^4 + 3(3)^3 - 97(3)^2 + 348(3)$
$M = 0.5(81) + 3(27) - 97(9) + 1044$
$M = 40.5 + 81 - 873 + 1044$
$M = 292.5$ mg

b. Less, the amount is decreasing.

c. The drug will wear off after 5 hours.

8

139. $F = \dfrac{kPQ}{d^2}$

   $F = kPQd^{-2}$

141. $\dfrac{5}{x^3} + \dfrac{3}{x^2} + \dfrac{2}{x^1} + 4 = 5x^{-3} + 3x^{-2} + 2x^{-1} + 4$

143. $R = (20 - 1x)(200 + 20x)$

   $R = 4000 + 400x - 200x - 20x^2$

   $R = 4000 + 200x - 20x^2$

   Let x represent the number of \$1 decreases.

x	R(x)
1	4180
2	4320
3	4420
4	4480
5	4500
6	4480
7	4420
8	4320
9	4180
10	4000

   Using the table, maximum revenue occurs at $x = 5$. Thus, $20 - 1x = 20 - 5 = 15$
   The most revenue will be earned when the price is \$15.

145. $\left(3x^2 + kx + 1\right) - \left(kx^2 + 5x - 7\right) + \left(2x^2 - 4x - k\right)$

   $\qquad = -x^2 - 3x + 2$

   $3x^2 + kx + 1 - kx^2 - 5x + 7 + 2x^2 - 4x - k$

   $\qquad = -x^2 - 3x + 2$

   $\left(kx - kx^2 - k\right) + 5x^2 + 8 - 9x = -x^2 - 3x + 2$

   $k\left(x - x^2 - 1\right) = -6x^2 + 6x - 6$

   $k = \dfrac{-6\left(x^2 - x + 1\right)}{x - x^2 - 1}$

   $k = \dfrac{-6\left(x^2 - x + 1\right)}{-\left(x^2 - x + 1\right)}$

   $k = 6$

R.4 Exercises

1. Product

3. Binomial; Conjugate

5. Answers will vary;
   $4x^2 - 36 = 4\left(x^2 - 9\right) = 4(x + 3)(x - 3)$

7. a. $-17x^2 + 51 = -17\left(x^2 - 3\right)$

   b. $21b^3 - 14b^2 + 56b = 7b\left(3b^2 - 2b + 8\right)$

   c. $-3a^4 + 9a^2 - 6a^3$
   $= -3a^2\left(a^2 - 3 + 2a\right)$
   $= -3a^2\left(a^2 + 2a - 3\right)$

9. a. $2a(a + 2) + 3(a + 2) = (a + 2)(2a + 3)$

   b. $\left(b^2 + 3\right)3b + \left(b^2 + 3\right)2 = \left(b^2 + 3\right)(3b + 2)$

   c. $4m(n + 7) - 11(n + 7) = (n + 7)(4m - 11)$

11. a. $9q^3 + 6q^2 + 15q + 10$
    $= \left(9q^3 + 6q^2\right) + (15q + 10)$
    $= 3q^2(3q + 2) + 5(3q + 2)$
    $= (3q + 2)\left(3q^2 + 5\right)$

    b. $h^5 - 12h^4 - 3h + 36$
    $= \left(h^5 - 12h^4\right) - (3h - 36)$
    $= h^4(h - 12) - 3(h - 12)$
    $= (h - 12)\left(h^4 - 3\right)$

    c. $k^5 - 7k^3 - 5k^2 + 35$
    $= \left(k^5 - 7k^3\right) - \left(5k^2 - 35\right)$
    $= k^3\left(k^2 - 7\right) - 5\left(k^2 - 7\right)$
    $= \left(k^2 - 7\right)\left(k^3 - 5\right)$

13. a. $-p^2 + 5p + 14 = -1\left(p^2 - 5p - 14\right)$
    $-p^2 + 5p + 14 = -1(p - 7)(p + 2)$

    b. $q^2 - 4q - 45 = (q - 9)(q + 5)$

    c. $n^2 + 20 - 9n$
    $= n^2 - 9n + 20 = (n - 4)(n - 5)$

15. a. $3p^2 - 13p - 10 = (3p + 2)(p - 5)$

    b. $4q^2 + 7q - 15 = (4q - 5)(q + 3)$

    c. $10u^2 - 19u - 15 = (5u + 3)(2u - 5)$

R.4 Exercises

17. a. $4s^2 - 25$
   $= (2s)^2 - (5)^2$
   $= (2s+5)(2s-5)$

   b. $9x^2 - 49$
   $= (3x)^2 - (7)^2$
   $= (3x+7)(3x-7)$

   c. $50x^2 - 72$
   $= 2(25x^2 - 36)$
   $= 2[(5x)^2 - (6)^2]$
   $= 2(5x+6)(5x-6)$

   d. $121h^2 - 144$
   $= (11h)^2 - (12)^2$
   $= (11h+12)(11h-12)$

   e. $b^2 - 5$
   $= (b+\sqrt{5})(b-\sqrt{5})$

19. a. $a^2 - 6a + 9 = (a-3)^2$

   b. $b^2 + 10b + 25 = (b+5)^2$

   c. $4m^2 - 20m + 25 = (2m-5)^2$

   d. $9n^2 - 42n + 49 = (3n-7)^2$

21. a. $8p^3 - 27 = (2p)^3 - (3)^3$
   $= (2p-3)(4p^2 + 6p + 9)$

   b. $m^3 + \dfrac{1}{8} = (m)^3 + \left(\dfrac{1}{2}\right)^3$
   $= \left(m + \dfrac{1}{2}\right)\left(m^2 - \dfrac{1}{2}m + \dfrac{1}{4}\right)$

   c. $g^3 - 0.027 = (g)^3 - (0.3)^3$
   $= (g-0.3)(g^2 + 0.3g + 0.09)$

   d. $-2t^4 + 54t = -2t(t^3 - 27)$
   $= -2t[(t)^3 - (3)^3]$
   $= -2t(t-3)(t^2 + 3t + 9)$

23. a. $x^4 - 10x^2 + 9$
   Let u represent x^2
   $= u^2 - 10u + 9$
   $= (u-9)(u-1)$
   $= (x^2 - 9)(x^2 - 1)$
   $= (x+3)(x-3)(x+1)(x-1)$

   b. $x^4 + 13x^2 + 36$
   Let u represent x^2
   $= u^2 + 13u + 36$
   $= (u+9)(u+4)$
   $= (x^2 + 9)(x^2 + 4)$

   c. $x^6 - 7x^3 - 8$
   Let u represent x^3
   $= u^2 - 7u - 8$
   $= (u-8)(u+1)$
   $= (x^3 - 8)(x^3 + 1)$
   $= (x-2)(x^2 + 2x + 4)(x+1)(x^2 - x + 1)$

25. a. $n^2 - 1 = (n+1)(n-1)$

   b. $n^3 - 1 = (n-1)(n^2 + n + 1)$

   c. $n^3 + 1 = (n+1)(n^2 - n + 1)$

   d. $28x^3 - 7x = 7x(4x^2 - 1)$
   $= 7x(2x+1)(2x-1)$

27. $a^2 + 7a + 10 = (a+5)(a+2)$

29. $2x^2 - 24x + 20$
   $= 2(x^2 - 12x + 20) = 2(x-2)(x-10)$

31. $64 - 9m^2 = -1(9m^2 - 64)$
   $= -1\left[(3m)^2 - 8^2\right] = -1(3m+8)(3m-8)$

33. $-9r + r^2 + 18 = r^2 - 9r + 18$
   $= (r-3)(r-6)$

35. $2h^2 + 7h + 6 = (2h+3)(h+2)$

37. $9k^2 - 24k + 16 = (3k-4)^2$

39. $-6x^3 + 39x^2 - 63x$
   $= -3x(2x^2 - 13x - 21)$
   $= -3x(2x-7)(x-3)$

41. $12m^2 - 40m + 4m^3 = 4m^3 + 12m^2 - 40m$
   $= 4m(m^2 + 3m - 10) = 4m(m+5)(m-2)$

43. $a^2 - 7a - 60 = (a+5)(a-12)$

45. $8x^3 - 125 = (2x)^3 - (5)^3$

$= (2x - 5)(4x^2 + 10x + 25)$

47. $m^2 + 9m - 24$, Prime

49. $x^3 - 5x^2 - 9x + 45$

$= (x^3 - 5x^2) - (9x - 45)$

$= x^2(x - 5) - 9(x - 5)$

$= (x - 5)(x^2 - 9)$

$= (x - 5)(x + 3)(x - 3)$

51. a. prime polynomial:

H. $x^2 + 9$

b. standard trinomial a = 1:

E. $x^2 - 3x - 10$

c. perfect square trinomial:

C. $x^2 - 10x + 25$

d. difference of cubes:

F. $8s^3 - 125t^3$

e. binomial square:

B. $(x + 3)^2$

f. sum of cubes:

A. $x^3 + 27$

g. binomial conjugates:

I. $(x - 7)$ and $(x + 7)$

h. difference of squares:

D. $x^2 - 144$

i. standard trinomial $a \neq 1$:

G. $2x^2 - x - 3$

53. $A = 2\pi r^2 + 2\pi rh$

$A = 2\pi r(r + h)$

$= 2\pi(35)(35 + 65)$

$= 2\pi(35)(100)$

$= 7000\pi$ cm^2 ; 21,991 cm^2

55. $V = \frac{1}{3}\pi R^2 h - \frac{1}{3}\pi r^2 h$

$V = \frac{1}{3}\pi h(R^2 - r^2) = \frac{1}{3}\pi h(R + r)(R - r)$

$V = \frac{1}{3}\pi(9)(5.1^2 - 4.9^2)$

$V = 3\pi(26.01 - 24.01)$

$V = 3\pi(2) = 6\pi$; 18.8 cm^3

57. $V = x^3 + 8x^2 + 15x$

$V = x(x^2 + 8x + 15)$

$= x(x + 3)(x + 5)$

a. If the height is x inches then the width is $x + 3$ which would make the width 3 inches more than the height.

b. If the height is x inches then the length is $x + 5$ which would make the length 5 inches more than the height.

c. 2 ft = 24 in;

$V = 24(24 + 5)(24 + 3) = 18,792$ ft^3

59. $L = L_0 \sqrt{1 - \left(\frac{v}{\iota}\right)^2}$

$L = L_0 \sqrt{\left(1 + \frac{v}{c}\right)\left(1 - \frac{v}{c}\right)}$

$L = 12\sqrt{(1 + 0.75)(1 - 0.75)}$

$L = 12\sqrt{(1.75)(0.25)}$

$L = 12\sqrt{0.4375} = 12\sqrt{0.0625(7)}$

$= 12(0.25)\sqrt{7} = 3\sqrt{7} \approx 7.94$ inches

61. a. $\frac{1}{2}x^4 + \frac{1}{8}x^3 - \frac{3}{4}x^2 + 4$

$= \frac{1}{8}(4x^4 + x^3 - 6x^2 + 32)$

b. $\frac{2}{3}b^5 - \frac{1}{6}b^3 + \frac{4}{9}b^2 - 1$

$= \frac{1}{18}(12b^5 - 3b^3 + 8b^2 - 18)$

63. $192x^3 - 164x^2 - 270x$

$= 2x(96x^2 - 82x - 135)$

$= 2x(16x - 27)(6x + 5)$

65. $x^4 - 81 = (x^2 - 9)(x^2 + 9)$

$= (x + 3)(x - 3)(x^2 + 9)$

67. $p^6 - 1 = (p^3)^2 - 1^2 = (p^3 + 1)(p^3 - 1)$

$= (p + 1)(p^2 - p + 1)(p - 1)(p^2 + p + 1)$

69. $q^4 - 28q^2 + 75 = (q^2 - 25)(q^2 - 3)$

$= (q + 5)(q - 5)(q + \sqrt{3})(q - \sqrt{3})$

R.5 Exercises

1. $1; -1$

3. common denominator

5. False; $x - (x+1) = x - x - 1 = -1$; numerator should be -1

7. a. $\dfrac{a-7}{-3a+21} = \dfrac{a-7}{-3(a-7)} = -\dfrac{1}{3}$

   b. $\dfrac{2x+6}{4x^2-8x} = \dfrac{2(x+3)}{4x(x-2)} = \dfrac{x+3}{2x(x-2)}$

9. a. $\dfrac{x^2-5x-14}{x^2+6x-7} = \dfrac{(x-7)(x+2)}{(x+7)(x-1)}$
   Simplified

   b. $\dfrac{a^2+3a-28}{a^2-49} = \dfrac{(a+7)(a-4)}{(a+7)(a-7)} = \dfrac{a-4}{a-7}$

11. a. $\dfrac{x-7}{7-x} = \dfrac{-(7-x)}{7-x} = -1$

    b. $\dfrac{5-x}{x-5} = \dfrac{-(x-5)}{x-5} = -1$

13. a. $\dfrac{-12a^3b^5}{4a^2b^{-4}} = -3a^{3-2}b^{5-(-4)} = -3ab^9$

    b. $\dfrac{7x+21}{63} = \dfrac{7(x+3)}{63} = \dfrac{x+3}{9}$

    c. $\dfrac{y^2-9}{3-y} = \dfrac{(y+3)(y-3)}{-(y-3)} = -1(y+3)$

    d. $\dfrac{m^3n-m^3}{m^4-m^4n} = \dfrac{m^3(n-1)}{m^4(1-n)}$
    $= \dfrac{m^3(n-1)}{-m^4(n-1)} = \dfrac{-1}{m}$

15. a. $\dfrac{2n^3+n^2-3n}{n^3-n^2} = \dfrac{n(2n^2+n-3)}{n^2(n-1)}$
    $= \dfrac{n(2n+3)(n-1)}{n^2(n-1)} = \dfrac{2n+3}{n}$

    b. $\dfrac{6x^2+x-15}{4x^2-9} = \dfrac{(2x-3)(3x+5)}{(2x-3)(2x+3)}$
    $= \dfrac{3x+5}{2x+3}$

    c. $\dfrac{x^3+8}{x^2-2x+4} = \dfrac{(x+2)(x^2-2x+4)}{x^2-2x+4}$
    $= x+2$

    d. $\dfrac{mn^2+n^2-4m-4}{mn+n+2m+2}$
    $= \dfrac{(mn^2+n^2)-(4m+4)}{(mn+n)+(2m+2)}$
    $= \dfrac{n^2(m+1)-4(m+1)}{n(m+1)+2(m+1)}$
    $= \dfrac{(m+1)(n^2-4)}{(m+1)(n+2)}$
    $= \dfrac{(m+1)(n+2)(n-2)}{(m+1)(n+2)}$
    $= n-2$

17. $\dfrac{a^2-4a+4}{a^2-9} \cdot \dfrac{a^2-2a-3}{a^2-4}$
    $= \dfrac{(a-2)(a-2)}{(a+3)(a-3)} \cdot \dfrac{(a-3)(a+1)}{(a+2)(a-2)}$
    $= \dfrac{(a-2)(a+1)}{(a+3)(a+2)}$

19. $\dfrac{x^2-7x-18}{x^2-6x-27} \cdot \dfrac{2x^2+7x+3}{2x^2+5x+2}$
    $= \dfrac{(x-9)(x+2)}{(x-9)(x+3)} \cdot \dfrac{(2x+1)(x+3)}{(2x+1)(x+2)} = 1$

21. $\dfrac{p^3-64}{p^3-p^2} \div \dfrac{p^2+4p+16}{p^2-5p+4}$

$= \dfrac{p^3-64}{p^3-p^2} \cdot \dfrac{p^2-5p+4}{p^2+4p+16}$

$= \dfrac{(p-4)\left(p^2+4p+16\right)}{p^2(p-1)} \cdot \dfrac{(p-4)(p-1)}{\left(p^2+4p+16\right)}$

$= \dfrac{(p-4)^2}{p^2}$

23. $\dfrac{3x-9}{4x+12} \div \dfrac{3-x}{5x+15} = \dfrac{3x-9}{4x+12} \cdot \dfrac{5x+15}{3-x}$

$= \dfrac{3(x-3)}{4(x+3)} \cdot \dfrac{5(x+3)}{-(x-3)} = \dfrac{-15}{4}$

25. $\dfrac{a^2+a}{a^2-3a} \cdot \dfrac{3a-9}{2a+2} = \dfrac{a(a+1)}{a(a-3)} \cdot \dfrac{3(a-3)}{2(a+1)} = \dfrac{3}{2}$

27. $\dfrac{8}{a^2-25} \cdot \left(a^2-2a-35\right)$

$= \dfrac{8}{(a+5)(a-5)} \cdot (a-7)(a+5) = \dfrac{8(a-7)}{a-5}$

29. $\dfrac{xy-3x+2y-6}{x^2-3x-10} \div \dfrac{xy-3x}{xy-5y}$

$= \dfrac{(xy-3x)+(2y-6)}{x^2-3x-10} \cdot \dfrac{xy-5y}{xy-3x}$

$= \dfrac{x(y-3)+2(y-3)}{(x-5)(x+2)} \cdot \dfrac{y(x-5)}{x(y-3)}$

$= \dfrac{(y-3)(x+2)}{(x-5)(x+2)} \cdot \dfrac{y(x-5)}{x(y-3)}$

$= \dfrac{y}{x}$

31. $\dfrac{m^2+2m-8}{m^2-2m} \div \dfrac{m^2-16}{m^2}$

$= \dfrac{m^2+2m-8}{m^2-2m} \cdot \dfrac{m^2}{m^2-16}$

$= \dfrac{(m+4)(m-2)}{m(m-2)} \cdot \dfrac{m^2}{(m+4)(m-4)}$

$= \dfrac{m}{m-4}$

33. $\dfrac{y+3}{3y^2+9y} \cdot \dfrac{y^2+7y+12}{y^2-16} \div \dfrac{y^2+4y}{y^2-4y}$

$= \dfrac{y+3}{3y(y+3)} \cdot \dfrac{(y+3)(y+4)}{(y+4)(y-4)} \cdot \dfrac{y^2-4y}{y^2+4y}$

$= \dfrac{y+3}{3y(y+3)} \cdot \dfrac{(y+3)(y+4)}{(y+4)(y-4)} \cdot \dfrac{y(y-4)}{y(y+4)}$

$= \dfrac{y+3}{3y(y+4)}$

35. $\dfrac{x^2-0.49}{x^2+0.5x-0.14} \div \dfrac{x^2-0.10x+0.21}{x^2-0.09}$

$= \dfrac{x^2-0.49}{x^2+0.5x-0.14} \cdot \dfrac{x^2-0.09}{x^2-0.10x+0.21}$

$= \dfrac{(x+0.7)(x-0.7)}{(x+0.7)(x-0.2)} \cdot \dfrac{(x+0.3)(x-0.3)}{(x-0.3)(x-0.7)}$

$= \dfrac{x+0.3}{x-0.2}$

37. $\dfrac{n^2-\dfrac{4}{9}}{n^2-\dfrac{13}{15}n+\dfrac{2}{15}} \div \dfrac{n^2+\dfrac{4}{3}n+\dfrac{4}{9}}{n^2-\dfrac{1}{25}}$

$= \dfrac{n^2-\dfrac{4}{9}}{n^2-\dfrac{13}{15}n+\dfrac{2}{15}} \cdot \dfrac{n^2-\dfrac{1}{25}}{n^2+\dfrac{4}{3}n+\dfrac{4}{9}}$

$= \dfrac{\left(n+\dfrac{2}{3}\right)\left(n-\dfrac{2}{3}\right)}{\left(n-\dfrac{1}{5}\right)\left(n-\dfrac{2}{3}\right)} \cdot \dfrac{\left(n+\dfrac{1}{5}\right)\left(n-\dfrac{1}{5}\right)}{\left(n+\dfrac{2}{3}\right)^2}$

$= \dfrac{n+\dfrac{1}{5}}{n+\dfrac{2}{3}}$

39. $\dfrac{3a^3-24a^2-12a+96}{a^2-11a+24} \div \dfrac{6a^2-24}{3a^3-81}$

$= \dfrac{3\left(a^3-8a^2-4a+32\right)}{a^2-11a+24} \cdot \dfrac{3a^3-81}{6a^2-24}$

$= \dfrac{3\left[\left(a^3-8a^2\right)-\left(4a-32\right)\right]}{(a-8)(a-3)} \cdot \dfrac{3\left(a^3-27\right)}{6\left(a^2-4\right)}$

$= \dfrac{3\left[a^2(a-8)-4(a-8)\right]}{(a-8)(a-3)} \cdot \dfrac{3(a-3)\left(a^2+3a+9\right)}{6(a+2)(a-2)}$

$= \dfrac{3(a-8)\left(a^2-4\right)}{(a-8)(a-3)} \cdot \dfrac{3(a-3)\left(a^2+3a+9\right)}{6(a+2)(a-2)}$

$= \dfrac{3(a-8)(a+2)(a-2)}{(a-8)(a-3)} \cdot \dfrac{3(a-3)\left(a^2+3a+9\right)}{6(a+2)(a-2)}$

$= \dfrac{3\left(a^2+3a+9\right)}{2}$

41. $\dfrac{4n^2-1}{12n^2-5n-3} \cdot \dfrac{6n^2+5n+1}{2n^2+n} \cdot \dfrac{12n^2-17n+6}{6n^2-7n+2}$

$= \dfrac{(2n+1)(2n-1)}{(4n-3)(3n+1)} \cdot \dfrac{(3n+1)(2n+1)}{n(2n+1)} \cdot \dfrac{(4n-3)(3n-2)}{(3n-2)(2n-1)}$

$= \dfrac{2n+1}{n}$

43. $\dfrac{3}{8x^2}+\dfrac{5}{2x} = \dfrac{3}{8x^2}+\dfrac{20x}{8x^2} = \dfrac{3+20x}{8x^2}$

45. $\dfrac{7}{4x^2y^3}-\dfrac{1}{8xy^4} = \dfrac{7(2y)}{8x^2y^4}-\dfrac{x}{8x^2y^4} = \dfrac{14y-x}{8x^2y^4}$

47. $\dfrac{4p}{p^2-36}-\dfrac{2}{p-6}$

$= \dfrac{4p}{(p-6)(p+6)}-\dfrac{2}{p-6}$

$= \dfrac{4p}{(p-6)(p+6)}-\dfrac{2(p+6)}{(p-6)(p+6)}$

$= \dfrac{4p-2p-12}{(p-6)(p+6)}$

$= \dfrac{2p-12}{(p-6)(p+6)}$

$= \dfrac{2(p-6)}{(p-6)(p+6)}$

$= \dfrac{2}{p+6}$

49. $\dfrac{m}{m^2-16}+\dfrac{4}{4-m}$

$= \dfrac{m}{(m+4)(m-4)}-\dfrac{4}{m-4}$

$= \dfrac{m}{(m+4)(m-4)}-\dfrac{4(m+4)}{(m+4)(m-4)}$

$= \dfrac{m-4m-16}{(m+4)(m-4)}$

$= \dfrac{-3m-16}{(m+4)(m-4)}$

51. $\dfrac{2}{m-7}-5 = \dfrac{2}{m-7}-\dfrac{5(m-7)}{m-7}$

$= \dfrac{2-5m+35}{m-7} = \dfrac{-5m+37}{m-7}$

53. $\dfrac{y+1}{y^2+y-30}-\dfrac{2}{y+6}$

$= \dfrac{y+1}{(y+6)(y-5)}-\dfrac{2(y-5)}{(y+6)(y-5)}$

$= \dfrac{y+1-2y+10}{(y+6)(y-5)} = \dfrac{-y+11}{(y+6)(y-5)}$

55. $\dfrac{1}{a+4}+\dfrac{a}{a^2-a-20}$

$= \dfrac{1}{a+4}+\dfrac{a}{(a+4)(a-5)}$

$= \dfrac{a-5}{(a+4)(a-5)}+\dfrac{a}{(a+4)(a-5)}$

$= \dfrac{a-5+a}{(a+4)(a-5)}$

$= \dfrac{2a-5}{(a+4)(a-5)}$

57. $\dfrac{3y-4}{y^2+2y+1} - \dfrac{2y-5}{y^2+2y+1}$

$= \dfrac{3y-4-(2y-5)}{y^2+2y+1}$

$= \dfrac{3y-4-2y+5}{y^2+2y+1}$

$= \dfrac{y+1}{(y+1)(y+1)}$

$= \dfrac{1}{y+1}$

59. $\dfrac{2}{m^2-9} + \dfrac{m-5}{m^2+6m+9}$

$= \dfrac{2}{(m+3)(m-3)} + \dfrac{m-5}{(m+3)^2}$

$= \dfrac{2(m+3)+(m-5)(m-3)}{(m+3)^2(m-3)}$

$= \dfrac{2m+6+m^2-8m+15}{(m+3)^2(m-3)}$

$= \dfrac{m^2-6m+21}{(m+3)^2(m-3)}$

61. $\dfrac{y+2}{5y^2+11y+2} + \dfrac{5}{y^2+y-6}$

$= \dfrac{y+2}{(5y+1)(y+2)} + \dfrac{5}{(y+3)(y-2)}$

$= \dfrac{1}{5y+1} + \dfrac{5}{(y+3)(y-2)}$

$= \dfrac{(y+3)(y-2)+5(5y+1)}{(5y+1)(y+3)(y-2)}$

$= \dfrac{y^2+y-6+25y+5}{(5y+1)(y+3)(y-2)}$

$= \dfrac{y^2+26y-1}{(5y+1)(y+3)(y-2)}$

63. a. $p^{-2}-5p^{-1} = \dfrac{1}{p^2} - \dfrac{5}{p}; \dfrac{1-5p}{p^2}$

b. $x^{-2}+2x^{-3} = \dfrac{1}{x^2} + \dfrac{2}{x^3}; \dfrac{x+2}{x^3}$

65. $\dfrac{\dfrac{5}{a}-\dfrac{1}{4}}{\dfrac{25}{a^2}-\dfrac{1}{16}} = \dfrac{\left(\dfrac{5}{a}-\dfrac{1}{4}\right)16a^2}{\left(\dfrac{25}{a^2}-\dfrac{1}{16}\right)16a^2} = \dfrac{80a-4a^2}{400-a^2}$

$= \dfrac{4a(20-a)}{(20+a)(20-a)} = \dfrac{4a}{20+a}$

67. $\dfrac{p+\dfrac{1}{p-2}}{1+\dfrac{1}{p-2}} = \dfrac{\left(p+\dfrac{1}{p-2}\right)(p-2)}{\left(1+\dfrac{1}{p-2}\right)(p-2)}$

$= \dfrac{p(p-2)+1}{p-2+1} = \dfrac{p^2-2p+1}{p-1}$

$= \dfrac{(p-1)(p-1)}{p-1}$

$= p-1$

69. $\dfrac{\dfrac{2}{3-x}+\dfrac{3}{x-3}}{\dfrac{4}{x}+\dfrac{5}{x-3}} = \dfrac{\dfrac{-2}{x-3}+\dfrac{3}{x-3}}{\dfrac{4}{x}+\dfrac{5}{x-3}}$

$= \dfrac{\left(\dfrac{-2}{x-3}+\dfrac{3}{x-3}\right)x(x-3)}{\left(\dfrac{4}{x}+\dfrac{5}{x-3}\right)x(x-3)}$

$= \dfrac{-2x+3x}{4(x-3)+5x} = \dfrac{x}{4x-12+5x} = \dfrac{x}{9x-12}$

71. $\dfrac{\dfrac{2}{y^2-y-20}}{\dfrac{3}{y+4}-\dfrac{4}{y-5}}=\dfrac{\dfrac{2}{(y-5)(y+4)}}{\dfrac{3}{y+4}-\dfrac{4}{y-5}}$

$=\dfrac{\left(\dfrac{2}{(y-5)(y+4)}\right)(y-5)(y+4)}{\left(\dfrac{3}{y+4}-\dfrac{4}{y-5}\right)(y-5)(y+4)}$

$=\dfrac{2}{3(y-5)-4(y+4)}$

$=\dfrac{2}{3y-15-4y-16}$

$=\dfrac{2}{-y-31}$

$=\dfrac{-2}{y+31}$

73. a. $\dfrac{1+3m^{-1}}{1-3m^{-1}}=\dfrac{1+\dfrac{3}{m}}{1-\dfrac{3}{m}}$

$=\dfrac{\left(1+\dfrac{3}{m}\right)m}{\left(1-\dfrac{3}{m}\right)m}$

$=\dfrac{m+3}{m-3}$

b. $\dfrac{1+2x^{-2}}{1-2x^{-2}}=\dfrac{1+\dfrac{2}{x^2}}{1-\dfrac{2}{x^2}}$

$=\dfrac{\left(1+\dfrac{2}{x^2}\right)x^2}{\left(1-\dfrac{2}{x^2}\right)x^2}$

$=\dfrac{x^2+2}{x^2-2}$

75. $\dfrac{1}{f_1}+\dfrac{1}{f_2}$

$\dfrac{f_2+f_1}{f_1f_2}$

77. $\dfrac{\dfrac{a}{x+h}-\dfrac{a}{x}}{h}$

$=\dfrac{\left(\dfrac{a}{x+h}-\dfrac{a}{x}\right)x(x+h)}{hx(x+h)}$

$=\dfrac{ax-a(x+h)}{hx(x+h)}$

$=\dfrac{ax-ax-ah}{hx(x+h)}$

$=\dfrac{-ah}{hx(x+h)}$

$=\dfrac{-a}{x(x+h)}$

79. $\dfrac{\dfrac{1}{2(x+h)^2}-\dfrac{1}{2x^2}}{h}$

$=\dfrac{\left(\dfrac{1}{2(x+h)^2}-\dfrac{1}{2x^2}\right)(x+h)^2(2x^2)}{h(x+h)^2(2x^2)}$

$=\dfrac{x^2-(x+h)^2}{h(x+h)^2(2x^2)}$

$=\dfrac{x^2-x^2-2xh-h^2}{h(x+h)^2(2x^2)}$

$=\dfrac{-2xh-h^2}{h(x+h)^2(2x^2)}$

$=\dfrac{-h(2x+h)}{h(x+h)^2(2x^2)}$

$=\dfrac{-(2x+h)}{2x^2(x+h)^2}$

81. $C = \dfrac{450P}{100 - P}$

P	$\dfrac{450P}{100 - P}$
40	$\dfrac{450(40)}{100 - 40} = 300$
60	$\dfrac{450(60)}{100 - 60} = 675$
80	$\dfrac{450(80)}{100 - 80} = 1800$
90	$\dfrac{450(90)}{100 - 90} = 4050$
93	$\dfrac{450(93)}{100 - 93} \approx 5979$
95	$\dfrac{450(95)}{100 - 95} = 8550$
98	$\dfrac{450(98)}{100 - 98} = 22050$
100	$\dfrac{450(100)}{100 - 100} = error$

a. $C = \dfrac{450(40)}{100 - 40} = \$300$ million;

  $C = \dfrac{450(85)}{100 - 85} = \$2550$ million

b. It would require many resources.

c. No

83. $P = \dfrac{50\left(7d^2 + 10\right)}{d^3 + 50}$

d	$P = \dfrac{50\left(7d^2 + 10\right)}{d^3 + 50}$
0	$P = \dfrac{50\left(7(0)^2 + 10\right)}{(0)^3 + 50} = 10$
1	$P = \dfrac{50\left(7(1)^2 + 10\right)}{(1)^3 + 50} = 16.67$
2	$P = \dfrac{50\left(7(2)^2 + 10\right)}{(2)^3 + 50} = 32.76$
3	$P = \dfrac{50\left(7(3)^2 + 10\right)}{(3)^3 + 50} = 47.40$
4	$P = \dfrac{50\left(7(4)^2 + 10\right)}{(4)^3 + 50} = 53.51$
5	$P = \dfrac{50\left(7(5)^2 + 10\right)}{(5)^3 + 50} = 52.86$
6	$P = \dfrac{50\left(7(6)^2 + 10\right)}{(6)^3 + 50} = 49.25$
7	$P = \dfrac{50\left(7(7)^2 + 10\right)}{(7)^3 + 50} = 44.91$
8	$P = \dfrac{50\left(7(8)^2 + 10\right)}{(8)^3 + 50} = 40.75$
9	$P = \dfrac{50\left(7(9)^2 + 10\right)}{(9)^3 + 50} = 37.03$
10	$P = \dfrac{50\left(7(10)^2 + 10\right)}{(10)^3 + 50} = 33.81$

Price rises rapidly for first four days, then begins a gradual decrease.
Yes, on the 35th day of trading.

85.

t	$N = \dfrac{60t - 120}{t}$
3	$N = \dfrac{60(3) - 120}{3} = 20$
4	$N = \dfrac{60(4) - 120}{4} = 30$
5	$N = \dfrac{60(5) - 120}{5} = 36$
6	$N = \dfrac{60(6) - 120}{6} = 40$
7	$N = \dfrac{60(7) - 120}{7} = 42.9$
8	$N = \dfrac{60(8) - 120}{8} = 45$

$t = 8$ weeks

87. (b); $\quad 20 \cdot n \div 10 \cdot n = \dfrac{20n}{10} \cdot n = 2n^2$

All the others equal 2.

89. $\dfrac{1}{\dfrac{5}{2} + \dfrac{4}{3}} = \dfrac{1(6)}{\left(\dfrac{5}{2} + \dfrac{4}{3}\right)(6)} = \dfrac{6}{15 + 8} = \dfrac{6}{23}$

The reciprocal of the sum of their

reciprocals is $\dfrac{6}{23}$.

$\dfrac{1}{\dfrac{b}{a} + \dfrac{d}{c}} = \dfrac{1(ac)}{\left(\dfrac{b}{a} + \dfrac{d}{c}\right)(ac)} = \dfrac{ac}{bc + ad}$

R.6 Exercises

1. Even

3. $\left(16^{\frac{1}{4}}\right)^3$

5. Answers will vary.

7. $\sqrt{x^2}$

   a. $\sqrt{(9)^2} = \sqrt{81} = 9$

   b. $\sqrt{(-10)^2} = \sqrt{100} = 10$

9. a. $\sqrt{49p^2} = \sqrt{(7p)^2} = 7|p|$

   b. $\sqrt{(x-3)^2} = |x-3|$

   c. $\sqrt{81m^4} = \sqrt{(9m^2)^2} = 9m^2$

   d. $\sqrt{x^2-6x+9} = \sqrt{(x-3)^2} = |x-3|$

11. a. $\sqrt[3]{64} = \sqrt[3]{(4)^3} = 4$

    b. $\sqrt[3]{-125x^3} = \sqrt[3]{(-5x)^3} = -5x$

    c. $\sqrt[3]{216z^{12}} = \sqrt[3]{(6z^4)^3} = 6z^4$

    d. $\sqrt[3]{\dfrac{v^3}{-8}} = \sqrt[3]{\left(\dfrac{v}{-2}\right)^3} = \dfrac{v}{-2}$

13. a. $\sqrt[6]{64} = \sqrt[6]{(2)^6} = 2$

    b. $\sqrt[6]{-64}$ Not a real number

    c. $\sqrt[5]{243x^{10}} = \sqrt[5]{(3x^2)^5} = 3x^2$

    d. $\sqrt[5]{-243x^5} = \sqrt[5]{(-3x)^5} = -3x$

    e. $\sqrt[5]{(k-3)^5} = k-3$

    f. $\sqrt[6]{(h+2)^6} = |h+2|$

15. a. $\sqrt[3]{-125} = \sqrt[3]{(-5)^3} = -5$

    b. $-\sqrt[4]{81n^{12}} = -\sqrt[4]{(3n^3)^4} = -3|n^3|$

    c. $\sqrt{-36}$ Not a real number

    d. $\sqrt{\dfrac{49v^{10}}{36}} = \sqrt{\left(\dfrac{7v^5}{6}\right)^2} = \dfrac{7|v^5|}{6}$

17. a. $8^{\frac{2}{3}} = \left(8^{1/3}\right)^2 = \sqrt[3]{8}^2 = 2^2 = 4$

    b. $\left(\dfrac{16}{25}\right)^{\frac{3}{2}} = \left[\left(\dfrac{16}{25}\right)^{\frac{1}{2}}\right]^3$
    $= \sqrt{\left(\dfrac{16}{25}\right)^3} = \left(\dfrac{4}{5}\right)^3 = \dfrac{64}{125}$

    c. $\left(\dfrac{4}{25}\right)^{-\frac{3}{2}} = \left(\dfrac{25}{4}\right)^{\frac{3}{2}} = \left[\left(\dfrac{25}{4}\right)^{\frac{1}{2}}\right]^3$
    $= \sqrt{\left(\dfrac{25}{4}\right)^3} = \left(\dfrac{5}{2}\right)^3 = \dfrac{125}{8}$

    d. $\left(\dfrac{-27p^6}{8q^3}\right)^{\frac{2}{3}} = \left[\left(\dfrac{-27p^6}{8q^3}\right)^{\frac{1}{3}}\right]^2$
    $= \sqrt[3]{\left(\dfrac{-27p^6}{8q^3}\right)^2} = \left(\dfrac{-3p^2}{2q}\right)^2 = \dfrac{9p^4}{4q^2}$

R.6 Exercises

19. a. $-144^{\frac{3}{2}} = -\left[(144)^{\frac{1}{2}}\right]^3 = -\sqrt{144}^3$

   $= -(12)^3 = -1728$

   b. $\left(-\frac{4}{25}\right)^{\frac{3}{2}} = \left[\left(-\frac{4}{25}\right)^{\frac{1}{2}}\right]^3 = \sqrt{\left(-\frac{4}{25}\right)}^3$

   Not a real number

   c. $(-27)^{\frac{-2}{3}} = \left(-\frac{1}{27}\right)^{\frac{2}{3}} = \left[\left(-\frac{1}{27}\right)^{\frac{1}{3}}\right]^2$

   $= \sqrt[3]{\left(-\frac{1}{27}\right)}^2 = \left(-\frac{1}{3}\right)^2 = \frac{1}{9}$

   d. $-\left(\frac{27x^3}{64}\right)^{-\frac{4}{3}} = -\left(\frac{64}{27x^3}\right)^{\frac{4}{3}}$

   $= -\left[\left(\frac{64}{27x^3}\right)^{\frac{1}{3}}\right]^4 = -\sqrt[3]{\left(\frac{64}{27x^3}\right)}^4$

   $= -\left(\frac{4}{3x}\right)^4 = \frac{-256}{81x^4}$

21. a. $\left(2n^2p^{\frac{-2}{5}}\right)^5 = 32n^{10}p^{-2} = \frac{32n^{10}}{p^2}$

   b. $\left(\frac{8y^{\frac{3}{4}}}{64y^{\frac{3}{2}}}\right)^{\frac{1}{3}} = \frac{8^{\frac{1}{3}}y^{\frac{1}{4}}}{64^{\frac{1}{3}}y^{\frac{1}{2}}}$

   $= \frac{\sqrt[3]{8}y^{\frac{1}{4}-\frac{1}{2}}}{\sqrt[3]{64}} = \frac{2y^{\frac{-1}{4}}}{4} = \frac{1}{2y^{\frac{1}{4}}}$

23. a. $\sqrt{18m^2} = \sqrt{(3m)^2 \cdot 2} = 3m\sqrt{2}$

   b. $-2\sqrt[3]{-125p^3q^7} = -2\sqrt[3]{\left(-5pq^2\right)^3 \cdot q}$

   $= -2\left(-5pq^2\right)\sqrt[3]{q} = 10pq^2\sqrt[3]{q}$

   c. $\frac{3}{8}\sqrt[3]{64m^3n^5} = \frac{3}{8}\sqrt[3]{(4mn)^3n^2}$

   $= \frac{3}{8}(4mn)\sqrt[3]{n^2} = \frac{3}{2}mn\sqrt[3]{n^2}$

   d. $\sqrt{32p^3q^6} = \sqrt{\left(16p^2q^6\right)\cdot 2p}$

   $\sqrt{\left(4pq^3\right)^2 \cdot 2p} = 4pq^3\sqrt{2p}$

   e. $\frac{-6+\sqrt{28}}{2} = \frac{-6+\sqrt{4\cdot 7}}{2} = \frac{-6+2\sqrt{7}}{2}$

   $= \frac{-6}{2} + \frac{2\sqrt{7}}{2} = -3 + \sqrt{7}$

   f. $\frac{27-\sqrt{72}}{6} = \frac{27-\sqrt{36\cdot 2}}{6} = \frac{27-6\sqrt{2}}{6}$

   $= \frac{27}{6} - \frac{6\sqrt{2}}{6} = \frac{9}{2} - \sqrt{2}$

25. a. $2.5\sqrt{18a}\sqrt{2a^3} = 2.5\sqrt{36a^4}$

   $= 2.5\sqrt{\left(6a^2\right)^2} = 2.5\left(6a^2\right) = 15a^2$

   b. $-\frac{2}{3}\sqrt{3b}\sqrt{12b^2} = -\frac{2}{3}\sqrt{36b^2}\sqrt{b}$

   $= -\frac{2}{3}(6b)\sqrt{b} = -4b\sqrt{b}$

   c. $\sqrt{\frac{x^3y}{3}}\sqrt{\frac{4x^5y}{12y}} = \sqrt{\frac{4x^8y^2}{36y}} = \sqrt{\frac{x^8y}{9}}$

   $= \sqrt{\frac{x^8}{9}}\sqrt{y} = \frac{x^4\sqrt{y}}{3}$

   d. $\sqrt[3]{9v^2u}\,\sqrt[3]{3u^5v^2} = \sqrt[3]{27v^4u^6}$

   $= \sqrt[3]{27u^6v^3}\sqrt[3]{v} = \sqrt[3]{\left(3u^2v\right)^3}\sqrt[3]{v} = 3u^2v\,\sqrt[3]{v}$

27. a. $\dfrac{\sqrt{8m^5}}{\sqrt{2m}} = \sqrt{\dfrac{8m^5}{2m}} = \sqrt{4m^4} = 2m^2$

b. $\dfrac{\sqrt[3]{108n^4}}{\sqrt[3]{4n}} = \sqrt[3]{\dfrac{108n^4}{4n}} = \sqrt[3]{27n^3} = 3n$

c. $\sqrt{\dfrac{45}{16x^2}} = \dfrac{\sqrt{9\cdot5}}{\sqrt{16x^2}} = \dfrac{3\sqrt{5}}{4x}$

d. $12\sqrt[3]{\dfrac{81}{8z^9}} = 12\dfrac{\sqrt[3]{81}}{\sqrt[3]{8z^9}} = 12\dfrac{\sqrt[3]{27\cdot3}}{2z^3}$

$= \dfrac{12(3)\sqrt[3]{3}}{2z^3} = \dfrac{18\sqrt[3]{3}}{z^3}$

29. a. $\sqrt[5]{32x^{10}y^{15}} = 2x^2y^3$

b. $x\sqrt[4]{x^5} = x\sqrt[4]{x^4\cdot x}$

$= x\cdot x\sqrt[4]{x} = x^2\sqrt[4]{x}$

c. $\sqrt[4]{\sqrt[3]{b}} = \sqrt[4]{b^{\frac{1}{3}}} = \left((b)^{\frac{1}{3}}\right)^{\frac{1}{4}}$

$= b^{\frac{1}{12}} = \sqrt[12]{b}$

d. $\dfrac{\sqrt[3]{6}}{\sqrt{6}} = \dfrac{6^{\frac{1}{3}}}{6^{\frac{1}{2}}} = 6^{\frac{1}{3}-\frac{1}{2}} = 6^{-\frac{1}{6}} = \dfrac{1}{\sqrt[6]{6}}$

or $\dfrac{\sqrt[6]{6^5}}{6}$

e. $\sqrt{b}\cdot\sqrt[4]{b} = b^{\frac{1}{2}}\cdot b^{\frac{1}{4}} = b^{\frac{2}{4}}\cdot b^{\frac{1}{4}} = b^{\frac{3}{4}}$

31. a. $12\sqrt{72} - 9\sqrt{98}$

$= 12\cdot6\sqrt{2} - 9\cdot7\sqrt{2}$

$= 72\sqrt{2} - 63\sqrt{2}$

$= 9\sqrt{2}$

b. $8\sqrt{48} - 3\sqrt{108}$

$= 8\cdot4\sqrt{3} - 3\cdot6\sqrt{3}$

$= 32\sqrt{3} - 18\sqrt{3}$

$= 14\sqrt{3}$

c. $7\sqrt{18m} - \sqrt{50m}$

$= 7\cdot3\sqrt{2m} - 5\sqrt{2m}$

$= 21\sqrt{2m} - 5\sqrt{2m}$

$= 16\sqrt{2m}$

d. $2\sqrt{28p} - 3\sqrt{63p}$

$= 2\cdot2\sqrt{7p} - 3\cdot3\sqrt{7p}$

$= 4\sqrt{7p} - 9\sqrt{7p}$

$= -5\sqrt{7p}$

33. a. $3x\sqrt[3]{54x} - 5\sqrt[3]{16x^4}$

$= 3x\cdot3\sqrt[3]{2x} - 5\cdot2x\sqrt[3]{2x}$

$= 9x\sqrt[3]{2x} - 10x\sqrt[3]{2x}$

$= -x\sqrt[3]{2x}$

b. $\sqrt{4} + \sqrt{3x} - \sqrt{12x} + \sqrt{45}$

$= 2 + \sqrt{3x} - 2\sqrt{3x} + 3\sqrt{5}$

$= 2 - \sqrt{3x} + 3\sqrt{5}$

c. $\sqrt{72x^3} + \sqrt{50} - \sqrt{7x} + \sqrt{27}$

$= 6x\sqrt{2x} + 5\sqrt{2} - \sqrt{7x} + 3\sqrt{3}$

35. a. $\left(7\sqrt{2}\right)^2 = 49\cdot2 = 98$

b. $\sqrt{3}\left(\sqrt{5} + \sqrt{7}\right) = \sqrt{15} + \sqrt{21}$

c. $\left(n + \sqrt{5}\right)\left(n - \sqrt{5}\right) = n^2 - 5$

d. $\left(6 - \sqrt{3}\right)^2 = 36 - 12\sqrt{3} + 3 = 39 - 12\sqrt{3}$

37. a. $\left(3+2\sqrt{7}\right)\left(3-2\sqrt{7}\right) = 9 - 4(7)$
    $= 9 - 28 = -19$

    b. $\left(\sqrt{5} - \sqrt{14}\right)\left(\sqrt{2} + \sqrt{13}\right)$
    $= \sqrt{10} + \sqrt{65} - \sqrt{28} - \sqrt{182}$
    $= \sqrt{10} + \sqrt{65} - 2\sqrt{7} - \sqrt{182}$

    c. $\left(2\sqrt{2} + 6\sqrt{6}\right)\left(3\sqrt{10} + \sqrt{7}\right)$
    $= 6\sqrt{20} + 2\sqrt{14} + 18\sqrt{60} + 6\sqrt{42}$
    $= 12\sqrt{5} + 2\sqrt{14} + 36\sqrt{15} + 6\sqrt{42}$

39. $x^2 - 4x + 1 = 0$

    a. $\left(2+\sqrt{3}\right)^2 - 4(2+\sqrt{3}) + 1 = 0$
    $4 + 4\sqrt{3} + 3 - 8 - 4\sqrt{3} + 1 = 0$
    $0 = 0$
    verified

    b. $\left(2-\sqrt{3}\right)^2 - 4\left(2-\sqrt{3}\right) + 1 = 0$
    $4 - 4\sqrt{3} + 3 - 8 + 4\sqrt{3} + 1 = 0$
    $0 = 0$
    verified

41. $x^2 + 2x - 9 = 0$

    a. $\left(-1+\sqrt{10}\right)^2 + 2\left(-1+\sqrt{10}\right) - 9 = 0$
    $1 - 2\sqrt{10} + 10 - 2 + 2\sqrt{10} - 9 = 0$
    $0 = 0$
    verified

    b. $\left(-1-\sqrt{10}\right)^2 + 2\left(-1-\sqrt{10}\right) - 9 = 0$
    $1 + 2\sqrt{10} + 10 - 2 - 2\sqrt{10} - 9 = 0$
    $0 = 0$
    verified

43. a. $\dfrac{3}{\sqrt{12}} = \dfrac{3}{2\sqrt{3}} \cdot \dfrac{\sqrt{3}}{\sqrt{3}} = \dfrac{3\sqrt{3}}{6} = \dfrac{\sqrt{3}}{2}$

    b. $\sqrt{\dfrac{20}{27x^3}} = \dfrac{2\sqrt{5}}{3x\sqrt{3x}} \cdot \dfrac{\sqrt{3x}}{\sqrt{3x}} = \dfrac{2\sqrt{15x}}{9x^2}$

    c. $\sqrt{\dfrac{27}{50b}} = \dfrac{3\sqrt{3}}{5\sqrt{2b}} \cdot \dfrac{\sqrt{2b}}{\sqrt{2b}} = \dfrac{3\sqrt{6b}}{10b}$

    d. $\sqrt[3]{\dfrac{1}{4p}} = \dfrac{1}{\sqrt[3]{4p}} \cdot \dfrac{\sqrt[3]{2p^2}}{\sqrt[3]{2p^2}} = \dfrac{\sqrt[3]{2p^2}}{2p}$

    e. $\dfrac{5}{\sqrt[3]{a}} = \dfrac{5}{\sqrt[3]{a}} \cdot \dfrac{\sqrt[3]{a^2}}{\sqrt[3]{a^2}} = \dfrac{5 \cdot \sqrt[3]{a^2}}{a}$

45. a. $\dfrac{8}{3+\sqrt{11}} \cdot \dfrac{3-\sqrt{11}}{3-\sqrt{11}}$
    $= \dfrac{8\left(3-\sqrt{11}\right)}{9-11} = \dfrac{8\left(3-\sqrt{11}\right)}{-2}$
    $= -4\left(3-\sqrt{11}\right) = -12 + 4\sqrt{11} \approx 1.27$

    b. $\dfrac{6}{\sqrt{x}-\sqrt{2}} \cdot \dfrac{\sqrt{x}+\sqrt{2}}{\sqrt{x}+\sqrt{2}}$
    $= \dfrac{6\sqrt{x} + 6\sqrt{2}}{x-2}$

47. a. $\dfrac{\sqrt{10}-3}{\sqrt{3}+\sqrt{2}} \cdot \dfrac{\sqrt{3}-\sqrt{2}}{\sqrt{3}-\sqrt{2}}$
    $= \dfrac{\sqrt{30} - \sqrt{20} - 3\sqrt{3} + 3\sqrt{2}}{3-2}$
    $= \sqrt{30} - 2\sqrt{5} - 3\sqrt{3} + 3\sqrt{2}$
    ≈ 0.05

    b. $\dfrac{7+\sqrt{6}}{3-3\sqrt{2}} \cdot \dfrac{3+3\sqrt{2}}{3+3\sqrt{2}}$
    $= \dfrac{21 + 21\sqrt{2} + 3\sqrt{6} + 3\sqrt{12}}{9-18}$
    $= \dfrac{21 + 21\sqrt{2} + 3\sqrt{6} + 6\sqrt{3}}{-9}$
    $= \dfrac{7 + 7\sqrt{2} + \sqrt{6} + 2\sqrt{3}}{-3}$
    ≈ -7.60

49. $L = 1.13(W)^{\frac{1}{3}}$

    $L = 1.13(400)^{\frac{1}{3}} \approx 8.33$ feet

51. $c^2 = a^2 + b^2$
    $c^2 = 8^2 + 24^2$
    $c^2 = 64 + 576$
    $c^2 = 640 ; c = \sqrt{640} = 8\sqrt{10}$ m ;
    About 25.3 m

53. $T = 0.407 R^{\frac{3}{2}}$

    a. $T = 0.407(93)^{\frac{3}{2}} \approx 365.02 \text{ days}$

    b. $T = 0.407(142)^{\frac{3}{2}} \approx 688.69 \text{ days}$

    c. $T = 0.407(36)^{\frac{3}{2}} \approx 87.91 \text{ days}$

55. $V = 2\sqrt{6L}$

    a. $V = 2\sqrt{6(54)} = 2\sqrt{324} = 36 \text{ mph}$

    b. $V = 2\sqrt{6(90)} = 2\sqrt{540} \approx 46.5 \text{ mph}$

57. $S = \pi r \sqrt{r^2 + h^2}$;

    $S = \pi(6)\sqrt{6^2 + 10^2}$

    $S = 6\pi\sqrt{136}$

    $S = 12\pi\sqrt{34}$

    $S \approx 219.82 \text{ m}^2$

59. a. $x^2 - 5$

       $x^2 - \left(\sqrt{5}\right)^2$

       $\left(x + \sqrt{5}\right)\left(x - \sqrt{5}\right)$

    b. $n^2 - 19$

       $n^2 - \left(\sqrt{19}\right)^2$

       $\left(n + \sqrt{19}\right)\left(n - \sqrt{19}\right)$

61. a.

    $\sqrt{3x} + \sqrt{9x} + \sqrt{27x} + \sqrt{81x} + \sqrt{243x} + \sqrt{729x}$

    $= \sqrt{3x} + 3\sqrt{x} + 3\sqrt{3x} + 9\sqrt{x} + 9\sqrt{3x} + 27\sqrt{x}$

    $= 13\sqrt{3x} + 39\sqrt{x}$

    b. Answers will vary.

63. $\left(x^{\frac{1}{2}} + x^{-\frac{1}{2}}\right)^2 = \frac{9}{2}$;

    $\sqrt{\left(x^{\frac{1}{2}} + x^{-\frac{1}{2}}\right)^2} = \sqrt{\frac{9}{2}}$

    $x^{\frac{1}{2}} + x^{-\frac{1}{2}} = \frac{3}{\sqrt{2}}$;

    $\frac{3}{\sqrt{2}} \cdot \frac{\sqrt{2}}{\sqrt{2}} = \frac{3\sqrt{2}}{2}$

Chapter R Practice Test

1. a. True
   b. True
   c. False; $\sqrt{2}$ cannot be expressed as a ratio of integers
   d. True

3. a. $\dfrac{7}{8} - \left(-\dfrac{1}{4}\right) = \dfrac{7}{8} + \dfrac{2}{8} = \dfrac{9}{8}$

   b. $-\dfrac{1}{3} - \dfrac{5}{6} = -\dfrac{2}{6} - \dfrac{5}{6} = -\dfrac{7}{6}$

   c. $-0.7 + 1.2 = 0.5$

   d. $1.3 + (-5.9) = 1.3 - 5.9 = -4.6$

5. $2000\left(1 + \dfrac{0.08}{12}\right)^{12\cdot10} \approx 4439.28$

7. a. $-2v^2 + 6v + 5$
   Terms: 3
   Coefficients: -2, 6, 5

   b. $\dfrac{c+2}{3} + c$
   Terms: 2
   Coefficients: $\dfrac{1}{3}$, 1

9. a. $x^3 - (2x - 9)$

   b. $2n - 3\left(\dfrac{n}{2}\right)^2$

11. a. $8v^2 + 4v - 7 + v^2 - v = 9v^2 + 3v - 7$
    b. $-4(3b - 2) + 5b$
       $= -12b + 8 + 5b = -7b + 8$
    c. $4x - \left(x - 2x^2\right) + x(3 - x)$
       $= 4x - x + 2x^2 + 3x - x^2$
       $= x^2 + 6x$

13. a. $\dfrac{5}{b^{-3}} = 5b^3$

    b. $\left(-2a^3\right)^2\left(a^2b^4\right)^3$
       $= \left(4a^6\right)\left(a^6b^{12}\right)$
       $= 4a^{12}b^{12}$

    c. $\left(\dfrac{m^2}{2n}\right)^3 = \dfrac{m^6}{8n^3}$

    d. $\left(\dfrac{5p^2q^3r^4}{-2pq^2r^4}\right)^2$

       $= \left(\dfrac{5pq}{-2}\right)^2$

       $= \dfrac{25}{4}p^2q^2$

15. a. $\left(3x^2 + 5y\right)\left(3x^2 - 5y\right) = 9x^4 - 25y^2$
    b. $(2a + 3b)^2 = 4a^2 + 12ab + 9b^2$

17. a. $\dfrac{x-5}{5-x} = \dfrac{-(5-x)}{5-x} = -1$

    b. $\dfrac{4 - n^2}{n^2 - 4n + 4}$

       $= \dfrac{-1(n^2 - 4)}{(n-2)(n-2)}$

       $= \dfrac{-1(n+2)(n-2)}{(n-2)(n-2)}$

       $= \dfrac{-1(n+2)}{(n-2)}$

       $= \dfrac{n+2}{2-n}$

    c. $\dfrac{x^3 - 27}{x^2 + 3x + 9}$

       $= \dfrac{(x-3)\left(x^2 + 3x + 9\right)}{x^2 + 3x + 9} = x - 3$

d. $\dfrac{3x^2 - 13x - 10}{9x^2 - 4}$

$= \dfrac{(3x + 2)(x - 5)}{(3x + 2)(3x - 2)}$

$\dfrac{x - 5}{3x - 2}$

e. $\dfrac{x^2 - 25}{3x^2 - 11x - 4} \div \dfrac{x^2 + x - 20}{x^2 - 8x + 16}$

$= \dfrac{x^2 - 25}{3x^2 - 11x - 4} \cdot \dfrac{x^2 - 8x + 16}{x^2 + x - 20}$

$= \dfrac{(x + 5)(x - 5)}{(3x + 1)(x - 4)} \cdot \dfrac{(x - 4)^2}{(x + 5)(x - 4)}$

$= \dfrac{x - 5}{3x + 1}$

f. $\dfrac{m + 3}{m^2 + m - 12} - \dfrac{2}{5(m + 4)}$

$= \dfrac{m + 3}{(m + 4)(m - 3)} - \dfrac{2}{5(m + 4)}$

$= \dfrac{5(m + 3) - 2(m - 3)}{5(m + 4)(m - 3)}$

$= \dfrac{5m + 15 - 2m + 6}{5(m + 4)(m - 3)}$

$= \dfrac{3m + 21}{5(m + 4)(m - 3)}$

$= \dfrac{3(m + 7)}{5(m + 4)(m - 3)}$

19. $R = (30 - 0.5x)(40 + x)$

$R = 1200 + 10x - 0.5x^2$

$R = -0.5x^2 + 10x + 1200$

a.

x	$R = 1200 + 10x - 0.5x^2$
1	$R = 1200 + 10(1) - 0.5(1)^2 = 1209.50$
2	$R = 1200 + 10(2) - 0.5(2)^2 = 1218$
3	$R = 1200 + 10(3) - 0.5(3)^2 = 1225.50$
4	$R = 1200 + 10(4) - 0.5(4)^2 = 1232$
5	$R = 1200 + 10(5) - 0.5(5)^2 = 1237.50$
6	$R = 1200 + 10(6) - 0.5(6)^2 = 1242$
7	$R = 1200 + 10(7) - 0.5(7)^2 = 1245.50$
8	$R = 1200 + 10(8) - 0.5(8)^2 = 1248$
9	$R = 1200 + 10(9) - 0.5(9)^2 = 1249.50$
10	$R = 1200 + 10(10) - 0.5(10)^2 = 1250$
11	$R = 1200 + 10(11) - 0.5(11)^2 = 1249.50$

Ten decreases of 0.50 or $5.00

b. Maximum revenue is $1250.

1.1 Technology Highlight

1. 12 pounds of premium ground beef;
   40 pounds of peanuts

1.1 Exercises

1. Identity, unknown

3. Literal, two

5. Answers will vary.

7. $4x + 3(x-2) = 18 - x$

   $4x + 3x - 6 = 18 - x$

   $7x - 6 = 18 - x$

   $8x - 6 = 18$

   $8x = 24$

   $x = 3$;

   Check:

   $4(3) + 3(3-2) = 18 - 3$

   $12 + 3(1) = 15$

   $12 + 3 = 15$

   $15 = 15$

9. $21 - (2v + 17) = -7 - 3v$

   $21 - 2v - 17 = -7 - 3v$

   $-2v + 4 = -7 - 3v$

   $v + 4 = -7$

   $v = -11$;

   Check:

   $21 - (2(-11) + 17) = -7 - 3(-11)$

   $21 - (-22 + 17) = -7 + 33$

   $21 - (-5) = 26$

   $21 + 5 = 26$

   $26 = 26$

11. $8 - (3b + 5) = -5 + 2(b + 1)$

    $8 - 3b - 5 = -5 + 2b + 2$

    $3 - 3b = -3 + 2b$

    $-5b = -6$

    $b = \dfrac{6}{5}$;

    Check:

    $8 - \left(3\left(\dfrac{6}{5}\right) + 5 \right) = -5 + 2\left(\dfrac{6}{5} + 1\right)$

    $8 - \left(\dfrac{18}{5} + \dfrac{25}{5} \right) = -5 + 2\left(\dfrac{6}{5} + \dfrac{5}{5}\right)$

    $8 - \left(\dfrac{43}{5} \right) = -5 + 2\left(\dfrac{11}{5}\right)$

    $\dfrac{40}{5} - \dfrac{43}{5} = \dfrac{-25}{5} + \dfrac{22}{5}$

    $\dfrac{-3}{5} = \dfrac{-3}{5}$

13. $\dfrac{1}{5}(b + 10) - 7 = \dfrac{1}{3}(b - 9)$

    $\dfrac{1}{5}b + 2 - 7 = \dfrac{1}{3}b - 3$

    $\dfrac{1}{5}b - 5 = \dfrac{1}{3}b - 3$

    $15\left(\dfrac{1}{5}b - 5\right) = 15\left(\dfrac{1}{3}b - 3\right)$

    $3b - 75 = 5b - 45$

    $-2b - 75 = -45$

    $-2b = 30$

    $b = -15$

15. $\dfrac{2}{3}(m + 6) = \dfrac{-1}{2}$

    $\dfrac{2}{3}m + 4 = \dfrac{-1}{2}$

    $6\left(\dfrac{2}{3}m + 4\right) = 6\left(\dfrac{-1}{2}\right)$

    $4m + 24 = -3$

    $4m = -27$

    $m = -\dfrac{27}{4}$

17. $\dfrac{1}{2}x + 5 = \dfrac{1}{3}x + 7$

$6\left(\dfrac{1}{2}x + 5\right) = 6\left(\dfrac{1}{3}x + 7\right)$

$3x + 30 = 2x + 42$

$x + 30 = 42$

$x = 12$

19. $\dfrac{x+3}{5} + \dfrac{x}{3} = 7$

$15\left(\dfrac{x+3}{5} + \dfrac{x}{3}\right) = 15(7)$

$3(x+3) + 5x = 105$

$3x + 9 + 5x = 105$

$8x + 9 = 105$

$8x = 96$

$x = 12$

21. $15 = -6 - \dfrac{3p}{8}$

$21 = -\dfrac{3p}{8}$

$\left(\dfrac{-8}{3}\right)(21) = \dfrac{-8}{3}\left(-\dfrac{3p}{8}\right)$

$-56 = p$

23. $0.2(24 - 7.5a) - 6.1 = 4.1$

$4.8 - 1.5a - 6.1 = 4.1$

$-1.5a - 1.3 = 4.1$

$-1.5a = 5.4$

$a = -3.6$

25. $6.2v - (2.1v - 5) = 1.1 - 3.7v$

$6.2v - 2.1v + 5 = 1.1 - 3.7v$

$4.1v + 5 = 1.1 - 3.7v$

$7.8v + 5 = 1.1$

$7.8v = -3.9$

$v = -0.5$

27. $\dfrac{n}{2} + \dfrac{n}{5} = \dfrac{2}{3}$

$30\left(\dfrac{n}{2} + \dfrac{n}{5}\right) = 30\left(\dfrac{2}{3}\right)$

$15n + 6n = 20$

$21n = 20$

$n = \dfrac{20}{21}$

29. $3p - \dfrac{p}{4} - 5 = \dfrac{p}{6} - 2p + 6$

$12\left(3p - \dfrac{p}{4} - 5\right) = 12\left(\dfrac{p}{6} - 2p + 6\right)$

$36p - 3p - 60 = 2p - 24p + 72$

$33p - 60 = -22p + 72$

$55p - 60 = 72$

$55p = 132$

$p = \dfrac{12}{5}$

31. $-3(4z + 5) = -15z - 20 + 3z$

$-12z - 15 = -15z - 20 + 3z$

$-12z - 15 = -12z - 20$

$-15 \neq -20$

Contradiction; $\{\ \}$

33. $8 - 8(3n + 5) = -5 + 6(1 + n)$

$8 - 24n - 40 = -5 + 6 + 6n$

$-24n - 32 = 1 + 6n$

$-30n = 33$

$n = -\dfrac{11}{10}$

Conditional; $n = -\dfrac{11}{10}$

35. $-4(4x + 5) = -6 - 2(8x + 7)$

$-16x - 20 = -6 - 16x - 14$

$-16x - 20 = -20 - 16x$

$0 = 0$

Identity; $\{x \mid x \in \mathbb{i}\ \}$

37. $P = C + CM$

    $P = C(1 + M)$

    $\dfrac{P}{1 + M} = C$

    $C = \dfrac{P}{1 + M}$

39. $C = 2\pi r$

    $\dfrac{C}{2\pi} = r$

    $r = \dfrac{C}{2\pi}$

41. $\dfrac{P_1 V_1}{T_1} = \dfrac{P_2 V_2}{T_2}$

    $T_1 T_2 \left(\dfrac{P_1 V_1}{T_1} \right) = T_1 T_2 \left(\dfrac{P_2 V_2}{T_2} \right)$

    $P_1 V_1 T_2 = T_1 P_2 V_2$

    $T_2 = \dfrac{T_1 P_2 V_2}{P_1 V_1}$

43. $V = \dfrac{4}{3} \pi r^2 h$

    $\dfrac{3}{4} V = \pi r^2 h$

    $\dfrac{3V}{4\pi r^2} = h$

    $h = \dfrac{3V}{4\pi r^2}$

45. $S_n = n \left(\dfrac{a_1 + a_n}{2} \right)$

    $\left(\dfrac{2}{a_1 + a_n} \right) \cdot S_n = \left(\dfrac{2}{a_1 + a_n} \right) \cdot n \left(\dfrac{a_1 + a_n}{2} \right)$

    $\dfrac{2S_n}{a_1 + a_n} = n$

    $n = \dfrac{2S_n}{a_1 + a_n}$

47. $S = B + \dfrac{1}{2} PS$

    $S - B = \dfrac{1}{2} PS$

    $2(S - B) = PS$

    $\dfrac{2(S - B)}{S} = P$

49. $Ax + By = C$

    $By = -Ax + C$

    $y = \dfrac{-A}{B} x + \dfrac{C}{B}$

51. $\dfrac{5}{6} x + \dfrac{3}{8} y = 2$

    $\dfrac{3}{8} y = -\dfrac{5}{6} x + 2$

    $\left(\dfrac{8}{3} \right) \left(\dfrac{3}{8} y \right) = \left(\dfrac{8}{3} \right) \left(-\dfrac{5}{6} x + 2 \right)$

    $y = -\dfrac{20}{9} x + \dfrac{16}{3}$

53. $y - 3 = \dfrac{-4}{5} (x + 10)$

    $y - 3 = \dfrac{-4}{5} x - 8$

    $y = \dfrac{-4}{5} x - 5$

55. $3x + 2 = -19$

    $a = 3, b = 2, c = -19$

    $x = \dfrac{-19 - 2}{3}$

    $x = -7$

57. $-6x + 1 = 33$

    $a = -6, b = 1, c = 33$

    $x = \dfrac{33 - 1}{-6}$

    $x = -\dfrac{16}{3}$

59. $7x - 13 = -27$

    $a = 7, b = -13, c = -27$

    $x = \dfrac{-27 - (-13)}{7}$

    $x = -2$

61. $SA = 2\pi r^2 + 2\pi rh$

    $1256 = 2(3.14)(8)^2 + 2(3.14)(8)h$

    $1256 = 401.92 + 50.24h$

    $854.08 = 50.24h$

    $17 = h$

    $h = 17 \text{cm}$

63. Let x represent the length of the second descent.
$$2x + 198 = 1218$$
$$2x = 1218 - 198$$
$$2x = 1020$$
$$x = 510$$
The second spelunker descended 510 feet.

65. Let L represent the length of the package.
$$2(14 + 12) + L = 108$$
$$2(26) + L = 108$$
$$52 + L = 108$$
$$L = 56$$
The package can be up to 56 inches long.

67. Let L represent the length of the Shimotsui bridge.
$$364 + 2L = 6532$$
$$2L = 6168$$
$$L = 3084$$
The Shimotsui bridge is 3084 feet long.

69. Let x represent the first consecutive even integer.
Let $x + 2$ represent the second consecutive even integer.
$$2x + x + 2 = 146$$
$$3x + 2 = 146$$
$$3x = 144$$
$$x = 48$$
The first integer is 48.
The second integer is 50.

71. Let x represent the first consecutive odd integer.
Let $x + 2$ represent the second consecutive odd integer.
$$7x = 5(x + 2)$$
$$7x = 5x + 10$$
$$2x = 10$$
$$x = 5$$
The first integer is 5.
The second integer is 7.

73. Let t represent the number of hours when Bruce overtakes Linda.
$$D_{Linda} = D_{Bruce}$$
$$60(t + 0.5) = 75t$$
$$60t + 30 = 75t$$
$$30 = 15t$$
$$2 = t$$
2 hours after 9:30am is 11:30am.

75. Let t represent the number of hours Jeff was driving in the construction zone.
$$D_1 + D_2 = 72$$
$$30t + 60(1.5 - t) = 72$$
$$30t + 90 - 60t = 72$$
$$-30t + 90 = 72$$
$$-30t = -18$$
$$t = 0.6 \text{ hour}$$
$$0.6(60) = 36 \text{ minutes.}$$

77. 2 quarts + 2 quarts = 4 quarts;
$$2(1.00) + 2(0.00) = 2$$
$$\frac{2}{4} = 50\% \text{ juice}$$
4 quart mixture of 50% orange juice

79. 8 lbs + 8 lbs = 16 lbs;
$$8(2.50) + 8(1.10) = 28.8$$
$$\frac{\$28.80}{16} = \$1.80 \text{ per pound}$$
Sixteen pound mixture at a cost of \$1.80 per pound

81. Let x represent the number of pounds of premium ground beef.
$$3.10(x) + 2.05(8) = 2.68(x + 8)$$
$$3.10x + 16.4 = 2.68x + 21.44$$
$$0.42x + 16.4 = 21.44$$
$$0.42x = 5.04$$
$$x = 12$$
12 pounds of premium ground beef

83. Let x represent the pounds of walnuts.

$$0.84x + 1.20(20) = 1.04(x + 20)$$
$$0.84x + 24 = 1.04x + 20.8$$
$$-0.2x = -3.2$$
$$x = 16$$

16 pounds of walnuts

85. Answers will vary.

87. $P + Q + S = 40$

$P + R + U = 34$
$S + T + U = 30$
$Q + R = 26$
$Q + T = 23$
$R + T = 19;$

$Q + R = 26$
$\underline{-Q - T = -23}$
$R - T = 3;$

$R - T = 3$
$R + T = 19$
$\overline{2R = 22}$
$R = 11;$

$Q + R = 26$
$Q + 11 = 26$
$Q = 15;$

$Q + T = 23$
$15 + T = 23$
$T = 8;$

$P + R + U = 34$
$P + 11 + U = 34$
$P + U = 23;$

$P + Q + S = 40$
$P + 15 + S = 40$
$P + S = 25;$

$P + U = 23$
$\underline{-P - S = -25}$
$U - S = -2;$

$S + T + U = 30$
$S + 8 + U = 30$
$S + U = 22;$

$U - S = -2$
$S + U = 22$
$\overline{2U = 20}$
$U = 10;$

$S + U = 22$
$S + 10 = 22$
$S = 12;$

$P + Q + S = 40$
$P + 15 + 12 = 40$
$P = 13;$

$P + Q + R + S + T + U$
$= 13 + 15 + 11 + 12 + 8 + 10 = 69$

89. $-2 - 6^2 \div 4 + 8$
$= -2 - 36 \div 4 + 8$
$= -2 - 9 + 8$
$= -11 + 8$
$= -3$

91. a. $4x^2 - 9$
$= (2x + 3)(2x - 3)$

   b. $x^3 - 27$
$= (x - 3)(x^2 + 3x + 9)$

1.2 Exercises

1. Set, interval

3. Intersection, union

5. Answers will vary.

7. $w \geq 45$

9. $250 < T < 450$

11. $y < 3$

13. $m \leq 5$

15. $x \neq 1$

17. $5 > x > 2$

19. $\{x | x \geq -2\}; [-2, \infty)$

21. $\{x | -2 \leq x \leq 1\}; [-2, 1]$

23. $5a - 11 \geq 2a - 5$
$3a \geq 6$
$a \geq 2$
$\{a | a \geq 2\}$, Interval notation: $a \in [2, \infty)$

25. $2(n+3) - 4 \leq 5n - 1$
$2n + 6 - 4 \leq 5n - 1$
$2n + 2 \leq 5n - 1$
$-3n \leq -3$
$n \geq 1$
$\{n | n \geq 1\}$, Interval notation: $n \in [1, \infty)$

27. $\dfrac{3x}{8} + \dfrac{x}{4} < -4$

$8\left(\dfrac{3x}{8} + \dfrac{x}{4}\right) < 8(-4)$
$3x + 2x < -32$
$5x < -32$
$x < -\dfrac{32}{5}$

$\left\{x \Big| x < \dfrac{-32}{5}\right\}$,

Interval notation: $x \in \left(-\infty, \dfrac{-32}{5}\right)$

29. $7 - 2(x+3) \geq 4x - 6(x-3)$
$7 - 2x - 6 \geq 4x - 6x + 18$
$-2x + 1 \geq -2x + 18$
$1 \geq 18$ false
$\{\ \}$

31. $4(3x-5) + 18 < 2(5x+1) + 2x$
$12x - 20 + 18 < 10x + 2 + 2x$
$12x - 2 < 12x + 2$
$-2 < 2$ true
$\{x | x \in \mathbb{R} \}$

33. $-6(p-1) + 2p \leq -2(2p-3)$
$-6p + 6 + 2p \leq -4p + 6$
$-4p + 6 \leq -4p + 6$
$6 \leq 6$ true
$\{p | p \in \mathbb{R} \}$

35. $A \cap B = \{2\}$
$A \cup B = \{-3, -2, -1, 0, 1, 2, 3, 4, 6, 8\}$

37. $A \cap D = \{\ \}$
$A \cup D = \{-3, -2, -1, 0, 1, 2, 3, 4, 5, 6, 7\}$

39. $B \cap D = \{4, 6\}$
$B \cup D = \{2, 4, 5, 6, 7, 8\}$

41. $x < -2$ or $x > 1$
$(-\infty, -2) \cup (1, \infty)$

43. $x < 5$ and $x \geq -2$

$[-2,5)$

45. $x \geq 3$ and $x \leq 1$

no solution

47. $4(x-1) \leq 20$ or $x + 6 > 9$

$4x - 4 \leq 20$ or $x > 3$

$4x \leq 24$

$x \leq 6$ or $x > 3$

$x \in (-\infty, \infty)$

49. $-2x - 7 \leq 3$ and $2x \leq 0$

$-2x \leq 10$ and $x \leq 0$

$x \geq -5$ and $x \leq 0$

$x \in [-5,0]$

51. $\dfrac{3}{5}x + \dfrac{1}{2} > \dfrac{3}{10}$ and $-4x > 1$

$10\left(\dfrac{3}{5}x + \dfrac{1}{2}\right) > \left(\dfrac{3}{10}\right)10$ and $-4x > 1$

$6x + 5 > 3$ and $x < -\dfrac{1}{4}$

$6x > -2$ and $x < -\dfrac{1}{4}$

$x > -\dfrac{1}{3}$ and $x < -\dfrac{1}{4}$

$x \in \left(\dfrac{-1}{3}, \dfrac{-1}{4}\right)$

53. $\dfrac{3x}{8} + \dfrac{x}{4} < -3$ or $x + 1 > -5$

$8\left(\dfrac{3x}{8} + \dfrac{x}{4}\right) < 8(-3)$ or $x > -6$

$3x + 2x < -24$ or $x > -6$

$5x < -24$ or $x > -6$

$x < -\dfrac{24}{5}$ or $x > -6$

$x \in (-\infty, \infty)$

55. $-3 \leq 2x + 5 < 7$

$-8 \leq 2x < 2$

$-4 \leq x < 1$

$x \in [-4,1)$

57. $-0.5 \leq 0.3 - x \leq 1.7$

$-0.8 \leq -x \leq 1.4$

$0.8 \geq x \geq -1.4$

$x \in [-1.4, 0.8]$

59. $-7 < -\dfrac{3}{4}x - 1 \leq 11$

$-6 < -\dfrac{3}{4}x \leq 12$

$\left(-\dfrac{4}{3}\right)(-6) > \left(-\dfrac{4}{3}\right)\left(-\dfrac{3}{4}x\right) \geq \left(-\dfrac{4}{3}\right)12$

$8 > x \geq -16$

$x \in [-16, 8)$

61. $\dfrac{12}{m}$

$m \neq 0$

$m \in (-\infty, 0) \cup (0, \infty)$

63. $\dfrac{5}{y + 7}$

$y + 7 \neq 0$

$y \neq -7$

$y \in (-\infty, -7) \cup (-7, \infty)$

65. $\dfrac{a + 5}{6a - 3}$

$6a - 3 \neq 0$

$6a \neq 3$

$a \neq \dfrac{1}{2}$

$a \in \left(-\infty, \dfrac{1}{2}\right) \cup \left(\dfrac{1}{2}, \infty\right)$

67. $\dfrac{15}{3x-12}$

$3x - 12 \neq 0$

$3x \neq 12$

$x \neq 4$

$x \in (-\infty, 4) \cup (4, \infty)$

69. $\sqrt{x-2}$

$x - 2 \geq 0$

$x \geq 2$

$x \in [2, \infty)$

71. $\sqrt{3n-12}$

$3n - 12 \geq 0$

$3n \geq 12$

$n \geq 4$

$n \in [4, \infty)$

73. $\sqrt{b - \dfrac{4}{3}}$

$b - \dfrac{4}{3} \geq 0$

$b \geq \dfrac{4}{3}$

$b \in \left[\dfrac{4}{3}, \infty\right)$

75. $\sqrt{8-4y}$

$8 - 4y \geq 0$

$-4y \geq -8$

$y \leq 2$

$y \in (-\infty, 2]$

77. a) $B = \dfrac{704W}{H^2}$

$BH^2 = 704W$

$\dfrac{BH^2}{704} = W$

$W = \dfrac{BH^2}{704}$

b) $W < \dfrac{BH^2}{704}$

$W < \dfrac{(27)(68)^2}{704}$

$W < 177.34$

Weight could be 177.34 pounds or less.

79. $\dfrac{82 + 76 + 65 + 71 + x}{5} \geq 75$

$82 + 76 + 65 + 71 + x \geq 375$

$294 + x \geq 375$

$x \geq 81$

81. $\dfrac{1125 + 850 + 625 + 400 + b}{5} \geq 1000$

$1125 + 850 + 625 + 400 + b \geq 5000$

$3000 + b \geq 5000$

$b \geq \$2000$

83. $0 < 20W < 150$

$0 < W < 7.5m$

85. $45 < \dfrac{9}{5}C + 32 < 85$

$13 < \dfrac{9}{5}C < 53$

$7.2° < C < 29.4°$

87. $20 + 4.50h < 11 + 6.00h$

$20 - 1.50h < 11$

$-1.50h < -9$

$h > 6$

89. Answers may vary.

91. $<$

93. $<$

95. $<$

97. $>$

99. $2n - 8$

101. $2\left(\dfrac{5}{9}x - 1\right) - \left(\dfrac{1}{6}x + 3\right)$

$= \dfrac{10}{9}x - 2 - \dfrac{1}{6}x - 3$

$= \dfrac{20}{18}x - 2 - \dfrac{3}{18}x - 3$

$= \dfrac{17}{18}x - 5$

1.3 Technology Highlight

1. Algebraic verification:

$$3|x+1| - 2 \geq 7$$
$$3|x+1| \geq 9$$
$$|x+1| \geq 3$$
$$x+1 \geq 3 \text{ or } x+1 \leq -3$$
$$x \geq 2 \text{ or } x \leq -4$$
$$x \in (-\infty, -4] \cup [2, \infty)$$

3. Algebraic verification:

$$-1 \leq 4|x-3| - 1$$
$$0 \leq 4|x-3|$$
$$0 \leq |x-3|$$
$$|x-3| \geq 0$$

Absolute value is always greater than or equal to zero.
$$x \in \mathbb{i}$$

1.3 Exercises

1. Reverse

3. $-7; 7$

5. No solution; answers will vary.

7. $2|m-1| - 7 = 3$

$$2|m-1| = 10$$
$$|m-1| = 5$$
$$m-1 = 5 \text{ or } m-1 = -5$$
$$m = 6 \text{ or } m = -4$$
$$\{-4, 6\}$$

9. $-3|x+5| + 6 = -15$

$$-3|x+5| = -21$$
$$|x+5| = 7$$
$$x+5 = 7 \text{ or } x+5 = -7$$
$$x = 2 \text{ or } x = -12$$
$$\{-12, 2\}$$

11. $2|4v+5| - 6.5 = 10.3$

$$2|4v+5| = 16.8$$
$$|4v+5| = 8.4$$
$$4v+5 = 8.4 \text{ or } 4v+5 = -8.4$$
$$4v = 3.4 \text{ or } 4v = -13.4$$
$$v = 0.85 \text{ or } v = -3.35$$
$$\{-3.35, 0.85\}$$

13. $-|7p-3| + 6 = -5$

$$-|7p-3| = -11$$
$$|7p-3| = 11$$
$$7p-3 = 11 \text{ or } 7p-3 = -11$$
$$7p = 14 \text{ or } 7p = -8$$
$$p = 2 \text{ or } p = \frac{-8}{7}$$
$$\left\{\frac{-8}{7}, 2\right\}$$

15. $-2|b| - 3 = -4$

$$-2|b| = -1$$
$$|b| = \frac{1}{2}$$
$$b = \frac{1}{2} \text{ or } b = -\frac{1}{2}$$
$$\left\{-\frac{1}{2}, \frac{1}{2}\right\}$$

17. $-2|3x| - 17 = -5$

$$-2|3x| = 12$$
$$|3x| = -6$$
$$\{ \}$$

19. $-3\left|\dfrac{w}{2} + 4\right| - 1 = -4$

$$-3\left|\frac{w}{2} + 4\right| = -3$$
$$\left|\frac{w}{2} + 4\right| = 1$$
$$\frac{w}{2} + 4 = 1 \text{ or } \frac{w}{2} + 4 = -1$$
$$\frac{w}{2} = -3 \text{ or } \frac{w}{2} = -5$$
$$w = -6 \text{ or } w = -10$$
$$\{-6, -10\}$$

21. $8.7|p-7.5|-26.6=8.2$

    $8.7|p-7.5|=34.8$

    $|p-7.5|=4$

    $p-7.5=4 \quad$ or $\quad p-7.5=-4$

    $p=11.5 \quad$ or $\quad p=3.5$

    $\{3.5, 11.5\}$

23. $8.7|-2.5x|-26.6=8.2$

    $8.7|-2.5x|=34.8$

    $|-2.5x|=4$

    $-2.5x=4 \quad$ or $\quad -2.5x=-4$

    $x=-1.6 \quad$ or $\quad x=1.6$

    $\{-1.6, 1.6\}$

25. $|x-2|\le 7$

    $-7\le x-2\le 7$

    $-5\le x\le 9$

    $x\in[-5,9]$

27. $-3|m|-2>4$

    $-3|m|>6$

    $|m|<-2$

    Absolute value is never less than a negative number.

    $\varnothing$

29. $\dfrac{|5v+1|}{4}+8<9$

    $\dfrac{|5v+1|}{4}<1$

    $|5v+1|<4$

    $-4<5v+1<4$

    $-5<5v<3$

    $-1<v<\dfrac{3}{5}$

    $v\in\left(-1,\dfrac{3}{5}\right)$

31. $3|p+4|+5<8$

    $3|p+4|<3$

    $|p+4|<1$

    $-1<p+4<1$

    $-5<p<-3$

    $p\in(-5,-3)$

33. $|3b-11|+6\le 9$

    $|3b-11|<3$

    $-3\le 3b-11\le 3$

    $8\le 3b\le 14$

    $\dfrac{8}{3}\le b\le\dfrac{14}{3}$

    $b\in\left[\dfrac{8}{3},\dfrac{14}{3}\right]$

35. $|4-3z|+12<7$

    $|4-3z|<-5$

    No solution

37. $\left|\dfrac{4x+5}{3}-\dfrac{1}{2}\right|\le\dfrac{7}{6}$

    $-\dfrac{7}{6}\le\dfrac{4x+5}{3}-\dfrac{1}{2}\le\dfrac{7}{6}$

    $6\left(-\dfrac{7}{6}\right)\le 6\left(\dfrac{4x+5}{3}-\dfrac{1}{2}\right)\le 6\left(\dfrac{7}{6}\right)$

    $-7\le 8x+10-3\le 7$

    $-7\le 8x+7\le 7$

    $-14\le 8x\le 0$

    $-\dfrac{14}{8}\le x\le 0$

    $-\dfrac{7}{4}\le x\le 0$

    $\left[-\dfrac{7}{4},0\right]$

39. $|n+3|>7$

    $n+3>7 \quad$ or $\quad n+3<-7$

    $n>4 \quad$ or $\quad n<-10$

    $n\in(-\infty,-10)\cup(4,\infty)$

41. $-2|w| - 5 \leq -11$

$-2|w| \leq -6$

$|w| \geq 3$

$w \geq 3$ or $w \leq -3$

$w \in (-\infty, -3] \cup [3, \infty)$

43. $\dfrac{|q|}{2} - \dfrac{5}{6} \geq \dfrac{1}{3}$

$6\left(\dfrac{|q|}{2} - \dfrac{5}{6}\right) \geq 6\left(\dfrac{1}{3}\right)$

$3|q| - 5 \geq 2$

$3|q| \geq 7$

$|q| \geq \dfrac{7}{3}$

$q \geq \dfrac{7}{3}$ or $q \leq -\dfrac{7}{3}$

$q \in \left(-\infty, -\dfrac{7}{3}\right] \cup \left[\dfrac{7}{3}, \infty\right)$

45. $3|5 - 7d| + 9 \geq 15$

$3|5 - 7d| \geq 6$

$|5 - 7d| \geq 2$

$5 - 7x \geq 2$ or $5 - 7x \leq -2$

$-7d \geq -3$

$\qquad\qquad -7d \leq -7$

$d \leq \dfrac{3}{7}$ $\qquad d \geq 1$

$d \in \left(-\infty, \dfrac{3}{7}\right] \cup [1, \infty)$

47. $|4z - 9| + 6 \geq 4$

$|4z - 9| \geq -2$

$z \in (-\infty, \infty)$

49. $4|5 - 2h| - 9 > 11$

$4|5 - 2h| > 20$

$|5 - 2h| > 5$

$5 - 2h > 5$ or $5 - 2h < -5$

$-2h > 0$ $\qquad -2h < -10$

$h < 0$ $\qquad\quad h > 5$

$h \in (-\infty, 0) \cup (5, \infty)$

51. $-3.9|4q - 5| + 8.7 \leq -22.5$

$-3.9|4q - 5| \leq -31.2$

$|4q - 5| \geq 8$

$4q - 5 \geq 8$ or $4q - 5 \leq -8$

$4q \geq 13$ $\qquad\quad 4q \leq -3$

$q \geq \dfrac{13}{4}$ $\qquad\quad q \leq -\dfrac{3}{4}$

$q \geq 3.25$ $\qquad\quad q \leq -0.75$

$q \in (-\infty, -0.75] \cup [3.25, \infty)$

53. $2 < \left|-3m + \dfrac{4}{5}\right| - \dfrac{1}{5}$

$\dfrac{11}{5} < \left|-3m + \dfrac{4}{5}\right|$

$\left|-3m + \dfrac{4}{5}\right| > \dfrac{11}{5}$

$-3m + \dfrac{4}{5} > \dfrac{11}{5}$ or $-3m + \dfrac{4}{5} < -\dfrac{11}{5}$

$-3m > \dfrac{7}{5}$

$\qquad\qquad -3m < -3$

$m < -\dfrac{7}{15}$

$\qquad\qquad m > 1$

$m \in \left(-\infty, \dfrac{-7}{15}\right) \cup (1, \infty)$

55. $|d - x| \leq L$

4 ft = 48 in.

$|d - 48| \leq 3$

$d - 48 \leq 3$ and $d - 48 \geq -3$

$d \leq 51$ and $\qquad d \geq 45$

$45 \leq d \leq 51$ in.

57. $|h - 35,050| \leq 2,550$

$h - 35050 \leq 2550$ and $h - 35050 \geq -2550$

$h \leq 37600$ $\qquad\qquad h \geq 32500$

$32,500 \leq h \leq 37,600$; yes, if between

32,500 feet and 37,600 feet, inclusive.

59. $|d - 394| - 20 > 164$

$|d - 394| > 184$

$d - 394 > 184$ or $d - 394 < -184$

$d > 578$ or $d < 210$;

$d < 210$ or $d > 578$

Less than 210 feet to go over the net, more than 578 feet to go under the net.

61. (a) $|s - 37.58| \leq 3.35$

(b) $s - 37.58 \leq 3.35$ and $s - 37.58 \geq -3.35$

$s \leq 40.93$ and $s \geq 34.23$

$34.23 \leq s \leq 40.93$

$[34.23, 40.93]$

63. (a) $|s - 125| \leq 23$

(b) $-23 \leq s - 125 \leq 23$

$102 \leq s \leq 148$

$[102, 148]$

65. a. $|d - 42.7| < 0.03$

b. $|d - 73.78| < 1.01$

c. $|d - 57.150| < 0.127$

d. $|d - 2171.05| < 12.05$

e. golf: $t = \dfrac{2(0.03)}{42.7} \approx 0.0014$;

baseball: $t = \dfrac{2(1.01)}{73.78} \approx 0.0274$;

billiard: $t = \dfrac{2(0.127)}{57.150} \approx 0.0044$;

bowling: $t = \dfrac{2(12.05)}{2171.05} \approx 0.0111$;

Golf balls.

67. a. $x = 4$

b. $\left[\dfrac{4}{3}, 4\right]$

c. $x = 0$

d. $\left[-\infty, \dfrac{3}{5}\right]$

e. $\{\ \}$

69. $18x^3 + 21x^2 - 60x$

$3x(6x^2 + 7x - 20)$

$3x(2x + 5)(3x - 4)$

71. $\dfrac{-1}{3 + \sqrt{3}}$

$\dfrac{-1}{3 + \sqrt{3}} g \dfrac{3 - \sqrt{3}}{3 - \sqrt{3}}$

$\dfrac{-3 + \sqrt{3}}{9 - 3\sqrt{3} + 3\sqrt{3} - 3}$

$\dfrac{-3 + \sqrt{3}}{6} \approx -0.21$

Chapter 1 Mid-Chapter Check

1. a. $\dfrac{r}{3}+5=2$

$\dfrac{r}{3}=-3$

$r=-9$

b. $5(2x-1)+4=9x-7$

$10x-5+4=9x-7$

$10x-1=9x-7$

$x=-6$

c. $m-2(m+3)=1-(m+7)$

$m-2m-6=1-m-7$

$-m-6=-m-6$

$0=0$

Identity; $x\in\mathbb{¡}$

d. $\dfrac{1}{5}y+3=\dfrac{3}{2}y-2$

$10\left(\dfrac{1}{5}y+3\right)=10\left(\dfrac{3}{2}y-2\right)$

$2y+30=15y-20$

$-13y+30=-20$

$-13y=-50$

$y=\dfrac{50}{13}$

e. $\dfrac{1}{2}(5j-2)=\dfrac{3}{2}(j-4)+j$

$\dfrac{5}{2}j-1=\dfrac{3}{2}j-6+j$

$\dfrac{5}{2}j-1=\dfrac{5}{2}j-6$

$-1=-6$

Contradiction; $\{\ \}$

f. $0.6(x-3)+0.3=1.8$

$0.6x-1.8+0.3=1.8$

$0.6x-1.5=1.8$

$0.6x=3.3$

$x=5.5$

3. $S=2\pi x^2+\pi x^2 y$

$S=x^2(2\pi+\pi y)$

$\dfrac{S}{2\pi+\pi y}=x^2$

$\dfrac{S}{\pi(2+y)}=x^2$

$\sqrt{\dfrac{S}{\pi(2+y)}}=x$

5. a. $\dfrac{3x+1}{2x-5}$

$2x-5\neq0$

$2x\neq5$

$x\neq\dfrac{5}{2}$

$x\in\left(-\infty,\dfrac{5}{2}\right)\cup\left(\dfrac{5}{2},0\right)$

b. $\sqrt{17-6x}$

$17-6x\geq0$

$-6x\geq-17$

$x\leq\dfrac{17}{6}$

$x\in\left(-\infty,\dfrac{17}{6}\right]$

7. a. $3|q+4|-2<10$

$3|q+4|<12$

$|q+4|<4$

$-4<q+4<4$

$-8<q<0$

$x\in(-8,0)$

b. $\left|\dfrac{x}{3}+2\right|+5\leq5$

$\left|\dfrac{x}{3}+2\right|\leq0$

$\dfrac{x}{3}+2=0$

$\dfrac{x}{3}=-2$

$x=-6$

$\{-6\}$

9. $\dfrac{x}{30} + \dfrac{115-x}{50} = 2 + \dfrac{50}{60}$

   $\dfrac{x}{30} + \dfrac{115-x}{50} = \dfrac{17}{6}$

   $150\left(\dfrac{x}{30} + \dfrac{115-x}{50}\right) = 150\left(\dfrac{17}{6}\right)$

   $5x + 3(115-x) = 425$

   $5x + 345 - 3x = 425$

   $2x = 80$

   $x = 40$ miles;

   $\dfrac{40}{30} = 1\dfrac{1}{3}$ hours or 1 hour 20 minutes

Reinforcing Basic Concepts

1. $|x-2| = 5$

   $x - 2 = -5$ or $x - 2 = 5$

   $x = -3$ or $x = 7$

3. $|2x-3| \geq 5$

   $2x - 3 \geq 5$ or $2x - 3 \leq -5$

   $2x \geq 8$ or $2x \leq -2$

   $x \geq 4$ or $x \leq -1$

   $x \in (-\infty, -1] \cup [4, \infty)$

1.4 Exercises

1. $3-2i$

3. $2, 3\sqrt{2}$

5. b is correct.

7. a. $\sqrt{-16} = 4i$
    b. $\sqrt{-49} = 7i$
    c. $\sqrt{27} = \sqrt{9(3)} = 3\sqrt{3}$
    d. $\sqrt{72} = \sqrt{36(2)} = 6\sqrt{2}$

9. a. $-\sqrt{-18} = -\sqrt{-1(9)(2)} = -3i\sqrt{2}$
    b. $-\sqrt{-50} = -\sqrt{-1(25)(2)} = -5i\sqrt{2}$
    c. $3\sqrt{-25} = 3(5i) = 15i$
    d. $2\sqrt{-9} = 2(3i) = 6i$

11. a. $\sqrt{-19} = i\sqrt{19}$
    b. $\sqrt{-31} = i\sqrt{31}$
    c. $\sqrt{\dfrac{-12}{25}} = \dfrac{\sqrt{-1(4)3}}{\sqrt{25}} = \dfrac{2\sqrt{3}}{5}i$
    d. $\sqrt{\dfrac{-9}{32}} = \dfrac{\sqrt{-9}}{\sqrt{32}} \cdot \dfrac{\sqrt{2}}{\sqrt{2}} = \dfrac{3\sqrt{2}}{\sqrt{64}}i = \dfrac{3\sqrt{2}}{8}i$

13. a. $\dfrac{2+\sqrt{-4}}{2} = \dfrac{2+2i}{2} = 1+i$
    $a = 1, b = 1$
    b. $\dfrac{6+\sqrt{-27}}{3} = \dfrac{6+3i\sqrt{3}}{3} = 2+\sqrt{3}\,i$
    $a = 2, b = \sqrt{3}$

15. a. $\dfrac{8+\sqrt{-16}}{2} = \dfrac{8+4i}{2} = 4+2i$
    $a = 4, b = 2$
    b. $\dfrac{10-\sqrt{-50}}{5} = \dfrac{10-5i\sqrt{2}}{5} = 2-\sqrt{2}\,i$
    $a = 2, b = -\sqrt{2}$

17. a. $5 = 5+0i$
    $a = 5, b = 0$
    b. $3i = 0+3i$
    $a = 0, b = 3$

19. a. $2\sqrt{-81} = 2(9i) = 0+18i = 18i$
    $a = 0, b = 18$
    b. $\dfrac{\sqrt{-32}}{8} = \dfrac{4\sqrt{2}}{8}i = 0 + \dfrac{\sqrt{2}}{2}i = \dfrac{\sqrt{2}}{2}i$
    $a = 0, b = \dfrac{\sqrt{2}}{2}$

21. a. $4+\sqrt{-50} = 4+5\sqrt{2}\,i$
    $a = 4, b = 5\sqrt{2}$
    b. $-5+\sqrt{-27} = -5+3\sqrt{3}\,i$
    $a = -5, b = 3\sqrt{3}$

23. a. $\dfrac{14+\sqrt{-98}}{8} = \dfrac{14+7i\sqrt{2}}{8} = \dfrac{7}{4} + \dfrac{7\sqrt{2}}{8}i$
    $a = \dfrac{7}{4}, b = \dfrac{7\sqrt{2}}{8}$
    b. $\dfrac{5+\sqrt{-250}}{10} = \dfrac{5+5i\sqrt{10}}{10} = \dfrac{1}{2} + \dfrac{\sqrt{10}}{2}i$
    $a = \dfrac{1}{2}, b = \dfrac{\sqrt{10}}{2}$

25. a. $\left(12-\sqrt{-4}\right)+\left(7+\sqrt{-9}\right)$
    $= (12-2i)+(7+3i)$
    $= 19+i$
    b. $\left(3+\sqrt{-25}\right)+\left(-1-\sqrt{-81}\right)$
    $= (3+5i)+(-1-9i)$
    $= 2-4i$
    c. $\left(11+\sqrt{-108}\right)-\left(2-\sqrt{-48}\right)$
    $= 11+\sqrt{-108} - 2+\sqrt{-48}$
    $= 11+\sqrt{-1(36)(3)} - 2+\sqrt{-1(16)(3)}$
    $= 9+6\sqrt{3}\,i + 4\sqrt{3}\,i$
    $= 9+10\sqrt{3}\,i$

27. a. $(2+3i)+(-5-i)$
    $= 2+3i-5-i$
    $= -3+2i$
    b. $(5-2i)+(3+2i)$
    $= 5-2i+3+2i$
    $= 8$
    c. $(6-5i)-(4+3i)$
    $= 6-5i-4-3i$
    $= 2-8i$

29. a. $(3.7 + 6.1i) - (1 + 5.9i)$
$= 3.7 + 6.1i - 1 - 5.9i$
$= 2.7 + 0.2i$

b. $\left(8 + \dfrac{3}{4}i\right) - \left(-7 + \dfrac{2}{3}i\right)$

$= 8 + \dfrac{3}{4}i + 7 - \dfrac{2}{3}i$
$= 15 + \dfrac{1}{12}i$

c. $\left(-6 - \dfrac{5}{8}i\right) + \left(4 + \dfrac{1}{2}i\right)$

$= -6 - \dfrac{5}{8}i + 4 + \dfrac{1}{2}i$
$= -2 - \dfrac{1}{8}i$

31. a. $5i \cdot (-3i)$

$= -15i^2$
$= 15$

b. $4i \cdot (-4i)$

$= -16i^2$
$= 16$

33. a. $-7i(5 - 3i)$

$= -35i + 21i^2$
$= -21 - 35i$

b. $6i(-3 + 7i)$

$= -18i + 42i^2$
$= -42 - 18i$

35. a. $(-3 + 2i)(2 + 3i)$

$= -6 - 9i + 4i + 6i^2$
$= -12 - 5i$

b. $(3 + 2i)(1 + i)$

$= 3 + 3i + 2i + 2i^2$
$= 1 + 5i$

37. a. conjugate $4 - 5i$

$= (4 + 5i)(4 - 5i)$
$= 16 - 20i + 20i - 25i^2$
$= 16 + 25$
$= 41$

b. conjugate $3 + i\sqrt{2}$

$= (3 + i\sqrt{2})(3 - i\sqrt{2})$
$= 9 - 3\sqrt{2}i + 3\sqrt{2}i - 2i^2$
$= 9 + 2$
$= 11$

39. a. conjugate $-7i$

$(7i)(-7i)$
$= -49i^2$
$= 49$

b. conjugate $\dfrac{1}{2} + \dfrac{2}{3}i$

$\left(\dfrac{1}{2} + \dfrac{2}{3}i\right)\left(\dfrac{1}{2} - \dfrac{2}{3}i\right)$
$= \dfrac{1}{4} - \dfrac{1}{3}i + \dfrac{1}{3}i - \dfrac{4}{9}i^2$
$= \dfrac{25}{36}$

41. a. $(4 - 5i)(4 + 5i)$

$= 16 + 20i - 20i - 25i^2$
$= 41$

b. $(7 - 5i)(7 + 5i)$

$= 49 + 35i - 35i - 25i^2$
$= 74$

43. a. $(3 - i\sqrt{2})(3 + i\sqrt{2})$

$= 9 + 3i\sqrt{2} - 3i\sqrt{2} - 2i^2$
$= 11$

b. $\left(\dfrac{1}{6} + \dfrac{2}{3}i\right)\left(\dfrac{1}{6} - \dfrac{2}{3}i\right)$

$= \dfrac{1}{36} - \dfrac{1}{9}i + \dfrac{1}{9}i - \dfrac{4}{9}i^2$
$= \dfrac{17}{36}$

45. a. $(2 + 3i)^2$

$= (2 + 3i)(2 + 3i)$
$= 4 + 6i + 6i + 9i^2$
$= -5 + 12i$

b. $(3 - 4i)^2$

$= (3 - 4i)(3 - 4i)$
$= 9 - 12i - 12i + 16i^2$
$= -7 - 24i$

47. a. $(-2 + 5i)^2$

$= (-2 + 5i)(-2 + 5i)$
$= 4 - 10i - 10i + 25i^2$
$= -21 - 20i$

b. $(3 + i\sqrt{2})^2$

$= (3 + i\sqrt{2})(3 + i\sqrt{2})$
$= 9 + 3i\sqrt{2} + 3i\sqrt{2} + 2i^2$
$= 7 + 6\sqrt{2}\ i$

49. $x^2 + 36 = 0, x = -6;$

$(-6)^2 + 36 = 0$
$36 + 36 = 0$
$72 \neq 0$ no

51. $x^2 + 49 = 0, x = -7i;$

$(-7i)^2 + 49 = 0$
$49i^2 + 49 = 0$
$-49 + 49 = 0$
$0 = 0$ yes

53. $(x-3)^2 = -9, x = 3 - 3i;$

$(3 - 3i - 3)^2 = -9$
$(-3i)^2 = -9$
$9i^2 = -9$
$-9 = -9$ yes

55. $x^2 - 2x + 5 = 0, x = 1 - 2i;$

$(1-2i)^2 - 2(1-2i) + 5 = 0$
$1 - 4i + 4i^2 - 2 + 4i + 5 = 0$
$4 + 4i^2 = 0$
$4 - 4 = 0$
$0 = 0$ yes

57. $x^2 - 4x + 9 = 0, x = 2 + i\sqrt{5};$

$(2 + i\sqrt{5})^2 - 4(2 + i\sqrt{5}) + 9 = 0$
$4 + 4i\sqrt{5} + 5i^2 - 8 - 4i\sqrt{5} + 9 = 0$
$5 + 5i^2 = 0$
$5 - 5 = 0$
$0 = 0$ yes

59. $x^2 - 2x + 17 = 0, x = 1 + 4i, 1 - 4i;$

$(1 + 4i)^2 - 2(1 + 4i) + 17 = 0$
$1 + 8i + 16i^2 - 2 - 8i + 17 = 0$
$16 + 16i^2 = 0$
$16 - 16 = 0$
$0 = 0$
$1 + 4i$ is a solution.
$(1 - 4i)^2 - 2(1 - 4i) + 17 = 0$
$1 - 8i + 16i^2 - 2 + 8i + 17 = 0$
$16 + 16i^2 = 0$
$16 - 16 = 0$
$0 = 0$
$1 - 4i$ is a solution.

61. a. $i^{48} = (i^4)^{12} = (1)^{12} = 1$

b. $i^{26} = (i^4)^6 i^2 = (1)^6(-1) = -1$

c. $i^{39} = (i^4)^9 i^3 = (1)^9(-i) = -i$

d. $i^{53} = (i^4)^{13} i^1 = (1)^{13}(i) = i$

63. a. $\dfrac{-2}{\sqrt{-49}} = \dfrac{-2}{7i} \cdot \dfrac{i}{i} = \dfrac{-2i}{7i^2} = \dfrac{2}{7}i$

b. $\dfrac{4}{\sqrt{-25}} = \dfrac{4}{5i} \cdot \dfrac{i}{i} = \dfrac{4i}{5i^2} = \dfrac{-4}{5}i$

65. a. $\dfrac{7}{3+2i} \cdot \dfrac{3-2i}{3-2i} = \dfrac{21-14i}{9-4i^2} = \dfrac{21-14i}{13}$

$= \dfrac{21}{13} - \dfrac{14}{13}i$

b. $\dfrac{-5}{2-3i} \cdot \dfrac{2+3i}{2+3i} = \dfrac{-10-15i}{4-9i^2} = \dfrac{-10-15i}{13}$

$= \dfrac{-10}{13} - \dfrac{15}{13}i$

67. a. $\dfrac{3+4i}{4i} \cdot \dfrac{i}{i} = \dfrac{3i+4i^2}{4i^2} = \dfrac{-4+3i}{-4} = 1 - \dfrac{3}{4}i$

b. $\dfrac{2-3i}{3i} \cdot \dfrac{i}{i} = \dfrac{2i-3i^2}{3i^2} = \dfrac{3+2i}{-3} = -1 - \dfrac{2}{3}i$

69. $|a + bi| = \sqrt{a^2 + b^2}$

a. $|2 + 3i| = \sqrt{(2)^2 + (3)^2} = \sqrt{13}$

b. $|4 - 3i| = \sqrt{(4)^2 + (-3)^2} = 5$

c. $|3 + \sqrt{2}\, i| = \sqrt{(3)^2 + (\sqrt{2})^2} = \sqrt{11}$

71. $5 + \sqrt{15}\, i + 5 - \sqrt{15}\, i = 10$
$10 = 10$
verified;

$(5 + \sqrt{15}\, i)(5 - \sqrt{15}\, i) = 40$
$25 - 5\sqrt{15}\, i + 5\sqrt{15}\, i - 15i^2 = 40$
$40 = 40$
verified

73. $Z = R + iX_L - iX_C$
$Z = 7 + i(6) - i(11) = 7 - 5i \ \Omega$

75. $V = IZ$

$V = (3 - 2i)(5 + 5i)$

$V = 15 + 15i - 10i - 10i^2$

$V = 25 + 5i$ volts

77. $Z = \dfrac{Z_1 Z_2}{Z_1 + Z_2}$

$Z = \dfrac{(1 + 2i)(3 - 2i)}{1 + 2i + 3 - 2i}$

$Z = \dfrac{3 - 2i + 6i - 4i^2}{4}$

$Z = \dfrac{7 + 4i}{4}$

$Z = \dfrac{7}{4} + i \ \Omega$

79. a. $x^2 + 36$

$(x + 6i)(x - 6i)$

b. $m^2 + 3$

$(m + i\sqrt{3})(m - i\sqrt{3})$

c. $n^2 + 12$

$(n + 2i\sqrt{3})(n - 2i\sqrt{3})$

d. $4x^2 + 49$

$(2x + 7i)(2x - 7i)$

81. $i^{17}(3 - 4i) - 3i^3(1 + 2i)^2$

$i(3 - 4i) + 3i(1 + 2i)^2$

$3i - 4i^2 + 3i(1 + 4i + 4i^2)$

$3i + 4 + 3i(1 + 4i - 4)$

$3i + 4 + 3i(4i - 3)$

$3i + 4 + 12i^2 - 9i$

$-6i + 4 - 12$

$-8 - 6i$

83. a. $P = 4s,\ A = s^2$

b. $P = 2L + 2W,\ A = LW$

c. $P = a + b + c,\ A = \dfrac{bh}{2}$

d. $C = \pi d,\ A = \pi r^2$

85. John takes $\dfrac{200}{10} = 20$ seconds.

Rick takes $\dfrac{200}{9} = 22.\overline{2}$ seconds.

Even with a 2 second head start, John will finish first.

86. a. $x^4 - 16$

$= (x^2 + 4)(x^2 - 4)$

$= (x^2 + 4)(x + 2)(x - 2)$

b. $n^3 - 27$

$= (n - 3)(n^2 + 3n + 9)$

c. $x^3 - x^2 - x + 1$

$= x^2(x - 1) - (x - 1)$

$= (x - 1)(x^2 - 1)$

$= (x - 1)(x + 1)(x - 1)$

$= (x - 1)^2(x + 1)$

d. $4n^2 m - 12nm^2 + 9m^3$

$= m(4n^2 - 12nm + 9m^2)$

$= m(2n - 3m)^2$

1.5 Exercises

1. Descending, 0

3. Quadratic, 1

5. GCF factoring;
$$4x^2 - 5x = 0$$
$$x(4x - 5) = 0$$
$$x = 0 \text{ or } 4x - 5 = 0$$
$$4x = 5$$
$$x = \frac{5}{4}$$

7. $2x - 15 - x^2 = 0$
$$-x^2 + 2x - 15 = 0$$
Quadratic; $a = -1, b = 2, c = -15$

9. $\frac{2}{3}x - 7 = 0$
not quadratic

11. $\frac{1}{4}x^2 = 6x$
$$\frac{1}{4}x^2 - 6x = 0$$
Quadratic; $a = \frac{1}{4}, b = -6, c = 0$

13. $2x^2 + 7 = 0$
Quadratic; $a = 2, b = 0, c = 7$

15. $-3x^2 + 9x - 5 + 2x^3 = 0$
Not quadratic

17. $(x-1)^2 + (x-1) + 4 = 9$
$$x^2 - 2x + 1 + x - 1 - 5 = 0$$
$$x^2 - x - 5 = 0$$
Quadratic; $a = 1, b = -1, c = -5$

19. $x^2 - 15 = 2x$
$$x^2 - 2x - 15 = 0$$
$$(x - 5)(x + 3) = 0$$
$$x - 5 = 0 \text{ or } x + 3 = 0$$
$$x = 5 \quad \text{ or } x = -3$$

21. $m^2 = 8m - 16$
$$m^2 - 8m + 16 = 0$$
$$(m - 4)(m - 4) = 0$$
$$m - 4 = 0$$
$$m = 4$$

23. $5p^2 - 10p = 0$
$$5p(p - 2) = 0$$
$$5p = 0 \text{ or } p - 2 = 0$$
$$p = 0 \quad \text{ or } p = 2$$

25. $-14h^2 = 7h$
$$-14h^2 - 7h = 0$$
$$-7h(2h + 1) = 0$$
$$-7h = 0 \text{ or } 2h + 1 = 0$$
$$h = 0 \quad \text{ or } 2h = -1$$
$$h = 0 \quad \text{ or } h = \frac{-1}{2}$$

27. $a^2 - 17 = -8$
$$a^2 - 9 = 0$$
$$(a + 3)(a - 3) = 0$$
$$a + 3 = 0 \text{ or } a - 3 = 0$$
$$a = -3 \quad \text{ or } a = 3$$

29. $g^2 + 18g + 70 = -11$
$$g^2 + 18g + 81 = 0$$
$$(g + 9)(g + 9) = 0$$
$$g + 9 = 0$$
$$g = -9$$

31. $m^3 + 5m^2 - 9m - 45 = 0$
$$m^2(m + 5) - 9(m + 5) = 0$$
$$(m + 5)(m^2 - 9) = 0$$
$$(m + 5)(m + 3)(m - 3) = 0$$
$$m + 5 = 0 \text{ or } m + 3 = 0 \text{ or } m - 3 = 0$$
$$m = -5 \quad \text{ or } m = -3 \quad \text{ or } m = 3$$

33. $(c - 12)c - 15 = 30$
$$c^2 - 12c - 15 = 30$$
$$c^2 - 12c - 45 = 0$$
$$(c - 15)(c + 3) = 0$$
$$c - 15 = 0 \text{ or } c + 3 = 0$$
$$c = 15 \quad \text{ or } c = -3$$

35. $9 + (r-5)r = 33$

$9 + r^2 - 5r = 33$

$r^2 - 5r - 24 = 0$

$(r-8)(r+3) = 0$

$r - 8 = 0$ or $r + 3 = 0$

$r = 8$ or $r = -3$

37. $(t+4)(t+7) = 54$

$t^2 + 11t + 28 = 54$

$t^2 + 11t - 26 = 0$

$(t+13)(t-2) = 0$

$t + 13 = 0$ or $t - 2 = 0$

$t = -13$ or $t = 2$

39. $2x^2 - 4x - 30 = 0$

$2(x^2 - 2x - 15) = 0$

$2(x-5)(x+3) = 0$

$x - 5 = 0$ or $x + 3 = 0$

$x = 5$ or $x = -3$

41. $2w^2 - 5w = 3$

$2w^2 - 5w - 3 = 0$

$(2w+1)(w-3) = 0$

$2w + 1 = 0$ or $w - 3 = 0$

$2w = -1$ or $w = 3$

$w = -\dfrac{1}{2}$ or $w = 3$

43. $m^2 = 16$

$m = \pm 4$

45. $y^2 - 28 = 0$

$y^2 = 28$

$y = \pm\sqrt{28}$

$y = \pm 2\sqrt{7} \approx \pm 5.29$

47. $p^2 + 36 = 0$

$p^2 = -36$

$p = \pm\sqrt{-36}$

No real solutions

49. $x^2 = \dfrac{21}{16}$

$x = \pm\dfrac{\sqrt{21}}{4} \approx \pm 1.15$

51. $(n-3)^2 = 36$

$n - 3 = \pm 6$

$n = 6 + 3$ or $n = -6 + 3$

$n = 9$ or $n = -3$

53. $(w+5)^2 = 3$

$w + 5 = \pm\sqrt{3}$

$w = -5 \pm\sqrt{3}$

$w \approx -3.27$ or $w \approx -6.73$

55. $(x-3)^2 + 7 = 2$

$(x-3)^2 = -5$

$x - 3 = \pm\sqrt{-5}$

No real solutions

57. $(m-2)^2 = \dfrac{18}{49}$

$m - 2 = \pm\dfrac{\sqrt{18}}{7}$

$m = 2 \pm\dfrac{3\sqrt{2}}{7}$

$m \approx 2.61$ or $m \approx 1.39$

59. $x^2 + 6x + \underline{9}$

$(x+3)^2$

61. $n^2 + 3n + \underline{\dfrac{9}{4}}$

$\left(n + \dfrac{3}{2}\right)^2$

63. $p^2 + \dfrac{2}{3}p + \underline{\dfrac{1}{9}}$

$\left(p + \dfrac{1}{3}\right)^2$

65. $x^2 + 6x = -5$

$x^2 + 6x + 9 = -5 + 9$

$(x+3)^2 = 4$

$x + 3 = \pm 2$

$x = -3 \pm 2$

$x = -1$ or $x = -5$

67. $p^2 - 6p + 3 = 0$

    $p^2 - 6p = -3$

    $p^2 - 6p + 9 = -3 + 9$

    $(p - 3)^2 = 6$

    $p - 3 = \pm\sqrt{6}$

    $p = 3 \pm \sqrt{6}$

    $p \approx 5.45$ or $p \approx 0.55$

69. $p^2 + 6p = -4$

    $p^2 + 6p + 9 = -4 + 9$

    $(p + 3)^2 = 5$

    $p + 3 = \pm\sqrt{5}$

    $p = -3 \pm \sqrt{5}$

    $p \approx -0.76$ or $p \approx -5.24$

71. $m^2 + 3m = 1$

    $m^2 + 3m + \dfrac{9}{4} = 1 + \dfrac{9}{4}$

    $\left(m + \dfrac{3}{2}\right)^2 = \dfrac{13}{4}$

    $m + \dfrac{3}{2} = \pm\dfrac{\sqrt{13}}{2}$

    $m = -\dfrac{3}{2} \pm \dfrac{\sqrt{13}}{2}$

    $m \approx 0.30$ or $m \approx -3.30$

73. $n^2 = 5n + 5$

    $n^2 - 5n = 5$

    $n^2 - 5n + \dfrac{25}{4} = 5 + \dfrac{25}{4}$

    $\left(n - \dfrac{5}{2}\right)^2 = \dfrac{45}{4}$

    $n - \dfrac{5}{2} = \pm\dfrac{\sqrt{45}}{2}$

    $n = \dfrac{5}{2} \pm \dfrac{3\sqrt{5}}{2}$

    $n \approx 5.85$ or $n \approx -0.85$

75. $2x^2 = -7x + 4$

    $2x^2 + 7x = 4$

    $x^2 + \dfrac{7}{2}x = 2$

    $x^2 + \dfrac{7}{2}x + \dfrac{49}{16} = 2 + \dfrac{49}{16}$

    $\left(x + \dfrac{7}{4}\right)^2 = \dfrac{81}{16}$

    $x + \dfrac{7}{4} = \pm\dfrac{9}{4}$

    $x = -\dfrac{7}{4} \pm \dfrac{9}{4}$

    $x = \dfrac{1}{2}$ or $x = -4$

77. $2n^2 - 3n - 9 = 0$

    $2n^2 - 3n = 9$

    $n^2 - \dfrac{3}{2}n = \dfrac{9}{2}$

    $n^2 - \dfrac{3}{2}n + \dfrac{9}{16} = \dfrac{9}{2} + \dfrac{9}{16}$

    $\left(n - \dfrac{3}{4}\right)^2 = \dfrac{81}{16}$

    $n - \dfrac{3}{4} = \pm\dfrac{9}{4}$

    $n = \dfrac{3}{4} \pm \dfrac{9}{4}$

    $n = 3$ or $n = -\dfrac{3}{2}$

79. $4p^2 - 3p - 2 = 0$

    $4p^2 - 3p = 2$

    $p^2 - \dfrac{3}{4}p = \dfrac{1}{2}$

    $p^2 - \dfrac{3}{4}p + \dfrac{9}{64} = \dfrac{1}{2} + \dfrac{9}{64}$

    $\left(p - \dfrac{3}{8}\right)^2 = \dfrac{41}{64}$

    $p - \dfrac{3}{8} = \pm\dfrac{\sqrt{41}}{8}$

    $p = \dfrac{3}{8} \pm \dfrac{\sqrt{41}}{8}$

    $p \approx 1.18$ or $p \approx -0.43$

81. $m^2 = 7m - 4$

$m^2 - 7m = -4$

$m^2 - 7m + \dfrac{49}{4} = \dfrac{49}{4} - 4$

$\left(m - \dfrac{7}{2}\right)^2 = \dfrac{33}{4}$

$m - \dfrac{7}{2} = \dfrac{\pm\sqrt{33}}{2}$

$m = \dfrac{7}{2} \pm \dfrac{\sqrt{33}}{2}$

$m \approx 6.37 \ \text{ or } \ m \approx 0.63$

83. $x^2 - 3x = 18$

$x^2 - 3x - 18 = 0$

$(x - 6)(x + 3) = 0$

$x - 6 = 0 \ \text{ or } \ x + 3 = 0$

$x = 6 \qquad \text{ or } \ x = -3$

85. $4m^2 - 25 = 0$

$4m^2 = 25$

$m^2 = \dfrac{25}{4}$

$m = \pm\dfrac{5}{2}$

87. $4n^2 - 8n - 1 = 0$

$a = 4, b = -8, c = -1$

$n = \dfrac{-(-8) \pm \sqrt{(-8)^2 - 4(4)(-1)}}{2(4)}$

$n = \dfrac{8 \pm \sqrt{80}}{8}$

$n = \dfrac{8 \pm 4\sqrt{5}}{8}$

$n = \dfrac{2 \pm \sqrt{5}}{2}$

$n \approx 2.12 \ \text{ or } \ n \approx -0.12$

89. $6w^2 - w = 2$

$6w^2 - w - 2 = 0$

$(2w + 1)(3w - 2) = 0$

$2w + 1 = 0 \ \text{ or } \ 3w - 2 = 0$

$2w = -1 \qquad \text{ or } \ 3w = 2$

$w = -\dfrac{1}{2} \ \text{ or } \ w = \dfrac{2}{3}$

91. $4m^2 = 12m - 15$

$4m^2 - 12m + 15 = 0$

$a = 4, b = -12, c = 15$

$m = \dfrac{-(-12) \pm \sqrt{(-12)^2 - 4(4)(15)}}{2(4)}$

$m = \dfrac{12 \pm \sqrt{-96}}{8}$

$m = \dfrac{12 \pm 4\sqrt{6}\,i}{8}$

$m = \dfrac{3 \pm \sqrt{6}\,i}{2}$

$m = \dfrac{3}{2} \pm \dfrac{\sqrt{6}}{2}\,i$

$m \approx 1.5 \pm 1.22i$

93. $4n^2 - 9 = 0$

$4n^2 = 9$

$n^2 = \dfrac{9}{4}$

$n = \pm\dfrac{3}{2}$

95. $5w^2 = 6w + 8$

$5w^2 - 6w - 8 = 0$

$(5w + 4)(w - 2) = 0$

$5w + 4 = 0 \ \text{ or } \ w - 2 = 0$

$5w = -4 \qquad \text{ or } \ w = 2$

$w = -\dfrac{4}{5} \ \text{ or } \ w = 2$

97. $3a^2 - a + 2 = 0$

$a = 3, b = -1, c = 2$

$a = \dfrac{-(-1) \pm \sqrt{(-1)^2 - 4(3)(2)}}{2(3)}$

$a = \dfrac{1 \pm \sqrt{-23}}{6}$

$a = \dfrac{1 \pm \sqrt{23}\,i}{6}$

$a = \dfrac{1}{6} \pm \dfrac{\sqrt{23}}{6}\,i$

$a \approx 0.1\overline{6} \pm 0.80i$

99. $5p^2 = 6p + 3$

$5p^2 - 6p - 3 = 0$
$a = 5, b = -6, c = -3$

$$p = \frac{-(-6) \pm \sqrt{(-6)^2 - 4(5)(-3)}}{2(5)}$$

$$p = \frac{6 \pm \sqrt{96}}{10}$$

$$p = \frac{6 \pm 4\sqrt{6}}{10}$$

$$p = \frac{3 \pm 2\sqrt{6}}{5}$$

$$p = \frac{3}{5} \pm \frac{2\sqrt{6}}{5}$$

$p \approx 1.58$ or $p \approx -0.38$

101. $5w^2 - w = 1$

$5w^2 - w - 1 = 0$
$a = 5, b = -1, c = -1$

$$w = \frac{-(-1) \pm \sqrt{(-1)^2 - 4(5)(-1)}}{2(5)}$$

$$w = \frac{1 \pm \sqrt{21}}{10}$$

$$w = \frac{1}{10} + \frac{\sqrt{21}}{10}$$

$w \approx 0.56$ or $w \approx -0.36$

103. $2a^2 + 5 = 3a$

$2a^2 - 3a + 5 = 0$
$a = 2, b = -3, c = 5$

$$a = \frac{-(-3) \pm \sqrt{(-3)^2 - 4(2)(5)}}{2(2)}$$

$$a = \frac{3 \pm \sqrt{-31}}{4}$$

$$a = \frac{3 \pm \sqrt{31}\, i}{4}$$

$$a = \frac{3}{4} \pm \frac{\sqrt{31}}{4} i$$

$a \approx 0.75 \pm 1.39i$

105. $2p^2 - 4p + 11 = 0$

$a = 2, b = -4, c = 11$

$$p = \frac{-(-4) \pm \sqrt{(-4)^2 - 4(2)(11)}}{2(2)}$$

$$p = \frac{4 \pm \sqrt{-72}}{4}$$

$$p = \frac{4 \pm 6\sqrt{2}\, i}{4}$$

$$p = 1 \pm \frac{3\sqrt{2}}{2} i$$

$p \approx 1 \pm 2.12i$

107. $w^2 + \dfrac{2}{3} w = \dfrac{1}{9}$

$$9\left[w^2 + \frac{2}{3} w = \frac{1}{9} \right]$$

$9w^2 + 6w = 1$
$9w^2 + 6w - 1 = 0$
$a = 9, b = 6, c = -1$

$$w = \frac{-(6) \pm \sqrt{(6)^2 - 4(9)(-1)}}{2(9)}$$

$$w = \frac{-6 \pm \sqrt{72}}{18}$$

$$w = \frac{-6 \pm 6\sqrt{2}}{18}$$

$$w = \frac{-1 \pm \sqrt{2}}{3}$$

$$w = \frac{-1}{3} \pm \frac{\sqrt{2}}{3}$$

$w \approx 0.14$ or $w \approx -0.80$

109. $0.2a^2 + 1.2a + 0.9 = 0$
$a = 0.2, b = 1.2, c = 0.9$

$$a = \frac{-(1.2) \pm \sqrt{(1.2)^2 - 4(0.2)(0.9)}}{2(0.2)}$$

$$a = \frac{-1.2 \pm \sqrt{0.72}}{0.4}$$

$$a = \frac{-1.2 \pm 0.6\sqrt{2}}{0.4}$$

$$a = \frac{-6 \pm 3\sqrt{2}}{2}$$

$$a = \frac{-6}{2} \pm \frac{3\sqrt{2}}{2}$$

$$a = -3 \pm \frac{3\sqrt{2}}{2}$$

$a \approx -0.88 \ \text{ or } \ a \approx -5.12$

111. $\dfrac{2}{7}p^2 - 3 = \dfrac{8}{21}p$

$$21\left[\frac{2}{7}p^2 - 3 = \frac{8}{21}p\right]$$

$6p^2 - 63 = 8p$
$6p^2 - 8p - 63 = 0$
$a = 6, b = -8, c = -63$

$$p = \frac{-(-8) \pm \sqrt{(-8)^2 - 4(6)(-63)}}{2(6)}$$

$$p = \frac{8 \pm \sqrt{1576}}{12}$$

$$p = \frac{8 \pm 2\sqrt{394}}{12}$$

$$p = \frac{4 \pm \sqrt{394}}{6}$$

$$p = \frac{4}{6} \pm \frac{\sqrt{394}}{6}$$

$$p = \frac{2}{3} \pm \frac{\sqrt{394}}{6}$$

$p \approx 3.97 \ \text{ or } \ p \approx -2.64$

113. $-3x^2 + 2x + 1 = 0$
$a = -3, b = 2, c = 1$
$(2)^2 - 4(-3)(1) = 16$
two rational solutions

115. $-4x + x^2 + 13 = 0$
$x^2 - 4x + 13 = 0$
$a = 1, b = -4, c = 13$
$(-4)^2 - 4(1)(13) = -36$
two complex solutions

117. $15x^2 - x - 6 = 0$
$a = 15, b = -1, c = -6$
$(-1)^2 - 4(15)(-6) = 361$
two rational solutions

119. $-4x^2 + 6x - 5 = 0$
$a = -4, b = 6, c = -5$
$(6)^2 - 4(-4)(-5) = -44$
two complex solutions

121. $2x^2 + 8 = -9x$
$2x^2 + 9x + 8 = 0$
$a = 2, b = 9, c = 8$
$(9)^2 - 4(2)(8) = 17$
two irrational solutions

123. $4x^2 + 12x = -9$
$4x^2 + 12x + 9 = 0$
$a = 4, b = 12, c = 9$
$(12)^2 - 4(4)(9) = 0$
one repeated solution

125. $-6x + 2x^2 + 5 = 0$
$2x^2 - 6x + 5 = 0$
$a = 2, b = -6, c = 5$

$$x = \frac{-(-6) \pm \sqrt{(-6)^2 - 4(2)(5)}}{2(2)}$$

$$x = \frac{6 \pm \sqrt{-4}}{4}$$

$$x = \frac{6 \pm 2i}{4}$$

$$x = \frac{3}{2} \pm \frac{1}{2}i$$

1.5 Exercises

127. $5x^2 + 5 = -5x$

$5x^2 + 5x + 5 = 0$
$a = 5, b = 5, c = 5$

$$x = \frac{-(5) \pm \sqrt{(5)^2 - 4(5)(5)}}{2(5)}$$

$$x = \frac{-5 \pm \sqrt{-75}}{10}$$

$$x = \frac{-5 \pm 5\sqrt{3}\, i}{10}$$

$$x = -\frac{1}{2} \pm \frac{\sqrt{3}}{2} i$$

129. $-2x^2 = -5x + 11$

$0 = 2x^2 - 5x + 11$
$a = 2, b = -5, c = 11$

$$x = \frac{-(-5) \pm \sqrt{(-5)^2 - 4(2)(11)}}{2(2)}$$

$$x = \frac{5 \pm \sqrt{-63}}{4}$$

$$x = \frac{5 \pm 3\sqrt{7}\, i}{4}$$

$$x = \frac{5}{4} \pm \frac{3\sqrt{7}}{4} i$$

131. $h = -16t^2 + vt$

$16t^2 - vt + h = 0$
$a = 16, b = -v, c = h$

$$t = \frac{-(-v) \pm \sqrt{(-v)^2 - 4(16)(h)}}{2(16)}$$

$$t = \frac{v \pm \sqrt{v^2 - 64h}}{32}$$

133. $-16t^2 + 96t + 408 = 0$

$16t^2 - 96t - 408 = 0$
$8(2t^2 - 12t - 51) = 0$
$a = 2, b = -12, c = -51$

$$t = \frac{-(-12) \pm \sqrt{(-12)^2 - 4(2)(-51)}}{2(2)}$$

$$t = \frac{12 \pm \sqrt{552}}{4}$$

$$t = \frac{12 \pm 2\sqrt{138}}{4}$$

$$t = \frac{6 \pm \sqrt{138}}{2}$$

$t \approx 8.87$ seconds

135. $R = x\left(40 - \frac{1}{3}x\right)$

$$900 = 40x - \frac{1}{3}x^2$$

$$\frac{1}{3}x^2 - 40x + 900 = 0$$

$x^2 - 120x + 2700 = 0$
$a = 1, b = -120, c = 2700$

$$x = \frac{-(-120) \pm \sqrt{(-120)^2 - 4(1)(2700)}}{2(1)}$$

$$x = \frac{120 \pm \sqrt{3600}}{2}$$

$$x = \frac{120 \pm 60}{2}$$

$x = 90$ or $x = 30$
30 thousand ovens

137.a. $P = -x^2 + 122x - 1965 - (2x + 35)$

$P = -x^2 + 120x - 2000$

b. $x^2 - 120x + 2000 = 0$
$(x - 100)(x - 20) = 0$
$x - 100 = 0$ or $x - 20 = 0$
$x = 100$ or $x = 20$
10,000 toys

139. $260 = -16t^2 + 144t$

$16t^2 - 144t + 260 = 0$

$4t^2 - 36t + 65 = 0$
$(2t - 13)(2t - 5) = 0$
$2t - 13 = 0$ or $2t - 5 = 0$
$2t = 13$ or $2t = 5$
$t = \dfrac{13}{2}$ seconds or $t = \dfrac{5}{2}$ seconds

$t = 6.5$ seconds or $t = 2.5$ seconds

141. $3750 = 17.4x^2 + 36.1x + 83.3$

$0 = 17.4x^2 + 36.1x - 3666.7$

$a = 17.4, b = 36.1, c = -3666.7$

$x = \dfrac{-36.1 + \sqrt{(36.1)^2 - 4(17.4)(-3666.7)}}{2(17.4)}$

$x = \dfrac{-36.1 + \sqrt{256505.53}}{34.8}$

$x = \dfrac{-36.1 + 506.4637499}{34.8}$

$x = \dfrac{470.3637499}{34.8}$

$x \approx 13.5$

$1995 + 13.5 = 2008.5$

In the year 2008.

143. Let w represent the width of the doubles court.

Let $2w + 6$ represent the length of the doubles court.

$w(2w + 6) = 2808$

$2w^2 + 6w - 2808 = 0$

$w^2 + 3w - 1404 = 0$
$(w + 39)(w - 36) = 0$
$w + 39 = 0$ or $w - 36 = 0$
$w = -39$ or $w = 36$
The width is 36 feet.
The length is 78 feet.

145.a. $x^2 + 6x - 16 = 0$

$b = 6, c = -16$

$b^2 - 4ac$

$6^2 - 4a(-16)$

$= 36 + 64a =$ perfect square

If $a = 7$,

$= 36 + 64(7) = 484$ (perfect square);

$7x^2 + 6x - 16 = 0$

b. $x^2 + 5x - 14 = 0$

$b = 5, c = -14$

$b^2 - 4ac$

$5^2 - 4a(-14)$

$= 25 + 56a =$ perfect square

If $a = 6$,

$= 25 + 56(6) = 361$ (perfect square);

$6x^2 + 5x - 14 = 0$

c. $x^2 - x - 6 = 0$

$b = -1, c = -6$

$b^2 - 4ac$

$(-1)^2 - 4a(-6)$

$= 1 + 24a =$ perfect square

If $a = 5$,

$= 1 + 24(5) = 121$ (perfect square);

$5x^2 + x - 6 = 0$

147. $z^2 - 3iz = -10$

$z^2 - 3iz + 10 = 0$
$a = 1, b = -3i, c = 10$

$z = \dfrac{-(-3i) \pm \sqrt{(-3i)^2 - 4(1)(10)}}{2(1)}$

$z = \dfrac{3i \pm \sqrt{9i^2 - 40}}{2}$

$z = \dfrac{3i \pm \sqrt{-49}}{2}$

$z = \dfrac{3i \pm 7i}{2}$

$z = 5i$ or $z = -2i$

149. $4iz^2 + 5z + 6i = 0$

$a = 4i, b = 5, c = 6i$

$$z = \frac{-(5) \pm \sqrt{(5)^2 - 4(4i)(6i)}}{2(4i)}$$

$$z = \frac{-5 \pm \sqrt{25 - 96i^2}}{8i}$$

$$z = \frac{-5 \pm \sqrt{121}}{8i}$$

$$z = \frac{-5 \pm 11}{8i}$$

$$z = \frac{6}{8i} \text{ or } z = \frac{-16}{8i} \quad \left(\text{Recall } \frac{1}{i} = -i \right)$$

$$z = -\frac{3}{4}i \text{ or } z = 2i$$

151. $0.5z^2 + (7 + i)z + (6 + 7i) = 0$

$a = 0.5, b = 7 + i, c = 6 + 7i$

$$z = \frac{-(7 + i) \pm \sqrt{(7 + i)^2 - 4(0.5)(6 + 7i)}}{2(0.5)}$$

$$z = \frac{-7 - i \pm \sqrt{49 + 14i + i^2 - 12 - 14i}}{1}$$

$z = -7 - i \pm \sqrt{36}$

$z = -7 - i \pm 6$

$z = -1 - i$ or $z = -13 - i$

153. a. $P = 2L + 2W, A = LW$

b. $P = 2\pi r, A = \pi r^2$

c. $A = \frac{1}{2}h(b_1 + b_2), P = c + h + b_1 + b_2$

d. $A = \frac{1}{2}bh, P = a + b + c$

155. Let x represent the number of good seats sold.

Let $900 - x$ represent the number of cheap seats sold.

$30x + 20(900 - x) = 25,000$

$30x + 18,000 - 20x = 25,000$

$10x = 7,000$

$x = 700$

700, $30 tickets

200, $20 tickets

1.6 Exercises

1. Excluded

3. Extraneous

5. Answers will vary.

7. $22x = x^3 - 9x^2$

$0 = x^3 - 9x^2 - 22x$

$0 = x(x^2 - 9x - 22)$

$0 = x(x - 11)(x + 2)$

$x = 0$ or $x - 11 = 0$ or $x + 2 = 0$

$x = 0, x = -2, x = 11$

9. $3x^3 = -7x^2 + 6x$

$3x^3 + 7x^2 - 6x = 0$

$x(3x^2 + 7x - 6) = 0$

$x(3x - 2)(x + 3) = 0$

$x = 0$ or $3x - 2 = 0$ or $x + 3 = 0$

$\qquad 3x = 2 \quad$ or $x = -3$

$\qquad x = \dfrac{2}{3}$

$x = 0, x = -3, x = \dfrac{2}{3}$

11. $2x^4 - 3x^3 = 9x^2$

$2x^4 - 3x^3 - 9x^2 = 0$

$x^2(2x^2 - 3x - 9) = 0$

$x^2(2x + 3)(x - 3) = 0$

$x^2 = 0$ or $2x + 3 = 0$ or $x - 3 = 0$

$x = 0 \qquad 2x = -3 \qquad x = 3$

$\qquad\qquad x = -\dfrac{3}{2}$

$x = 0, x = -\dfrac{3}{2}, x = 3$

13. $2x^4 - 16x = 0$

$2x(x^3 - 8) = 0$

$2x(x - 2)(x^2 + 2x + 4) = 0$

$2x = 0$ or $x - 2 = 0$ or $x^2 + 2x + 4 = 0$

$x = 0$ or $x = 2$ or $x = \dfrac{-2 \pm \sqrt{(2)^2 - 4(1)(4)}}{2(1)}$

$\qquad\qquad x = \dfrac{-2 \pm \sqrt{-12}}{2}$

$\qquad\qquad x = \dfrac{-2 \pm 2\sqrt{3}i}{2}$

$\qquad\qquad x = -1 \pm i\sqrt{3}$

$x = 0, x = 2, x = -1 \pm i\sqrt{3}$

15. $x^3 - 4x = 5x^2 - 20$

$x^3 - 5x^2 - 4x + 20 = 0$

$x^2(x - 5) - 4(x - 5) = 0$

$(x - 5)(x^2 - 4) = 0$

$(x - 5)(x + 2)(x - 2) = 0$

$x - 5 = 0$ or $x + 2 = 0$ or $x - 2 = 0$

$\qquad x = 5$ or $\quad x = -2 \quad$ or $x = 2$

$x = 5, x = 2, x = -2$

17. $4x - 12 = 3x^2 - x^3$

$x^3 - 3x^2 + 4x - 12 = 0$

$x^2(x - 3) + 4(x - 3) = 0$

$(x - 3)(x^2 + 4) = 0$

$x - 3 = 0$ or $x^2 + 4 = 0$

$\qquad x = 3$ or $x^2 = -4$

$x = 3, x = \pm 2i$

19. $2x^3 - 12x^2 = 10x - 60$

$2x^3 - 12x^2 - 10x + 60 = 0$

$2(x^3 - 6x^2 - 5x + 30) = 0$

$2[x^2(x - 6) - 5(x - 6)] = 0$

$2(x - 6)(x^2 - 5) = 0$

$x - 6 = 0$ or $x^2 - 5 = 0$

$x = 6$ or $x^2 = 5$

$x = 6, x = \pm\sqrt{5}$

1.6 Exercises

21. $x^4 - 7x^3 + 4x^2 = 28x$

    $x^4 - 7x^3 + 4x^2 - 28x = 0$

    $x(x^3 - 7x^2 + 4x - 28) = 0$

    $x[x^2(x-7) + 4(x-7)] = 0$

    $x(x-7)(x^2+4) = 0$

    $x = 0$ or $x - 7 = 0$ or $x^2 + 4 = 0$

    $\qquad x = 7$ or $x^2 = -4$

    $x = 0, x = 7, x = \pm 2i$

23. $x^4 - 81 = 0$

    $(x^2 + 9)(x^2 - 9) = 0$

    $x^2 + 9 = 0$ or $x^2 - 9 = 0$

    $x^2 = -9$ or $x^2 = 9$

    $x = \pm 3i, x = \pm 3$

25. $x^4 - 256 = 0$

    $(x^2 + 16)(x^2 - 16) = 0$

    $x^2 + 16 = 0$ or $x^2 - 16 = 0$

    $x^2 = -16$ or $x^2 = 16$

    $x = \pm 4i, x = \pm 4$

27. $x^6 - 2x^4 - x^2 + 2 = 0$

    $x^4(x^2-2) - 1(x^2-2) = 0$

    $(x^2-2)(x^4-1) = 0$

    $(x^2-2)(x^2+1)(x^2-1) = 0$

    $(x^2-2)(x^2+1)(x+1)(x-1) = 0$

    $x^2 - 2 = 0$ or $x^2 + 1 = 0$

    $\qquad$ or $x + 1 = 0$ or $x - 1 = 0$

    $x^2 = 2$ or $x^2 = -1$

    $\qquad$ or $x = -1$ or $x = 1$

    $x = \pm\sqrt{2}, x = \pm i, x = -1, x = 1$

29. $x^5 - x^3 - 8x^2 + 8 = 0$

    $x^3(x^2-1) - 8(x^2-1) = 0$

    $(x^2-1)(x^3-8) = 0$

    $(x+1)(x-1)(x-2)(x^2+2x+4) = 0$

    $x + 1 = 0$ or $x - 1 = 0$

    $\qquad$ or $x - 2 = 0$ or $x^2 + 2x + 4 = 0$

    $x = -1$ or $x = 1$ or $x = 2$

    $\qquad$ or $x = \dfrac{-2 \pm \sqrt{(2)^2 - 4(1)(4)}}{2(1)}$

    $x = \dfrac{-2 \pm \sqrt{-12}}{2}$

    $x = \dfrac{-2 \pm 2i\sqrt{3}}{2}$

    $x = \pm 1, x = 2, x = -1 \pm i\sqrt{3}$

31. $x^6 - 1 = 0$

    $(x^3+1)(x^3-1) = 0$

    $(x+1)(x^2-x+1)(x-1)(x^2+x+1) = 0$

    $x + 1 = 0$ or $x^2 - x + 1 = 0$

    or $x - 1 = 0$ or $x^2 - x + 1 = 0$;

    $x = -1$

    or $x = \dfrac{-(-1) \pm \sqrt{(-1)^2 - 4(1)(1)}}{2(1)} = \dfrac{1 \pm \sqrt{-3}}{2} = \dfrac{1 \pm i\sqrt{3}}{2}$

    or $x = 1$

    or $x = \dfrac{-(1) \pm \sqrt{1^2 - 4(1)(1)}}{2(1)} = \dfrac{-1 \pm \sqrt{-3}}{2} = \dfrac{-1 \pm i\sqrt{3}}{2}$;

    $x = \pm 1, x = \dfrac{1}{2} \pm \dfrac{i\sqrt{3}}{2}, x = -\dfrac{1}{2} \pm \dfrac{i\sqrt{3}}{2}$

33. $\dfrac{2}{x} + \dfrac{1}{x+1} = \dfrac{5}{x^2+x}$

    $\dfrac{2}{x} + \dfrac{1}{x+1} = \dfrac{5}{x(x+1)}$

    $x(x+1)\left[\dfrac{2}{x} + \dfrac{1}{x+1} = \dfrac{5}{x(x+1)}\right]$

    $2x + 2 + x = 5$

    $3x = 3$

    $x = 1$

35. $\dfrac{21}{a+2} = \dfrac{3}{a-1}$

$(a+2)(a-1)\left[\dfrac{21}{a+2} = \dfrac{3}{a-1}\right]$

$21a - 21 = 3a + 6$

$18a = 27$

$a = \dfrac{27}{18} = \dfrac{3}{2}$

37. $\dfrac{1}{3y} - \dfrac{1}{4y} = \dfrac{1}{y^2}$

$12y^2\left[\dfrac{1}{3y} - \dfrac{1}{4y} = \dfrac{1}{y^2}\right]$

$4y - 3y = 12$

$y = 12$

39. $x + \dfrac{14}{x-7} = 1 + \dfrac{2x}{x-7}$

$(x-7)\left[x + \dfrac{14}{x-7} = 1 + \dfrac{2x}{x-7}\right]$

$x(x-7) + 14 = 1(x-7) + 2x$

$x^2 - 7x + 14 = x - 7 + 2x$

$x^2 - 7x + 14 = 3x - 7$

$x^2 - 10x + 21 = 0$

$(x-7)(x-3) = 0$

$x - 7 = 0 \text{ or } x - 3 = 0$

$x = 7 \quad \text{ or } \quad x = 3$

$x = 3; x = 7 \text{ is extraneous}$

41. $\dfrac{6}{n+3} + \dfrac{20}{n^2+n-6} = \dfrac{5}{n-2}$

$\dfrac{6}{n+3} + \dfrac{20}{(n+3)(n-2)} = \dfrac{5}{n-2}$

$(n+3)(n-2)\left[\dfrac{6}{n+3} + \dfrac{20}{(n+3)(n-2)} = \dfrac{5}{n-2}\right]$

$6(n-2) + 20 = 5(n+3)$

$6n - 12 + 20 = 5n + 15$

$6n + 8 = 5n + 15$

$n = 7$

43. $\dfrac{a}{2a+1} - \dfrac{2a^2+5}{2a^2-5a-3} = \dfrac{3}{a-3}$

$\dfrac{a}{2a+1} - \dfrac{2a^2+5}{(2a+1)(a-3)} = \dfrac{3}{a-3}$

$(2a+1)(a-3)\left[\dfrac{a}{2a+1} - \dfrac{2a^2+5}{(2a+1)(a-3)} = \dfrac{3}{a-3}\right]$

$a(a-3) - (2a^2+5) = 3(2a+1)$

$a^2 - 3a - 2a^2 - 5 = 6a + 3$

$-a^2 - 3a - 5 = 6a + 3$

$0 = a^2 + 9a + 8$

$(a+8)(a+1) = 0$

$a + 8 = 0 \text{ or } a + 1 = 0$

$a = -8 \quad \text{ or } a = -1$

45. $\dfrac{1}{f} = \dfrac{1}{f_1} + \dfrac{1}{f_2}$

$f\,f_1 f_2\left[\dfrac{1}{f} = \dfrac{1}{f_1} + \dfrac{1}{f_2}\right]$

$f_1 f_2 = f\,f_2 + f\,f_1$

$f_1 f_2 = f(f_2 + f_1)$

$\dfrac{f_1 f_2}{f_1 + f_2} = f$

47. $I = \dfrac{E}{R+r}$

$(R+r)\left[I = \dfrac{E}{R+r}\right]$

$IR + Ir = E$

$Ir = E - IR$

$r = \dfrac{E - IR}{I} \quad \text{ or } r = \dfrac{E}{I} - R$

49. $V = \dfrac{1}{3}\pi r^2 h$

$3V = \pi r^2 h$

$\dfrac{3V}{\pi r^2} = h$

1.6 Exercises

51. $V = \dfrac{4}{3}\pi r^3$

    $3V = 4\pi r^3$

    $\dfrac{3V}{4\pi} = r^3$

53. a. $-3\sqrt{3x-5} = -9$

    $\sqrt{3x-5} = 3$
    $3x-5 = 9$
    $3x = 14$
    $x = \dfrac{14}{3}$

    b. $x = \sqrt{3x+1}+3$

    $x-3 = \sqrt{3x+1}$

    $(x-3)^2 = \left(\sqrt{3x+1}\right)^2$

    $x^2 -6x+9 = 3x+1$
    $x^2 -9x+8 = 0$
    $(x-8)(x-1) = 0$
    $x-8 = 0 \ \text{ or } \ x-1 = 0$
    $x = 8 \qquad \text{ or } \ x = 1$
    $x = 8; \ x = 1$ is extraneous.

55. a. $2 = \sqrt[3]{3m-1}$

    $8 = 3m-1$
    $9 = 3m$
    $3 = m$

    b. $2\sqrt[3]{7-3x} - 3 = -7$

    $2\sqrt[3]{7-3x} = -4$

    $\sqrt[3]{7-3x} = -2$

    $7-3x = -8$
    $-3x = -15$
    $x = 5$

    c. $\dfrac{\sqrt[3]{2m+3}}{-5} + 2 = 3$

    $\dfrac{\sqrt[3]{2m+3}}{-5} = 1$

    $\sqrt[3]{2m+3} = -5$
    $2m+3 = -125$
    $2m = -128$
    $m = -64$

    d. $\sqrt[3]{2x-9} = \sqrt[3]{3x+7}$

    $2x-9 = 3x+7$
    $-x = 16$
    $x = -16$

57. a. $\sqrt{x-9}+\sqrt{x} = 9$

    $\sqrt{x-9} = 9-\sqrt{x}$

    $\left(\sqrt{x-9}\right)^2 = \left(9-\sqrt{x}\right)^2$

    $x-9 = 81-18\sqrt{x}+x$
    $-90 = -18\sqrt{x}$
    $5 = \sqrt{x}$
    $25 = x$

    b. $\sqrt{x+9}+\sqrt{x-7} = 8$

    $\sqrt{x+9} = 8-\sqrt{x-7}$

    $\left(\sqrt{x+9}\right)^2 = \left(8-\sqrt{x-7}\right)^2$

    $x+9 = 64-16\sqrt{x-7}+x-7$
    $x+9 = 57-16\sqrt{x-7}+x$
    $-48 = -16\sqrt{x-7}$
    $3 = \sqrt{x-7}$
    $9 = x-7$
    $16 = x$

    c. $\sqrt{x-2}-\sqrt{2x} = -2$

    $\sqrt{x-2} = \sqrt{2x}-2$

    $\left(\sqrt{x-2}\right)^2 = \left(\sqrt{2x}-2\right)^2$

    $x-2 = 2x-4\sqrt{2x}+4$
    $-x-6 = -4\sqrt{2x}$
    $x+6 = 4\sqrt{2x}$

    $(x+6)^2 = \left(4\sqrt{2x}\right)^2$

    $x^2 +12x+36 = 16(2x)$
    $x^2 +12x+36 = 32x$
    $x^2 -20x+36 = 0$
    $(x-2)(x-18) = 0$
    $x-2 = 0 \ \text{ or } \ x-18 = 0$
    $x = 2 \qquad \text{ or } \ x = 18$

d. $\sqrt{12x+9} - \sqrt{24x} = -3$

$\sqrt{12x+9} = \sqrt{24x} - 3$

$\left(\sqrt{12x+9}\right)^2 = \left(\sqrt{24x}-3\right)^2$

$12x+9 = 24x - 6\sqrt{24x} + 9$

$-12x = -6\sqrt{24x}$

$2x = \sqrt{24x}$

$(2x)^2 = \left(\sqrt{24x}\right)^2$

$4x^2 = 24x$

$4x^2 - 24x = 0$

$4x(x-6) = 0$

$4x = 0$ or $x - 6 = 0$

$x = 0$ or $x = 6$

$x = 6; x = 0$ is extraneous

59. $x^{\frac{3}{5}} + 17 = 9$

$x^{\frac{3}{5}} = -8$

$\left(x^{\frac{3}{5}}\right)^{\frac{5}{3}} = (-8)^{\frac{5}{3}}$

$x = -32$

61. $0.\overline{3}x^{\frac{5}{2}} - 39 = 42$

$\frac{1}{3}x^{\frac{5}{2}} = 81$

$x^{\frac{5}{2}} = 243$

$\left(x^{\frac{5}{2}}\right)^{\frac{2}{5}} = (243)^{\frac{2}{5}}$

$x = 9$

63. $2(x+5)^{\frac{2}{3}} - 11 = 7$

$2(x+5)^{\frac{2}{3}} = 18$

$(x+5)^{\frac{2}{3}} = 9$

$\left((x+5)^{\frac{2}{3}}\right)^{\frac{3}{2}} = 9^{\frac{3}{2}}$

$x+5 = 27$ or $x+5 = -27$

$x = 22$ $x = -32$

65. $x^{\frac{2}{3}} - 2x^{\frac{1}{3}} - 15 = 0$

Let $u = x^{\frac{1}{3}}$

$u^2 = x^{\frac{2}{3}}$;

$u^2 - 2u - 15 = 0$

$(u-5)(u+3) = 0$

$u - 5 = 0$ or $u + 3 = 0$

$u = 5$ or $u = -3$

$x^{\frac{1}{3}} = 5$ or $x^{\frac{1}{3}} = -3$

$\left(x^{\frac{1}{3}}\right)^3 = (5)^3$ or $\left(x^{\frac{1}{3}}\right)^3 = (-3)^3$

$x = 125$ or $x = -27$

67. $x^4 - 24x^2 - 25 = 0$

Let $u = x^2$, then $u^2 = x^4$;

$u^2 - 24u - 25 = 0$

$(u-25)(u+1) = 0$

$u - 25 = 0$ or $u + 1 = 0$

$u = 25$ or $u = -1$;

$x^2 = 25$ or $x^2 = -1$

$x = \pm 5$ or $x = \pm i$

69. $(x^2-3)^2 + (x^2-3) - 2 = 0$

Let $u = x^2 - 3$

$u^2 = (x^2-3)^2$;

$u^2 + u - 2 = 0$

$(u-1)(u+2) = 0$

$u - 1 = 0$ or $u + 2 = 0$

$u = 1$ or $u = -2$

$x^2 - 3 = 1$ or $x^2 - 3 = -2$

$x^2 = 4$ or $x^2 = 1$

$x = \pm 2$ or $x = \pm 1$

71. $x^{-2} - 3x^{-1} - 4 = 0$

Let $u = x^{-1}$

$u^2 = x^{-2}$;

$u^2 - 3u - 4 = 0$

$(u - 4)(u + 1) = 0$

$u - 4 = 0$ or $u + 1 = 0$

$u = 4$ or $u = -1$

$x^{-1} = 4$ or $x^{-1} = -1$

$\left(x^{-1}\right)^{-1} = (4)^{-1}$ or $\left(x^{-1}\right)^{-1} = (-1)^{-1}$

$x = \dfrac{1}{4}$ or $x = -1$

73. $x^{-4} - 13x^{-2} + 36 = 0$

Let $u = x^{-2}$

$u^2 = x^{-4}$;

$u^2 - 13u + 36 = 0$

$(u - 9)(u - 4) = 0$

$u - 9 = 0$ or $u - 4 = 0$

$u = 9$ or $u = 4$

$x^{-2} = 9$ or $x^{-2} = 4$

$\dfrac{1}{x^2} = 9$ or $\dfrac{1}{x^2} = 4$

$x^2 = \dfrac{1}{9}$ or $x^2 = \dfrac{1}{4}$

$x = \pm\dfrac{1}{3}$ or $x = \pm\dfrac{1}{2}$

75. $x + 4 = 7\sqrt{x+4}$

Let $u = (x+4)^{\frac{1}{2}}$

$u^2 = x + 4$;

$u^2 = 7u$

$u^2 - 7u = 0$

$u(u - 7) = 0$

$u = 0$ or $u - 7 = 0$

$u = 0$ or $u = 7$

$\sqrt{x+4} = 0$ or $\sqrt{x+4} = 7$

$x + 4 = 0$ or $x + 4 = 49$

$x = -4$ or $x = 45$

77. $2\sqrt{x+10} + 8 = 3(x+10)$

Let $u = (x+10)^{\frac{1}{2}}$

$u^2 = x + 10$;

$2u + 8 = 3u^2$

$0 = 3u^2 - 2u - 8$

$0 = (3u + 4)(u - 2)$

$3u + 4 = 0$ or $u - 2 = 0$

$3u = -4$ or $u = 2$

$u = -\dfrac{4}{3}$ or $u = 2$

$\sqrt{x+10} = -\dfrac{4}{3}$ or $\sqrt{x+10} = 2$

$x + 10 = \dfrac{16}{9}$ or $x + 10 = 4$

$x = -\dfrac{74}{9}$ or $x = -6$

$x = -6$; $x = -\dfrac{74}{9}$ is extraneous

79. a. $S = \pi r\sqrt{r^2 + h^2}$

$\dfrac{S}{\pi r} = \sqrt{r^2 + h^2}$

$\left(\dfrac{S}{\pi r}\right)^2 = r^2 + h^2$

$\left(\dfrac{S}{\pi r}\right)^2 - r^2 = h^2$

$\sqrt{\left(\dfrac{S}{\pi r}\right)^2 - r^2} = h$

b. $S = \pi(6)\sqrt{(6)^2 + (10)^2}$

$S = 6\pi\sqrt{36 + 100}$

$S = 6\pi\sqrt{136}$

$S = 12\pi\sqrt{34} \text{ m}^2$

81. Let x represent the number.
$$x^3 + 2x^2 = 18 + 9x$$
$$x^3 + 2x^2 - 9x - 18 = 0$$
$$x^2(x+2) - 9(x+2) = 0$$
$$(x+2)(x^2 - 9) = 0$$
$$(x+2)(x+3)(x-3) = 0$$
$$x+2 = 0 \text{ or } x+3 = 0 \text{ or } x-3 = 0$$
$$x = -2 \text{ or } x = -3 \text{ or } x = 3$$
$$x = -2, \ x = \pm 3$$

83. Let x represent the first integer.
Let $x + 2$ represent the second integer.
Let $x + 4$ represent the third integer.
$$4(x+4) + x^4 = (x+2)^2 + 24$$
$$4x + 16 + x^4 = x^2 + 4x + 4 + 24$$
$$4x + 16 + x^4 = x^2 + 4x + 28$$
$$x^4 - x^2 - 12 = 0$$
$$(x^2 - 4)(x^2 + 3) = 0$$
$$x^2 - 4 = 0 \text{ or } x^2 + 3 = 0$$
$$x^2 = 4 \quad \text{ or } \quad x^2 = -3$$
$$x = \pm 2 \quad \text{ or } \quad \text{Not Real;}$$
if $x = 2, x+2 = 4, x+4 = 6;$
if $x = -2, x+2 = 0, x+4 = 2;$
$$x = 2, 4, 6 \text{ or } x = -2, 0, 2$$

85. Let w represent the width.
$$w(w+2) = 143$$
$$w^2 + 2w = 143$$
$$w^2 + 2w - 143 = 0$$
$$(w-11)(w+13) = 0$$
$$w - 11 = 0 \text{ or } w + 13 = 0$$
$$w = 11 \quad \text{ or } w = -13$$
11 inches by 13 inches

87. $24\pi r = \dfrac{2}{3}\pi r^3 + \pi r^2(6)$
$$0 = \frac{2}{3}\pi r^3 + 6\pi r^2 - 24\pi r$$
$$3(0) = 3\left(\frac{2}{3}\pi r^3 + 6\pi r^2 - 24\pi r\right)$$
$$0 = 2\pi r^3 + 18\pi r^2 - 72\pi r$$
$$0 = 2\pi r(r^2 + 9r - 36)$$
$$0 = 2\pi r(r-3)(r+12)$$
$$2\pi r = 0 \text{ or } r - 3 = 0 \text{ or } r + 12 = 0$$
$$r = 0 \quad \text{ or } \quad r = 3 \text{ or } \quad r = -12$$
$$r = 3 \text{ m;}$$
$r = 0$ m and $r = 12$ m do not fit the context

89. Let x represent the number of decreases in price.
$$(70 - 2x)(15 + 3x) = 2250$$
$$1050 + 210x - 30x - 6x^2 = 2250$$
$$6x^2 - 180x + 1200 = 0$$
$$x^2 - 30x + 200 = 0$$
$$(x - 10)(x - 20) = 0$$
$$x - 10 = 0 \text{ or } x - 20 = 0$$
$$x = 10 \text{ or } x = 20$$
10, \$2 decreases results in a price of \$50 and a sale of 45 shoes.
20, \$2 decreases results in a price of \$30 and a sale of 75 shoes.

91. $h = -16t^2 + vt + k$
$$h = -16t^2 + 176t - 480$$

   a. $h = -16(4)^2 + 176(4) - 480$
$$h = -32 \text{ feet}$$
32 feet below the rim.

   b. $-480 = -16t^2 + 176t - 480$
$$0 = -16t^2 + 176t$$
$$0 = -16t(t - 11)$$
$$-16t = 0 \text{ or } t - 11 = 0$$
$$t = 0 \quad \text{ or } t = 11$$
The pebble returns after 11 seconds.

   c. $h = -16(5)^2 + 176(5) - 480 = 0;$
$$h = -16(6)^2 + 176(6) - 480 = 0;$$
The pebble is at the canyon's rim.

93. $\dfrac{1}{20} + \dfrac{1}{30} = \dfrac{1}{x}$

$60x\left(\dfrac{1}{20} + \dfrac{1}{30}\right) = 60x\left(\dfrac{1}{x}\right)$

$3x + 2x = 60$

$5x = 60$

$x = 12$ minutes

95. Let v represent the rate Tom can row in still water.
Then $v + 4$ is the rate downstream,
$v - 4$ is the rate upstream.

$t_{up} + t_{down} = 3$

$\dfrac{5}{v-4} + \dfrac{5}{v+4} = 3$

$(v+4)(v-4)\left(\dfrac{5}{v-4} + \dfrac{5}{v+4}\right) = 3(v+4)(v-4)$

$5(v+4) + 5(v-4) = 3(v^2 - 16)$

$5v + 20 + 5v - 20 = 3v^2 - 48$

$10v = 3v^2 - 48$

$0 = 3v^2 - 10v - 48$

$0 = (3v+8)(v-6)$

$3v + 8 = 0 \;\text{ or }\; v - 6 = 0$

$v = -\dfrac{8}{3} \qquad v = 6;$

$v = 6$ mph

97. $C = \dfrac{92P}{100 - P}$

$100 = \dfrac{92P}{100 - P}$

$(100 - P)\left[100 = \dfrac{92P}{100 - P}\right]$

$10000 - 100P = 92P$

$10000 = 192P$

$52.1\% = P$

99. $T = 0.407R^{\frac{3}{2}}$

a. $88 = 0.407R^{\frac{3}{2}}$

$216.22 \approx R^{\frac{3}{2}}$

$(216.22)^{\frac{2}{3}} \approx \left(R^{\frac{3}{2}}\right)^{\frac{2}{3}}$

$R \approx 36$ million miles

b. $225 = 0.407R^{\frac{3}{2}}$

$552.83 \approx R^{\frac{3}{2}}$

$(552.83)^{\frac{2}{3}} \approx \left(R^{\frac{3}{2}}\right)^{\frac{2}{3}}$

$R \approx 67$ million miles

c. $365 = 0.407R^{\frac{3}{2}}$

$896.81 \approx R^{\frac{3}{2}}$

$(896.81)^{\frac{2}{3}} \approx \left(R^{\frac{3}{2}}\right)^{\frac{2}{3}}$

$R \approx 93$ million miles

d. $687 = 0.407R^{\frac{3}{2}}$

$1687.96 \approx R^{\frac{3}{2}}$

$(1687.96)^{\frac{2}{3}} \approx \left(R^{\frac{3}{2}}\right)^{\frac{2}{3}}$

$R \approx 142$ million miles

e. $4333 = 0.407R^{\frac{3}{2}}$

$10646.19 \approx R^{\frac{3}{2}}$

$(10646.19)^{\frac{2}{3}} \approx \left(R^{\frac{3}{2}}\right)^{\frac{2}{3}}$

$R \approx 484$ million miles

f. $10759 = 0.407R^{\frac{3}{2}}$

$26434.89 \approx R^{\frac{3}{2}}$

$(26434.89)^{\frac{2}{3}} \approx \left(R^{\frac{3}{2}}\right)^{\frac{2}{3}}$

$R \approx 887$ million miles

101. The constant "3" was not multiplied by the LCD;

$$3 - \frac{8}{x+3} = \frac{1}{x}$$
$$3x(x+3) - 8x = x+3$$
$$3x^2 + 9x - 8x = x+3$$
$$3x^2 + x = x+3$$
$$3x^2 = 3$$
$$x^2 = 1$$
$$x = \pm 1$$

103. $\dfrac{\sqrt{x-1}}{x^2-4}$

$$x - 1 > 0 \text{ and } x^2 - 4 \neq 0$$
$$x \geq 1 \quad \text{and } (x+2)(x-2) \neq 0$$
$$x \geq 1 \quad \text{and } x + 2 \neq 0 \text{ and } x - 2 \neq 0$$
$$x \geq 1 \quad \text{and } x \neq -2 \text{ and } x \neq 2$$
$$x \in [1,2) \cup (2,\infty)$$

105. a. $|x^2 - 2x - 25| = 10$

$$x^2 - 2x - 25 = 10 \quad \text{or } x^2 - 2x - 25 = -10$$
$$x^2 - 2x - 35 = 0 \quad \text{or } x^2 - 2x - 15 = 0$$
$$(x-7)(x+5) = 0 \quad \text{or } (x-5)(x+3) = 0$$
$$x - 7 = 0 \text{ or } x + 5 = 0 \quad \text{or } x - 5 = 0 \text{ or } x + 3 = 0$$
$$x = 7 \text{ or } x = -5 \quad \text{or} \quad x = 5 \text{ or } x - 3$$
$$x = -5, -3, 5, 7$$

b. $|x^2 - 5x - 10| = 4$

$$x^2 - 5x - 10 = 4 \quad \text{or } x^2 - 5x - 10 = -4$$
$$x^2 - 5x - 14 = 0 \quad \text{or } x^2 - 5x - 6 = 0$$
$$(x-7)(x+2) = 0 \quad \text{or } (x-6)(x+1) = 0$$
$$x - 7 = 0 \text{ or } x + 2 = 0 \quad \text{or } x - 6 = 0 \text{ or } x + 1 = 0$$
$$x = 7 \text{ or } x = -2 \quad \text{or } x = 6 \text{ or } x = 1$$
$$x = -2, -1, 6, 7$$

c. $|x^2 - 4| = x + 2$

$$x^2 - 4 = x + 2 \quad \text{or} \quad x^2 - 4 = -(x+2)$$
$$x^2 - x - 6 = 0 \quad\quad x^2 - 4 = -x - 2$$
$$(x-3)(x+2) = 0 \quad\quad x^2 + x - 2 = 0$$
$$x - 3 = 0 \text{ or } x + 2 = 0 \quad (x+2)(x-1) = 0$$
$$x = 3 \text{ or } x = -2 \quad\quad x + 2 = 0 \text{ or } x - 1 = 0$$
$$x = -2, x = 1$$
$$x = -2, 1, 3$$

d. $|x^2 - 9| = -x + 3$

$$x^2 - 9 = -x + 3 \quad \text{or} \quad x^2 - 9 = -(-x+3)$$
$$x^2 + x - 12 = 0 \quad\quad x^2 - 9 = x - 3$$
$$(x+4)(x-3) = 0 \quad\quad x^2 - x - 6 = 0$$
$$x + 4 = 0 \text{ or } x - 3 = 0 \quad (x-3)(x+2) = 0$$
$$x = -4 \quad x = 3 \quad\quad x - 3 = 0 \text{ or } x + 2 = 0$$
$$x = 3, x = -2$$
$$x = -4, -2, 3$$

e. $|x^2 - 7x| = -x + 7$

$$x^2 - 7x = -x + 7 \quad \text{or} \quad x^2 - 7x = -(-x+7)$$
$$x^2 - 6x - 7 = 0 \quad\quad x^2 - 7x = x - 7$$
$$(x-7)(x+1) = 0 \quad\quad x^2 - 8x + 7 = 0$$
$$x - 7 = 0 \text{ or } x + 1 = 0 \quad (x-7)(x-1) = 0$$
$$x = 7 \quad x = -1 \quad\quad x - 7 = 0 \quad x - 1 = 0$$
$$x = 7, x = 1$$
$$x = -1, 1, 7$$

f. $|x^2 - 5x - 2| = x + 5$

$$x^2 - 5x - 2 = x + 5 \quad \text{or} \quad x^2 - 5x - 2 = -(x+5)$$
$$x^2 - 6x - 7 = 0 \quad\quad x^2 - 5x - 2 = -x - 5$$
$$(x-7)(x+1) = 0 \quad\quad x^2 - 4x + 3 = 0$$
$$x - 7 = 0 \text{ or } x + 1 = 0 \quad (x-3)(x-1) = 0$$
$$x = 7 \quad x = -1 \quad\quad x - 3 = 0 \quad x - 1 = 0$$
$$x = 3, x = 1$$
$$x = -1, 1, 3, 7$$

107. $x^2 + 10^2 = 12^2$

$$x^2 + 100 = 144$$
$$x^2 = 44$$
$$x = \pm\sqrt{44}$$
$$x = \pm 2\sqrt{11}$$
$$x = 2\sqrt{11} \text{ cm}$$

109. $2x - 3 < 7$ and $x + 2 > 1$

$$2x < 10 \quad \text{and } x > -1$$
$$x < 5 \quad \text{and } x > -1$$
$$-1 < x < 5$$

Chapter 1 Summary and Concept Review

1. a. $6x - (2 - x) = 4(x - 5)$

    $6(-6) - (2 - (-6)) = 4(-6 - 5)$

    $-36 - (2 + 6) = 4(-11)$

    $-36 - 8 = -44$

    $-44 = -44;$ yes

    b. $\dfrac{3}{4}b + 2 = \dfrac{5}{2}b + 16$

    $\dfrac{3}{4}(-8) + 2 = \dfrac{5}{2}(-8) + 16$

    $-6 + 2 = -20 + 16$

    $-4 = -4$

    yes

    c. $4d - 2 = -\dfrac{1}{2} + 3d$

    $4\left(\dfrac{3}{2}\right) - 2 = -\dfrac{1}{2} + 3\left(\dfrac{3}{2}\right)$

    $6 - 2 = -\dfrac{1}{2} + \dfrac{9}{2}$

    $4 = 4$

    yes

3. $3(2n - 6) + 1 = 7$

    $6n - 18 + 1 = 7$

    $6n - 17 = 7$

    $6n = 24$

    $n = 4$

5. $\dfrac{1}{2}x + \dfrac{2}{3} = \dfrac{3}{4}$

    $12\left[\dfrac{1}{2}x + \dfrac{2}{3} = \dfrac{3}{4}\right]$

    $6x + 8 = 9$

    $6x = 1$

    $x = \dfrac{1}{6}$

7. $-\dfrac{g}{6} = 3 - \dfrac{1}{2} - \dfrac{5g}{12}$

    $12\left[-\dfrac{g}{6} = 3 - \dfrac{1}{2} - \dfrac{5g}{12}\right]$

    $-2g = 36 - 6 - 5g$

    $3g = 30$

    $g = 10$

9. $P = 2L + 2W$

    $P - 2W = 2L$

    $\dfrac{P - 2W}{2} = L$

11. $2x - 3y = 6$

    $-3y = -2x + 6$

    $y = \dfrac{2}{3}x - 2$

13. $3(4) + \dfrac{1}{2}\pi(1.5)^2$

    $12 + \dfrac{9}{8}\pi \ \text{ft}^2 \approx 15.5 \ \text{ft}^2$

15. $a \geq 35$

17. $s \leq 65$

19. $7x > 35$

    $x > 5$

    $(5, \infty)$

21. $2(3m - 2) \leq 8$

    $6m - 4 \leq 8$

    $6m \leq 12$

    $m \leq 2$

    $(-\infty, 2]$

23. $-4 < 2b + 8$ and $3b - 5 > -32$

    $-12 < 2b$ and $3b > -27$

    $-6 < b$ and $b > -9$

    $b > -6$ and $b > -9$

    $(-6, \infty)$

25. a. $\dfrac{7}{n-3}$

$n - 3 \neq 0$

$n \neq 3$

$(-\infty, 3) \cup (3, \infty)$

b. $\dfrac{5}{2x-3}$

$2x - 3 \neq 0$

$2x \neq 3$

$x \neq \dfrac{3}{2}$

$\left(-\infty, \dfrac{3}{2}\right) \cup \left(\dfrac{3}{2}, \infty\right)$

c. $\sqrt{x+5}$

$x + 5 \geq 0$

$x \geq -5$

$[-5, \infty)$

d. $\sqrt{-3n+18}$

$-3n + 18 \geq 0$

$-3n \geq -18$

$n \leq 6$

$(-\infty, 6]$

27. $7 = |x - 3|$

$x - 3 = 7$ or $x - 3 = -7$

$x = 10$ or $x = -4$

$\{-4, 10\}$

29. $|-2x + 3| = 13$

$-2x + 3 = 13$ or $-2x + 3 = -13$

$-2x = 10$ or $-2x = -16$

$x = -5$ $\qquad$ $x = 8$

$\{-5, 8\}$

31. $-3|x + 2| - 2 < -14$

$-3|x + 2| < -12$

$|x + 2| > 4$

$x + 2 > 4$ or $x + 2 < -4$

$x > 2$ or $x < -6$

$x \in (-\infty, -6) \cup (2, \infty)$

33. $|3x + 5| = -4$

Absolute value can never be negative.

$\{\ \}$

35. $2|x + 1| > -4$

$|x + 1| > -2$

Absolute value is always greater than a negative number.

$(-\infty, \infty)$

37. $\dfrac{|3x - 2|}{2} + 6 \geq 10$

$\dfrac{|3x - 2|}{2} \geq 1$

$|3x - 2| \geq 8$

$3x - 2 \geq 8$ or $3x - 2 \leq -8$

$3x \geq 10$ or $3x \leq -6$

$x \geq \dfrac{10}{3}$ or $x \leq -2$

$(-\infty, -2] \cup \left[\dfrac{10}{3}, \infty\right)$

39. $\sqrt{-72} = \sqrt{-1(36)(2)} = 6\sqrt{2}\, i$

41. $\dfrac{-10 + \sqrt{-50}}{5} = \dfrac{-10 + \sqrt{-1(25)(2)}}{5}$

$= \dfrac{-10 + 5\sqrt{2}\, i}{5} = -2 + \sqrt{2}\, i$

43. $i^{57} = \left(i^4\right)^{14} i = i$

45. $\dfrac{5i}{1 - 2i} \cdot \dfrac{1 + 2i}{1 + 2i} = \dfrac{5i + 10i^2}{1 + 2i - 2i - 4i^2}$

$= \dfrac{-10 + 5i}{5} = -2 + i$

47. $(2 + 3i)(2 - 3i) = 4 - 6i + 6i - 9i^2 = 13$

49. $x^2 - 9 = -34, x = 5i$;

$(5i)^2 - 9 = -34$

$25i^2 - 9 = -34$
$-25 - 9 = -34$
$-34 = -34$ verified;

$(-5i)^2 - 9 = -34$

$25i^2 - 9 = -34$
$-25 - 9 = -34$
$-34 = -34$ verified

51. a. $-3 = 2x^2$

$2x^2 + 3 = 0$

$a = 2, b = 0, c = 3$

b. $7 = -2x + 11$ is not quadratic

c. $99 = x^2 - 8x$

$x^2 - 8x - 99 = 0$

$a = 1, b = -8, c = -99$

d. $20 = 4 - x^2$

$x^2 + 16 = 0$

$a = 1, b = 0, c = 16$

53. a. $x^2 - 9 = 0$

$x^2 = 9$
$x = \pm 3$

b. $2(x-2)^2 + 1 = 11$

$2(x-2)^2 = 10$

$(x-2)^2 = 5$

$x - 2 = \pm\sqrt{5}$

$x = 2 \pm \sqrt{5}$

c. $3x^2 + 15 = 0$

$3x^2 = -15$

$x^2 = -5$

$x = \pm\sqrt{5}\, i$

d. $-2x^2 + 4 = -46$

$-2x^2 = -50$
$x^2 = 25$
$x = \pm 5$

55. a. $x^2 - 4x = -9$

$x^2 - 4x + 9 = 0$

$a = 1, b = -4, c = 9$

$x = \dfrac{-(-4) \pm \sqrt{(-4)^2 - 4(1)(9)}}{2(1)}$

$x = \dfrac{4 \pm \sqrt{-20}}{2}$

$x = \dfrac{4 \pm 2\sqrt{5}\, i}{2}$

$x = 2 \pm \sqrt{5}\, i$

$x \approx 2 \pm 2.24\, i$

b. $4x^2 + 7 = 12x$

$4x^2 - 12x + 7 = 0$

$a = 4, b = -12, c = 7$

$x = \dfrac{-(-12) \pm \sqrt{(-12)^2 - 4(4)(7)}}{2(4)}$

$x = \dfrac{12 \pm \sqrt{32}}{8}$

$x = \dfrac{12 \pm 4\sqrt{2}}{8}$

$x = \dfrac{3 \pm \sqrt{2}}{2}$

$x \approx 2.21$ or $x \approx 0.79$

c. $2x^2 - 6x + 5 = 0$

$a = 2, b = -6, c = 5$

$x = \dfrac{-(-6) \pm \sqrt{(-6)^2 - 4(2)(5)}}{2(2)}$

$x = \dfrac{6 \pm \sqrt{-4}}{4}$

$x = \dfrac{6 \pm 2i}{4}$

$x = \dfrac{3}{2} \pm \dfrac{1}{2}i$

57. a. $120 = -16t^2 + 64t + 80$

$16t^2 - 64t + 40 = 0$

$2t^2 - 8t + 5 = 0$

$a = 2, b = -8, c = 5$

$t = \dfrac{-(-8) \pm \sqrt{(-8)^2 - 4(2)(5)}}{2(2)}$

$t = \dfrac{8 \pm \sqrt{24}}{4}$

$t = \dfrac{8 \pm 2\sqrt{6}}{4}$

$t = \dfrac{4 \pm \sqrt{6}}{2}$

$t \approx 3.22$ or $t \approx 0.78$

0.8 seconds

b. 3.2 seconds

c. $0 = -16t^2 + 64t + 80$

$0 = -16(t^2 - 4t - 5)$

$0 = (t - 5)(t + 1)$

$t - 5 = 0$ or $t + 1 = 0$

$t = 5$ or $t = -1$

5 seconds

59. Let x represent the time for the smaller pump.

$\dfrac{1}{x-3} + \dfrac{1}{x} = \dfrac{1}{2}$

$2x(x-3)\left[\dfrac{1}{x-3} + \dfrac{1}{x} = \dfrac{1}{2}\right]$

$2x + 2(x-3) = x(x-3)$

$2x + 2x - 6 = x^2 - 3x$

$0 = x^2 - 7x + 6$

$0 = (x-6)(x-1)$

$x - 6 = 0$ or $x - 1 = 0$

$x = 6$ or $x = 1$

It takes the smaller pump 6 hours.

61. $3x^3 + 5x^2 = 2x$

$3x^3 + 5x^2 - 2x = 0$

$x(3x^2 + 5x - 2) = 0$

$x(3x-1)(x+2) = 0$

$x = 0$ or $3x - 1 = 0$ or $x + 2 = 0$

         $3x = 1$ or $x = -2$

            $x = \dfrac{1}{3}$

$x = -2, x = 0, x = \dfrac{1}{3}$

63. $x^4 - \dfrac{1}{16} = 0$

$\left(x^2 + \dfrac{1}{4}\right)\left(x^2 - \dfrac{1}{4}\right) = 0$

$x^2 + \dfrac{1}{4} = 0$ or $x^2 - \dfrac{1}{4} = 0$

$x^2 = -\dfrac{1}{4}$ or $x^2 = \dfrac{1}{4}$

$x = \pm\dfrac{1}{2}i$ or $x = \pm\dfrac{1}{2}$

$x = \pm\dfrac{1}{2}, x = \pm\dfrac{1}{2}i$

65. $\dfrac{3h}{h+3} - \dfrac{7}{h^2 + 3h} = \dfrac{1}{h}$

$\dfrac{3h}{h+3} - \dfrac{7}{h(h+3)} = \dfrac{1}{h}$

$h(h+3)\left[\dfrac{3h}{h+3} - \dfrac{7}{h(h+3)} = \dfrac{1}{h}\right]$

$3h^2 - 7 = h + 3$

$3h^2 - h - 10 = 0$

$(3h + 5)(h - 2) = 0$

$3h + 5 = 0$ or $h - 2 = 0$

$h = -\dfrac{5}{3}$ or $h = 2$

67. $\dfrac{\sqrt{x^2 + 7}}{2} + 3 = 5$

$\dfrac{\sqrt{x^2 + 7}}{2} = 2$

$\sqrt{x^2 + 7} = 4$

$x^2 + 7 = 16$

$x^2 = 9$

$x = \pm 3$

Chapter 1 Summary and Concept Review

69. $\sqrt{3x+4} = 2 - \sqrt{x+2}$

$\left(\sqrt{3x+4}\right)^2 = \left(2 - \sqrt{x+2}\right)^2$

$3x + 4 = 4 - 4\sqrt{x+2} + x + 2$

$2x - 2 = -4\sqrt{x+2}$

$2(x-1) = -4\sqrt{x+2}$

$x - 1 = -2\sqrt{x+2}$

$(x-1)^2 = \left(-2\sqrt{x+2}\right)^2$

$x^2 - 2x + 1 = 4(x+2)$

$x^2 - 2x + 1 = 4x + 8$

$x^2 - 6x - 7 = 0$

$(x-7)(x+1) = 0$

$x - 7 = 0$ or $x + 1 = 0$

$x = 7 \quad$ or $x = -1$

$x = -1$; $x = 7$ is extraneous.

71. $-2(5x+2)^{\frac{2}{3}} + 17 = -1$

$-2(5x+2)^{\frac{2}{3}} = -18$

$(5x+2)^{\frac{2}{3}} = 9$

$\left[(5x+2)^{\frac{2}{3}}\right]^{\frac{3}{2}} = 9^{\frac{3}{2}}$

$5x + 2 = 27$ or $5x + 2 = -27$

$5x = 25 \quad$ or $5x = -29$

$x = 5 \quad$ or $x = -5.8$

73. $x^4 - 7x^2 = 18$

$x^4 - 7x^2 - 18 = 0$

$(x^2 - 9)(x^2 + 2) = 0$

$x^2 - 9 = 0$ or $x^2 + 2 = 0$

$x^2 = 9 \qquad x^2 = -2$

$x = \pm 3 \qquad x = \pm i\sqrt{2}$

$x = -3, x = 3, x = -i\sqrt{2}, x = i\sqrt{2}$

75. Let w represent the width.
Let $w + 3$ represent the length.

$w(w+3) = 54$

$w^2 + 3w - 54 = 0$

$(w+9)(w-6) = 0$

$w + 9 = 0$ or $w - 6 = 0$

$w = -9 \quad$ or $w = 6$

6 inches by 9 inches
width, 6 in; length, 9 in

77. Let x represent the number of \$2 decreases.

$(50 - 2x)(40 + 5x) = 2520$

$2000 + 250x - 80x - 10x^2 = 2520$

$0 = 10x^2 - 170x + 520$

$0 = x^2 - 17x + 52$

$0 = (x-4)(x-13)$

$x - 4 = 0$ or $x - 13 = 0$

$x = 4 \quad$ or $x = 13$

$50 - 2(4) = \$42$ or $50 - 2(13) = \$24$

Chapter 1 Mixed Review

1. a. $\dfrac{10}{\sqrt{x-8}}$

   $x-8>0$

   $x>8$

   $x \in (8, \infty)$

   b. $\dfrac{-5}{3x+4}$

   $3x+4 \ne 0$

   $3x \ne -4$

   $x \ne -\dfrac{4}{3}$

   $x \in \left(-\infty, \dfrac{-4}{3}\right) \cup \left(\dfrac{-4}{3}, \infty\right)$

3. a. $-2x^3 + 4x^2 = 50x - 100$

   $0 = 2x^3 - 4x^2 + 50x - 100$

   $0 = 2(x^3 - 2x^2 + 25x - 50)$

   $0 = 2\left[x^2(x-2) + 25(x-2)\right]$

   $0 = 2(x-2)(x^2 + 25)$

   $x - 2 = 0$ or $x^2 + 25 = 0$

   $x = 2$ or $x^2 = -25$

   $\qquad$ or $x = \pm 5i$;

   $x = 2, \quad x = \pm 5i$

   b. $-3x^4 - 375x = 0$

   $-3x(x^3 + 125) = 0$

   $-3x(x+5)(x^2 - 5x + 25) = 0$

   $-3x = 0$ or $x + 5 = 0$ or $x^2 - 5x + 25 = 0$

   $x = 0$ or $x = -5$

   $\qquad$ or $x = \dfrac{-(-5) \pm \sqrt{(-5)^2 - 4(1)(25)}}{2(1)}$

   $x = \dfrac{5 \pm \sqrt{-75}}{2}$

   $x = \dfrac{5 \pm 5i\sqrt{3}}{2} = \dfrac{5}{2} \pm \dfrac{5i\sqrt{3}}{2}$;

   $x = 0, x = -5, x = \dfrac{5}{2} \pm \dfrac{5i\sqrt{3}}{2}$

c. $-2|3x+1| = -12$

   $|3x+1| = 6$

   $3x+1 = 6$ or $3x+1 = -6$

   $3x = 5 \qquad 3x = -7$

   $x = \dfrac{5}{3} \qquad x = -\dfrac{7}{3}$;

   $x = -\dfrac{7}{3}, x = \dfrac{5}{3}$

d. $-3\left|\dfrac{x}{3} - 5\right| \le -12$

   $\left|\dfrac{x}{3} - 5\right| \ge 4$

   $\dfrac{x}{3} - 5 \ge 4$ or $\dfrac{x}{3} - 5 \le -4$

   $\dfrac{x}{3} \ge 9$ or $\dfrac{x}{3} \le 1$

   $x \ge 27$ or $x \le 3$

   $(-\infty, 3] \cup [27, \infty)$

e. $v^{\frac{4}{3}} = 81$

   $\left(v^{\frac{4}{3}}\right)^{\frac{3}{4}} = 81^{\frac{3}{4}}$

   $v = \pm 27$

f. $-2(x+1)^{\frac{1}{4}} = -6$

   $(x+1)^{\frac{1}{4}} = 3$

   $\left[(x+1)^{\frac{1}{4}}\right]^4 = 3^4$

   $x + 1 = 81$

   $x = 80$

5. $3x + 4y = -12$

   $4y = -3x - 12$

   $y = -\dfrac{3}{4}x - 3$

7. a. $5x - (2x - 3) + 3x = -4(5 + x) + 3$
    $5x - 2x + 3 + 3x = -20 - 4x + 3$
    $3 + 6x = -17 - 4x$
    $10x = -20$
    $x = -2$

    b. $\dfrac{n}{5} - 2 = 2 - \dfrac{5}{3} - \dfrac{4}{15}n$

    $15\left[\dfrac{n}{5} - 2 = 2 - \dfrac{5}{3} - \dfrac{4}{15}n\right]$
    $3n - 30 = 30 - 25 - 4n$
    $3n - 30 = 5 - 4n$
    $7n = 35$
    $n = 5$

9. $x^2 - 18x + 77 = 0$
    $(x - 7)(x - 11) = 0$
    $x - 7 = 0$ or $x - 11 = 0$
    $x = 7$ or $x = 11$

11. $4x^2 - 5 = 19$
    $4x^2 = 24$
    $x^2 = 6$
    $x = \pm\sqrt{6}$

13. $25x^2 + 16 = 40x$
    $25x^2 - 40x + 16 = 0$
    $(5x - 4)(5x - 4) = 0$
    $5x - 4 = 0$
    $5x = 4$
    $x = \dfrac{4}{5}$

15. $2x^4 - 50 = 0$
    $2\left(x^4 - 25\right) = 0$
    $2\left(x^2 + 5\right)\left(x^2 - 5\right) = 0$

    $x^2 + 5 = 0$ or $x^2 - 5 = 0$
    $x^2 = -5$ or $x^2 = 5$
    $x = \pm\sqrt{5}\,i$ or $x = \pm\sqrt{5}$

17. a. $\sqrt{2v - 3} + 3 = v$
    $\sqrt{2v - 3} = v - 3$
    $2v - 3 = v^2 - 6v + 9$
    $0 = v^2 - 8v + 12$
    $0 = (v - 6)(v - 2)$
    $v - 6 = 0$ or $v - 2 = 0$
    $v = 6$ or $v = 2$
    $v = 6; v = 2$ is extraneous

    b. $\sqrt[3]{x^2 - 9} + \sqrt[3]{x - 11} = 0$

    $\sqrt[3]{x^2 - 9} = -\sqrt[3]{x - 11}$
    $x^2 - 9 = -(x - 11)$
    $x^2 + x - 20 = 0$
    $(x - 4)(x + 5) = 0$
    $x - 4 = 0$ or $x + 5 = 0$
    $x = 4$ or $x = -5$

    c. $\sqrt{x + 7} - \sqrt{2x} = 1$
    $\sqrt{x + 7} = 1 + \sqrt{2x}$

    $\left(\sqrt{x + 7}\right)^2 = \left(1 + \sqrt{2x}\right)^2$

    $x + 7 = 1 + 2\sqrt{2x} + 2x$

    $-x + 6 = 2\sqrt{2x}$

    $\left(-x + 6\right)^2 = \left(2\sqrt{2x}\right)^2$

    $x^2 - 12x + 36 = 4(2x)$

    $x^2 - 12x + 36 = 8x$

    $x^2 - 20x + 36 = 0$

    $(x - 2)(x - 18) = 0$

    $x - 2 = 0$ or $x - 18 = 0$

    $x = 2$ or $x = -18$;

    $x = 2; x = 18$ is extraneous

19. $\dfrac{2(75) + 2(79) + x}{5} = 78$

    $2(75) + 2(79) + x = 390$
    $150 + 158 + x = 390$
    $x = 82$ inches
    $6'10''$

Chapter 1 Practice Test

1. a. $-\dfrac{2}{3}x - 5 = 7 - (x+3)$

$-\dfrac{2}{3}x - 5 = 7 - x - 3$

$-\dfrac{2}{3}x - 5 = 4 - x$

$\dfrac{1}{3}x = 9$

$x = 27$

b. $-5.7 + 3.1x = 14.5 - 4(x+1.5)$

$-5.7 + 3.1x = 14.5 - 4x - 6$

$-5.7 + 3.1x - 8.5 = 4x$

$7.1x = 14.2$

$x = 2$

c. $P = C + kC$

$P = C(1+k)$

$\dfrac{P}{1+k} = C$

d. $2|2x+5| - 17 = -11$

$2|2x+5| = 6$

$|2x+5| = 3$

$2x+5 = 3$ or $2x+5 = -3$

$2x = -2$ or $2x = -8$

$x = -1$ or $x = -4$

3. a. $-\dfrac{2}{5}x + 7 < 19$

$-\dfrac{2}{5}x < 12$

$\left(-\dfrac{5}{2}\right)\left(-\dfrac{2}{5}x\right) > \left(-\dfrac{5}{2}\right)(12)$

$x > -30$

b. $-1 < 3 - x \le 8$

$-4 < -x \le 5$

$4 > x \ge -5$

$-5 \le x < 4$

c. $\dfrac{1}{2}x + 3 < 9$ or $\dfrac{2}{3}x - 1 \ge 3$

$\dfrac{1}{2}x < 6$ or $\dfrac{2}{3}x \ge 4$

$x < 12$ or $x \ge 6$

$x \in \mathbb{R}$

d. $\dfrac{1}{2}|x-3| + \dfrac{5}{4} = \dfrac{7}{4}$

$\dfrac{1}{2}|x-3| = \dfrac{2}{4}$

$|x-3| = 1$

$x-3 = 1$ or $x-3 = -1$

$x = 4$ or $x = 2$

e. $-\dfrac{2}{3}|x+1| - 5 < -7$

$-\dfrac{2}{3}|x+1| < -2$

$|x+1| > 3$

$x+1 > 3$ or $x+1 < -3$

$x > 2$ or $x < -4$

5. $z^2 - 7z - 30 = 0$

$(z-10)(z+3) = 0$

$z - 10 = 0$ or $z + 3 = 0$

$z = 10$ or $z = -3$

7. $(x-1)^2 + 3 = 0$

$(x-1)^2 = -3$

$x - 1 = \pm i\sqrt{3}$

$x = 1 \pm i\sqrt{3}$

9. $3x^2 - 20x = -12$

$3x^2 - 20x + 12 = 0$

$(3x-2)(x-6) = 0$

$3x - 2 = 0$ or $x - 6 = 0$

$3x = 2$ or $x = 6$

$x = \dfrac{2}{3}$ or $x = 6$

11. $\dfrac{2}{x-3}+\dfrac{2x}{x+2}=\dfrac{x^2+16}{x^2-x-16}$

$(x-3)(x+2)\left[\dfrac{2}{x-3}+\dfrac{2x}{x+2}=\dfrac{x^2+16}{(x-3)(x+2)}\right]$

$2(x+2)+2x(x-3)=x^2+16$

$2x+4+2x^2-6x=x^2+16$

$2x^2-4x+4=x^2+16$

$x^2-4x-12=0$

$(x-6)(x+2)=0$

$x-6=0$ or $x+2=0$

$x=6$ or $x=-2$

$x=6, x=-2$ is extraneous

13. $\sqrt{x}+1=\sqrt{2x-7}$

$\left(\sqrt{x}+1\right)^2=\left(\sqrt{2x-7}\right)^2$

$x+2\sqrt{x}+1=2x-7$

$2\sqrt{x}=x-8$

$\left(2\sqrt{x}\right)^2=(x-8)^2$

$4x=x^2-16x+64$

$0=x^2-20x+64$

$0=(x-4)(x-16)$

$x-4=0$ or $x-16=0$

$x=4$ $x=16$

Check:

$\sqrt{4}+1=\sqrt{2(4)-7}$

$2+1=\sqrt{1}$

$3\neq 1$

Check:

$\sqrt{16}+1=\sqrt{2(16)-7}$

$4+1=\sqrt{32-7}$

$5=\sqrt{25}$

$5=5$

$x=16; x=4$ is extraneous

15. $P=(120-2x)(3+0.10x)$

$P=360+12x-6x-0.2x^2$

$P=-0.2x^2+6x+360$

a. $405=-0.2x^2+6x+360$

      $0.2x^2-6x+45=0$

      $2x^2-60x+450=0$

      $x^2-30x+225=0$

      $(x-15)^2=0$

      $x=15$

      $3+0.10(15)=\$4.50$ per tin

b. $120-2(15)=90$ tins

17. $\dfrac{-8+\sqrt{-20}}{6}=\dfrac{-8+\sqrt{-4(5)}}{6}=\dfrac{-8+2\sqrt{5}\,i}{6}$

    $=-\dfrac{4}{3}+\dfrac{\sqrt{5}}{3}i$

19. a. $\left(\dfrac{1}{2}+\dfrac{\sqrt{3}}{2}i\right)+\left(\dfrac{1}{2}-\dfrac{\sqrt{3}}{2}i\right)=1$

   b. $\left(\dfrac{1}{2}+\dfrac{\sqrt{3}}{2}i\right)-\left(\dfrac{1}{2}-\dfrac{\sqrt{3}}{2}i\right)$

      $=\dfrac{1}{2}+\dfrac{\sqrt{3}}{2}i-\dfrac{1}{2}+\dfrac{\sqrt{3}}{2}i=i\sqrt{3}$

   c. $\left(\dfrac{1}{2}+\dfrac{\sqrt{3}}{2}i\right)\cdot\left(\dfrac{1}{2}-\dfrac{\sqrt{3}}{2}i\right)$

      $=\dfrac{1}{4}-\dfrac{\sqrt{3}}{4}i+\dfrac{\sqrt{3}}{4}i-\dfrac{3}{4}i^2=1$

21. $(3i+5)(5-3i)=15i-9i^2+25-15i=34$

23. a. $\quad 2x^2 - 20x + 49 = 0$

$$2x^2 - 20x = -49$$

$$x^2 - 10x = -\frac{49}{2}$$

$$x^2 - 10x + 25 = -\frac{49}{2} + 25$$

$$(x-5)^2 = \frac{1}{2}$$

$$x - 5 = \pm\sqrt{\frac{1}{2}}$$

$$x - 5 = \pm\sqrt{\frac{1}{2}}\sqrt{\frac{2}{2}}$$

$$x = 5 \pm \frac{\sqrt{2}}{2}$$

b. $\quad 2x^2 - 5x = -4$

$$x^2 - \frac{5}{2}x = -2$$

$$x^2 - \frac{5}{2}x + \frac{25}{16} = -2 + \frac{25}{16}$$

$$\left(x - \frac{5}{4}\right)^2 = \frac{-7}{16}$$

$$x - \frac{5}{4} = \pm\frac{\sqrt{7}}{4}i$$

$$x = \frac{5}{4} \pm \frac{\sqrt{7}}{4}i$$

25. $F \approx 0.3W^{\frac{3}{4}}$

   a. $\quad F \approx 0.3(1296)^{\frac{3}{4}}$

   $\quad F \approx 64.8$ g

   b. $\quad 19.2 \approx 0.3W^{\frac{3}{4}}$

   $$64 \approx W^{\frac{3}{4}}$$

   $$(64)^{\frac{4}{3}} \approx \left(W^{\frac{3}{4}}\right)^{\frac{4}{3}}$$

   $$256 \text{ g} \approx W$$

Chapter 1 Calculator Exploration and Discovery

1. They differ by 0.2.

3. They differ by $\sqrt{2}$ or ≈ 1.41.

Strengthening Core Skills

1. $2x^2 - 5x - 7 = 0$
   $a = 2, b = -5, c = -7$

   $x_1 = \dfrac{7}{2}, x_2 = -1$

   $\dfrac{7}{2} + (-1) = \dfrac{5}{2} = -\dfrac{b}{a};$

   $\dfrac{7}{2} \cdot (-1) = \dfrac{-7}{2} = \dfrac{c}{a}$

3. $x^2 - 10x + 37 = 0$
   $a = 1, b = -10, c = 37$

   $x_1 = 5 + 2\sqrt{3}\,i, x_2 = 5 - 2\sqrt{3}\,i$

   $\left(5 + 2\sqrt{3}\,i\right) + \left(5 - 2\sqrt{3}\,i\right) = 10 = \dfrac{-b}{a};$

   $\left(5 + 2\sqrt{3}\,i\right)\left(5 - 2\sqrt{3}\,i\right) = 25 + 12 = 37 = \dfrac{c}{a}$

Technology Highlight

1. $y = \pm 4.8; y = \pm 3.6$, Answers will vary.

2.1 Exercises

1. First, second

3. Radius, center

5. Answers will vary.

7.

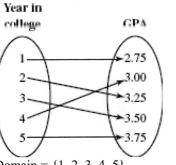

Domain = {1, 2, 3, 4, 5}
Range = {2.75, 3.00, 3.25, 3.50, 3.75}

9. D = {1, 3, 5, 7, 9}
   R = {2, 4, 6, 8, 10}

11. D = {4, -1, 2, -3}
    R = {0, 5, 4, 2, 3}

13. $y = -\dfrac{2}{3}x + 1$

x	y
-6	$-\dfrac{2}{3}(-6) + 1 = 4 + 1 = 5$
-3	$-\dfrac{2}{3}(-3) + 1 = 2 + 1 = 3$
0	$-\dfrac{2}{3}(0) + 1 = 0 + 1 = 1$
3	$-\dfrac{2}{3}(3) + 1 = -2 + 1 = -1$
6	$-\dfrac{2}{3}(6) + 1 = -4 + 1 = -3$
8	$-\dfrac{2}{3}(8) + 1 = -\dfrac{16}{3} + 1 = -\dfrac{13}{3}$

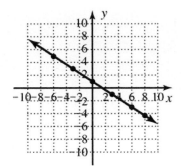

15. $x + 2 = |y|$

x	y
-2	0
0	2, -2
1	3, -3
3	5, -5
6	8, -8
7	9, -9

$-2 + 2 = |y|$ $0 + 2 = |y|$
$0 = |y|$ $2 = |y|$
$0 = y;$ $\pm 2 = y;$

$1 + 2 = |y|$ $3 + 2 = |y|$
$3 = |y|$ $5 = |y|$
$\pm 3 = y;$ $\pm 5 = y;$

$6 + 2 = |y|$ $7 + 2 = |y|$
$8 = |y|$ $9 = |y|$
$\pm 8 = y;$ $\pm 9 = y;$

17. $y = x^2 - 1$

x	y
-3	$(-3)^2 - 1 = 9 - 1 = 8$
-2	$(-2)^2 - 1 = 4 - 1 = 3$
0	$(0)^2 - 1 = 0 - 1 = -1$
2	$(2)^2 - 1 = 4 - 1 = 3$
3	$(3)^2 - 1 = 9 - 1 = 8$
4	$(4)^2 - 1 = 16 - 1 = 15$

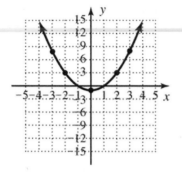

19. $y = \sqrt{25 - x^2}$

x	y
-4	$\sqrt{25 - (-4)^2} = \sqrt{25 - 16} = \sqrt{9} = 3$
-3	$\sqrt{25 - (-3)^2} = \sqrt{25 - 9} = \sqrt{16} = 4$
0	$\sqrt{25 - (0)^2} = \sqrt{25} = 5$
2	$\sqrt{25 - (2)^2} = \sqrt{25 - 4} = \sqrt{21}$
3	$\sqrt{25 - (3)^2} = \sqrt{25 - 9} = \sqrt{16} = 4$
4	$\sqrt{25 - (4)^2} = \sqrt{25 - 16} = \sqrt{9} = 3$

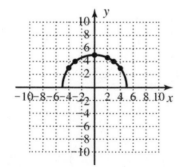

21. $x - 1 = y^2$

$y = \pm\sqrt{x - 1}$

x	y
10	$\sqrt{(10) - 1} = \sqrt{9} = \pm 3$
5	$\sqrt{(5) - 1} = \sqrt{4} = \pm 2$
4	$\sqrt{(4) - 1} = \pm\sqrt{3}$
2	$\sqrt{(2) - 1} = \sqrt{1} = \pm 1$
1.25	$\sqrt{(1.25) - 1} = \sqrt{0.25} = \pm 0.5$
1	$\sqrt{(1) - 1} = \sqrt{0} = 0$

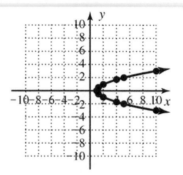

23. $y = \sqrt[3]{x + 1}$

x	y
-9	$\sqrt[3]{(-9) + 1} = \sqrt[3]{-8} = -2$
-2	$\sqrt[3]{(-2) + 1} = \sqrt[3]{-1} = -1$
-1	$\sqrt[3]{(-1) + 1} = \sqrt[3]{0} = 0$
0	$\sqrt[3]{(0) + 1} = \sqrt[3]{1} = 1$
4	$\sqrt[3]{(4) + 1} = \sqrt[3]{5}$
7	$\sqrt[3]{(7) + 1} = \sqrt[3]{8} = 2$

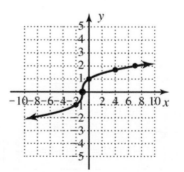

25. $M = \left(\dfrac{x_1 + x_2}{2}, \dfrac{y_1 + y_2}{2} \right)$

$M = \left(\dfrac{1+5}{2}, \dfrac{8+(-6)}{2} \right)$

$M = \left(\dfrac{6}{2}, \dfrac{2}{2} \right)$

$M = (3, 1)$

27. $M = \left(\dfrac{x_1 + x_2}{2}, \dfrac{y_1 + y_2}{2} \right)$

$M = \left(\dfrac{-4.5 + 3.1}{2}, \dfrac{9.2 + (-9.8)}{2} \right)$

$M = \left(\dfrac{-1.4}{2}, \dfrac{-0.6}{2} \right)$

$M = (-0.7, -0.3)$

29. $M = \left(\dfrac{x_1 + x_2}{2}, \dfrac{y_1 + y_2}{2} \right)$

$M = \left(\dfrac{\frac{1}{5} + \left(\frac{-1}{10} \right)}{2}, \dfrac{\frac{-2}{3} + \frac{3}{4}}{2} \right)$

$M = \left(\dfrac{\frac{1}{10}}{2}, \dfrac{\frac{1}{12}}{2} \right)$

$M = \left(\dfrac{1}{20}, \dfrac{1}{24} \right)$

31. $(-5, -4)\ (5, 2)$

$M = \left(\dfrac{x_1 + x_2}{2}, \dfrac{y_1 + y_2}{2} \right)$

$M = \left(\dfrac{-5+5}{2}, \dfrac{-4+2}{2} \right)$

$M = \left(\dfrac{0}{2}, \dfrac{-2}{2} \right)$

$M = (0, -1)$

33. $(-4, -4)\ (2, 4)$

$M = \left(\dfrac{x_1 + x_2}{2}, \dfrac{y_1 + y_2}{2} \right)$

$M = \left(\dfrac{-4+2}{2}, \dfrac{-4+4}{2} \right)$

$M = \left(\dfrac{-2}{2}, \dfrac{0}{2} \right)$

$M = (-1, 0)$

The center of the circle is $(-1, 0)$.

35. $(-5, -4)\ (5, 2)$

$d = \sqrt{(x_2 - x_1)^2 + (y_2 - y_1)^2}$

$d = \sqrt{(5 - (-5))^2 + (2 - (-4))^2}$

$d = \sqrt{(10)^2 + (6)^2}$

$d = \sqrt{100 + 36}$

$d = \sqrt{136}$

$d = 2\sqrt{34}$

37. $(-4, -4)\ (2, 4)$

$d = \sqrt{(x_2 - x_1)^2 + (y_2 - y_1)^2}$

$d = \sqrt{(2 - (-4))^2 + (4 - (-4))^2}$

$d = \sqrt{6^2 + 8^2}$

$d = \sqrt{36 + 64}$

$d = \sqrt{100}$

$d = 10$

39. $(5, 2)\ (0, -3)$

$m = \dfrac{-3 - 2}{0 - 5} = \dfrac{-5}{-5} = 1;$

$(0, -3)\ (4, -4)$

$m = \dfrac{-4 - (-3)}{4 - 0} = \dfrac{-4 + 3}{4} = \dfrac{-1}{4};$

$(5, 2)\ (4, -4)$

$m = \dfrac{-4 - 2}{4 - 5} = \dfrac{-6}{-1} = 6$

Not a right triangle. Lines are not

perpendicular. Slopes: 1; $\dfrac{-1}{4}$; 6

41. $(-4, 3) (-7, -1)$

$$m = \frac{-1-3}{-7-(-4)} = \frac{-4}{-7+4} = \frac{-4}{-3} = \frac{4}{3};$$

$(-7, -1) (3, -2)$

$$m = \frac{-2-(-1)}{3-(-7)} = \frac{-2+1}{3+7} = \frac{-1}{10};$$

$(-4, 3) (3, -2)$

$$m = \frac{-2-3}{3-(-4)} = \frac{-5}{7}$$

Not a right triangle. Lines are not

perpendicular. Slopes: $\frac{4}{3}; \frac{-1}{10}; \frac{-5}{7}$

43. $(-3, 2) (-1, 5)$

$$m = \frac{5-2}{-1-(-3)} = \frac{3}{-1+3} = \frac{3}{2};$$

$(-3, 2) (-6, 4)$

$$m = \frac{4-2}{-6-(-3)} = \frac{2}{-6+3} = -\frac{2}{3}$$

Right triangle because these two lines are

perpendicular. Slopes: $\frac{3}{2}; \frac{-2}{3}$

45. Center (0,0), radius 3

$$x^2 + y^2 = 9$$

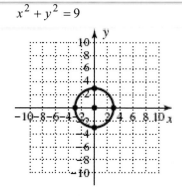

47. Center (5,0), radius $\sqrt{3}$

$$(x-5)^2 + y^2 = 3$$

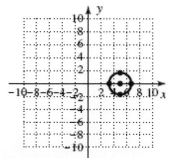

49. Center $(4, -3)$, radius 2

$$(x-4)^2 + (y+3)^2 = 4$$

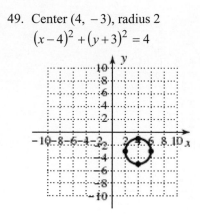

51. Center $(-7, -4)$, radius $\sqrt{7}$

$$(x+7)^2 + (y+4)^2 = 7$$

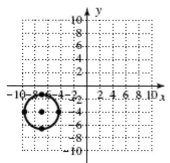

53. Center $(1, -2)$, radius $2\sqrt{3}$

$$(x-1)^2 + (y+2)^2 = 12$$

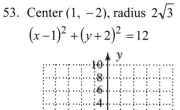

55. Center (4,5), diameter $4\sqrt{3}$

$$\text{radius} = \frac{1}{2} \cdot \text{diameter}$$

$$r = \frac{1}{2}\left(4\sqrt{3}\right) = 2\sqrt{3}$$

$$(x-4)^2 + (y-5)^2 = \left(2\sqrt{3}\right)^2$$

$$(x-4)^2 + (y-5)^2 = 12$$

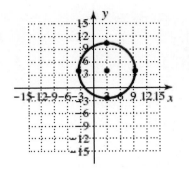

57. Center at (7,1),
graph contains the point $(1, -7)$

$$(x-7)^2 + (y-1)^2 = r^2;$$

$$(1-7)^2 + (-7-1)^2 = r^2$$

$$36 + 64 = r^2$$

$$100 = r^2;$$

$$(x-7)^2 + (y-1)^2 = 100$$

59. Center at (3,4),
graph contains the point (7,9)

$$(x-3)^2 + (y-4)^2 = r^2;$$

$$(7-3)^2 + (9-4)^2 = r^2$$

$$16 + 25 = r^2$$

$$41 = r^2;$$

$$(x-3)^2 + (y-4)^2 = 41$$

61. Diameter has endpoints (5,1) and (5,7);
midpoint of diameter = center of circle

$$\left(\frac{5+5}{2}, \frac{1+7}{2}\right) = (5,4);$$

radius = distance from center to endpt

$$r = \sqrt{(5-5)^2 + (1-4)^2} = 3;$$

$$(x-5)^2 + (y-4)^2 = 9$$

63. Center: $(2,3), r = 2$

$$D : x \in [0,4]$$

$$R : y \in [1,5]$$

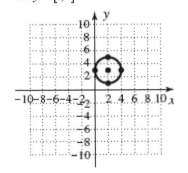

65. Center: $(-1,2), r = 2\sqrt{3}$

$D: x \in \left[-1-2\sqrt{3}, -1+2\sqrt{3}\right]$

$R: y \in \left[2-2\sqrt{3}, 2+2\sqrt{3}\right]$

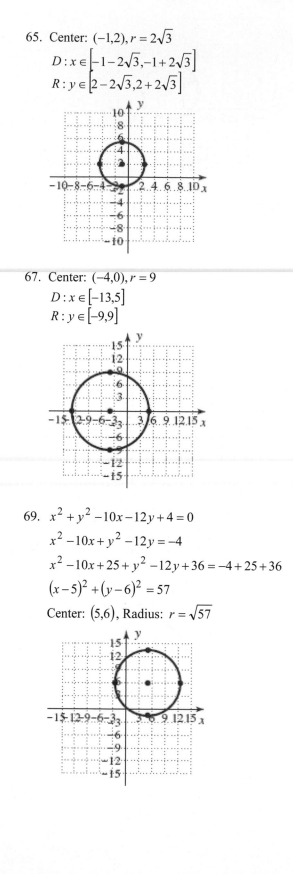

67. Center: $(-4,0), r = 9$

$D: x \in \left[-13,5\right]$

$R: y \in \left[-9,9\right]$

69. $x^2 + y^2 - 10x - 12y + 4 = 0$

$x^2 - 10x + y^2 - 12y = -4$

$x^2 - 10x + 25 + y^2 - 12y + 36 = -4 + 25 + 36$

$(x-5)^2 + (y-6)^2 = 57$

Center: $(5,6)$, Radius: $r = \sqrt{57}$

71. $x^2 + y^2 - 10x + 4y + 4 = 0$

$x^2 - 10x + y^2 + 4y = -4$

$x^2 - 10x + 25 + y^2 + 4y + 4 = -4 + 25 + 4$

$(x-5)^2 + (y+2)^2 = 25$

Center: $(5,-2)$, Radius: $r = 5$

73. $x^2 + y^2 + 6y - 5 = 0$

$x^2 + y^2 + 6y = 5$

$x^2 + y^2 + 6y + 9 = 5 + 9$

$x^2 + (y+3)^2 = 14$

Center: $(0,-3)$, Radius: $r = \sqrt{14}$

75. $x^2 + y^2 + 4x + 10y + 18 = 0$

$x^2 + 4x + y^2 + 10y = -18$

$x^2 + 4x + 4 + y^2 + 10y + 25 = -18 + 4 + 25$

$(x+2)^2 + (y+5)^2 = 11$

Center: $(-2,-5)$, Radius: $r = \sqrt{11}$

77. $x^2 + y^2 + 14x + 12 = 0$

$x^2 + 14x + y^2 = -12$

$x^2 + 14x + 49 + y^2 = -12 + 49$

$(x+7)^2 + y^2 = 37$

Center: $(-7,0)$, Radius: $r = \sqrt{37}$

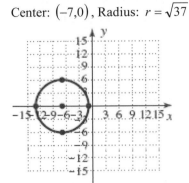

79. $2x^2 + 2y^2 - 12x + 20y + 4 = 0$

$x^2 + y^2 - 6x + 10y + 2 = 0$

$x^2 - 6x + y^2 + 10y = -2$

$x^2 - 6x + 9 + y^2 + 10y + 25 = -2 + 9 + 25$

$(x-3)^2 + (y+5)^2 = 32$

Center: $(3,-5)$, Radius: $r = 4\sqrt{2}$

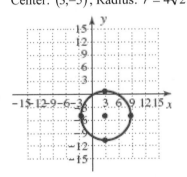

81. $s = 12.5t + 59$

a. Let $t = 1, s = 12.5(1) + 59 = 71.5$;

Let $t = 2, s = 12.5(2) + 59 = 84$;

Let $t = 3, s = 12.5(3) + 59 = 96.5$;

Let $t = 5, s = 12.5(5) + 59 = 121.5$;

Let $t = 7, s = 12.5(7) + 59 = 146.5$;

$(1, 71.5), (2, 84), (3, 96.5), (5, 121.5), (7, 146.5)$

b. Let $t = 8, s = 12.5(8) + 59 = 159$

Average amount spend in 2008 is $159.

c. Let $s = 196$,

$196 = 12.5t + 59$

$137 = 12.5t$

$10.96 = t$

In 2011, annual spending surpasses $196.

d.

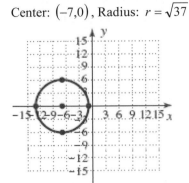

83. a. $(x-5)^2 + (y-12)^2 = 25^2$

$(x-5)^2 + (y-12)^2 = 625$

b. $d = \sqrt{(15-5)^2 + (36-12)^2}$

$d = \sqrt{10^2 + 24^2} = \sqrt{676} = 26$

No, radar cannot pick up the liner's sister ship.

85. Red: $(x-2)^2 + (y-2)^2 = 4$;

Center: (2,2), Radius: 2

Blue: $(x-2)^2 + y^2 = 16$;

Center: (2,0), Radius: 4

Area of blue: $\pi(16) - \pi(4) = 12\pi$ units^2

87. $x^2 + y^2 + 8x - 6y = 0$

$x^2 + 8x + y^2 - 6y = 0$

$x^2 + 8x + 16 + y^2 - 6y + 9 = 0 + 16 + 9$

$(x+4)^2 + (y-3)^2 = 25$;

$x^2 + y^2 - 10x + 4y = 0$

$x^2 - 10x + y^2 + 4y = 0$

$x^2 - 10x + 25 + y^2 + 4y + 4 = 0 + 25 + 4$

$(x-5)^2 + (y+2)^2 = 29$;

Distance between centers: $(-4,3), (5, -2)$

$d = \sqrt{(-4-5)^2 + (3-(-2))^2}$

$= \sqrt{81+25} = \sqrt{106} \approx 10.30$;

Sum of the radii: $5 + \sqrt{29} \approx 10.39$

No, Distance between the centers is less than the sum of the radii.

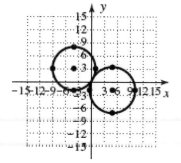

89. Answers will vary.

91. a. $x^2 + y^2 - 12x + 4y + 40 = 0$

$x^2 - 12x + y^2 + 4y = -40$

$x^2 - 12x + 36 + y^2 + 4y + 4 = -40 + 36 + 4$

$(x-6)^2 + (y+2)^2 = 0$

Center $(6, -2)$, $r = 0$, degenerate case

b. $x^2 + y^2 - 2x - 8y - 8 = 0$

$x^2 - 2x + y^2 - 8y = 8$

$x^2 - 2x + 1 + y^2 - 8y + 16 = 8 + 1 + 16$

$(x-1)^2 + (y-4)^2 = 25$

Center $(1, 4)$, $r = 5$

c. $x^2 + y^2 - 6x - 10y + 35 = 0$

$x^2 - 6x + y^2 - 10y = -35$

$x^2 - 6x + 9 + y^2 - 10y + 25 = -35 + 9 + 25$

$(x-3)^2 + (y-5)^2 = -1$

Center $(3, 5)$, $r^2 = -1$, degenerate case

93. a. 0

b. not possible

c. 0.3 ;many answers possible

d. not possible

e. not possible

f. $\sqrt{3}$;many answers possible

95. $1 - \sqrt{n+3} = -n$

$-\sqrt{n+3} = -n-1$

$\sqrt{n+3} = n+1$

$\left(\sqrt{n+3}\right)^2 = (n+1)^2$

$n+3 = n^2 + 2n + 1$

$0 = n^2 + n - 2$

$0 = (n+2)(n-1)$

$n+2 = 0$ or $n-1 = 0$

$n = -2$ or $n = 1$

Check: $n = -2$

$1 - \sqrt{-2+3} = -(-2)$

$1 - \sqrt{1} = 2$

$0 \neq 2$;

Check: $n = 1$

$1 - \sqrt{1+3} = -1$

$1 - \sqrt{4} = -1$

$-1 = -1$;

$n = 1$ is a solution, $n = -2$ is extraneous.

2.2 Technology Highlight

Exercise 1: $Y_1 = \dfrac{2}{3}x + 1$; $(-1.5, 0)$, $(0, 1)$

2.2 Exercises

1. 0; 0.

3. negative, downward

5. yes; slopes are not equal $m_1 \neq m_2$;
    No; $m_1 \cdot m_2 \neq -1$

7. $2x + 3y = 6$
    $3y = -2x + 6$

    $y = -\dfrac{2}{3}x + 2$

x	y
-6	$-\dfrac{2}{3}(-6) + 2 = 4 + 2 = 6$
-3	$-\dfrac{2}{3}(-3) + 2 = 2 + 2 = 4$
0	$-\dfrac{2}{3}(0) + 2 = 0 + 2 = 2$
3	$-\dfrac{2}{3}(3) + 2 = -2 + 2 = 0$

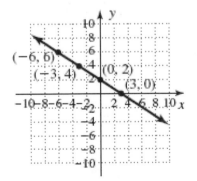

9. $y = \dfrac{3}{2}x + 4$

x	y
-2	$\dfrac{3}{2}(-2) + 4 = -3 + 4 = 1$
0	$\dfrac{3}{2}(0) + 4 = 0 + 4 = 4$
2	$\dfrac{3}{2}(2) + 4 = 3 + 4 = 7$
4	$\dfrac{3}{2}(4) + 4 = 6 + 4 = 10$

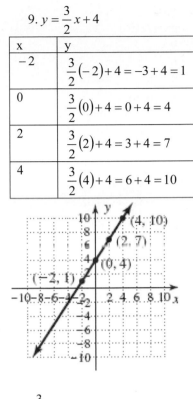

11. $y = \dfrac{3}{2}x + 4$

    $-0.5 = \dfrac{3}{2}(-3) + 4$

    $-0.5 = -\dfrac{9}{2} + 4$

    $-0.5 = -0.5$;

    $\dfrac{19}{4} = \dfrac{3}{2}\left(\dfrac{1}{2}\right) + 4$

    $\dfrac{19}{4} = \dfrac{3}{4} + 4$

    $\dfrac{19}{4} = \dfrac{19}{4}$

13. $3x + y = 6$

    x-intercept: $(2, 0)$
    $3x + 0 = 6$
    $3x = 6$
    $x = 2$
    y-intercept: $(0, 6)$
    $3(0) + y = 6$
    $y = 6$

17. $-5x + 2y = 6$

    x-intercept: $\left(-\dfrac{6}{5}, 0\right)$
    $-5x + 2(0) = 6$
    $-5x = 6$
    $x = -\dfrac{6}{5}$
    y-intercept: $(0, 3)$
    $-5(0) + 2y = 6$
    $2y = 6$
    $y = 3$

15. $5y - x = 5$

    x-intercept: $(-5, 0)$
    $5(0) - x = 5$
    $-x = 5$
    $x = -5$
    y-intercept: $(0, 1)$
    $5y - 0 = 5$
    $5y = 5$
    $y = 1$

19. $2x - 5y = 4$

    x-intercept: $(2, 0)$
    $2x - 5(0) = 4$
    $2x = 4$
    $x = 2$
    y-intercept: $\left(0, -\dfrac{4}{5}\right)$
    $2(0) - 5y = 4$
    $-5y = 4$
    $y = -\dfrac{4}{5}$

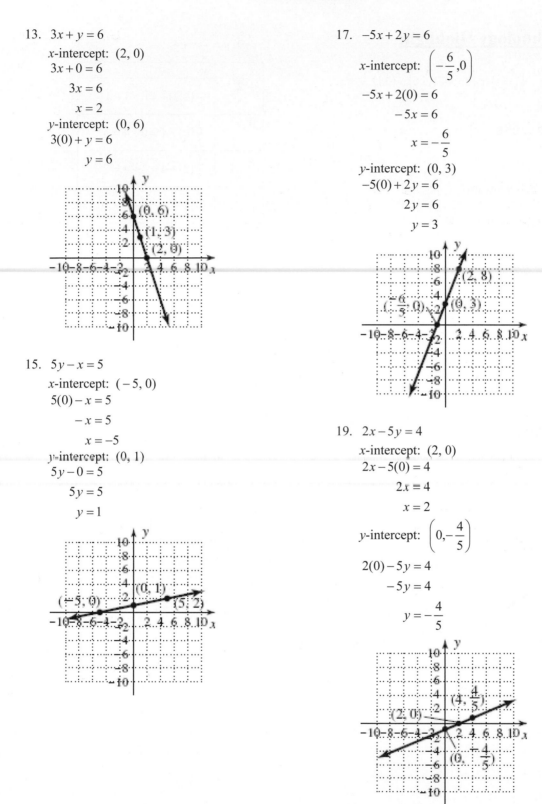

Chapter 2: Relations, Functions and Graphs

21. $2x + 3y = -12$

 x-intercept: $(-6, 0)$

 $2x + 3(0) = -12$

 $2x = -12$

 $x = -6$

 y-intercept: $(0, -4)$

 $2(0) + 3y = -12$

 $3y = -12$

 $y = -4$

25. $y - 25 = 50x$

 $y - 25 = 50(-1)$

 $y - 25 = -50$

 $y = -25$

 $(-1, -25);$

 $y - 25 = 50x$

 $y - 25 = 50(1)$

 $y - 25 = 50$

 $y = 75$

 $(1, 75)$

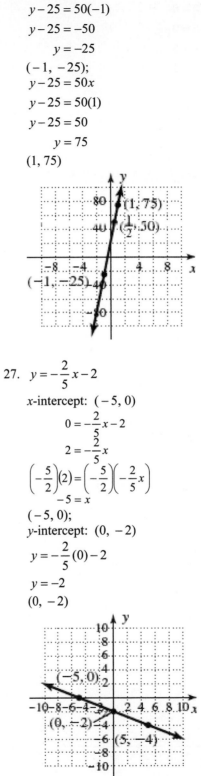

23. $y = -\dfrac{1}{2}x$

 $y = -\dfrac{1}{2}(2)$

 $y = -1$

 $(2, -1);$

 $y = -\dfrac{1}{2}x$

 $y = -\dfrac{1}{2}(4)$

 $y = -2$

 $(4, -2);$

 $y = -\dfrac{1}{2}x$

 $y = -\dfrac{1}{2}(0)$

 $y = 0$

 $(0, 0)$

27. $y = -\dfrac{2}{5}x - 2$

 x-intercept: $(-5, 0)$

 $0 = -\dfrac{2}{5}x - 2$

 $2 = -\dfrac{2}{5}x$

 $\left(-\dfrac{5}{2}\right)(2) = \left(-\dfrac{5}{2}\right)\left(-\dfrac{2}{5}x\right)$

 $-5 = x$

 $(-5, 0);$

 y-intercept: $(0, -2)$

 $y = -\dfrac{2}{5}(0) - 2$

 $y = -2$

 $(0, -2)$

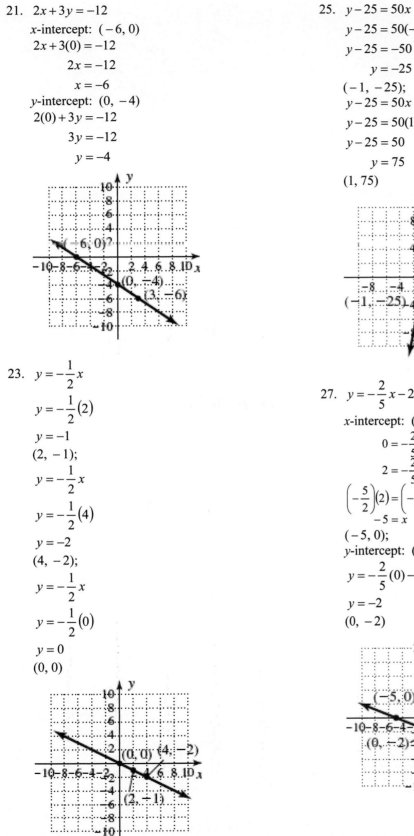

29. $2y - 3x = 0$
    $2y - 3(2) = 0$
    $2y - 6 = 0$
    $2y = 6$
    $y = 3$
    $(2, 3)$;
    $2y - 3x = 0$
    $2y - 3(4) = 0$
    $2y - 12 = 0$
    $2y = 12$
    $y = 6$
    $(4, 6)$;
    $2y - 3x = 0$
    $2y - 3(0) = 0$
    $2y = 0$
    $y = 0$
    $(0, 0)$

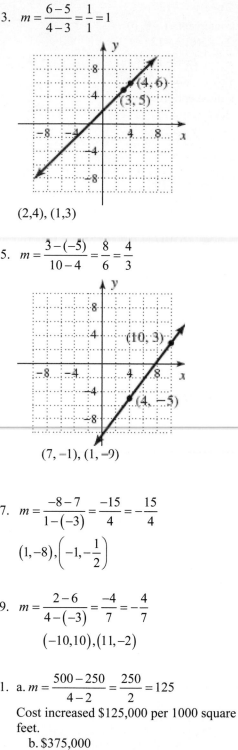

31. $3y + 4x = 12$
    x-intercept: $(3, 0)$
    $3(0) + 4x = 12$
    $4x = 12$
    $x = 3$
    y-intercept: $(0, 4)$
    $3y + 4(0) = 12$
    $3y = 12$
    $y = 4$

33. $m = \dfrac{6-5}{4-3} = \dfrac{1}{1} = 1$

    $(2, 4), (1, 3)$

35. $m = \dfrac{3-(-5)}{10-4} = \dfrac{8}{6} = \dfrac{4}{3}$

    $(7, -1), (1, -9)$

37. $m = \dfrac{-8-7}{1-(-3)} = \dfrac{-15}{4} = -\dfrac{15}{4}$

    $(1, -8), \left(-1, -\dfrac{1}{2}\right)$

39. $m = \dfrac{2-6}{4-(-3)} = \dfrac{-4}{7} = -\dfrac{4}{7}$

    $(-10, 10), (11, -2)$

41. a. $m = \dfrac{500-250}{4-2} = \dfrac{250}{2} = 125$
    Cost increased \$125,000 per 1000 square feet.
    b. \$375,000

43. a. $m = \dfrac{270-90}{12-4} = \dfrac{180}{8} = 22.5$
    Distance increases 22.5 miles per hour.
    b. 186 miles

45.

a. $m = \dfrac{165-142}{70-64} = \dfrac{23}{6}$

A person weighs 23 pounds more for each additional 6 inches in height.

b. $\dfrac{23}{6} \approx 3.8$ pounds

47. Convert 48 feet to inches: $48(12) = 576$;

$(0, -6)$ represents position of the sewer line at edge of house;

$(576, -18)$ represents position of sewer line at the main line.

$m = \dfrac{-18-(-6)}{576-0} = \dfrac{-12}{576} = -\dfrac{1}{48}$

The sewer line is one inch deeper for each 48 inches in length.

49. $x = -3$

$x + 0y = -3$

$x + 0(4) = -3$

$x = -3$

$(-3, 4)$;

$x + 0y = -3$

$x + 0(-4) = -3$

$x = -3$

$(-3, -4)$

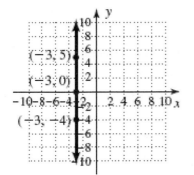

51. $x = 2$

$x + 0y = 2$

$x + 0(2) = 2$

$x = 2$

$(2, 0)$

$x + 0y = 2$

$x + 0(-2) = 2$

$x = 2$

$(2, 0)$

53. $L_1 : x = 2$

$L_2 : y = 4$

Point of intersection: $(2, 4)$

55. a. Choose any two points (t, j).

$(0, 9), (10, 9)$

$m = \dfrac{9-9}{10-0} = \dfrac{0}{10} = 0$

Which indicates there is no increase or decrease in the number of Supreme Court justices.

b. Choose any two points (t, n).

$(0, 0), (10, 1)$

$m = \dfrac{1-0}{10-0} = \dfrac{1}{10}$

Which indicates that over the last 5 decades, one non-white or non-female justice has been added to the court every ten years.

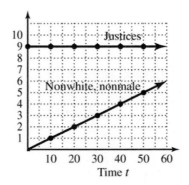

2.2 Exercises

57. $L_1: m = \dfrac{6-0}{0-(-2)} = \dfrac{6}{2} = 3$

    $L_2: m = \dfrac{5-8}{0-1} = \dfrac{-3}{-1} = 3$

    Parallel

59. $L_1: m = \dfrac{-4-1}{-3-0} = \dfrac{-5}{-3} = \dfrac{5}{3}$

    $L_2: m = \dfrac{4-0}{-4-0} = \dfrac{4}{-4} = -1$

    Neither

61. $L_1: m = \dfrac{7-3}{8-6} = \dfrac{4}{2} = 2$

    $L_2: m = \dfrac{2-0}{7-6} = \dfrac{2}{1} = 2$

    Parallel

63. $(5, 2)\,(0, -3)$

    $m = \dfrac{-3-2}{0-5} = \dfrac{-5}{-5} = 1$;

    $(0, -3)\,(4, -4)$

    $m = \dfrac{-4-(-3)}{4-0} = \dfrac{-4+3}{4} = \dfrac{-1}{4}$;

    $(5, 2)\,(4, -4)$

    $m - \dfrac{-4-2}{4-5} - \dfrac{-6}{-1} - 6$

    Not a right triangle. Lines are not

    perpendicular. Slopes: $1;\ \dfrac{-1}{4};\ 6$

65. $(-4, 3)\,(-7, -1)$

    $m = \dfrac{-1-3}{-7-(-4)} = \dfrac{-4}{-7+4} = \dfrac{-4}{-3} = \dfrac{4}{3}$;

    $(-7, -1)\,(3, -2)$

    $m = \dfrac{-2-(-1)}{3-(-7)} = \dfrac{-2+1}{3+7} = \dfrac{-1}{10}$;

    $(-4, 3)\,(3, -2)$

    $m = \dfrac{-2-3}{3-(-4)} = \dfrac{-5}{7}$

    Not a right triangle. Lines are not

    perpendicular. Slopes: $\dfrac{4}{3};\ \dfrac{-1}{10};\ \dfrac{-5}{7}$

67. $(-3, 2)\,(-1, 5)$

    $m = \dfrac{5-2}{-1-(-3)} = \dfrac{3}{-1+3} = \dfrac{3}{2}$;

    $(-3, 2)\,(-6, 4)$

    $m = \dfrac{4-2}{-6-(-3)} = \dfrac{2}{-6+3} = -\dfrac{2}{3}$

    Right triangle because these two lines are

    perpendicular. Slopes: $\dfrac{3}{2};\ \dfrac{-2}{3}$

69. $L = 0.11T + 74.2$

    a. $L(20) = 0.11(20) + 74.2 = 76.4$ years

    b. $77.5 = 0.11T + 74.2$

    $\quad 3.3 = 0.11T$

    $\quad 30 = T$

    $\qquad\qquad\qquad 1980 + 30 = 2010$

71. $V = 8500 - 1250y$

    a. $V = 8500 - 1250(4) = \$3500$

    b. $2250 = 8500 - 1250y$

    $-6250 = -1250y$

    $\quad 5 = y$

    $\quad 5$ years

73. Let h represent the water level, in inches.

    Let t represent the time, in months.

    $h - -3t + 300$

    a. $\quad h = -3(9) + 300 = 273$ in .

    b. $\quad$ Convert feet to inches: $20(12) = 240$;

    $\quad 240 = -3t + 300$

    $\quad -60 = -3t$

    $\quad 20 = t$

    $\quad 20$ months

75. Slope of FM 1960: $\dfrac{38}{12}$;

    Slope of FM 380: $\dfrac{30}{9.5}$;

    Since $\dfrac{38}{12} \neq \dfrac{30}{9.5}$, the roads are not parallel

    and yes, the roads will meet.

77. $y = 144x + 621$

   a. $y = 144(22) + 621$

      $y = 3789$

        $\$3,789$

   b. $5250 = 144x + 621$

      $4629 = 144x$

      $32.15 \approx x$

      $1980 + 32 = 2012$

      Year 2012

79. $y = -\dfrac{7}{15}x + 32$

   a. $y = -\dfrac{7}{15}(20) + 32$

      $y = \dfrac{-28}{3} + 32$

      $y = 22\dfrac{2}{3}$

        23%

   b. $20 = -\dfrac{7}{15}x + 32$

      $-12 = -\dfrac{7}{15}x$

      $-180 = -7x$

      $25.7 = x$

      $1980 + 25.7 = 2005.7$

      During the year 2005

81. $4y + 2x = -5$

    $4y = -2x - 5$

    $y = -\dfrac{1}{2}x - \dfrac{5}{4}$;

    $3y + ax = -2$

    $3y = -ax - 2$

    $y = -\dfrac{a}{3}x - \dfrac{2}{3}$;

    $-\dfrac{a}{3} \cdot -\dfrac{1}{2} = -1$

      $\dfrac{a}{6} = -1$

      $a = -6$

83. $t_n = t_1 + (n-1)d$

   a. $n = 21, t_1 = 2, d = 9 - 2 = 7$

      $t_{21} = 2 + (21-1)7 = 142$

   b. $n = 31, t_1 = 7, d = 4 - 7 = -3$

      $t_{31} = 7 + (31-1)(-3) = -83$

   c. $n = 27, t_1 = 5.10, d = 5.25 - 5.10 = 0.15$

      $t_{27} = 5.10 + (27-1)(0.15) = 9$

   d. $n = 17, t_1 = \dfrac{3}{2}, d = \dfrac{9}{4} - \dfrac{3}{2} = \dfrac{3}{4}$

    $t_{17} = \dfrac{3}{2} + (17-1)\left(\dfrac{3}{4}\right) = \dfrac{27}{2}$

85. $P = 2L + 2W$

    Perimeter of a rectangle;

    $V = LWH$

    Volume of a rectangular prism;

    $V = \pi r^2 h$

    Volume of a cylinder;

    $C = 2\pi r$

    Circumference of a circle

87.

	Distance	Rate	Time
Westbound Boat	D	15	t
Eastbound Boat	$70 - D$	20	t

$$\begin{cases} D = 15t \\ 70 - D = 20t \end{cases}$$

$70 - 15t = 20t$

$70 = 35t$

$2 = t$

2 hours

2.3 Exercises

1. $-\dfrac{7}{4}$; $(0, 3)$

3. 2.5

5. Answers will vary.

7. $4x + 5y = 10$

$5y = -4x + 10$

$y = -\dfrac{4}{5}x + 2$

x	$y = -\dfrac{4}{5}x + 2$
-5	$y = -\dfrac{4}{5}(-5) + 2 = 4 + 2 = 6$
-2	$y = -\dfrac{4}{5}(-2) + 2 = \dfrac{8}{5} + 2 = \dfrac{18}{5}$
0	$y = -\dfrac{4}{5}(0) + 2 = 0 + 2 = 2$
1	$y = -\dfrac{4}{5}(1) + 2 = -\dfrac{4}{5} + 2 = \dfrac{6}{5}$
3	$y = -\dfrac{4}{5}(3) + 2 = -\dfrac{12}{5} + 2 = -\dfrac{2}{5}$

9. $-0.4x + 0.2y = 1.4$

$0.2y = 0.4x + 1.4$

$y = 2x + 7$

x	$y = 2x + 7$
-5	$y = 2(-5) + 7 = -10 + 7 = -3$
-2	$y = 2(-2) + 7 = -4 + 7 = 3$
0	$y = 2(0) + 7 = 0 + 7 = 7$
1	$y = 2(1) + 7 = 2 + 7 = 9$
3	$y = 2(3) + 7 = 6 + 7 = 13$

11. $\dfrac{1}{3}x + \dfrac{1}{5}y = -1$

$\dfrac{1}{5}y = -\dfrac{1}{3}x - 1$

$y = -\dfrac{5}{3}x - 5$

x	$y = -\dfrac{5}{3}x - 5$
-5	$y = -\dfrac{5}{3}(-5) - 5 = \dfrac{25}{3} - 5 = \dfrac{10}{3}$
-2	$y = -\dfrac{5}{3}(-2) - 5 = \dfrac{10}{3} - 5 = -\dfrac{5}{3}$
0	$y = -\dfrac{5}{3}(0) - 5 = 0 - 5 = -5$
1	$y = -\dfrac{5}{3}(1) - 5 = -\dfrac{5}{3} - 5 = -\dfrac{20}{3}$
3	$y = -\dfrac{5}{3}(3) - 5 = -5 - 5 = -10$

13. $6x - 3y = 9$

$-3y = -6x + 9$

$y = 2x - 3$

New Coefficient: 2
New Constant: -3

15. $-0.5x - 0.3y = 2.1$

$-0.3y = 0.5x + 2.1$

$y = \dfrac{-5}{3}x - 7$

New Coefficient: $\dfrac{-5}{3}$

New Constant: -7

17. $\dfrac{5}{6}x + \dfrac{1}{7}y = -\dfrac{4}{7}$

$\dfrac{1}{7}y = -\dfrac{5}{6}x - \dfrac{4}{7}$

$y = -\dfrac{35}{6}x - 4$

New Coefficient: $-\dfrac{35}{6}$

New Constant: -4

19. $y = -\dfrac{4}{3}x + 5$

x	$y = -\dfrac{4}{3}x + 5$
0	$y = -\dfrac{4}{3}(0) + 5 = 0 + 5 = 5$
3	$y = -\dfrac{4}{3}(3) + 5 = -4 + 5 = 1$
6	$y = -\dfrac{4}{3}(6) + 5 = -8 + 5 = -3$

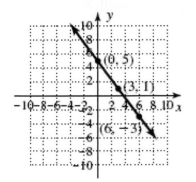

23. $y = -\dfrac{1}{6}x + 4$

x	$y = -\dfrac{1}{6}x + 4$
-6	$y = -\dfrac{1}{6}(-6) + 4 = 1 + 5 = 5$
0	$y = -\dfrac{1}{6}(0) + 4 = 0 + 4 = 4$
6	$y = -\dfrac{1}{6}(6) + 4 = -1 + 4 = 3$

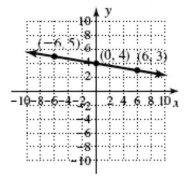

21. $y = -\dfrac{3}{2}x - 2$

x	$y = -\dfrac{3}{2}x - 2$
0	$y = -\dfrac{3}{2}(0) - 2 = 0 - 2 = -2$
2	$y = -\dfrac{3}{2}(2) - 2 = -3 - 2 = -5$
4	$y = -\dfrac{3}{2}(4) - 2 = -6 - 2 = -8$

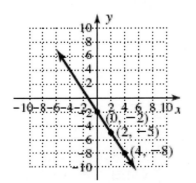

25. $3x + 4y = 12$

x-intercept: (4, 0)   y-intercept: (0, 3)

$3x + 4(0) = 12$ $3(0) + 4y = 12$

$\qquad 3x = 12$ $\qquad 4y = 12$

$\qquad x = 4$ $\qquad y = 3$

a. $m = \dfrac{0-3}{4-0} = -\dfrac{3}{4}$

b. $y = -\dfrac{3}{4}x + 3$

c. The coefficient of x is the slope and the constant is the y-intercept.

27. $2x - 5y = 10$

x-intercept: (5, 0)　　　y-intercept: (0, −2)

$2x - 5(0) = 10$　　$2(0) - 5y = 10$

$\quad 2x = 10$　　　　$-5y = 10$

$\quad\quad x = 5$　　　　　$y = -2$

a. $m = \dfrac{0 - (-2)}{5 - 0} = \dfrac{2}{5}$

b. $y = \dfrac{2}{5}x - 2$

c. The coefficient of x is the slope and the constant is the y-intercept.

29. $4x - 5y = -15$

x-intercept: $\left(-\dfrac{15}{4}, 0\right)$　　　y-intercept: (0, 3)

$4x - 5(0) = -15$　　　$4(0) - 5y = -15$

$\quad 4x = -15$　　　　　$-5y = -15$

$\quad\quad x = -\dfrac{15}{4}$　　　　$y = 3$

a. $m = \dfrac{0 - 3}{-\dfrac{15}{4} - 0} = \dfrac{-3}{-\dfrac{15}{4}} = \dfrac{12}{15} = \dfrac{4}{5}$

b. $y = \dfrac{4}{5}x + 3$

c. The coefficient of x is the slope and the constant is the y-intercept.

31. $2x + 3y = 6$

$3y = -2x + 6$

$y = -\dfrac{2}{3}x + 2$

$m = -\dfrac{2}{3}$; y-intercept (0, 2)

33. $5x + 4y = 20$

$4y = -5x + 20$

$y = -\dfrac{5}{4}x + 5$

$m = -\dfrac{5}{4}$; y-intercept (0, 5)

35. $x = 3y$

$y = \dfrac{1}{3}x$

$m = \dfrac{1}{3}$; y-intercept (0, 0)

37. $3x + 4y - 12 = 0$

$4y = -3x + 12$

$y = -\dfrac{3}{4}x + 3$

$m = -\dfrac{3}{4}$; y-intercept (0, 3)

39. $m = \dfrac{2}{3}$; y-intercept (0, 1)

$y = mx + b$

$y = \dfrac{2}{3}x + 1$

41. $m = 3$; y-intercept (0, 3)

$y = mx + b$

$y = 3x + 3$

43. $m = 3$; y-intercept (0, 2)

$y = mx + b$

$y = 3x + 2$

45. $m = 250$; (14, 4000)

$y - y_1 = m(x - x_1)$

$y - 4000 = 250(x - 14)$

$y - 4000 = 250x - 3500$

$\quad\quad y = 250x + 500$

$f(x) = 250x + 500$

47. $m = \dfrac{75}{2}$; (24, 1050)

$y - y_1 = m(x - x_1)$

$y - 1050 = \dfrac{75}{2}(x - 24)$

$y - 1050 = \dfrac{75}{2}x - 900$

$\quad\quad y = \dfrac{75}{2}x + 150$

$f(x) = \dfrac{75}{2}x + 150$

49. $m = 2; (5, -3)$

$$y - y_1 = m(x - x_1)$$
$$y + 3 = 2(x - 5)$$
$$y + 3 = 2x - 10$$
$$y = 2x - 13$$

51. $3x + 5y = 20$

$$5y = -3x + 20$$
$$y = -\frac{3}{5}x + 4$$

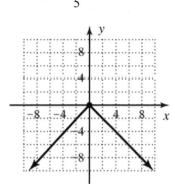

53. $2x - 3y = 15$

$$-3y = -2x + 15$$
$$y = \frac{2}{3}x - 5$$

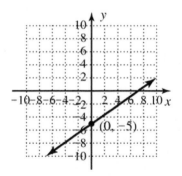

55. $y = \frac{2}{3}x + 3$

$m = \frac{2}{3}$; y-intercept $(0, 3)$

57. $y = -\frac{1}{3}x + 2$

$m = \frac{-1}{3}$; y-intercept $(0, 2)$

59. $y = 2x - 5$

$m = 2$; y-intercept $(0, -5)$

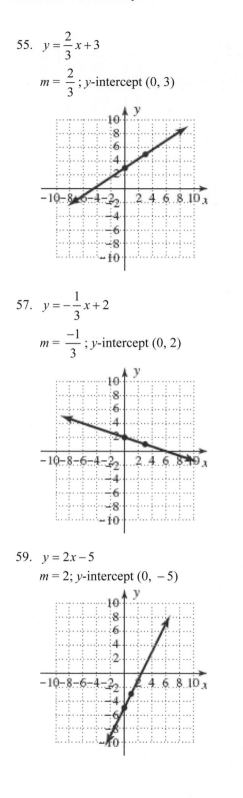

61. $f(x) = \dfrac{1}{2}x - 3$

$m = \dfrac{1}{2}$; y-intercept $(0, -3)$

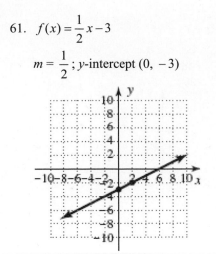

63. $2x - 5y = 10$

$-5y = -2x + 10$

$y = \dfrac{2}{5}x - 2$

$m = \dfrac{2}{5}$; $(-5, 2)$

$y - y_1 = m(x - x_1)$

$y - 2 = \dfrac{2}{5}\big(x - (-5)\big)$

$y - 2 = \dfrac{2}{5}x + 2$

$y = \dfrac{2}{5}x + 4$

65. $5y - 3x = 9$

$5y = 3x + 9$

$y = \dfrac{3}{5}x + \dfrac{9}{5}$

$m = -\dfrac{5}{3}$; $(6, -3)$;

$y - y_1 = m(x - x_1)$

$y - (-3) \equiv -\dfrac{5}{3}(x - 6)$

$y + 3 = -\dfrac{5}{3}x + 10$

$y = -\dfrac{5}{3}x + 7$

67. $12x + 5y = 65$

$5y = -12x + 65$

$y = -\dfrac{12}{5}x + 13$

$m = -\dfrac{12}{5}$; $(-2, -1)$

$y - y_1 = m(x - x_1)$

$y + 1 = -\dfrac{12}{5}(x + 2)$

$y + 1 = -\dfrac{12}{5}x - \dfrac{24}{5}$

$y = -\dfrac{12}{5}x - \dfrac{29}{5}$

69. $y = -3$ has slope of zero.

Slope of any line parallel to this line has the same slope, 0.

$y = mx + b$

$5 = 0(2) + b$

$5 = b$;

$y = 0x + 5$

$y = 5$

71. $4y - 5x = 8$

$4y = 5x + 8$

$y = \dfrac{5}{4}x + 2$;

$5y + 4x = -15$

$5y = -4x - 15$

$y = -\dfrac{4}{5}x - 3$

perpendicular

73. $2x - 5y = 20$

$-5y = -2x + 20$

$y = \dfrac{2}{5}x - 4$;

$4x - 3y = 18$

$-3y = -4x + 18$

$y = \dfrac{4}{3}x - 6$

Neither

75. $-4x + 6y = 12$

$6y = 4x + 12$

$y = \dfrac{2}{3}x + 2;$

$2x + 3y = 6$

$3y = -2x + 6$

$y = -\dfrac{2}{3}x + 2$

Neither

77. $(0,1),(4,-2)$

$m = \dfrac{-2-1}{4-0} = -\dfrac{3}{4}$

a. $y - (-4) = -\dfrac{3}{4}(x-2)$

$y + 4 = -\dfrac{3}{4}x + \dfrac{3}{2}$

$y = -\dfrac{3}{4}x - \dfrac{5}{2}$

b. $y - (-4) = \dfrac{4}{3}(x-2)$

$y + 4 = \dfrac{4}{3}x - \dfrac{8}{3}$

$y = \dfrac{4}{3}x - \dfrac{20}{3}$

79. $(-4,0),(5,4)$

$m = \dfrac{4-0}{5-(-4)} = \dfrac{4}{9}$

a. $y - 3 = \dfrac{4}{9}(x - (-1))$

$y - 3 = \dfrac{4}{9}(x + 1)$

$y - 3 = \dfrac{4}{9}x + \dfrac{4}{9}$

$y = \dfrac{4}{9}x + \dfrac{31}{9}$

b. $y - 3 = \dfrac{-9}{4}(x - (-1))$

$y - 3 = \dfrac{-9}{4}(x + 1)$

$y - 3 = \dfrac{-9}{4}x - \dfrac{9}{4}$

$y = \dfrac{-9}{4}x + \dfrac{3}{4}$

81. $(-2,3),(4,0)$

$m = \dfrac{0-3}{4-(-2)} = \dfrac{-3}{6} = \dfrac{-1}{2}$

a. $y - (-2) = \dfrac{-1}{2}(x - 0)$

$y + 2 = \dfrac{-1}{2}x$

$y = \dfrac{-1}{2}x - 2$

b. $y - (-2) = 2(x - 0)$

$y + 2 = 2x$

$y = 2x - 2$

83. $m = 2; \ P_1 = (2,-5)$

$y - y_1 = m(x - x_1)$

$y + 5 = 2(x - 2)$

$y + 5 = 2x - 4$

$y = 2x - 9$

2.3 Exercises

85. $P_1(3,-4), P_2(11,-1)$

$$m = \frac{-1-(-4)}{11-3} = \frac{3}{8};$$

$$y - y_1 = m(x - x_1)$$

$$y - (-4) = \frac{3}{8}(x-3)$$

$$y + 4 = \frac{3}{8}x - \frac{9}{8}$$

$$y = \frac{3}{8}x - \frac{41}{8}$$

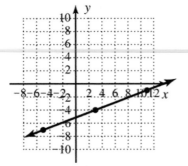

87. $m = 0.5;\ P_1 = (1.8, -3.1)$

$$y - y_1 = m(x - x_1)$$

$$y + 3.1 = 0.5(x - 1.8)$$

$$y + 3.1 = 0.5x - 0.9$$

$$y = 0.5x - 4$$

89. $m = \frac{6}{5};\ (4, 2)$

$$y - y_1 = m(x - x_1)$$

$$y - 2 = \frac{6}{5}(x - 4)$$

For each 5000 additional sales, income rises $6000.

91. $m = -20;\ (0.5, 100)$

$$y - y_1 = m(x - x_1)$$

$$y - 100 = -20(x - 0.5)$$

For every hour of television, a student's final grade falls 20%.

93. $m = \frac{35}{2};\ (0.5, 10)$

$$y - y_1 = m(x - x_1)$$

$$y - 10 = \frac{35}{2}(x - 0.5)$$

Every 2 inches of rainfall increases the number of cattle raised per acre by 35.

95. C

97. A

99. B

101. D

103. $ax + by = c$

$$by = -ax + c$$

$$y = -\frac{a}{b}x + \frac{c}{b};$$

Slope $-\frac{a}{b}$, y-intercept $\left(0, \frac{c}{b}\right)$

a. $3x + 4y = 8$

$$m = -\frac{a}{b} = -\frac{3}{4};$$

$$y - \text{int} = \frac{c}{b} = \frac{8}{4} = 2, (0,2)$$

b. $2x + 5y = -15$

$$m = -\frac{a}{b} = -\frac{2}{5};$$

$$y - \text{int} = \frac{c}{b} = -\frac{15}{5} = -3, (0,-3)$$

c. $5x - 6y = -12$

$$m = -\frac{5}{-6} = \frac{5}{6};$$

$$y - \text{int} = \frac{c}{b} = \frac{-12}{-6} = 2, (0,2)$$

d. $3y - 5x = 9$

$$m = -\frac{a}{b} = -\frac{-5}{3} = \frac{5}{3};$$

$$y - \text{int} = \frac{c}{b} = \frac{9}{3} = 3, (0,3)$$

105.a. As the temperature increases 5°C, the velocity of sound waves increases 3 m/s. At a temperature of 0°C, the velocity is 331 m/s.

b. $V(20) = \dfrac{3}{5}(20) + 331 = 343$ m/s

c. $361 = \dfrac{3}{5}C + 331$

$30 = \dfrac{3}{5}C$

$50 = C$

$50°C$

107.a. $m = \dfrac{190-150}{6-0} = \dfrac{40}{6} = \dfrac{20}{3}$

$V(t) = \dfrac{20}{3}t + 150$

b. Every three years, the coin increased in value by $20. The initial value was $150.

109.a. $m = \dfrac{51-9}{2001-1995} = \dfrac{42}{6} = 7$

$N(t) = 7t + 9$

b. Every 1 year, the number of homes hooked to the internet increases by 7 million.

c. $0 = 7t + 9$

$-9 = 7t$

$-\dfrac{9}{7} = t$

$-1.29 = t$

1.29 years prior to 1995 is 1993.

111 $m = \dfrac{1320000 - 740000}{2000 - 1990}$

$= \dfrac{580000}{10} = 58000$

$P(t) = 58000t + 740000$

Grows 58,000 every year.

$P(17) = 58000(17) + 740000 = 1726000$

113. Answers will vary.

115.a. $ax + by = c$

Find x-intercept by letting $y = 0$.

$ax + b(0) = c$

$x = \dfrac{c}{a}$

$\left(\dfrac{c}{a}, 0\right)$;

Find y-intercept by letting $x = 0$.

$a(0) + by = c$

$y = \dfrac{c}{b}$

$\left(0, \dfrac{c}{b}\right)$

The intercept method works most efficiently when a and b are factors of c.

b. Solve $ax + by = c$ for y.

$ax + by = c$

$by = -ax + c$

$y = -\dfrac{a}{b}x + \dfrac{c}{b}$;

$m = -\dfrac{a}{b}; y-\text{intercept}\left(0, \dfrac{c}{b}\right)$

The slope-intercet method works most efficiently when b is a factor of c.

117. $3x^2 - 10x = 9$

$3x^2 - 10 - 9 = 0$

$x = \dfrac{10 \pm \sqrt{(-10)^2 - 4(3)(-9)}}{2(3)}$

$x = \dfrac{10 \pm \sqrt{100 + 108}}{6}$

$x = \dfrac{10 \pm \sqrt{208}}{6}$

$x = \dfrac{10 \pm 4\sqrt{13}}{6}$

$x = \dfrac{5 \pm 2\sqrt{13}}{3}$

$x \approx 4.07$ or $x \approx -0.74$

119. $A = \pi r^2$

Larger circle: Smaller Circle

$A = \pi(10)^2$ $A = \pi(8)^2$

$A = 100\pi$ $A = 64\pi$

$100\pi - 64\pi = 36\pi \approx 113.10$ yds^2

2.4 Exercises

1. First

3. Range

5. Answers will vary.

7. Function

9. Not a function. The Shaq is paired with two heights.

11. Not a function, 4 is paired with 2 and -5.

13. Function

15. Function

17. Not a function, -2 is paired with 3 and -4.

19. Function

21. Function

23. Not a function, 0 is paired with 4 and -4.

25. Function

27. Not a function, 5 is paired with -1 and 1

29. Function

31.

x	$y = x$
-2	$y = -2$
-1	$y = -1$
0	$y = 0$
1	$y = 1$
2	$y = 2$

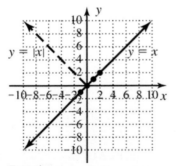

Function

33.

x	$y = (x+2)^2$
-4	$y = (-4+2)^2 = 4$
-3	$y = (-3+2)^2 = 1$
-2	$y = (-2+2)^2 = 0$
-1	$y = (-1+2)^2 = 1$
0	$y = (0+2)^2 = 4$
1	$y = (1+2)^2 = 9$

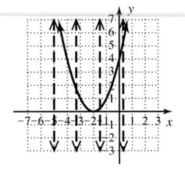

Function

35. Function; $x \in [-4,-5]$ $y \in [-2,3]$

37. Function; $x \in [-4,\infty)$ $y \in [-4,\infty)$

39. Function; $x \in [-4,4]$ $y \in [-5,-1]$

41. Function; $x \in (-\infty,\infty)$ $y \in (-\infty,\infty)$

43. Not a function; $x \in [-3,5]$ $y \in [-3,3]$

45. Not a function; $x \in (-\infty,3]$ $y \in (-\infty,\infty)$

47. $f(x) = \dfrac{3}{x-5}$

$x - 5 = 0$

$x = 5$

$x \in (-\infty,5) \cup (5,\infty)$

49. $h(a) = \sqrt{3a+5}$

$3a + 5 \geq 0$

$3a \geq -5$

$a \geq -\dfrac{5}{3}$

$a \in \left[-\dfrac{5}{3}, \infty\right)$

51. $v(x) = \dfrac{x+2}{x^2 - 25}$

$x^2 - 25 = 0$

$x^2 = 25$

$x = \pm 5$

$x \in (-\infty, -5) \cup (-5, 5) \cup (5, \infty)$

53. $u = \dfrac{v-5}{v^2 - 18}$

$v^2 - 18 = 0$

$v^2 = 18$

$v = \pm 3\sqrt{2}$

$v \in \left(-\infty, -3\sqrt{2}\right) \cup \left(-3\sqrt{2}, 3\sqrt{2}\right) \cup \left(3\sqrt{2}, \infty\right)$

55. $y = \dfrac{17}{25}x + 123$

$x \in (-\infty, \infty)$

57. $m = n^2 - 3n - 10$

$n \in (-\infty, \infty)$

59. $y = 2|x| + 1$

$x \in (-\infty, \infty)$

61. $y_1 = \dfrac{x}{x^2 - 3x - 10}$

$x^2 - 3x - 10 = 0$

$(x-5)(x+2) = 0$

$x = 5$ or $x = -2$

$x \in (-\infty, -2) \cup (-2, 5) \cup (5, \infty)$

63. $y = \dfrac{\sqrt{x-2}}{2x-5}$, $x \geq 2$

$2x - 5 = 0$

$2x = 5$

$x = \dfrac{5}{2}$

$x \in \left[2, \dfrac{5}{2}\right) \cup \left(\dfrac{5}{2}, \infty\right)$

65. $f(x) = \sqrt{\dfrac{5}{x-2}}$

Since the radicand must be non-negative,

solve the inequality: $\dfrac{5}{x-2} > 0$, $x \neq 2$

Use test points to each side of 2.

If $x = 0, \dfrac{5}{0-2} \geq 0$ false

If $x = 3, \dfrac{5}{3-2} \geq 0$ true

Domain: $x \in (2, \infty)$

67. $h(x) = \dfrac{-2}{\sqrt{4+x}}$

Since the radicand must be non-negative and the denominator cannot equal zero, solve the inequality: $4 + x > 0, x > -4$.

Domain: $x \in (-4, \infty)$

69. $f(x) = \dfrac{1}{2}x + 3$

$f(-6) = \dfrac{1}{2}(-6) + 3 = -3 + 3 = 0$;

$f\left(\dfrac{3}{2}\right) = \dfrac{1}{2}\left(\dfrac{3}{2}\right) + 3 = \dfrac{3}{4} + 3 = \dfrac{15}{4}$;

$f(2c) = \dfrac{1}{2}(2c) + 3 = c + 3$

71. $f(x) = 3x^2 - 4x$

$f(-6) = 3(-6)^2 - 4(-6) = 108 + 24 = 132$;

$f\left(\dfrac{3}{2}\right) = 3\left(\dfrac{3}{2}\right)^2 - 4\left(\dfrac{3}{2}\right) = 3\left(\dfrac{9}{4}\right) - 6$

$= \dfrac{27}{4} - 6 = \dfrac{3}{4}$;

$f(2c) = 3(2c)^2 - 4(2c) = 3\left(4c^2\right) - 8c$

$= 12c^2 - 8c$

73. $h(x) = \dfrac{3}{x}$

$h(3) = \dfrac{3}{(3)} = 1$;

$h\left(-\dfrac{2}{3}\right) = \dfrac{3}{\left(-\dfrac{2}{3}\right)} = -\dfrac{9}{2}$;

$h(3a) = \dfrac{3}{3a} = \dfrac{1}{a}$

75. $h(x) = \dfrac{5|x|}{x}$

$h(3) = \dfrac{5|3|}{3} = \dfrac{5(3)}{3} = 5$;

$h\left(-\dfrac{2}{3}\right) = \dfrac{5\left|-\dfrac{2}{3}\right|}{-\dfrac{2}{3}} = \dfrac{5\left(\dfrac{2}{3}\right)}{-\dfrac{2}{3}} = -5$;

$h(3a) = \dfrac{5|3a|}{3a} = \dfrac{15|a|}{3a} = \dfrac{5|a|}{a}$;

-5 if $a < 0$; 5 if $a > 0$

77. $g(r) = 2\pi r$

$g(0.4) = 2\pi(0.4) = 0.8\pi$;

$g\left(\dfrac{9}{4}\right) = 2\pi\left(\dfrac{9}{4}\right) = \dfrac{9}{2}\pi$;

$g(h) = 2\pi(h) = 2\pi h$;

79. $g(r) = \pi r^2$

$g(0.4) = \pi(0.4)^2 = 0.16\pi$;

$g\left(\dfrac{9}{4}\right) = \pi\left(\dfrac{9}{4}\right)^2 = \dfrac{81}{16}\pi$;

$g(h) = \pi(h)^2 = \pi h^2$

81. $p(x) = \sqrt{2x+3}$

$p(0.5) = \sqrt{2(0.5)+3} = \sqrt{1+3} = \sqrt{4} = 2$;

$p\left(\dfrac{9}{4}\right) = \sqrt{2\left(\dfrac{9}{4}\right)+3} = \sqrt{\dfrac{9}{2}+3} = \sqrt{\dfrac{15}{2}} = \dfrac{\sqrt{30}}{2}$;

$p(a) = \sqrt{2(a)+3} = \sqrt{2a+3}$

83. $p(x) = \dfrac{3x^2-5}{x^2}$

$p(0.5) = \dfrac{3(0.5)^2-5}{(0.5)^2} = \dfrac{3(0.25)-5}{0.25}$

$= \dfrac{0.75-5}{0.25} = \dfrac{-4.25}{0.25} = -17$

$p\left(\dfrac{9}{4}\right) = \dfrac{3\left(\dfrac{9}{4}\right)^2-5}{\left(\dfrac{9}{4}\right)^2} = \dfrac{3\left(\dfrac{81}{16}\right)-5}{\dfrac{81}{16}}$

$= \dfrac{\dfrac{243}{16}-5}{\dfrac{81}{16}} = \dfrac{\dfrac{163}{16}}{\dfrac{81}{16}} = \dfrac{163}{81}$;

$p(a) = \dfrac{3(a)^2-5}{(a)^2} = \dfrac{3a^2-5}{a^2}$

85. a. D: $\{-1, 0, 1, 2, 3, 4, 5\}$
b. R: $\{-2, -1, 0, 1, 2, 3, 4\}$
c. $f(2) = 1$
d. $f(-1) = 4$

87. a. $x \in [-5,5]$
b. $y \in [-3,4]$
c. $f(2) = -2$
d. when $y = 1$, $x = 0$ and $x = -4$.

89. a. $x \in [-3,\infty)$
b. $y \in (-\infty,4]$
c. $f(2) = 2$
d. when $y = 2$, $x = 2$ and $x = -2$

91. $W(H) = \dfrac{9}{2}H - 151$

a. $W(75) = \dfrac{9}{2}(75) - 151 = 186.5$ lb

b. $W(72) = \dfrac{9}{2}(72) - 151 = 173$ lb
$210 - 173 = 37$ lb

93. $A = \dfrac{1}{2}B + I - 1$

$\triangle PQR$

$P(-3,1), Q(3,9), R(7,6)$

$m = \dfrac{9-1}{3-(-3)} = \dfrac{4}{3}$

$y - 1 = \dfrac{4}{3}(x+3)$

$y - 1 = \dfrac{4}{3}x + 4$

$y = \dfrac{4}{3}x + 5$

(0,5) lies on PQ;
Lattice points are points that join vertical and horizontal grids in a Cartesian coordinate system.
There are four lattice points on the boundary; three vertices and point (0,5), thus $B = 8$. There are 24 lattice points in the interior of the triangle, thus $I = 24$.

$A = \dfrac{1}{2}(8) + 22 - 1 = 25$ units^2

95. a. $N(g) = 2.5g$

b. $g \in [0,5];\quad N \in [0,12.5]$

97. a. $D \in [0,\infty)$

b. $V(7.5) = 100\pi(7.5) = 750\pi$

c. $V\left(\dfrac{8}{\pi}\right) = 100\pi\left(\dfrac{8}{\pi}\right) = 800$ cm^3

99. a. $c(t) = 42.50t + 50$

b. $c(2.5) = 42.50(2.5) + 50 = \$156.25$

c. $262.50 = 42.50t + 50$
$212.50 = 42.50t$
$5\ \text{hr} = t$

d. $500 = 42.50t + 50$
$450 = 42.50t$
$10.6\ \text{hr} \approx t$
$t \in [0,10.6];\quad c \in [0,500]$

101. a.   Yes.
Each "x" is paired with exactly one "y".

b.   10 P.M.

c.   0.9 m

d. 7 P.M. and 1 A.M.

103. a.   Average rate of change from 1920 to 1940, use (20,3.2) and (40,2.2).
$\dfrac{\Delta fertility}{\Delta time} = \dfrac{2.2-3.2}{40-20} = -\dfrac{1}{20}$; Negative;
Fertility is decreasing by one child every 20 years.

b.   Average rate of change from 1940 to 1950, use (40,2.2).and (50,3.0).
$\dfrac{\Delta fertility}{\Delta time} = \dfrac{3.0-2.2}{50-40} = \dfrac{0.8}{10}$; Positive;
Fertility is increasing by less than one child every 10 years.

c.   from 1980 to 1990, use (80,1.8).and (90,2.0).
$\dfrac{\Delta fertility}{\Delta time} = \dfrac{2.0-1.8}{90-80} = \dfrac{0.2}{10}$; The fertility rate was increasing four times as fast from 1940 to 1950.

105. The y-values of the negative x integers would become positive.
All points would be in Quadrants I and III.

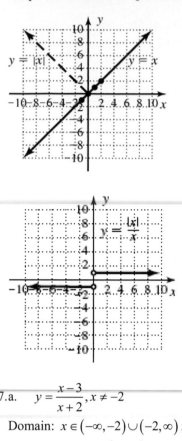

107.a. $y = \dfrac{x-3}{x+2}, x \neq -2$

Domain: $x \in (-\infty, -2) \cup (-2, \infty)$;

$(x+2)y = (x+2)\left(\dfrac{x-3}{x+2}\right)$

$xy + 2y = x - 3$

$xy - x = -2y - 3$

$x(y-1) = -2y - 3$

$x = \dfrac{-2y-3}{y-1} = \dfrac{2y+3}{1-y}, y \neq 1$

Range: $y \in (-\infty, 1) \cup (1, \infty)$

b. $y = x^2 - 3$

Domain: $x \in \mathbb{R}$;

$y = x^2 - 3$

$y + 3 = x^2$

$\pm\sqrt{y+3} = x$;

$y + 3 \geq 0$

$y \geq -3$

Range: $y \in [-3, \infty)$

109.a. $\sqrt{24} + 6\sqrt{54} - \sqrt{6}$

$= 2\sqrt{6} + 6 \cdot 3\sqrt{6} - \sqrt{6}$

$= 2\sqrt{6} + 18\sqrt{6} - \sqrt{6}$

$= 19\sqrt{6}$

b. $(2+\sqrt{3})(2-\sqrt{3})$

$= 4 - 2\sqrt{3} + 2\sqrt{3} - 3$

$= 1$

111.a. $x^3 - 3x^2 - 25x + 75$

$= (x^3 - 3x^2) - (25x - 75)$

$= x^2(x-3) - 25(x-3)$

$= (x-3)(x^2 - 25)$

$= (x-3)(x-5)(x+5)$

b. $2x^2 - 13x - 24 = (2x+3)(x-8)$

c. $8x^3 - 125 = (2x-5)(4x^2 + 10x + 25)$

Chapter 2 Mid-Chapter Check

1. $4x - 3y = 12$

   $-3y = -4x + 12$

   $y = \dfrac{4}{3}x - 4$

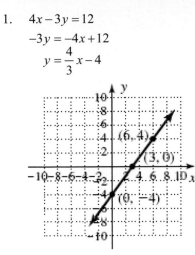

3. $m = \dfrac{-0.5 - (-2)}{2003 - 2002} = \dfrac{1.5}{1} = 1.5$;

   Positive, loss is decreasing, profit is increasing.
   Data.com's loss decreases by 1.5 million dollars per year.

5. $x = -3$; not a function. Input -3 is paired with more than one output.

7. a. $h(2) = 0$

   b. $x \in [-3,5]$

   c. $x = -1$ when $h(x) = -3$

   d. $y \in [-4,5]$

9. $m = \dfrac{3}{4}$; $(1, 2)$

   $y - y_1 = m(x - x_1)$

   $y - 2 = \dfrac{3}{4}(x - 1)$

   $y - 2 = \dfrac{3}{4}x - \dfrac{3}{4}$

   $y = \dfrac{3}{4}x + \dfrac{5}{4}$

   $F(p) = \dfrac{3}{4}p + \dfrac{5}{4}$

   For every 4000 pheasants, the fox population increases by 300.

   $F(20) = \dfrac{3}{4}(20) + \dfrac{5}{4} = 15 + 1.25 = 16.25$

   Fox population is 1625 when the pheasant population is 20,000.

Chapter 2 Reinforcing Basic Concepts

1. $P_1(0,5)$; $P_2(6,7)$

    a. $m = \dfrac{7-5}{6-0} = \dfrac{2}{6} = \dfrac{1}{3}$; increasing

    b. $y - 5 = \dfrac{1}{3}(x - 0)$

    c. $y = \dfrac{1}{3}x + 5$

    d. $y = \dfrac{1}{3}x + 5$

    $-\dfrac{1}{3}x + y = 5$

    $x - 3y = -15$

    e. x-intercept: $(-15, 0)$ y-intercept: $(0, 5)$

    $\begin{array}{ll} x - 3(0) = -15 & 0 - 3y = -15 \\ x = -15 & -3y = -15 \\ & y = 5 \end{array}$

3. $P_1(3,2)$; $P_2(9,5)$

    a. $m = \dfrac{5-2}{9-3} = \dfrac{3}{6} = \dfrac{1}{2}$; increasing

    b. $y - 2 = \dfrac{1}{2}(x - 3)$

    c. $y - 2 = \dfrac{1}{2}x - \dfrac{3}{2}$

    $y = \dfrac{1}{2}x + \dfrac{1}{2}$

    d. $y = \dfrac{1}{2}x + \dfrac{1}{2}$

    $-\dfrac{1}{2}x + y = \dfrac{1}{2}$

    $x - 2y = -1$

    e. x-intercept: $(-1, 0)$ y-intercept: $\left(0, \dfrac{1}{2}\right)$

    $\begin{array}{ll} x - 2(0) = -1 & 0 - 2y = -1 \\ x = -1 & y = \dfrac{1}{2} \end{array}$

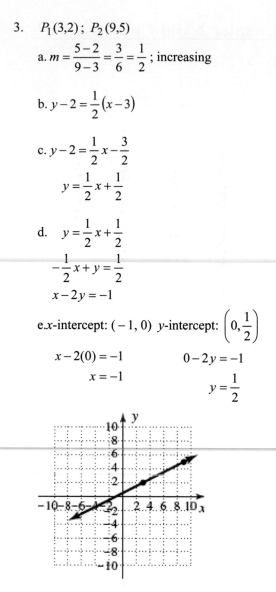

5. $P_1(-2,5)$; $P_2(6,-1)$

a. $m = \dfrac{-1-5}{6-(-2)} = \dfrac{-6}{8} = -\dfrac{3}{4}$; decreasing

b. $y - 5 = -\dfrac{3}{4}(x+2)$

c. $y - 5 = -\dfrac{3}{4}x - \dfrac{3}{2}$

$ y = -\dfrac{3}{4}x + \dfrac{7}{2}$

d. $y = -\dfrac{3}{4}x + \dfrac{7}{2}$

$ \dfrac{3}{4}x + y = \dfrac{7}{2}$

$ 3x + 4y = 14$

e. x-intercept: $\left(\dfrac{14}{3}, 0\right)$ y-intercept: $\left(0, \dfrac{7}{2}\right)$

$3x + 4(0) = 14$

$ 3x = 14$

$ x = \dfrac{14}{3}$

$3(0) + 4y = 14$

$ 4y = 14$

$ y = \dfrac{7}{2}$

2.5 Technology Highlight

Exercise 1: $x \approx -2.87$, $x \approx 0.87$,
  $\min : y = -7$ at $(-1, -7)$, no max
Exercise 3: $x \approx 1.35$, $x \approx 6.65$,
  $\min : y = -7$ at $(4, -7)$, no max
Exercise 5: $x = -2$, $x = 0$, $x \approx 2.41$,
  $\min : y = -3.20$ at $(-1.47, -3.20)$,
    $\min : y \approx -9.51$ at $(1.67, -9.51)$,
  $\max : y = 0$ at $(0, 0)$

2.5 Exercises

1. Linear; bounce

3. Increasing

5. Answers will vary.

7.

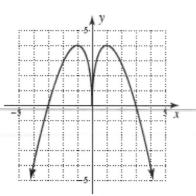

9. $f(x) = -7|x| + 3x^2 + 5$
  $f(k) = -7|k| + 3(k)^2 + 5;$
  $f(-k) = -7|-k| + 3(-k)^2 + 5$
  $= -7|k| + 3(k)^2 + 5 = f(k);$
  Even

11. $g(x) = \frac{1}{3}x^4 - 5x^2 + 1$
  $g(k) = \frac{1}{3}(k)^4 - 5(k)^2 + 1$
  $= \frac{1}{3}k^4 - 5k^2 + 1;$
  $g(-k) = \frac{1}{3}(-k)^4 - 5(-k)^2 + 1$
  $= \frac{1}{3}k^4 - 5k^2 + 1;$
  $g(k) = g(-k)$
  Even

13.

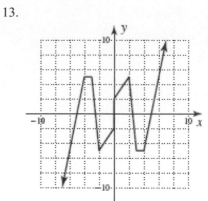

15. $f(x) = 4\sqrt[3]{x} - x$
  $f(k) = 4\sqrt[3]{k} - k$
  $f(-k) = 4\sqrt[3]{-k} - (-k)$
  $= -4\sqrt[3]{k} + k = -\left(4\sqrt[3]{k} - k\right);$
  $f(k) = -f(k)$
  Odd

17. $p(x) = 3x^3 - 5x^2 + 1$
  $p(k) = 3(k)^3 - 5(k)^2 + 1$
  $= 3k^3 - 5k^2 + 1$
  $p(-k) = 3(-k)^3 - 5(-k)^2 + 1$
  $= -3k^3 - 5k^2 + 1$
  $p(k) \neq -p(k)$; Not Odd

19. $w(x) = x^3 - x^2$
  $w(-x) = (-x)^3 - (-x)^2$
  $= -x^3 - x^2$; neither

21. 20. $p(x) = 2\sqrt[3]{x} - \frac{1}{4}x^3$
  $p(-x) = 2\sqrt[3]{(-x)} - \frac{1}{4}(-x)^3$
  $= -2\sqrt[3]{x} + \frac{1}{4}x^3 = -\left(2\sqrt[3]{x} - \frac{1}{4}x^3\right);$ odd

23. $v(x) = x^3 + 3|x|$
  $v(-x) = (-x)^3 + 3|-x|$
  $= -x^3 + 3|x|$; neither

25. $f(x) = x^3 - 3x^2 - x + 3$

   Verify Zeros: Let $f(x) = 0$

   $0 = x^2(x-3) - (x-3)$

   $0 = (x-3)(x^2-1)$

   $0 = (x-3)(x+1)(x-1)$

   Zeros: $(-1,0), (1,0), (3,0)$

   For $f(x) \geq 0$, $x \in [-1,1], [3,\infty)$

27. $f(x) = x^4 - 2x^2 + 1$

   Verify Zeros: Let $f(x) = 0$

   $0 = x^4 - 2x^2 + 1$

   $0 = (x^2-1)(x^2-1)$

   $0 = (x+1)(x-1)(x+1)(x-1)$

   Zeros: $(-1,0), (1,0)$

   For $f(x) > 0$, $x \in (-\infty,-1) \cup (-1,1) \cup (1,\infty)$

29. $p(x) = \sqrt[3]{x-1} - 1$

   $p(x) \geq 0$ for $x \in [2,\infty)$

31. $f(x) = (x-1)^3 - 1$

   $f(x) \leq 0$ for $x \in (-\infty, 2]$

33. $f(x)\uparrow: (-3,1) \cup (4,6)$

   $f(x)\downarrow: (-\infty,-3), (1,4)$

   Constant : None

35. $f(x)\uparrow: (1,4)$

   $f(x)\downarrow: (-2,1) \cup (4,\infty)$

   Constant: $(-\infty,-2)$

37. $p(x) = 0.5(x+2)^3$

   a. $p(x)\uparrow: x \in (-\infty,\infty)$

     $p(x)\downarrow$: None

   b. down, up

39. $y = p(x)$

   a. $p(x)\uparrow: x \in (-3,0) \cup (3,\infty)$

     $p(x)\downarrow: x \in (-\infty,-3) \cup (0,3)$

   b. up, up

41. $H(x) = -5|x-2| + 5$

   a. $x \in (-\infty,\infty)$

     $y \in (-\infty,5]$

   b. $(1, 0), (3, 0)$

   c. $H(x) \geq 0: x \in [1,3]$

     $H(x) \leq 0: x \in (-\infty,1] \cup [3,\infty)$

   d. $H(x)\uparrow: x \in (-\infty,2)$

     $H(x)\downarrow: x \in (2,\infty)$

   e. local maximum: $y = 2$ at $(2, 5)$

43. $y = g(x)$

   a. $x \in (-\infty,\infty)$

     $y \in (-\infty,\infty)$

   b. $(-1,0), (5, 0)$

   c. $g(x) \geq 0: x \in [-1,\infty)$

     $g(x) \leq 0: x \in (-\infty,-1] \cup \{5\}$

   d. $g(x)\uparrow: x \in (-\infty,1) \cup (5,\infty)$

     $g(x)\downarrow: x \in (1,5)$

   e. local maximum: $y = 6$ at $(1,6)$

     local minimum: $y = 0$ at $(5, 0)$

45. $y = Y_2$

   a. $x \in (-\infty,\infty)$

     $y \in (-\infty,3]$

   b. $(0, 0), (2, 0)$

   c. $Y_2 \geq 0: x \in [0,2]$

     $Y_2 \leq 0: x \in (-\infty,0] \cup [2,\infty)$

   d. $Y_2 \uparrow: x \in (-\infty,1)$

     $Y_2 \downarrow: x \in (1,\infty)$

   e. local maximum: $y = 3$ at $(1, 3)$

47. $p(x) = (x+3)^3 + 1$

   a. $x \in \mathbb{R}$, $y \in \mathbb{R}$

   b. $x = -4$

   c. $p(x) \geq 0: x \in [-4,\infty)$;

     $p(x) \leq 0: x \in (-\infty,-4]$

   d. $p(x)\uparrow: x \in (-\infty,-3) \cup (-3,\infty)$

     $p(x)\downarrow$: never decreasing

   e. Local max: none

     Local min: none

49. $y = \frac{1}{3}\sqrt{4x^2 - 36}$

    a. $x \in (-\infty, -3] \cup [3, \infty)$

         $y \in [0, \infty)$

    b. $(-3, 0), (3, 0)$

    c. $f(x)\uparrow : x \in (3, \infty)$

         $f(x)\downarrow : x \in (-\infty, -3)$

    d.   Even

    $y = \cos(x)$

    a. $y \in [-1, 1]$

    b. $(-90, 0), (90, 0), (270, 0)$

    c. $y\uparrow : x \in (-180, 0) \cup (180, 360)$

         $y\downarrow : x \in (0, 180)$

    d.   Minimum: $(-180, -1)$; $(180, -1)$

         Maximum: $(0, 1)$; $(360, 1)$

    e.   Even

51. a. $x \in [0, 260]$

         $y \in [0, 80]$

    b.   80 feet

    c.   120 feet

    d.   Yes

    e. $(0, 120)$

    f. $(120, 260)$

53. $f(x) = x^{\frac{2}{3}} - 1$

    a. $x \in (-\infty, \infty)$

         $y \in [-1, \infty)$

    b. $(-1, 0), (1, 0)$

    c. $f(x) \geq 0 : x \in (-\infty, -1] \cup [1, \infty)$

         $f(x) < 0 : x \in (-1, 1)$

    d. $f(x)\uparrow : x \in (0, \infty)$

         $f(x)\downarrow : x \in (-\infty, 0)$

    e.   Minimum: $(0, -1)$

55. a. $D : t \in [72, 96]$

         $R : I \in [7.25, 16]$

    b. $I(t)\uparrow$ for $t \in$ $(72, 74) \cup (77, 81) \cup (83, 84) \cup (93, 94)$

         $I(t)\downarrow$ for $t \in (74, 75) \cup (81, 83) \cup (84, 86) \cup (90, 93) \cup (94, 95)$

         $I(t)$ constant for $t \in (75, 77) \cup (86, 90) \cup (95, 96)$

    c.   Maximum: $(74, 9.25), (81, 16)$ (global max), $(84, 13), (94, 8.5)$

         Minimum: $(72, 7.5), (83, 12.75), (93, 7.2)$

    d.   Increase: 1980 to 1981

         Decrease: 1982 to 1983 or 1985 to 1986

57.

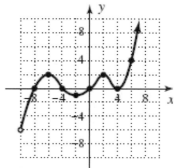

Zeroes: $(-8, 0), (-4, 0), (0, 0), (4, 0)$

Maximum: $(-6, 2), (2, 2)$

Minimum: $(-2, -1), (4, 0)$

59. $f(x) = x^3$

    a. $\dfrac{\Delta f}{\Delta x} = \dfrac{f(-1) - f(-2)}{-1 - (-2)} = \dfrac{-1 - (-8)}{1} = 7$

    b. $\dfrac{\Delta f}{\Delta x} = \dfrac{f(2) - f(1)}{2 - 1} = \dfrac{8 - 1}{1} = 7$

    c.   They are the same.

    d.   Slopes of the lines are the same.

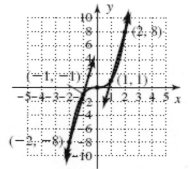

61. $h(t) = -16t^2 + 192t$

   a. $h(1) = -16(1)^2 + 192(1) = 176$ ft

   b. $h(2) = -16(2)^2 + 192(2) = 320$ ft

   c. $\dfrac{\Delta h}{\Delta t} = \dfrac{h(2) - h(1)}{2 - 1} = \dfrac{320 - 176}{1}$
     $= 144$ ft/sec

   d. $\dfrac{\Delta h}{\Delta t} = \dfrac{h(11) - h(10)}{11 - 10} = \dfrac{176 - 320}{1}$
     $= -144$ ft/sec
     The arrow is going down.

63. $v = \sqrt{2gs}$, $v = \sqrt{2(32)s} = 8\sqrt{s}$

   a. $v = \sqrt{2(32)(5)} = \sqrt{320} = 17.89$ ft/sec ;
     $v = \sqrt{2(32)(10)} = \sqrt{640} = 25.30$ ft/sec

   b. $v = \sqrt{2(32)(15)} = \sqrt{960} = 30.98$ ft/sec ;
     $v = \sqrt{2(32)(20)} = \sqrt{1280} = 35.78$ ft/sec

   c. Between $s = 5$ and $s = 10$

   d. $\dfrac{\Delta v}{\Delta s} = \dfrac{v(10) - v(5)}{10 - 5} = \dfrac{25.3 - 17.89}{5}$
     $= 1.482$ ft/sec;
     $\dfrac{\Delta v}{\Delta s} = \dfrac{v(20) - v(15)}{20 - 15} = \dfrac{35.78 - 30.98}{5}$
     $= 0.96$ ft/sec

65. $f(x) = 2x - 3$

   $\dfrac{\Delta f}{\Delta x} = \dfrac{f(x+h) - f(x)}{h}$

   $= \dfrac{[2(x+h) - 3] - (2x - 3)}{h}$

   $= \dfrac{2x + 2h - 3 - 2x + 3}{h}$

   $= \dfrac{2h}{h} = 2$

67. $h(x) = x^2 + 3$

   $\dfrac{\Delta f}{\Delta x} = \dfrac{h(x+h) - h(x)}{h} = \dfrac{[(x+h)^2 + 3] - (x^2 + 3)}{h}$

   $= \dfrac{x^2 + 2xh + h^2 + 3 - x^2 - 3}{h}$

   $= \dfrac{2xh + h^2}{h} = \dfrac{h(2x + h)}{h} = 2x + h$

69. $g(x) = x^2 + 2x - 3$

   $\dfrac{\Delta g}{\Delta x} = \dfrac{g(x+h) - g(x)}{h}$

   $= \dfrac{[(x+h)^2 + 2(x+h) - 3] - (x^2 + 2x - 3)}{h}$

   $= \dfrac{x^2 + 2xh + h^2 + 2x + 2h - 3 - x^2 - 2x + 3}{h}$

   $= \dfrac{2xh + h^2 + 2h}{h} = \dfrac{h(2x + h + 2)}{h} = 2x + 2 + h$

71. $f(x) = \dfrac{2}{x}$

   $\dfrac{\Delta f}{\Delta x} = \dfrac{f(x+h) - f(x)}{h}$

   $= \dfrac{\dfrac{2}{x+h} - \dfrac{2}{x}}{h} = \dfrac{\dfrac{2x - 2(x+h)}{x(x+h)}}{h}$

   $= \dfrac{\dfrac{2x - 2x - 2h}{x(x+h)}}{h} = \dfrac{-2h}{x(x+h)} \cdot \dfrac{1}{h} = \dfrac{-2}{x(x+h)}$

73. a. $g(x) = x^2 + 2x$

   $\dfrac{\Delta g}{\Delta x} = \dfrac{g(x+h) - g(x)}{h}$

   $= \dfrac{[(x+h)^2 + 2(x+h)] - (x^2 + 2x)}{h}$

   $= \dfrac{x^2 + 2xh + h^2 + 2x + 2h - x^2 - 2x}{h}$

   $= \dfrac{2xh + h^2 + 2h}{h} = \dfrac{h(2x + h + 2)}{h} = 2x + 2 + h$

   b. For [–3.0, –2.9], $x = -3.0$ and $h = 0.1$
     Rate of change:
     $2(-3.0) + 2 + 0.1 = -3.9$

   c. For [0.50, 0.51], $x = 0.50$ and $h = 0.01$
     Rate of change:
     $2(0.50) + 2 + 0.01 = 3.01$

   d.

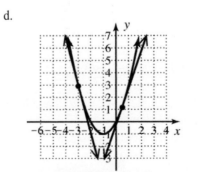

     The rates of change have opposite signs, with the secant line to the left being more steep.

75. a. $g(x) = x^3 + 1$

$$\frac{\Delta g}{\Delta x} = \frac{g(x+h) - g(x)}{h}$$

$$= \frac{[(x+h)^3 + 1] - (x^3 + 1)}{h}$$

$$= \frac{x^3 + 3x^2h + 3xh^2 + h^3 + 1 - x^3 - 1}{h}$$

$$= \frac{3x^2h + 3xh^2 + h^3}{h}$$

$$= \frac{h(3x^2 + 3xh + h^2)}{h} = 3x^2 + 3xh + h^2$$

b. For $[-2.1, -2]$, $x = -2.1$ and $h = 0.1$
Rate of change:

$$3(-2.1)^2 + 3(-2.1)(0.1) + (0.1)^2 = 12.61$$

c. For $[0.40, 0.41]$, $x = 0.40$ and $h = 0.01$
Rate of change:

$$3(0.40)^2 + 3(0.40)(0.01) + (0.01)^2 \approx 0.49$$

d.

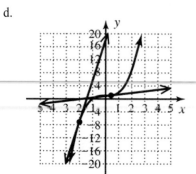

Both lines have a positive slope, but the line at $x = -2$ is much steeper.

77. $d(x) = 1.5\sqrt{x}$

$$\frac{\Delta d}{\Delta x} = \frac{d(x+h) - d(x)}{h}$$

$$= \frac{1.5\sqrt{x+h} - 1.5\sqrt{x}}{h}$$

a. For $[9, 9.01]$, $x = 9$ and $h = 0.01$
Rate of change:

$$\frac{1.5\sqrt{9 + 0.01} - 1.5\sqrt{9}}{0.01} \approx 0.25$$

b. For $[225, 225.01]$, $x = 225$ and $h = 0.01$
Rate of change:

$$\frac{1.5\sqrt{225 + 0.01} - 1.5\sqrt{225}}{0.01} \approx 0.05$$

c.

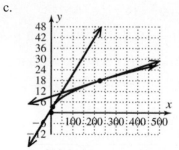

As height increases, you can see farther, the sight distance is increasing much slower.

79. No; No; Answers will vary.

81. Answers will vary.

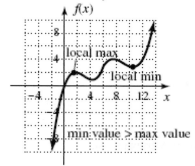

min value > max value

83. $x^2 - 8x - 20 = 0$

a. $(x - 10)(x + 2) = 0$

$x = 10$; $x = -2$

b. $(x^2 - 8x) - 20 = 0$

$(x^2 - 8x + 16) - 20 - 16 = 0$

$(x - 4)^2 - 36 = 0$

$(x - 4)^2 = 36$

$x - 4 = \pm 6$

$x = 4 \pm 6$

$x = 10$; $x = -2$

c. $x = \dfrac{8 \pm \sqrt{(-8)^2 - 4(1)(-20)}}{2(1)}$

$x = \dfrac{8 \pm \sqrt{64 + 80}}{2}$

$x = \dfrac{8 \pm \sqrt{144}}{2}$

$x = \dfrac{8 \pm 12}{2}$

$x = 10$; $x = -2$

85. $y = \dfrac{2}{3}x - 1$

2.6 Technology Highlight

Exercise 1: Shifted right 3 units; answers will vary.

2.6 Exercises

1. Stretch; compression

3. (–5, –9); upward

5. Answers will vary.

7. $f(x) = x^2 + 4x$
   a. quadratic
   b. up/up, Vertex (–2,–4),
      Axis of symmetry $x = -2$,
      x–intercepts (–4, 0) and (0,0),
      y–intercept (0,0)
   c. D: $x \in \mathbb{R}$, R: $y \in [-4, \infty)$

9. $p(x) = x^2 - 2x - 3$
   a. quadratic
   b. up/up, Vertex (1,–4),
      Axis of symmetry $x = 1$,
      x–intercepts (–1, 0) and (3,0),
      y–intercept (0,–3)
   c. D: $x \in \mathbb{R}$, R: $y \in [-4, \infty)$

11. $f(x) = x^2 - 4x - 5$
    a. quadratic
    b. up/up, Vertex (2,–9),
       Axis of symmetry $x = 2$,
       x–intercepts (–1, 0) and (5,0),
       y–intercept (0,–5)
    c. D: $x \in \mathbb{R}$, R: $y \in [-9, \infty)$

13. $p(x) = 2\sqrt{x+4} - 2$
    a. square root
    b. up to the right, Initial point (–3,–4),
       x–intercept (–3, 0),
       y–intercept (0,2)
    c. D: $x \in [-4, \infty)$, R: $y \in [-2, \infty)$

15. $r(x) = -3\sqrt{4-x} + 3$
    a. square root
    b. down to the left, Initial point (4,3),
       x–intercept (3, 0),
       y–intercept (0,–3)
    c. D: $x \in (-\infty, 4]$, R: $y \in (-\infty, 3]$

17. $g(x) = 2\sqrt{4-x}$
    a. square root
    b. up to the left, Initial point (4,0),
       x–intercept (4, 0),
       y–intercept (0,4)
    c. D: $x \in (-\infty, 4]$, R: $y \in [0, \infty)$

19. $p(x) = 2|x+1| - 4$
    a. absolute value
    b. up/up, Vertex (–1,–4),
       Axis of symmetry $x = -1$,
       x–intercepts (–3, 0) and (1,0),
       y–intercept (0,–2)
    c. D: $x \in \mathbb{R}$, R: $y \in [-4, \infty)$

21. $r(x) = -2|x+1| + 6$
    a. absolute value
    b. down/down, Vertex (–1,6),
       Axis of symmetry $x = -1$,
       x–intercepts (–4, 0) and (2,0),
       y–intercept (0,4)
    c. D: $x \in \mathbb{R}$, R: $y \in (-\infty, 6]$

23. $g(x) = -3|x| + 6$
    a. absolute value
    b. down/down, Vertex (0,6),
       Axis of symmetry $x = 0$,
       x–intercepts (–2, 0) and (2,0),
       y–intercept (0,6)
    c. D: $x \in \mathbb{R}$, R: $y \in (-\infty, 6]$

25. $f(x) = -(x-1)^3$
    a. cubic
    b. up/down, Inflection point (1,0),
       x–intercept (1, 0),
       y–intercept (0,1)
    c. D: $x \in \mathbb{R}$, R: $y \in \mathbb{R}$

27. $h(x) = x^3 + 1$
    a. cubic
    b. down/up, Inflection point (0,1),
       x–intercept (–1, 0),
       y–intercept (0,1)
    c. D: $x \in \mathbb{R}$, R: $y \in \mathbb{R}$

29. $q(x) = \sqrt[3]{x-1} - 1$
    a. cube root
    b. down/up, Inflection point (1,–1),
       x–intercept (2, 0),
       y–intercept (0,–2)
    c. D: $x \in \mathbb{R}$, R: $y \in \mathbb{R}$

31. Function family: Square root
    x–intercept: (–3, 0)
    y–intercept: (0, 2)
    Initial point: (–4, –2)
    End behavior: Up on right

33. Function family: Cubic
    x–intercept: (–2, 0)
    y–intercept: (0, –2)
    Inflection point: (–1, –1)
    End behavior: Up/down

35. $f(x) = \sqrt{x}$; $g(x) = \sqrt{x} + 2$; $h(x) = \sqrt{x} - 3$

x	f(x)	g(x)	h(x)
0	0	2	–3
4	2	4	–1
9	3	5	0
16	4	6	1
25	5	7	2

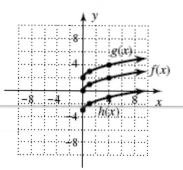

From the parent graph $f(x) = \sqrt{x}$, $g(x)$ shifts up 2 units and $h(x)$ shifts down 3 units.

37. $p(x) = |x|$; $q(x) = |x| - 5$; $r(x) = |x| + 2$

x	p(x)	q(x)	r(x)
–2	2	–3	4
–1	1	–4	3
0	0	–5	2
1	1	–4	3
2	2	–3	4

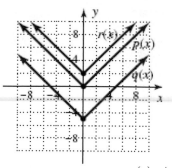

From the parent graph $p(x) = |x|$, $q(x)$ shifts down 5 units and $r(x)$ shifts up 2 units.

39. $f(x) = x^3 - 2$
    Shifts down 2 units.

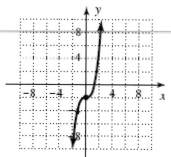

41. $h(x) = x^2 + 3$
    Shifts up 3 units.

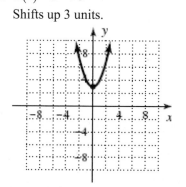

43. $p(x) = x^2$; $q(x) = (x+3)^2$

x	$p(x) = x^2$	$q(x) = (x+3)^2$
−5	25	4
−3	9	0
−1	1	4
1	1	16
3	9	36

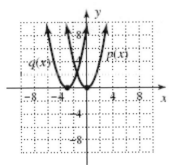

From the parent graph $p(x) = x^2$, $q(x)$ shifts
   left 3 units.

45. $Y_1 = |x|$; $Y_2 = |x-1|$

x	$Y_1 = |x|$	$Y_2 = |x-1|$
−2	2	3
−1	1	2
0	0	1
1	1	0
2	2	1

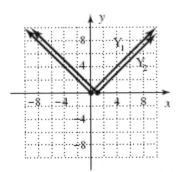

From the parent graph $Y_1 = |x|$, Y_2 shifts
   right 1 unit.

47. $p(x) = (x-3)^2$
   Shifts right 3 units.

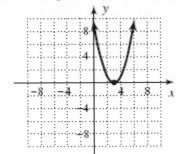

49. $h(x) = |x+3|$
   Shifts left 3 units

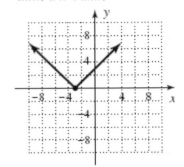

51. $g(x) = -|x|$
   Reflects across the x–axis.

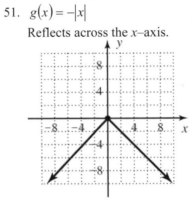

2.6 Exercises

53. $f(x) = \sqrt[3]{-x}$

Reflects across the y–axis.

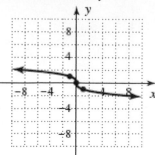

55. $p(x) = x^2$; $q(x) = 2x^2$; $r(x) = \frac{1}{2}x^2$

x	p(x)	q(x)	r(x)
−2	4	8	2
−1	1	2	½
0	0	0	0
1	1	2	½
2	4	8	2

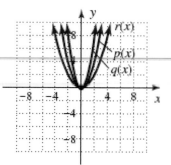

From the parent graph $p(x) = x^2$, $q(x)$ stretches upward and $r(x)$ compresses downward.

57. $Y_1 = |x|$; $Y_2 = 3|x|$; $Y_3 = \frac{1}{3}|x|$

x	Y_1	Y_2	Y_3
−2	2	6	2/3
−1	1	3	1/3
0	0	0	0
1	1	3	1/3
2	2	6	2/3

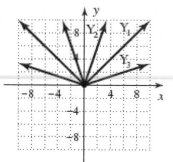

From the parent graph $Y_1 = |x|$, Y_2 stretches upward and Y_3 compresses downward.

59. $f(x) = 4\sqrt[3]{x}$

Stretches upward and downward.

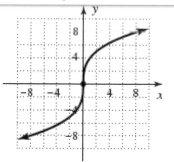

61. $p(x) = \frac{1}{3}x^3$

Compresses downward.

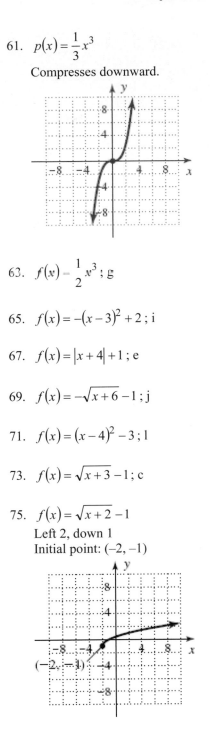

63. $f(x) = \frac{1}{2}x^3 \; ; \; g$

65. $f(x) = -(x-3)^2 + 2 \; ; \; i$

67. $f(x) = |x+4| + 1 \; ; \; e$

69. $f(x) = -\sqrt{x+6} - 1 \; ; \; j$

71. $f(x) = (x-4)^2 - 3 \; ; \; l$

73. $f(x) = \sqrt{x+3} - 1 \; ; \; c$

75. $f(x) = \sqrt{x+2} - 1$

Left 2, down 1
Initial point: $(-2, -1)$

77. $h(x) = -(x+3)^2 - 2$

Left 3, reflected across x–axis, down 2
Vertex: $(-3, -2)$

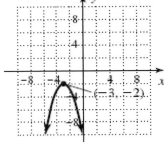

79. $p(x) = (x+3)^3 - 1$

Left 3, down 1
Inflection point: $(-3, -1)$

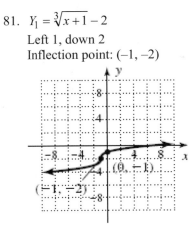

81. $Y_1 = \sqrt[3]{x+1} - 2$

Left 1, down 2
Inflection point: $(-1, -2)$

2.6 Exercises

83. $f(x) = -|x + 3| - 2$

Left 3, reflected across x–axis, down 2
Vertex: (–3, –2)

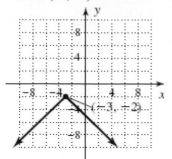

85. $h(x) = -2(x + 1)^2 - 3$

Left 1, stretched vertically, reflected across
x–axis, down 3
Vertex: (–1, –3)

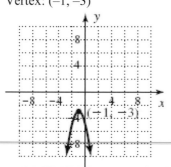

87. $p(x) = -\frac{1}{3}(x + 2)^3 - 1$

Left 2, compressed vertically, reflected
across x–axis, down 1
Inflection point: (–2, –1)

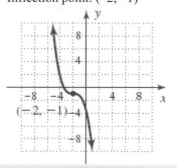

89. $Y_1 = -2\sqrt{-x - 1} + 3$

Reflected across y–axis, left 1, reflected
across x–axis, stretched vertically, up 3
Initial point: (–1, 3)

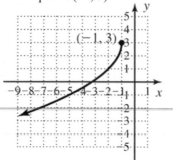

91. $h(x) = \frac{1}{5}(x - 3)^2 + 1$

Right 3, compressed vertically, up 1
Vertex: (3, 1)

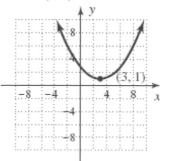

93. a. $f(x-2)$

95. a. $h(x)+3$

b. $-f(x)-3$

b. $-h(x-2)$

c. $\dfrac{1}{2}f(x+1)$

c. $h(x-2)-1$

d. $f(-x)+1$

d. $\dfrac{1}{4}h(x)+5$

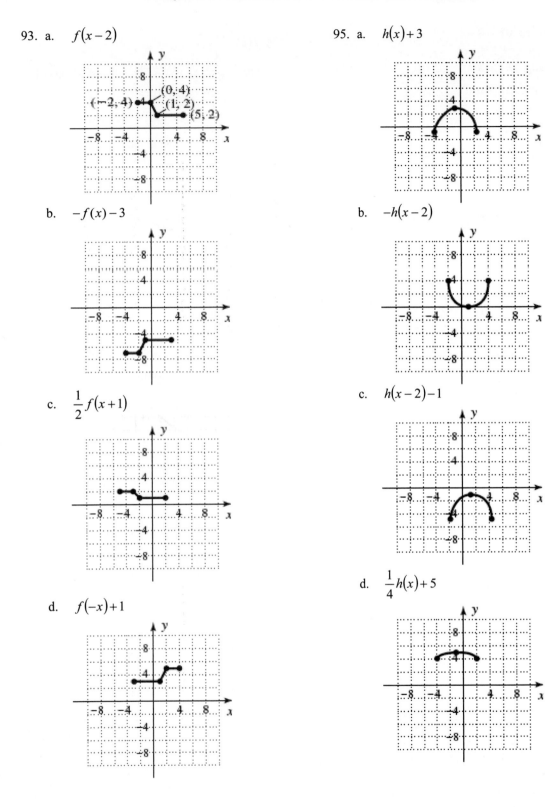

97. Vertex: (2, 0)
Point: (0, –4)

$$y = a(x-h)^2 + k$$
$$-4 = a(0-2)^2 + 0$$
$$-4 = 4a$$
$$-1 = a;$$
$$y = -(x-2)^2$$

99. Node: (–3, 0)
Point: (6, 4.5)

$$y = a\sqrt{x-h} + k$$
$$4.5 = a\sqrt{6-(-3)} + 0$$
$$4.5 = 3a$$
$$1.5 = a;$$
$$y = 1.5\sqrt{x+3}$$

101. Vertex: (–4, 0)
Point: (1, 4)

$$y = a|x-h| + k$$
$$4 = a|1+4| + 0$$
$$4 = 5a$$
$$\frac{4}{5} = a;$$
$$y = \frac{4}{5}|x+4|$$

103. $V = \frac{4}{3}\pi r^3$

$$\frac{4}{3}\pi \approx 4.2$$

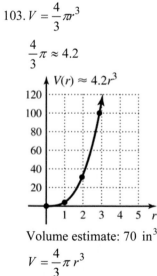

Volume estimate: 70 in³

$$V = \frac{4}{3}\pi r^3$$
$$V = \frac{4}{3}\pi(2.5)^3$$
$$V = \frac{4}{3}\pi(15.625) \approx 65.4 \text{ in}^3$$

Yes

105. $T(x) = \frac{1}{4}\sqrt{x}$

The graph can be obtained from $y = \sqrt{x}$ if it is compressed vertically.

$$T(81) = \frac{1}{4}\sqrt{81} = \frac{1}{4}(9) = 2.25 \text{ sec}$$

This point is on the graph.

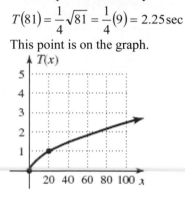

107. $P(v) = \frac{8}{125}v^3$

a. The graph can be obtained from $y = v^3$ if it is compressed vertically.

b.

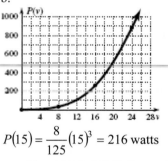

$$P(15) = \frac{8}{125}(15)^3 = 216 \text{ watts}$$

c. About 15.6, 161.5, Power increases dramatically at higher windspeeds.

109. $d(t) = 2t^2$

    a.   Vertical stretch by a factor of 2

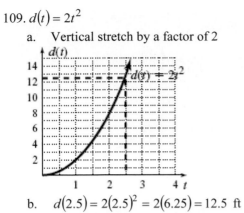

    b. $d(2.5) = 2(2.5)^2 = 2(6.25) = 12.5$ ft

    c.   5, 13, distance fallen per unit time increases very fast.

111. $f(x) = |x|$ and $g(x) = 2\sqrt{x}$

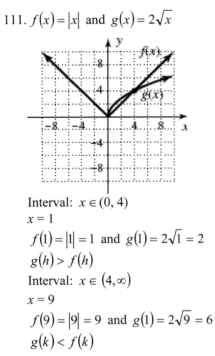

Interval: $x \in (0, 4)$

$x = 1$

$f(1) = |1| = 1$ and $g(1) = 2\sqrt{1} = 2$

$g(h) > f(h)$

Interval: $x \in (4, \infty)$

$x = 9$

$f(9) = |9| = 9$ and $g(1) = 2\sqrt{9} = 6$

$g(k) < f(k)$

113. $f(x) = x^2 - 4$

    $F(x) = |x^2 - 4|$

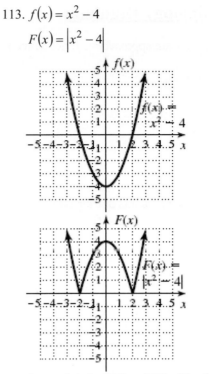

Any points in QIII and IV will reflected across the x–axis and thus move to QI and II.

115. $P = 32 + 32 + 38 + 24 + 6 + 8 = 140$ in.

    $A = 32(32) + 24(6) = 1024 + 144 = 1168$ in^2

117. $f(x) = (x - 4)^2 + 3$

    Quadratic, opens upward, Vertex (4,3)

    $f(x) \downarrow: (-\infty, 4)$;

    $f(x) \uparrow: (4, \infty)$

2.7 Technology Highlight

Exercise 1: They are approaching 4; not defined.

2.7 Exercises

1. Continuous

3. Smooth

5. Each piece must be continuous on the corresponding interval, and the function values at the endpoints of each interval must be equal. Answers will vary.

7. a. $f(x) = \begin{cases} x^2 - 6x + 10 & 0 \le x \le 5 \\ \dfrac{3}{2}x - \dfrac{5}{2} & 5 < x \le 9 \end{cases}$

   b. $y \in [1, 11]$

9. $h(x) = \begin{cases} -2 & x < -2 \\ |x| & -2 \le x < 3 \\ 5 & x \ge 3 \end{cases}$

   $h(-5) = -2$;

   $h(-2) = |-2| = 2$;

   $h\left(-\dfrac{1}{2}\right) = \left|-\dfrac{1}{2}\right| = \dfrac{1}{2}$;

   $h(0) = |0| = 0$;

   $h(2.999) = |2.999| = 2.999$;

   $h(3) = 5$

11. $p(x) = \begin{cases} 5 & x < -3 \\ x^2 - 4 & -3 \le x \le 3 \\ 2x + 1 & x > 3 \end{cases}$

    $p(-5) = 5$;

    $p(-3) = (-3)^2 - 4 = 9 - 4 = 5$;

    $p(-2) = (-2)^2 - 4 = 4 - 4 = 0$;

    $p(0) = (0)^2 - 4 = 0 - 4 = -4$;

    $p(3) = (3)^2 - 4 = 9 - 4 = 5$;

    $p(5) = 2(5) + 1 = 10 + 1 = 11$

13. $p(x) = \begin{cases} x + 2 & -6 \le x \le 2 \\ 2|x - 4| & x > 2 \end{cases}$

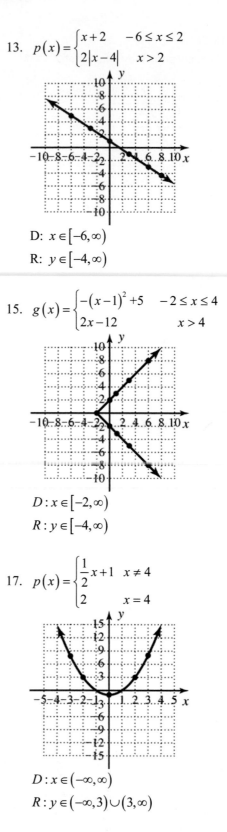

   D: $x \in [-6, \infty)$

   R: $y \in [-4, \infty)$

15. $g(x) = \begin{cases} -(x-1)^2 + 5 & -2 \le x \le 4 \\ 2x - 12 & x > 4 \end{cases}$

   $D : x \in [-2, \infty)$

   $R : y \in [-4, \infty)$

17. $p(x) = \begin{cases} \dfrac{1}{2}x + 1 & x \ne 4 \\ 2 & x = 4 \end{cases}$

   $D : x \in (-\infty, \infty)$

   $R : y \in (-\infty, 3) \cup (3, \infty)$

19. $H(x) = \begin{cases} -x+3 & x < 1 \\ -|x-5|+6 & 1 \le x < 9 \end{cases}$

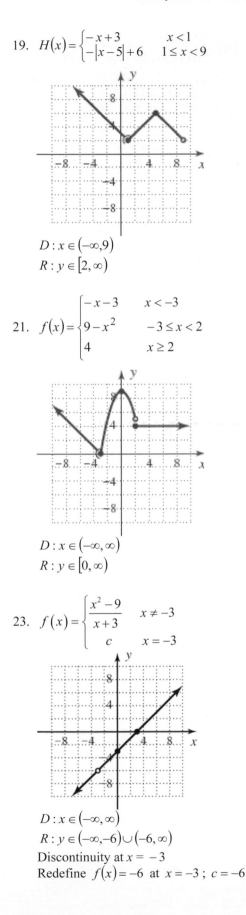

$D : x \in (-\infty, 9)$
$R : y \in [2, \infty)$

21. $f(x) = \begin{cases} -x-3 & x < -3 \\ 9-x^2 & -3 \le x < 2 \\ 4 & x \ge 2 \end{cases}$

$D : x \in (-\infty, \infty)$
$R : y \in [0, \infty)$

23. $f(x) = \begin{cases} \dfrac{x^2-9}{x+3} & x \ne -3 \\ c & x = -3 \end{cases}$

$D : x \in (-\infty, \infty)$
$R : y \in (-\infty, -6) \cup (-6, \infty)$
Discontinuity at $x = -3$
Redefine $f(x) = -6$ at $x = -3$; $c = -6$

25. $f(x) = \begin{cases} \dfrac{x^3-1}{x-1} & x \ne 1 \\ c & x = 1 \end{cases}$

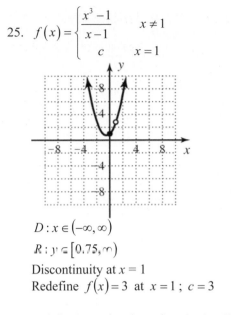

$D : x \in (-\infty, \infty)$
$R : y \in [0.75, \infty)$
Discontinuity at $x = 1$
Redefine $f(x) = 3$ at $x = 1$; $c = 3$

27. Left line contains the points $(-4, -3)$ and $(2, 0)$.

$$m = \frac{0-(-3)}{2-(-4)} = \frac{1}{2} \,;$$

$$y - 0 = \frac{1}{2}(x-2)$$

$$y = \frac{1}{2}x - 1;$$

Right line contains the points $(2, 0)$ and $(3, 3)$.

$$m = \frac{3-0}{3-2} = 3 \,;$$

$$y - 0 = 3(x-2)$$

$$y = 3x - 6;$$

$$f(x) = \begin{cases} \dfrac{1}{2}x - 1 & -4 \le x < 2 \\ 3x - 6 & x \ge 2 \end{cases}$$

29. The first equation is a quadratic with vertex $(-1, -4)$, opening up.

$$y = (x+1)^2 - 4$$

$$y = x^2 + 2x - 3;$$

The line is bounded by $(1, 2)$ and contains $(4, 5)$.

$$m = \frac{5-2}{4-1} = 1$$

$$y - 2 = 1(x-1)$$

$$y = x + 1;$$

$$p(x) = \begin{cases} x^2 + 2x - 3 & x \le 1 \\ x + 1 & x > 1 \end{cases}$$

2.7 Exercises

31. $|x| = \begin{cases} -x & x < 0 \\ x & x \geq 0 \end{cases}$

$f(x) = \dfrac{|x|}{x}$

Graph is discontinuous at $x = 0$.

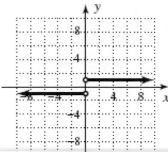

If $x < 0$, $f(x) = -1$.

If $x > 0$, $f(x) = 1$.

33.a. $S(t) = \begin{cases} -t^2 + 6t & 0 \leq t \leq 5 \\ 500 & t > 5 \end{cases}$

   b. $S(t) \in [0, 9]$

35. $P(t) = \begin{cases} -0.03t^2 + 1.28t + 1.68 & 0 \leq t \leq 30 \\ 1.89t - 43.5 & t > 30 \end{cases}$

   a. $P(5) = -0.03(5)^2 + 1.28(5) + 1.68 = 7.33$

   $P(15) = -0.03(15)^2 + 1.28(15) + 1.68 = 14.13$

   $P(25) = -0.03(25)^2 + 1.28(25) + 1.68 = 14.93$

   $P(35) = 1.89(35) - 43.5 = 22.65$

   $P(45) = 1.89(45) - 43.5 = 41.55$

   $P(55) = 1.89(55) - 43.5 = 60.45$

   b. Each piece gives a slightly different value due to rounding of coefficients in each model. At $t = 30$ we use the "first" piece: $P(30) = 13.08$.

37. $C(h) = \begin{cases} 0.09h & 0 \leq h \leq 1000 \\ 0.18h - 90 & h > 1000 \end{cases}$

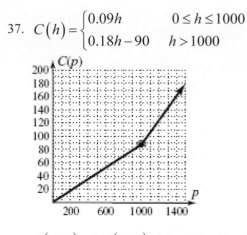

$C(1200) = 0.18(1200) - 90 = 216 - 90 = \$126$

39. $C(t) = \begin{cases} 0.75t & 0 \leq t \leq 25 \\ 1.5t - 18.75 & t > 25 \end{cases}$

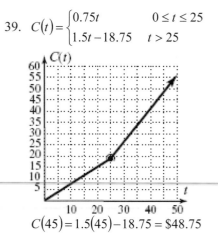

$C(45) = 1.5(45) - 18.75 = \$48.75$

41. $S(t) = \begin{cases} -1.35t^2 + 31.9t + 152 & 0 \le t \le 12 \\ 2.5t^2 - 80.6t + 950 & 12 < t \le 22 \end{cases}$

$S(25) = 2.5(25)^2 - 80.6(25) + 950$

$= 2.5(625) - 2015 + 950 = 497.5$

$\approx \$498$ billion;

$S(28) = 2.5(28)^2 - 80.6(28) + 950$

$= 2.5(784) - 2256.8 + 950 = 653.2$

$\approx \$653$ billion;

$S(30) = 2.5(30)^2 - 80.6(30) + 950$

$= 2.5(900) - 2418 + 950 = 782$

$\approx \$782$ billion

43. $C(m) = \begin{cases} 3.3m & 0 \le m \le 30 \\ 3.3(30) + 7(m-30) & m > 30 \end{cases}$

$C(m) = \begin{cases} 3.3m & 0 \le m \le 30 \\ 7m - 111 & m > 30 \end{cases}$

$C(46) = 7(46) - 111 = \$2.11$

45. $C(a) = \begin{cases} 0 & a < 2 \\ 2 & 2 \le a < 13 \\ 5 & 13 \le a < 20 \\ 7 & 20 \le a < 65 \\ 5 & a \ge 65 \end{cases}$

One grandparent:

$C(70) = 5$;

Two adults:

$C(44) = 7; C(45) = 7$;

Three teenagers:

$3 \cdot 5 = 15$;

Two children:

$2 \cdot 2 = 4$;

One infant: 0

Total Cost: $5 + 7 + 7 + 15 + 4 + 0 = \$38$

47. a. $C(w) = 17\lceil w - 1 \rceil + 80$

   For an envelope weighing between 0 and 1 oz, the cost is $0.80. Each step interval increases by 0.17.

   b. $0 < w \le 13$

   c. 80 cents

   d. 165 cents

   e. 165 cents

   f. 165 cents

   g. 182 cents

49. $h(x) = |x-2| - |x+3|$

x	$h(x) = |x-2| - |x+3|$
-5	$h(-5) = |-5-2| - |-5+3| = 7 - 2 = 5$
-4	$h(-4) = |-4-2| - |-4+3| = 6 - 1 = 5$
-3	$h(-3) = |-3-2| - |-3+3| = 5 - 0 = 5$
-2	$h(-2) = |-2-2| - |-2+3| = 4 - 1 = 3$
-1	$h(-1) = |-1-2| - |-1+3| = 3 - 2 = 1$
0	$h(0) = |0-2| - |0+3| = 2 - 3 = -1$
1	$h(1) = |1-2| - |1+3| = 1 - 4 = -3$
2	$h(2) = |2-2| - |2+3| = 0 - 5 = -5$
3	$h(3) = |3-2| - |3+3| = 1 - 6 = -5$
4	$h(4) = |4-2| - |4+3| = 2 - 7 = -5$
5	$h(5) = |5-2| - |5+3| = 3 - 8 = -5$

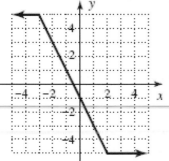

The function is continuous.

$$h(x) = \begin{cases} 5 & x \le -3 \\ -2x-1 & -3 < x < 2 \\ -5 & x \ge 2 \end{cases}$$

51. $Y_1 = \dfrac{x+2}{x+2}$, $Y_2 = \dfrac{|x+2|}{x+2}$

Y_1 has a removable discontinuity at $x = -2$.

Y_2 is discontinuous at $x = -2$.

53. $\dfrac{3}{x-2} + 1 = \dfrac{30}{x^2-4}$

$$\left(\dfrac{3}{x-2} + 1 \right)(x-2)(x+2) = \dfrac{30}{(x-2)(x+2)}(x-2)(x+2)$$

$3(x+2) + 1(x-2)(x+2) = 30$

$3x + 6 + x^2 - 4 = 30$

$x^2 + 3x - 28 = 0$

$(x+7)(x-4) = 0$

$x = -7; \quad x = 4$

55. a. $a^2 + b^2 = c^2$

$8^2 + b^2 = 12^2$

$64 + b^2 = 144$

$b^2 = 80$

$b = 4\sqrt{5}$ cm

b. $A = \dfrac{1}{2}bh$

$A = \dfrac{1}{2}(4\sqrt{5})(8)$

$A = 16\sqrt{5}$ cm^2

c. $V = \left(\dfrac{bh}{2} \right)h$

$V = (16\sqrt{5})(20) = 320\sqrt{5}$ cm^3

2.8 Technology Highlight

Exercise 1: $Y_1 = \sqrt{x}$ and $Y_2 = x + 7$

Yes, graph shifts 7 units to the left.

2.8 Exercises

1. $(f+g)(x);\ A \cap B$

3. Intersection; $g(x)$

5. Answers will vary.

7. a. Domain:

$f(x) = 2x^2 - x - 3; x \in \mathbb{R}$;

$g(x) = x^2 + 5x; x \in \mathbb{R}$;

$h(x) = f(x) - g(x); x \in \mathbb{R}$

$h(-2) = f(-2) - g(-2)$

b. $= 2(-2)^2 - (-2) - 3 - \left((-2)^2 + 5(-2)\right)$

$= 7 - (-6) = 13$

9. $h(x) = f(x) - g(x)$

a. $h(x) = 2x^2 - x - 3 - \left(x^2 + 5x\right)$

$= 2x^2 - x - 3 - x^2 - 5x$

$= x^2 - 6x - 3$

b. $h(-2) = (-2)^2 - 6(-2) - 3 = 13$

c. Same result

11. a. Domain of $f(x) = \sqrt{x-3}$

$x - 3 \geq 0$

$x \geq 3;\ [3, \infty)$

Domain of $g(x): x \in \mathbb{R}$;

Domain of $h(x): x \in [3, \infty)$

b. $h(x) = (f+g)(x)$

$= f(x) + g(x)$

$= \sqrt{x-3} + 2x^3 - 54$

c. $h(4) = \sqrt{4-3} + 2(4)^3 - 54 = 75$;

$h(2) = \sqrt{2-3} + 2(2)^3 - 54$

$= \sqrt{-1} + 16 - 54$

$\sqrt{-1}$ is not a real number;

2 is not in the domain of $h(x)$.

13. a. Domain of $p(x) = \sqrt{x+5}$

$x + 5 \geq 0$

$x \geq -5;\ x \in [-5, \infty)$

Domain of $q(x) = \sqrt{3-x}$

$3 - x \geq 0$

$-x \geq -3$

$x \leq 3;\ x \in (-\infty, 3]$

Domain of $r(x): x \in [-5, 3]$

b. $r(x) = (p+q)(x)$

$= p(x) + q(x)$

$= \sqrt{x+5} + \sqrt{3-x}$

c. $r(2) = \sqrt{2+5} + \sqrt{3-2} = \sqrt{7} + 1$

$r(4) = \sqrt{4+5} + \sqrt{3-4} = \sqrt{9} + \sqrt{-1}$

$\sqrt{-1}$ is not a real number;

4 is not in the domain of $r(x)$.

15. a. Domain of $f(x) = \sqrt{x+4}$

$x + 4 \geq 0$

$x \geq -4;\ x \in [-4, \infty)$

Domain of $g(x) = 2x + 3 : x \in \mathbb{R}$

Domain of $h(x): x \in [-4, \infty)$

b. $h(x) = (f \cdot g)(x)$

$= f(x) \cdot g(x)$

$= \sqrt{x+4}(2x+3)$

c. $h(-4) = \sqrt{-4+4}(2(-4)+3) = 0$;

$h(21) = \sqrt{21+4}(2(21)+3) = 225$

17. a. Domain of $p(x) = \sqrt{x+1}$

    $x + 1 \geq 0$

    $x \geq -1;\ x \in [-1, \infty)$

    Domain of $q(x) = \sqrt{7-x}$

    $7 - x \geq 0$

    $-x \geq -7$

    $x \leq 7;\ x \in (-\infty, 7]$

    Domain of $r(x): x \in [-1, 7]$

    b. $r(x) = (p \cdot q)(x)$

    $= p(x) \cdot q(x)$

    $= \sqrt{x+1} \cdot \sqrt{7-x}$

    $= \sqrt{-x^2 + 6x + 7}$

    c. $r(15) = \sqrt{-(15)^2 + 6(15) + 7} = \sqrt{-128}$

    $\sqrt{-128}$ is not a real number;

    15 is not in the domain of $r(x)$.

    $r(3) = \sqrt{-(3)^2 + 6(3) + 7} = \sqrt{16} = 4$

19. a. Domain of $f(x) = x^2 - 16 : x \in \mathbb{R}$

    Domain of $g(x) = x + 4 : x \in \mathbb{R}$

    Domain of $h(x) = \dfrac{x^2 - 16}{x+4}, x \neq -4$

    $x \in (-\infty, -4) \cup (-4, \infty)$

    b. $h(x) = \dfrac{f}{g}(x) = \dfrac{x^2 - 16}{x+4}$

    $h(x) = \dfrac{(x+4)(x-4)}{x+4} = x - 4;\ x \neq -4$

21. a. Domain of

    $f(x) = x^3 + 4x^2 - 2x - 8 : x \in \mathbb{R}$

    Domain of $g(x) = x + 4, x \in \mathbb{R}$

    Domain of

    $h(x) = \dfrac{x^3 + 4x^2 - 2x - 8}{x+4}, x \neq -4$

    $x \in (-\infty, -4) \cup (-4, \infty)$

    b. $h(x) = \dfrac{f}{g}(x) = \dfrac{x^3 + 4x^2 - 2x - 8}{x+4}$

    $h(x) = \dfrac{x^2(x+4) - 2(x+4)}{x+4}$

    $= \dfrac{(x+4)(x^2 - 2)}{x+4} = x^2 - 2;\ x \neq -4$

23. a. Domain of $f(x) = x^3 - 7x^2 + 6x : x \in \mathbb{R}$

    Domain of $g(x) = x - 1 : x \in \mathbb{R}$

    Domain of

    $h(x) = \dfrac{x^3 - 7x^2 + 6x}{x-1}, x \neq 1$

    $x \in (-\infty, 1) \cup (1, \infty)$

    b. $h(x) = \dfrac{f}{g}(x) = \dfrac{x^3 - 7x^2 + 6x}{x-1}$

    $h(x) = \dfrac{x(x^2 - 7x + 6)}{x-1}$

    $= \dfrac{x(x-6)(x-1)}{x-1} = x(x-6)$

    $= x^2 - 6x;\ x \neq 1$

25. a. Domain of $f(x) = x + 1 : x \in \mathbb{R}$

    Domain of $g(x) = x - 5 : x \in \mathbb{R}$

    Domain of

    $h(x) = \dfrac{x+1}{x-5}, x \neq 5$

    $x \in (-\infty, 5) \cup (5, \infty)$

    b. $h(x) = \dfrac{f}{g}(x) = \dfrac{x+1}{x-5};\ x \neq 1$

27. a. Domain of $p(x) = 2x - 3 : x \in \mathbb{R}$

    Domain of $q(x) = \sqrt{-2 - x}$,

    $-2 - x \geq 0$

    $-x \geq 2$

    $x \leq -2;\ x \in (-\infty, -2]$

    Domain of $r(x) = \dfrac{2x-3}{\sqrt{-2-x}}$,

    $-2 - x > 0$

    $-x > 2$

    $x < -2;\ x \in (-\infty, -2)$

    b. $r(x) = \dfrac{p}{q}(x) = \dfrac{2x-3}{\sqrt{-2-x}}$

    c. $r(6) = \dfrac{2(6)-3}{\sqrt{-2-6}} = \dfrac{9}{\sqrt{-8}}$

    $\sqrt{-8}$ is not a real number;

    6 is not in the domain of $r(x)$.

    $r(-6) = \dfrac{2(-6)-3}{\sqrt{-2+6}} = \dfrac{-15}{\sqrt{4}} = -\dfrac{15}{2}$

Chapter 2: Relations, Functions and Graphs

29. a. Domain of $p(x) = x - 5 : x \in \mathbb{R}$

   Domain of $q(x) = \sqrt{x-5}$,

   $x - 5 \geq 0$

   $x \geq 5; \; x \in [5, \infty)$

   Domain of $r(x) = \dfrac{x-5}{\sqrt{x-5}}$,

   $x - 5 > 0$

   $x > 5; \; x \in (5, \infty)$

   b. $r(x) = \dfrac{p}{q}(x) = \dfrac{x-5}{\sqrt{x-5}}$

   c. $r(6) = \dfrac{6-5}{\sqrt{6-5}} = \dfrac{1}{\sqrt{1}} = 1$

   $r(-6) = \dfrac{-6-5}{\sqrt{-6-5}} = \dfrac{-11}{\sqrt{-11}}$

   $\sqrt{-11}$ is not a real number;

   -6 is not in the domain of $r(x)$.

31. a. Domain of $p(x) = x^2 - 36 : x \in \mathbb{R}$

   Domain of $q(x) = \sqrt{2x+13}$,

   $2x + 13 \geq 0$

   $2x \geq -13$

   $x \geq -\dfrac{13}{2}; \; x \in \left[-\dfrac{13}{2}, \infty\right)$

   Domain of $r(x) = \dfrac{x^2 - 36}{\sqrt{2x+13}}$,

   $2x + 13 > 0$

   $2x > -13$

   $x > -\dfrac{13}{2}; \; x \in \left(-\dfrac{13}{2}, \infty\right)$

   b. $r(x) = \dfrac{p}{q}(x) = \dfrac{x^2-36}{\sqrt{2x+13}}$

   c. $r(6) = \dfrac{6^2 - 36}{\sqrt{2(6)+13}} = \dfrac{0}{\sqrt{25}} = 0$

   $r(-6) = \dfrac{(-6)^2 - 36}{\sqrt{2(-6)+13}} = \dfrac{0}{\sqrt{1}} = 0$

33. a. $f(x) = \dfrac{6x}{x-3}, \; g(x) = \dfrac{3x}{x+2}$

   $h(x) = \dfrac{f(x)}{g(x)} = \dfrac{\dfrac{6x}{x-3}}{\dfrac{3x}{x+2}}$

   $= \dfrac{6x}{x-3} \div \dfrac{3x}{x+2} = \dfrac{6x}{x-3} \cdot \dfrac{x+2}{3x}$

   $= \dfrac{2(x+2)}{x-3} = \dfrac{2x+4}{x-3}$

   b. Domain of $h(x) = \dfrac{2x+4}{x-3}, \; x \neq 3$

   $x \in (-\infty, 3) \cup (3, \infty)$

   c. $x + 2 \neq 0$

   $x \neq -2;$

   $\dfrac{3x}{x+2} \neq 0$

   $x \neq 0$

35. $f(x) = 2x + 3$ and $g(x) = x - 2$

   Sum:

   $f(x) + g(x) = 2x + 3 + x - 2 = 3x + 1$

   Domain contains all values of x.

   $D : x \in (-\infty, \infty)$

   Difference:

   $f(x) - g(x) = 2x + 3 - (x-2)$

   $= 2x + 3 - x + 2 = x + 5$

   Domain contains all values of x.

   $D : x \in (-\infty, \infty)$

   Product:

   $f(x) \cdot g(x) = (2x+3)(x-2)$

   $= 2x^2 - 4x + 3x - 6$

   $= 2x^2 - x - 6$

   Domain contains all values of x.

   $D : x \in (-\infty, \infty)$

   Quotient:

   $\dfrac{f(x)}{g(x)} = \dfrac{2x+3}{x-2}$

   $x - 2 \neq 0$

   $x \neq 2$

   $D : x \in (-\infty, 2) \cup (2, \infty)$

2.8 Exercises

37. $f(x) = x^2 + 7$ and $g(x) = 3x - 2$

Sum:

$f(x) + g(x) = x^2 + 7 + 3x - 2 = x^2 + 3x + 5$

Domain contains all values of x.

$D : x \in (-\infty, \infty)$

Difference:

$f(x) - g(x) = x^2 + 7 - (3x - 2)$

$= x^2 + 7 - 3x + 2$

$= x^2 - 3x + 9$

Domain contains all values of x.

$D : x \in (-\infty, \infty)$

Product:

$f(x) \cdot g(x) = (x^2 + 7)(3x - 2)$

$= 3x^3 - 2x^2 + 21x - 14$

Domain contains all values of x.

$D : x \in (-\infty, \infty)$

Quotient:

$\dfrac{f(x)}{g(x)} = \dfrac{x^2 + 7}{3x - 2}$

$3x - 2 \neq 0$

$3x \neq 2$

$x \neq \dfrac{2}{3}$

$D : x \in \left(-\infty, \dfrac{2}{3}\right) \cup \left(\dfrac{2}{3}, \infty\right)$

39. $f(x) = x^2 + 2x - 3$ and $g(x) = x - 1$

Sum:

$f(x) + g(x) = x^2 + 2x - 3 + x - 1$

$= x^2 + 3x - 4$

Domain contains all values of x.

$D : x \in (-\infty, \infty)$

Difference:

$f(x) - g(x) = x^2 + 2x - 3 - (x - 1)$

$= x^2 + 2x - 3 - x + 1$

$= x^2 + x - 2$

Domain contains all values of x.

$D : x \in (-\infty, \infty)$

Product:

$f(x) \cdot g(x) = (x^2 + 2x - 3)(x - 1)$

$= x^3 - x^2 + 2x^2 - 2x - 3x + 3$

$= x^3 + x^2 - 5x + 3$

Domain contains all values of x.

$D : x \in (-\infty, \infty)$

Quotient:

$\dfrac{f(x)}{g(x)} = \dfrac{x^2 + 2x - 3}{x - 1}$

$= \dfrac{(x + 3)(x - 1)}{x - 1} = x + 3$

$x - 1 \neq 0$

$x \neq 1$

$D : x \in (-\infty, 1) \cup (1, \infty)$

41. $f(x) = 3x + 1$ and $g(x) = \sqrt{x - 3}$

Sum:

$f(x) + g(x) = 3x + 1 + \sqrt{x - 3}$

$x - 3 \geq 0$

$x \geq 3$

$D : x \in [3, \infty)$

Difference:

$f(x) - g(x) = 3x + 1 - \sqrt{x - 3}$

$x - 3 \geq 0$

$x \geq 3$

$D : x \in [3, \infty)$

Product:

$f(x) \cdot g(x) = (3x + 1)\sqrt{x - 3}$

$x - 3 \geq 0$

$x \geq 3$

$D : x \in [3, \infty)$

Quotient:

$\dfrac{f(x)}{g(x)} = \dfrac{3x + 1}{\sqrt{x - 3}}$

$x - 3 > 0$

$x > 3$

$D : x \in (3, \infty)$

43. $f(x) = 2x^2$ and $g(x) = \sqrt{x+1}$

Sum:

$f(x) + g(x) = 2x^2 + \sqrt{x+1}$

$x + 1 \geq 0$

$x \geq -1$

$D : x \in [-1, \infty)$

Difference:

$f(x) - g(x) = 2x^2 - \sqrt{x+1}$

$x + 1 \geq 0$

$x \geq -1$

$D : x \in [-1, \infty)$

Product:

$f(x) \cdot g(x) = 2x^2 \sqrt{x+1}$

$x + 1 \geq 0$

$x \geq -1$

$D : x \in [-1, \infty)$

Quotient:

$\dfrac{f(x)}{g(x)} = \dfrac{2x^2}{\sqrt{x+1}}$

$x + 1 > 0$

$x > -1$

$D : x \in (-1, \infty)$

45. $f(x) = \dfrac{2}{x-3}$ and $g(x) = \dfrac{5}{x+2}$

Sum:

$f(x) + g(x) = \dfrac{2}{x-3} + \dfrac{5}{x+2}$

$= \dfrac{2(x+2) + 5(x-3)}{(x-3)(x+2)}$

$= \dfrac{2x+4+5x-15}{(x-3)(x+2)}$

$= \dfrac{7x-11}{(x-3)(x+2)}$

$x - 3 \neq 0 \quad x + 2 \neq 0$

$x \neq 3 \qquad x \neq -2$

$D : x \in (-\infty, -2) \cup (-2, 3) \cup (3, \infty)$

Difference:

$f(x) - g(x) = \dfrac{2}{x-3} - \dfrac{5}{x+2}$

$= \dfrac{2(x+2) - 5(x-3)}{(x-3)(x+2)}$

$= \dfrac{2x+4-5x+15}{(x-3)(x+2)}$

$= \dfrac{-3x+19}{(x-3)(x+2)}$

$x - 3 \neq 0 \quad x + 2 \neq 0$

$x \neq 3 \qquad x \neq -2$

$D : x \in (-\infty, -2) \cup (-2, 3) \cup (3, \infty)$

Product:

$f(x) \cdot g(x) = \left(\dfrac{2}{x-3}\right)\left(\dfrac{5}{x+2}\right)$

$= \dfrac{10}{(x-3)(x+2)}$

$= \dfrac{10}{x^2 - x - 6}$

$x - 3 \neq 0 \quad x + 2 \neq 0$

$x \neq 3 \qquad x \neq -2$

$D : x \in (-\infty, -2) \cup (-2, 3) \cup (3, \infty)$

Quotient:

$\dfrac{f(x)}{g(x)} = \dfrac{\dfrac{2}{x-3}}{\dfrac{5}{x+2}} = \left(\dfrac{2}{x-3}\right)\left(\dfrac{x+2}{5}\right)$

$= \dfrac{2(x+2)}{5(x-3)} = \dfrac{2x+4}{5x-15}$

$x - 3 \neq 0 \quad x + 2 \neq 0$

$x \neq 3 \qquad x \neq -2$

$D : x \in (-\infty, -2) \cup (-2, 3) \cup (3, \infty)$

47. $f(x) = x^2 - 5x - 14$

$f(-2) = (-2)^2 - 5(-2) - 14 = 4 + 10 - 14 = 0;$

$f(7) = (7)^2 - 5(7) - 14 = 49 - 35 - 14 = 0;$

$f(2) = (2)^2 - 5(2) - 14 = 4 - 10 - 14 = -20;$

$f(a-2) = (a-2)^2 - 5(a-2) - 14$

$= a^2 - 4a + 4 - 5a + 10 - 14$

$= a^2 - 9a$

49. $f(x) = \sqrt{x+3}$ and $g(x) = 2x - 5$
    (a) $h(x) = (f \circ g)(x) = f[g(x)]$
    $\quad = \sqrt{g(x) + 3}$
    $\quad = \sqrt{(2x - 5) + 3}$
    $\quad = \sqrt{2x - 2}$
    (b) $H(x) = (g \circ f)(x) = g[f(x)]$
    $\quad = 2(f(x)) - 5$
    $\quad = 2\sqrt{x+3} - 5$
    (c) $2x - 2 \geq 0$
    $\quad 2x \geq 2$
    $\quad x \geq 1$
    Domain of $h: x \in [1, \infty)$
    $x + 3 \geq 0$
    $\quad x \geq -3$
    Domain of $H: x \in [-3, \infty)$

51. $f(x) = \sqrt{x - 3}$ and $g(x) = 3x + 4$
    (a) $h(x) = (f \circ g)(x)$
    $\quad h(x) = f[g(x)]$
    $\quad h(x) = \sqrt{g(x) - 3}$
    $\quad = \sqrt{3x + 4 - 3}$
    $\quad = \sqrt{3x + 1}$
    (b) $H(x) = (g \circ f)(x)$
    $\quad H(x) = g[f(x)]$
    $\quad H(x) = 3(f(x)) + 4$
    $\quad = 3\sqrt{x - 3} + 4$
    (c) $3x + 1 \geq 0$
    $\quad 3x \geq -1$
    $\quad x \geq -\dfrac{1}{3}$
    Domain of h: $\left\{ x \middle| x \geq -\dfrac{1}{3} \right\}$
    or $\left[-\dfrac{1}{3}, \infty \right)$;
    $x - 3 \geq 0$
    $\quad x \geq 3$
    Domain of H: $\{ x | x \geq 3 \}$
    or $[3, \infty)$

53. $f(x) = x^2 - 3x$ and $g(x) = x + 2$
    (a) $h(x) = (f \circ g)(x)$
    $\quad h(x) = f[g(x)]$
    $\quad h(x) = (g(x))^2 - 3(g(x))$
    $\quad = (x + 2)^2 - 3(x + 2)$
    $\quad = x^2 + 4x + 4 - 3x - 6$
    $\quad = x^2 + x - 2$
    (b) $H(x) = (g \circ f)(x)$
    $\quad H(x) = g[f(x)]$
    $\quad H(x) = (f(x)) + 2$
    $\quad = x^2 - 3x + 2$
    (c) Domain of h: $(-\infty, \infty)$
    Domain of H: $(-\infty, \infty)$

55. $f(x) = x^2 + x - 4$ and $g(x) = x + 3$
    (a) $h(x) = (f \circ g)(x)$
    $\quad h(x) = f[g(x)]$
    $\quad h(x) = (g(x))^2 + g(x) - 4$
    $\quad = (x + 3)^2 + x + 3 - 4$
    $\quad = x^2 + 6x + 9 + x - 1$
    $\quad = x^2 + 7x + 8$
    (b) $H(x) = (g \circ f)(x)$
    $\quad H(x) = g[f(x)]$
    $\quad H(x) = f(x) + 3$
    $\quad = x^2 + x - 4 + 3$
    $\quad = x^2 + x - 1$
    (c) Domain of h: $(-\infty, \infty)$
    Domain of H: $(-\infty, \infty)$

57. $f(x) = |x| - 5$ and $g(x) = -3x + 1$
    (a) $h(x) = (f \circ g)(x)$
    $\quad h(x) = f[g(x)]$
    $\quad h(x) = |g(x)| - 5$
    $\quad = |-3x + 1| - 5$
    (b) $H(x) = (g \circ f)(x)$
    $\quad H(x) = g[f(x)]$
    $\quad H(x) = -3(f(x)) + 1$
    $\quad = -3(|x| - 5) + 1$
    $\quad = -3|x| + 15 + 1$
    $\quad = -3|x| + 16$
    (c) Domain of h: $(-\infty, \infty)$
    Domain of H: $(-\infty, \infty)$

59. $f(x) = \dfrac{2x}{x+3}$ and $g(x) = \dfrac{5}{x}$

(a)

$(f \circ g)(x)$: For $g(x)$ to be defined, $x \neq 0$.

For $f[g(x)] = \dfrac{2g(x)}{g(x)+3}$,

$g(x) \neq -3$ so $x \neq -\dfrac{5}{3}$.

Domain: $\left\{ x \middle| x \neq 0, x \neq -\dfrac{5}{3} \right\}$

(b)

$(g \circ f)(x)$: For $f(x)$ to be defined, $x \neq -3$.

For $g[f(x)] = \dfrac{5}{f(x)}$,

$f(x) \neq 0$ so $x \neq 0$.

Domain: $\{ x | x \neq 0, x \neq -3 \}$

(c) $(f \circ g)(x) = f[g(x)]$

$= \dfrac{2(g(x))}{g(x)+3} = \dfrac{2\left(\dfrac{5}{x}\right)}{\dfrac{5}{x}+3} = \dfrac{\dfrac{10}{x}}{\dfrac{5+3x}{x}}$

$= \dfrac{10x}{x(5+3x)} = \dfrac{10}{5+3x}$

$(g \circ f)(x) = g[f(x)]$

$= \dfrac{5}{f(x)} = \dfrac{5}{\dfrac{2x}{x+3}} = \dfrac{5(x+3)}{2x} = \dfrac{5x+15}{2x}$

61. $f(x) = \dfrac{4}{x}$ and $g(x) = \dfrac{1}{x-5}$

(a)

$(f \circ g)(x)$: For $g(x)$ to be defined, $x \neq 5$.

For $f[g(x)] = \dfrac{4}{g(x)}$,

$g(x) \neq 0$ and $g(x)$ is never zero.

Domain: $\{ x | x \neq 5 \}$

(b)

$(g \circ f)(x)$: For $f(x)$ to be defined, $x \neq 0$.

For $g[f(x)] = \dfrac{1}{f(x)-5}$,

$f(x) \neq 5$ so $x \neq \dfrac{4}{5}$.

Domain: $\left\{ x \middle| x \neq 0, x \neq \dfrac{4}{5} \right\}$

(c) $h(x) = (f \circ g)(x)$

$h(x) = f[g(x)]$

$h(x) = \dfrac{4}{g(x)}$

$= \dfrac{4}{\dfrac{1}{x-5}}$

$= 4(x-5)$

$= 4x - 20$

$H(x) = (g \circ f)(x)$

$H(x) = g[f(x)]$

$H(x) = \dfrac{1}{f(x)-5}$

$= \dfrac{1}{\dfrac{4}{x}-5}$

$= \dfrac{1}{\dfrac{4-5x}{x}}$

$= \dfrac{x}{4-5x}$

63. $f(x) = x^2 - 8$ and $g(x) = x+2$

$h(x) = (f \circ g)(x)$

a. $(f \circ g)(x) = f[g(x)]$

$= \left(g(x)^2\right) - 8$

$= (x+2)^2 - 8$

$= x^2 + 4x + 4 - 8$

$= x^2 + 4x - 4$;

$h(x) = x^2 + 4x - 4$

$h(5) = (5)^2 + 4(5) - 4 = 25 + 20 - 4 = 41$

b. $g(5) = 5 + 2 = 7$

$f[g(5)] = f(7)$

$= (7)^2 - 8 = 49 - 8 = 41$

2.8 Exercises

65. $h(x) = \left(\sqrt{x-2}+1\right)^3 - 5$

Answers may vary.

$g(x) = \sqrt{x-2}+1, f(x) = x^3 - 5$

67. $f(x) = 2x-1, g(x) = x^2 - 1,$

$h(x) = x+4$

   a. $p(x) = f\left[g\left(\left[h(x)\right]\right)\right]$

      $p(x) = f\left[(x+4)^2 - 1\right]$

      $= 2\left[(x+4)^2 - 1\right] - 1$

      $= 2(x+4)^2 - 2 - 1$

      $= 2(x+4)^2 - 3$

   b. $q(x) = g\left[f\left(\left[h(x)\right]\right)\right]$

      $q(x) = g\left[2(x+4)-1\right]$

      $= g\left[2x+8-1\right] = g\left[2x+7\right]$

      $= (2x+7)^2 - 1$

69. a. $C(5) = 6000$

   b. $T(8) = 3000$

   c. $C(9) + T(9) = 6000 + 2000 = 8000$

   d. $C(9) - T(9) = 6000 - 2000 = 4000$

71. a. $R(2) = \$1$ billion

   b. $C(8) = \$5$ billion

   c. $R(t) = C(t)$

      Broke even 2003, 2007, 2010

      $C(t) > R(t)$:

   d. $t \in (2000, 2003) \cup (2007, 2010)$

   e. $R(t) > C(t), t \in (2003, 2007)$

   f. $R(5) - C(5) = 5 - 1 = \$4$ billion

73. a. $(f+g)(-4) = f(-4) + g(-4)$

      $= 5 + (-1) = 4$

   b. $(f \cdot g)(1) = f(1) \cdot g(1)$

      $= 0(3) = 0$

   c. $(f-g)(4) = f(4) - g(4)$

      $= 5 - 3 = 2$

   d. $(f+g)(0) = f(0) + g(0)$

      $= 1 + 2 = 3$

   e. $\left(\dfrac{f}{g}\right)(2) = \dfrac{f(2)}{g(2)} = \dfrac{-1}{3}$

   f. $(f \cdot g)(-2) = f(-2) \cdot g(-2)$

      $= 3(2) = 6$

   g. $(g \cdot f)(2) = g(2) \cdot f(2)$

      $= 3(-1) = -3$

   h. $(f-g)(-1) = f(-1) - g(-1)$

      $= 2 - 1 = 1$

   i. $(f+g)(8) = f(8) + g(8)$

      $= -1 + 2 = 1$

   j. $\left(\dfrac{f}{g}\right)(7) = \dfrac{f(7)}{g(7)} = \dfrac{1}{0}$ undefined

   k. $(g \circ f)(4) = g(f(4))$

      $= g(5) = \dfrac{1}{2} = 0.5$

   l. $(f \circ g)(4) = f(g(4))$

      $= f(3) = 2$

75. $h(x) = f(x) - g(x)$

    $= 5 - \left(\dfrac{2}{3}x + 1\right) = 5 - \dfrac{2}{3}x - 1$

    $= -\dfrac{2}{3}x + 4$

77. $h(x) = f(x) - g(x)$

    $= \left(5x - x^2\right) - x = 4x - x^2$

79. $A = 40\pi r + 2\pi r^2$

    $A = 2\pi r(20 + r)$

    $A(r) = (f \cdot g)(r)$

    $f(r) = 2\pi r, g(r) = 20 + r$

    $A(5) = 2\pi(5)(20+5) = 10\pi(25) = 250\pi$ units^2

81. Revenue: $R(x) = 40,000x$

    Cost: $C(x) = 108,000 + 28,000x$

   a. $P(x) = R(x) - C(x)$

      $= 40,000x - 108,000 - 28,000x$

      $= 12,000x - 108,000$

   b. Break even when $P(x) = 0$

      $12,000x - 108,000 = 0$

      $12,000x = 108,000$

      $x = 9$

      9 boats must be sold to break even.

83. a. $P(n) = R(n) - C(n)$

    $P(n) = 11.45n - 0.1n^2$

    b. $P(12) = 11.45(12) - 0.1(12)^2$

    $= 137.4 - 14.4 = \$123$

    c. $P(60) = 11.45(60) - 0.1(60)^2$

    $= 687 - 360 = \$327$

    d. At $n = 115$, costs exceed revenue, $C(115) > R(115)$.

85. $f(x) = 0.5x - 14$; $g(x) = 2x + 23$

    $h(x) = (f \circ g)(x) = f[g(x)]$

    $h(x) = 0.5(g(x)) - 14$

    $= 0.5(2x + 23) - 14$

    $= x + 11.5 - 14$

    $= x - 2.5;$

    $h(13) = 13 - 2.5 = 10.5$

87. $T(x) = 41.6x$; $R(x) = 10.9x$

    (a) $T(100) = 41.6(100) = 4160 \text{ baht}$

    (b) $R(4160) = 10.9(4160) = 45,344$

    rRinggit

    (c) $M(x) = (R \circ T)(x) = R[T(x)]$

    $M(x) = 10.9(T(x))$

    $= 10.9(41.6x)$

    $= 453.44x;$

    $M(100) = 453.44(100) = 45344 \text{ ringgit}$

    Parts B and C agree.

89. $r(t) = 3t$; $A = \pi r^2$

    (a) $r(2) = 3(2) = 6 \text{ ft}$

    (b) $A(6) = \pi(6)^2 = 36\pi \text{ ft}^2$

    (c) $A(t) = (A \circ r)(t) = A[r(t)]$

    $A(t) = \pi(r(t))^2$

    $= \pi(3t)^2$

    $= 9\pi t^2;$

    $A(2) = 9\pi(2)^2 = 36\pi \text{ ft}^2$

    The answers do agree.

91. $C(x) = 0.0345x^4 - 0.8996x^3 + 7.5383x^2 - 21.7215x + 40$

    $L(x) = -0.0345x^4 + 0.8996x^3 - 7.5383x^2 + 21.7215x + 10$

    (a) Using the grapher, 1995 to 1996; 1999 to 2004

    (b) Using the grapher, 30 seats; 1995

    (c) Using the grapher, 20 seats; 1997

    (d) Using the grapher, the total number in the senate (50); the number of additional seats held by the majority.

93. $f(x) = \sqrt{1 - x}$ and $g(x) = \sqrt{x - 2}$

    Using the grapher,

    $(f + g)(x)$ cannot be found because their domains do not overlap.

95. $f(x) = (x - 3)^2 + 2, g(x) = 4|x - 3| - 5$

x	$f(x)$	$g(x)$	$(f–g)(x)$
−2	27	15	12
−1	18	11	7
0	11	7	4
1	6	3	3
2	3	−1	4
3	2	−5	7
4	3	−1	4
5	6	3	3
6	11	7	4
7	18	11	7
8	27	15	12

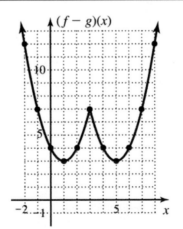

97. $f(x) = \sqrt{x}$; $g(x) = \sqrt[3]{x}$; $h(x) = |x|$

(a)

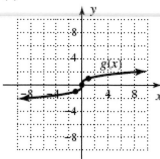

(b)

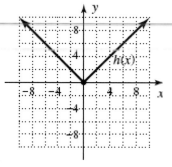

(c)

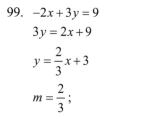

Chapter 2 Summary and Review

1.

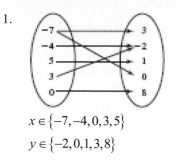

$$x \in \{-7, -4, 0, 3, 5\}$$
$$y \in \{-2, 0, 1, 3, 8\}$$

3. $(19, 25), (-14, -31)$
$$d = \sqrt{(-14 - 19)^2 + (-31 - 25)^2}$$
$$= \sqrt{1089 + 3136} = \sqrt{4225} = 65$$
65 miles

5. $x^2 + y^2 = 16$
Center (0,0), Radius 4

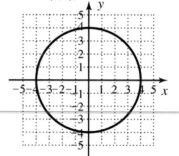

99. $-2x + 3y = 9$
$$3y = 2x + 9$$
$$y = \frac{2}{3}x + 3$$
$$m = \frac{2}{3};$$

Slope of a line perpendicular is $-\frac{3}{2}$.

y– intercept (0,0);

Equation: $y = -\frac{3}{2}x$

7. $(-3,0)$ and $(0,4)$

To find the center, find the midpoint.

$$\left(\frac{-3+0}{2}, \frac{0+4}{2}\right) = \left(-\frac{3}{2}, 2\right)$$

To find the radius, find the distance between

$\left(-\frac{3}{2}, 2\right)$ and $(0,4)$

$$d = \sqrt{\left(0-\left(-\frac{3}{2}\right)\right)^2 + (4-2)^2}$$

$$= \sqrt{\frac{9}{4}+4} = \sqrt{\frac{25}{4}} = \frac{5}{2} = 2.5$$

Radius: 2.5

Equation: $\left(x+\frac{3}{2}\right)^2 + (y-2)^2 = 6.25$

9. a. L_1: $(-2, 0)$ and $(0, 6)$

$$m = \frac{6-0}{0-(-2)} = \frac{6}{2} = 3$$

L_2: $(1, 8)$ and $(0, 5)$

$$m = \frac{5-8}{0-1} = \frac{-3}{-1} = 3$$

Parallel

b. L_1: $(1, 10)$ and $(-1, 7)$

$$m = \frac{7-10}{-1-1} = \frac{-3}{-2} = \frac{3}{2}$$

L_2: $(-2, -1)$ and $(1, -3)$

$$m = \frac{-3-(-1)}{1-(-2)} = \frac{-2}{3}$$

Perpendicular

11. a. $2x+3y=6$

x–intercept: $(3, 0)$ y–intercept: $(0, 2)$

$2x+3(0)=6$ $\qquad$ $2(0)+3y=6$

$\qquad 2x=6$ $\qquad\qquad 3y=6$

$\qquad\quad x=3$ $\qquad\qquad\quad y=2$

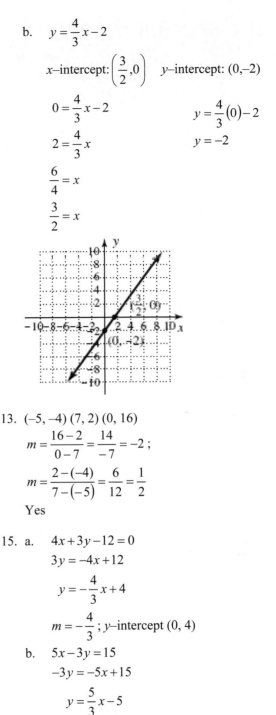

b. $y = \frac{4}{3}x - 2$

x–intercept: $\left(\frac{3}{2}, 0\right)$ $\qquad$ y–intercept: $(0, -2)$

$0 = \frac{4}{3}x - 2$ $\qquad\qquad$ $y = \frac{4}{3}(0) - 2$

$2 = \frac{4}{3}x$ $\qquad\qquad\qquad$ $y = -2$

$\frac{6}{4} = x$

$\frac{3}{2} = x$

13. $(-5, -4)$ $(7, 2)$ $(0, 16)$

$$m = \frac{16-2}{0-7} = \frac{14}{-7} = -2 \; ;$$

$$m = \frac{2-(-4)}{7-(-5)} = \frac{6}{12} = \frac{1}{2}$$

Yes

15. a. $4x+3y-12=0$

$3y = -4x+12$

$$y = -\frac{4}{3}x + 4$$

$m = -\frac{4}{3}$; y–intercept $(0, 4)$

b. $5x-3y=15$

$-3y = -5x+15$

$$y = \frac{5}{3}x - 5$$

$m = \frac{5}{3}$; y–intercept $(0, -5)$

Summary and Review

17. a. $m = \dfrac{2}{3}$; (1, 4)

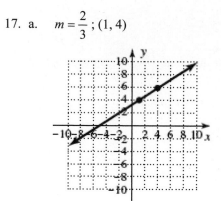

b. $m = -\dfrac{1}{2}$; (–2, 3)

19. (1, 2) and (–3, 5)

$$m = \frac{5-2}{-3-1} = -\frac{3}{4}$$

$$y - 2 = -\frac{3}{4}(x-1)$$

$$y - 2 = -\frac{3}{4}x + \frac{3}{4}$$

$$y = -\frac{3}{4}x + \frac{11}{4}$$

21. $m = \dfrac{2}{5}$; y–intercept (0, 2)

$$y = \frac{2}{5}x + 2$$

When the rabbit population increases by 500, the wolf population increases by 200.

23. a. $f(x) = \sqrt{4x+5}$

$$4x + 5 \geq 0$$

$$4x \geq -5$$

$$x \geq -\frac{5}{4}$$

$$x \in \left[-\frac{5}{4}, \infty \right)$$

b. $g(x) = \dfrac{x-4}{x^2 - x - 6}$

$$x^2 - x - 6 = 0$$

$$(x-3)(x+2) = 0$$

$$x - 3 = 0 \text{ or } x + 2 = 0$$

$$x = 3 \quad \text{or } x = -2$$

These values must be excluded because they cause division by zero.

$$x \in (-\infty, -2) \cup (-2, 3) \cup (3, \infty)$$

25. It is a function.

27. $D : x \in (-\infty, \infty)$

$R : y \in [-5, \infty)$

$f(x) \uparrow : x \in (2, \infty)$

$f(x) \downarrow : x \in (-\infty, 2)$

$f(x) > 0 : x \in (-\infty, -1) \cup (5, \infty)$

$f(x) < 0 : x \in (-1, 5)$

29. $D : x \in (-\infty, \infty)$

$R : y \in (-\infty, \infty)$

$f(x) \uparrow : x \in (-\infty, -3) \cup (1, \infty)$

$f(x) \downarrow : x \in (-3, 1)$

$f(x) > 0 : x \in (-5, -1) \cup (4, \infty)$

$f(x) < 0 : x \in (-\infty, -5) \cup (-1, 4)$

31. a. $\dfrac{f(x_2)-f(x_1)}{x_2-x_1} = \dfrac{\sqrt{5+4}-\sqrt{-3+4}}{5-(-3)}$

   $= \dfrac{3-1}{8} = \dfrac{1}{4}$

   Graph is rising to the right.

   b. $\dfrac{j(x+h)-j(x)}{h}$

   $= \dfrac{(x+h)^2-(x+h)-(x^2-x)}{h}$

   $= \dfrac{x^2+2xh+h^2-x-h-x^2+x}{h}$

   $= \dfrac{2xh+h^2-h}{h} = 2x-1+h$

   $x = 2, h = 0.01$

   $2(2)-1+0.01 = 3.01$

33. Squaring function
   a. up on left/up on the right
   b. x–intercepts: $(-4,0), (0,0)$
      y–intercept: $(0,0)$
   c. Vertex: $(-2,-4)$
   d. $x \in (-\infty, \infty), y \in [-4, \infty)$

35. Cubing function
   a. down on left/up on right
   b. x–intercepts: $(-2,0), (-1,0), (4,0)$
      y–intercept: $(0,2)$
   c. Inflection point: $(1,0)$
   d. $x \in (-\infty, \infty), y \in (-\infty, \infty)$

37. Cube root
   a. up on left/ down on right
   b. x–intercept: $(1,0)$
      y–intercept: $(0,1)$
   c. Inflection: $(1,0)$
   d. $x \in (-\infty, \infty), y \in (-\infty, \infty)$

39. $f(x) = 2|x+3|$; Absolute Value

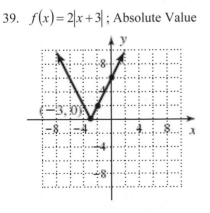

41. $f(x) = \sqrt{x-5}+2$; Square Root

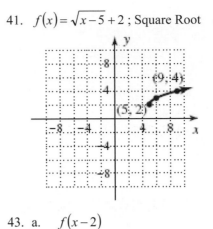

43. a. $f(x-2)$
      Right 2

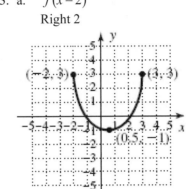

   b. $-f(x)+4$
      Reflect, up 4

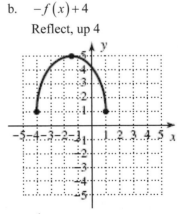

   c. $\dfrac{1}{2}f(x)$
      Compressed down

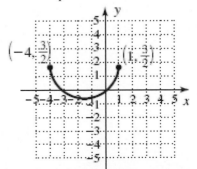

45. $h(x) = \begin{cases} \dfrac{x^2 - 2x - 15}{x+3} & x \neq -3 \\ -6 & x = -3 \end{cases}$

$x \in (-\infty, \infty)$

$y \in (-\infty, -8) \cup (-8, \infty)$

Discontinuity at $x = -3$

Define $h(x) = -8$ at $x = -3$

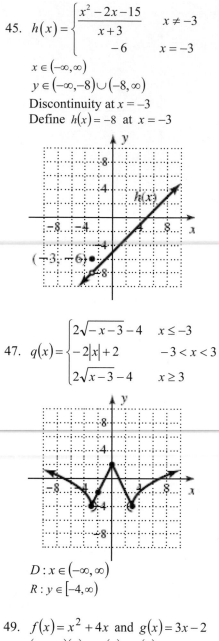

47. $q(x) = \begin{cases} 2\sqrt{-x-3} - 4 & x \leq -3 \\ -2|x| + 2 & -3 < x < 3 \\ 2\sqrt{x-3} - 4 & x \geq 3 \end{cases}$

$D : x \in (-\infty, \infty)$

$R : y \in [-4, \infty)$

49. $f(x) = x^2 + 4x$ and $g(x) = 3x - 2$

$(f + g)(a) = f(a) + g(a)$

$= a^2 + 4a + 3a - 2$

$= a^2 + 7a - 2$

51. $f(x) = x^2 + 4x$ and $g(x) = 3x - 2$

$\left(\dfrac{f}{g}\right)(x) = \dfrac{x^2 + 4x}{3x - 2}$

$D : x \in \left(-\infty, \dfrac{2}{3}\right) \cup \left(\dfrac{2}{3}, \infty\right)$

53. $p(x) = 4x - 3$; $q(x) = x^2 + 2x$;

$(q \circ p)(3) = q[p(3)]$

$p(3) = 4(3) - 3 = 12 - 3 = 9$

$q(9) = (9)^2 + 2(9) = 81 + 18 = 99$

55. $h(x) = \sqrt{3x - 2} + 1$;

$f(x) = \sqrt{x} + 1$;

$g(x) = 3x - 2$

57. $r(t) = 2t + 3$

$A(t) = \pi(2t + 3)^2$

Chapter 2 Mixed Review

1. $4x + 3y = 12$

   $3y = -4x + 12$

   $y = -\dfrac{4}{3}x + 4$

3. a. $f(x) = \dfrac{x+1}{x^2 - 5x + 4}$

   $x^2 - 5x + 4 = 0$

   $(x-4)(x-1) = 0$

   $x - 4 = 0$ or $x - 1 = 0$

   $x = 4$ or $x = 1$

   These values are restricted because they cause division by zero.

   Domain: $(-\infty, 1) \cup (1,4) \cup (4,\infty)$

   b. $g(x) = \dfrac{1}{\sqrt{2x - 3}}$

   Set the radicand greater than zero. (Zero must be excluded because the radical is in the denominator.

   $2x - 3 > 0$

   $2x > 3$

   $x > \dfrac{3}{2}$

   Domain: $\left(\dfrac{3}{2}, \infty \right)$

5. $m = -\dfrac{3}{2}$; y–intercept $(0,-2)$

   $y = -\dfrac{3}{2}x - 2$

7. $L_1 : (-3,7), (2,2)$

   Slope: $\dfrac{2-7}{2-(-3)} = -1$;

   $L_2 : (2,2), (5,5)$

   Slope: $\dfrac{5-2}{5-2} = 1$;

   Lines are perpendicular. Vertex of right angle is (2, 2).

   Radius: distance between (–3,7) and (2,2).

   $d = \sqrt{(2-(-3))^2 + (2-7)^2} = \sqrt{50}$;

   Center (2,2), radius $\sqrt{50}$

   Equation: $(x-2)^2 + (y-2)^2 = 50$

9. $y = \dfrac{3}{5}x - 2$

   y–intercept $(0,-2)$

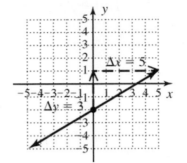

11. a. $p(x) = -2x^2 + 8x$

   Rate of change is positive in $[-2, -1]$ since p is increasing in $(-\infty, 2)$.

   The rate of change in [1, 2] will be less than the rate of change in [–2, –1].

   $\dfrac{\Delta p}{\Delta x} = \dfrac{p(2) - p(1)}{2-1} = \dfrac{8-6}{1} = 2$;

   $\dfrac{\Delta p}{\Delta x} = \dfrac{p(-1) - p(-2)}{-1-(-2)} = \dfrac{-10-(-24)}{1} = 14$

   b. $A(t) = 1000e^{0.07t}$

   For [10,10.01],

   $\dfrac{1000e^{0.07(10.01)} - 1000e^{0.07(10)}}{10.01 - 10} \approx 141.0$;

   For [15,15.01],

   $\dfrac{1000e^{0.07(15.01)} - 1000e^{0.07(15)}}{15.01 - 15} \approx 200.1$;

   For [20, 20.01],

   $\dfrac{1000e^{0.07(20.01)} - 1000e^{0.07(20)}}{20.01 - 20} \approx 284.0$;

   In the interval: [15, 15.01]

Chapter 2 Mixed Review

13. $f(x) = \dfrac{3}{x^2 - 1}, g(x) = 3x - 2$

$(f \circ g)(x) = \dfrac{3}{(3x-2)^2 - 1}$

$= \dfrac{3}{9x^2 - 12x + 4 - 1} = \dfrac{3}{9x^2 - 12x + 3}$

$= \dfrac{3}{3(3x^2 - 4x + 1)} = \dfrac{1}{3x^2 - 4x + 1}$

To find domain, $3x^2 - 4x + 1 = 0$

$(3x - 1)(x - 1) = 0$

$x = \dfrac{1}{3}, x = 1$

Domain: $\left(-\infty, \dfrac{1}{3}\right) \cup \left(\dfrac{1}{3}, 1\right) \cup (1, \infty)$

15. $f(x) = x^2 + 1, g(x) = 3x - 2$

$\dfrac{f(x+h) - f(x)}{h}$

$\dfrac{\left[(x+h)^2 + 1\right] - \left[x^2 + 1\right]}{h}$

$= \dfrac{x^2 + 2xh + h^2 + 1 - x^2 - 1}{h}$

$= \dfrac{2xh + h^2}{h} = \dfrac{h(2x + h)}{h} = 2x + h \;;$

$\dfrac{g(x+h) - g(x)}{h}$

$\dfrac{\left[3(x+h) - 2\right] - \left[3x - 2\right]}{h}$

$= \dfrac{3x + 3h - 2 - 3x + 2}{h} = \dfrac{3h}{h} = 3 \;;$

For small h, $2x + h = 3$

when $x \approx \dfrac{3}{2}$

17. a. $D : x \in (-\infty, 6]$
       $R : y \in (-\infty, 3]$

    b. Min: $(3, -3)$
       Max: $y = 3$ for $x \in (-6, -3)$; $(6, 0)$

    c. $g(x)\uparrow : x \in (-\infty, -6) \cup (3, 6)$
       $g(x)\downarrow : x \in (-3, 3)$
       $g(x)$ constant : $x \in (-6, -3)$

    d. $g(x) > 0 : x \in (-7, -1)$
       $g(x) < 0 : x \in (-\infty, -7) \cup (-1, 6)$

19. x–intercepts: $(-1, 0)$, $(1.5, 0)$
    y–intercept: $(0, 3)$

    $f(x) = a(x+1)(x-1.5)$

    $f(x) = a\left(x^2 - \dfrac{1}{2}x - \dfrac{3}{2}\right)$

    $3 = a\left(0^2 - \dfrac{1}{2}(0) - \dfrac{3}{2}\right)$

    $3 = -\dfrac{3}{2}a$

    $-2 = a;$

    $f(x) = -2x^2 + x + 3$

Chapter 2 Practice Test

1. a. $x = y^2 + 2y$

   b. $y = \sqrt{5 - 2x}$

   c. $|y| + 1 = x$

   d. $y = x^2 + 2x$

   a and c are non–functions, do not pass the vertical line test.

3. $x + 4y = 8$

   $4y = -x + 8$

   $y = -\dfrac{1}{4}x + 2$

5. $6x + 5y = 3$

   $6x + 5y = 3$

   $5y = -6x + 3$

   $y = -\dfrac{6}{5}x + \dfrac{3}{5}$

   Slope: $-\dfrac{6}{5}$

   Point $(2,-2)$, slope $-\dfrac{6}{5}$

   $y - (-2) = -\dfrac{6}{5}(x - 2)$

   $y + 2 = -\dfrac{6}{5}x + \dfrac{12}{5}$

   $y = -\dfrac{6}{5}x + \dfrac{2}{5}$

7. $L_1: x = -3$

   $L_2: y = 4$

9. a. $W(24) = 300$

   b. $h = 30$ when $W(h) = 375$

   c. $(20, 250)$ and $(40, 500)$

   $$m = \frac{500 - 250}{40 - 20} = \frac{250}{20} = \frac{25}{2}$$

   $$W(h) = \frac{25}{2}h$$

   d. Wages are \$12.50 per hour.

   e. $h \in [0, 40]$

   $w \in [0, 500]$

11. $f(x) = \dfrac{2 - x^2}{x^2}$

   a. $f\left(\dfrac{2}{3}\right) = \dfrac{2 - \left(\dfrac{2}{3}\right)^2}{\left(\dfrac{2}{3}\right)^2} = \dfrac{2 - \left(\dfrac{4}{9}\right)}{\dfrac{4}{9}}$

   $= \dfrac{\dfrac{14}{9}}{\dfrac{4}{9}} = \dfrac{14}{9} \div \dfrac{4}{9} = \dfrac{7}{2}$

   b. $f(a + 3) = \dfrac{2 - (a + 3)^2}{(a + 3)^2}$

   $= \dfrac{2 - (a^2 + 6a + 9)}{a^2 + 6a + 9} = \dfrac{2 - a^2 - 6a - 9}{a^2 + 6a + 9}$

   $= \dfrac{-a^2 - 6a - 7}{a^2 + 6a + 9}$

   c. $f(1 + 2i) = \dfrac{2 - (1 + 2i)^2}{(1 + 2i)^2}$

   $= \dfrac{2 - (1 + 4i + 4i^2)}{1 + 4i + 4i^2} = \dfrac{2 - 1 - 4i - 4i^2}{1 + 4i - 4}$

   $= \dfrac{1 - 4i + 4}{-3 + 4i} = \dfrac{5 - 4i}{-3 + 4i}$

   $= \dfrac{5 - 4i}{-3 + 4i} \cdot \dfrac{-3 - 4i}{-3 - 4i} = \dfrac{-15 - 20i + 12i + 16i^2}{9 - 16i^2}$

   $= \dfrac{-15 - 8i - 16}{9 + 16} = \dfrac{-31 - 8i}{25}$

   $= -\dfrac{31}{25} - \dfrac{8}{25}i$

13. $S(t) = 2t^2 - 3t$

   a. No, new company and sales should be growing.

   b. For $[5,6]$, $S(5) = 2(5)^2 - 3(5) = 35$

$S(6) = 2(6)^2 - 3(6) = 54$

Rate of Change: $\dfrac{54-35}{6-5} = 19$

For $[6,7]$, $S(7) = 2(7)^2 - 3(7) = 77$

Rate of Change: $\dfrac{77-54}{7-6} = 23$

   c. $\dfrac{2(t+h)^2 - 3(t+h) - (2t^2 - 3t)}{h}$

$= \dfrac{2(t^2 + 2th + h^2) - 3t - 3h - 2t^2 + 3t}{h}$

$= \dfrac{2t^2 + 4th + 2h^2 - 3h - 2t^2}{h}$

$= \dfrac{4th + 2h^2 - 3h}{h} = 4t - 3 + 2h$

For small h:

$4(10) - 3 = 37, 4(18) - 3 = 69,$

$4(24) - 3 = 93$

For small h, sales volume is approximately

$\dfrac{37{,}000 \text{ units}}{1 \text{ mo}}$ in month 10,

$\dfrac{69{,}000 \text{ units}}{1 \text{ mo}}$ in month 18,

$\dfrac{93{,}000 \text{ units}}{1 \text{ mo}}$ in month 24

15. $g(x) = -(x+3)^2 - 2$

Left 3, reflected across x–axis, down 2

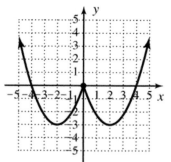

17. a. $D: x \in [-4, \infty)$

      $R: y \in [-3, \infty)$

   b. $f(-1) \approx 2.2$

   c. $f(x) < 0 : x \in (-4, -3)$

      $f(x) > 0 : x \in (-3, \infty)$

   d. $f(x) \uparrow : (-4, \infty)$

      $f(x) \downarrow :$ none

   e. Parent graph: $y = \sqrt{x}$

      Graph shifts left 4, down 3

      $y = a\sqrt{x+4} - 3$

      $3 = a\sqrt{0+4} - 3$

      $3 = a\sqrt{4} - 3$

      $3 = 2a - 3$

      $6 = 2a$

      $3 = a$

      $y = 3\sqrt{x+4} - 3$

19.

Ch 2 Calculator Exploration

Exercise 1: $y = -5(x+4)^2 + 6; (-4, 6)$

Ch. 2 Strengthening Core Skills

Exercise 1: $f(x) = x^2 - 8x - 12$

$\dfrac{b}{2a} = \dfrac{-8}{2(1)} = -4$

$g(x) = x + 4$

$h(x) = f[g(x)] = f(x+4)$

$= (x+4)^2 - 8(x+4) - 12$

$= x^2 + 8x + 16 - 8x - 32 - 12$

$= x^2 - 28$

$x^2 - 28 = 0$

$x^2 = 28$

$x = \pm 2\sqrt{7}$;

$4 \pm 2\sqrt{7}$

Exercise 3: $f(x) = 2x^2 - 10x + 11$

$\dfrac{b}{2a} = \dfrac{-10}{2(2)} = -\dfrac{5}{2}$

$g(x) = x + \dfrac{5}{2}$

$h(x) = f[g(x)] = f\left(x + \dfrac{5}{2}\right)$

$= 2\left(x + \dfrac{5}{2}\right)^2 - 10\left(x + \dfrac{5}{2}\right) + 11$

$= 2\left(x^2 + 5x + \dfrac{25}{4}\right) - 10x - \dfrac{50}{2} + 11$

$= 2x^2 + 10x + \dfrac{50}{4} - 10x - \dfrac{50}{2} + 11$

$= 2x^2 - \dfrac{3}{2}$

$2x^2 - \dfrac{3}{2} = 0$

$2x^2 = \dfrac{3}{2}$

$x^2 = \dfrac{3}{4}$

$x = \pm \dfrac{\sqrt{3}}{2}$

$\dfrac{5}{2} \pm \dfrac{\sqrt{3}}{2}$

Cumulative Review Chapters 1–2

1. $\left(x^3 - 5x^2 + 2x - 10\right) \div (x - 5)$

$= \dfrac{x^2(x-5) + 2(x-5)}{x-5}$

$= \dfrac{(x-5)(x^2+2)}{x-5}$

$= x^2 + 2$

3. $A = \pi r^2$

$69 = \pi r^2$

$\dfrac{69}{\pi} = r^2$

$21.96 \approx r^2$

$4.686 \approx r;$

$C = 2\pi r$

$C = 2\pi(4.686)$

$C \approx 29.45$ cm

5. $-2(3 - x) + 5x = 4(x+1) - 7$

$-6 + 2x + 5x = 4x + 4 - 7$

$7x - 6 = 4x - 3$

$3x = 3$

$x = 1$

7. a. $(-4, 7)$ and $(2, 5)$

$m = \dfrac{7-5}{-4-2} = \dfrac{2}{-6} = -\dfrac{1}{3}$

b. $3x - 5y = 20$

$-5y = -3x + 20$

$y = \dfrac{3}{5}x - 4$

$m = \dfrac{3}{5}$

9. $(-3, 2)$; $m = \dfrac{1}{2}$

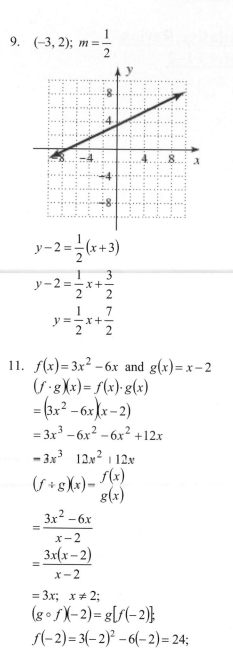

$$y - 2 = \frac{1}{2}(x + 3)$$

$$y - 2 = \frac{1}{2}x + \frac{3}{2}$$

$$y = \frac{1}{2}x + \frac{7}{2}$$

11. $f(x) = 3x^2 - 6x$ and $g(x) = x - 2$

$$(f \cdot g)(x) = f(x) \cdot g(x)$$

$$= \left(3x^2 - 6x\right)(x - 2)$$

$$= 3x^3 - 6x^2 - 6x^2 + 12x$$

$$= 3x^3 - 12x^2 + 12x$$

$$(f \div g)(x) = \frac{f(x)}{g(x)}$$

$$= \frac{3x^2 - 6x}{x - 2}$$

$$= \frac{3x(x - 2)}{x - 2}$$

$$= 3x; \quad x \neq 2;$$

$(g \circ f)(-2) = g[f(-2)];$

$f(-2) = 3(-2)^2 - 6(-2) = 24;$

$g(24) = 24 - 2 = 22$

13. $f(x) = \begin{cases} x^2 - 4 & x < 2 \\ x - 1 & 2 \leq x \leq 8 \end{cases}$

a. $D: x \in (-\infty, 8]$

   $R: y \in [-4, \infty)$

b. $f(-3) = (-3)^2 - 4 = 9 - 4 = 5$;

   $f(-1) = (-1)^2 - 4 = 1 - 4 = -3$;

   $f(1) = (1)^2 - 4 = 1 - 4 = -3$;

   $f(2) = 2 - 1 = 1$;

   $f(3) = 3 - 1 = 2$

c. $(-2, 0)$

d. $f(x) < 0: x \in (-2, 2)$

   $f(x) > 0: x \in (-\infty, -2) \cup [2, 8]$

e. Max: $(8, 7)$

   Min: $(0, -4)$

f. $f(x) \uparrow: x \in (0, 8)$

   $f(x) \downarrow: x \in (-\infty, 0)$

15. a. $\dfrac{-2}{x^2 - 3x - 10} + \dfrac{1}{x + 2}$

$$= \frac{-2}{(x - 5)(x + 2)} + \frac{1}{x + 2}$$

$$= \frac{-2}{(x - 5)(x + 2)} + \frac{1(x - 5)}{(x - 5)(x + 2)}$$

$$= \frac{-2 + x - 5}{(x - 5)(x + 2)}$$

$$= \frac{x - 7}{(x - 5)(x + 2)}$$

b. $\dfrac{b^2}{4a^2} - \dfrac{c}{a} = \dfrac{b^2}{4a^2} - \dfrac{4ac}{4a^2} = \dfrac{b^2 - 4ac}{4a^2}$

17. a. $N \subset Z \subset W \subset Q \subset R$

    False

    b. $W \subset N \subset Z \subset Q \subset R$

    False

    c. $N \subset W \subset Z \subset Q \subset R$

    True

    d. $N \subset R \subset Z \subset Q \subset W$

    False

19. $2x^2 + 49 = -20x$

    $2x^2 + 20x + 49 = 0$

    $2x^2 + 20x = -49$

    $x^2 + 10x = \dfrac{-49}{2}$

    $x^2 + 10x + 25 = -\dfrac{49}{2} + 25$

    $(x+5)^2 = \dfrac{1}{2}$

    $x + 5 = \pm\sqrt{\dfrac{1}{2}}$

    $x + 5 = \pm\dfrac{\sqrt{2}}{2}$

    $x = -5 \pm \dfrac{\sqrt{2}}{2};$

    $x \approx -4.293$

    $x \approx -5.707$

21. Let w represent the width.
    Let l represent the length.

    $A = lw$

    $1457 = (w+16)w$

    $0 = w^2 + 16w - 1457$

    $0 = (w-31)(w+47)$

    $w = 31$ cm; $l = 47$ cm

23. a. $6x^2 - 7x = 20$

    $6x^2 - 7x - 20 = 0$

    $(3x+4)(2x-5) = 0$

    $x = -\dfrac{4}{3}; \quad x = \dfrac{5}{2}$

    b. $x^3 + 5x^2 - 15 = 3x$

    $x^3 + 5x^2 - 3x - 15 = 0$

    $(x^3 + 5x^2) - (3x + 15) = 0$

    $x^2(x+5) - 3(x+5) = 0$

    $(x+5)(x^2 - 3) = 0$

    $x = -5; \quad x = \sqrt{3}; \quad x = -\sqrt{3}$

25. $(-4, 5), (4, -1), (0, 8)$

    $d = \sqrt{(-4-4)^2 + (5+1)^2}$

    $d = \sqrt{(-8)^2 + (6)^2}$

    $d = \sqrt{100}$

    $d = 10;$

    $d = \sqrt{(4-0)^2 + (-1-8)^2}$

    $d = \sqrt{(4)^2 + (-9)^2}$

    $d = \sqrt{97}$

    $d \approx 9.85;$

    $d = \sqrt{(-4-0)^2 + (5-8)^2}$

    $d = \sqrt{(-4)^2 + (-3)^2}$

    $d = \sqrt{25}$

    $d = 5;$

    $P = 10 + \sqrt{97} + 5 = 15 + \sqrt{97}$

    $\approx 15 + 9.85 \approx 24.85$ units
    No it is not a right triangle.

    $5^2 + (\sqrt{97})^2 \neq 10^2$

MWT 1 Exercises

MWT I

1. Positive

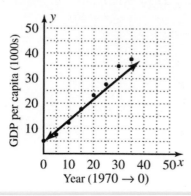

3. a. Linear
   b. Negative

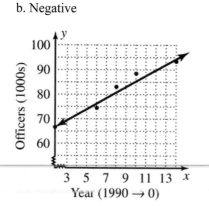

5. a.

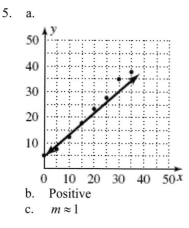

   b. Positive
   c. $m \approx 1$

7. a.

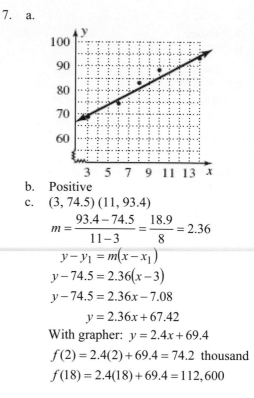

   b. Positive
   c. $(3, 74.5)$ $(11, 93.4)$

   $$m = \frac{93.4 - 74.5}{11 - 3} = \frac{18.9}{8} = 2.36$$
   $$y - y_1 = m(x - x_1)$$
   $$y - 74.5 = 2.36(x - 3)$$
   $$y - 74.5 = 2.36x - 7.08$$
   $$y = 2.36x + 67.42$$

   With grapher: $y = 2.4x + 69.4$

   $$f(2) = 2.4(2) + 69.4 = 74.2 \text{ thousand}$$
   $$f(18) = 2.4(18) + 69.4 = 112,600$$

9. a.

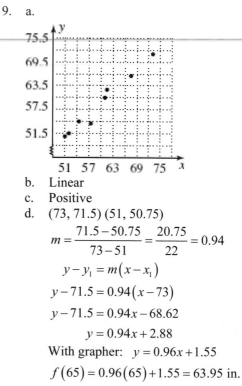

   b. Linear
   c. Positive
   d. $(73, 71.5)$ $(51, 50.75)$

   $$m = \frac{71.5 - 50.75}{73 - 51} = \frac{20.75}{22} = 0.94$$
   $$y - y_1 = m(x - x_1)$$
   $$y - 71.5 = 0.94(x - 73)$$
   $$y - 71.5 = 0.94x - 68.62$$
   $$y = 0.94x + 2.88$$

   With grapher: $y = 0.96x + 1.55$

   $$f(65) = 0.96(65) + 1.55 = 63.95 \text{ in.}$$

144

Modeling with Technology 1: Linear and Quadratic Equation Models

11. a.

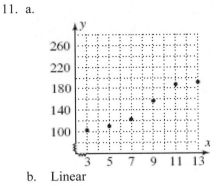

    b. Linear
    c. Positive
    d. With grapher: $y = 9.55x + 70.42$

$$f(21) = 9.55(21) + 70.42 = 271,000$$

    The number of patents are up because the slope is larger.

13. a.

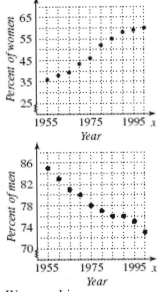

    b. Women: Linear
       Men: Linear
    c. Women: Positive
       Men: Negative
    d. The percentage of females in the work force is increasing faster than the percentage of males because the absolute value of the slope is greater.

15. a. Linear

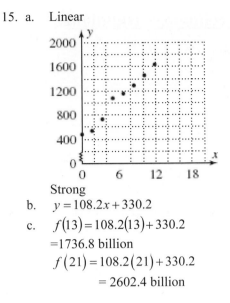

    Strong
    b. $y = 108.2x + 330.2$
    c. $f(13) = 108.2(13) + 330.2$
       $=1736.8$ billion

$$f(21) = 108.2(21) + 330.2$$
$$= 2602.4 \text{ billion}$$

17. a. Quadratic

$$h(t) = -14.5t^2 + 90t$$

    b. $v = 90$ ft/sec
    c. $-14.5 = -\dfrac{1}{2}g$

$$29 = g \text{ ; Venus}$$

3.1 Technology Highlight

Exercise 1: 1.35, 6.65

Exercise 3: $-2.87, 0.87$

3.1 Exercises

1. $\dfrac{25}{2}$

3. $0, f(x)$

5. Answers will vary.

7. $f(x) = x^2 + 4x - 5$

$f(x) = \left(x^2 + 4x + 4\right) - 5 - 4$

$f(x) = (x+2)^2 - 9;$

$x = \dfrac{-4 \pm \sqrt{(4)^2 - 4(1)(-5)}}{2(1)}$

$x = \dfrac{-4 \pm \sqrt{36}}{2}$

$x = \dfrac{-4 \pm 6}{2}$

$x = 1; \quad x = -5$

Left 2, down 9
x-intercepts: $(1, 0), (-5, 0)$
y-intercept: $(0, -5)$
Vertex: $(-2, -9)$

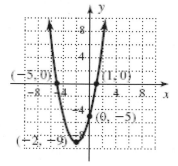

9. $h(x) = -x^2 + 2x + 3$

$h(x) = -\left(x^2 - 2x + 1\right) + 3 + 1$

$h(x) = -(x-1)^2 + 4;$

$x = \dfrac{-2 \pm \sqrt{(2)^2 - 4(-1)(3)}}{2(-1)}$

$x = \dfrac{-2 \pm \sqrt{16}}{-2}$

$x = \dfrac{-2 \pm 4}{-2}$

$x = -1; \quad x = 3$

Reflected in x-axis, right 1, up 4
x-intercepts: $(-1, 0), (3, 0)$
y-intercept: $(0, 3)$
Vertex: $(1, 4)$

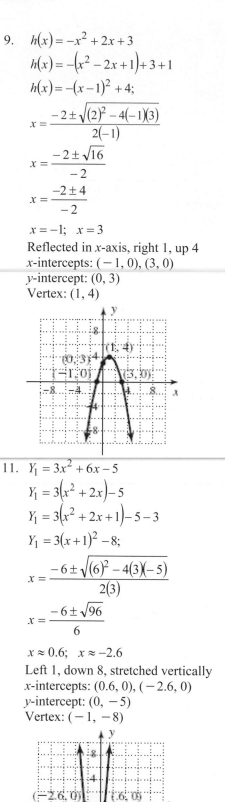

11. $Y_1 = 3x^2 + 6x - 5$

$Y_1 = 3\left(x^2 + 2x\right) - 5$

$Y_1 = 3\left(x^2 + 2x + 1\right) - 5 - 3$

$Y_1 = 3(x+1)^2 - 8;$

$x = \dfrac{-6 \pm \sqrt{(6)^2 - 4(3)(-5)}}{2(3)}$

$x = \dfrac{-6 \pm \sqrt{96}}{6}$

$x \approx 0.6; \quad x \approx -2.6$

Left 1, down 8, stretched vertically
x-intercepts: $(0.6, 0), (-2.6, 0)$
y-intercept: $(0, -5)$
Vertex: $(-1, -8)$

13. $f(x) = -2x^2 + 8x + 7$

$f(x) = -2(x^2 - 4x) + 7$

$f(x) = -2(x^2 - 4x + 4) + 7 + 8$

$f(x) = -2(x - 2)^2 + 15;$

$x = \dfrac{-8 \pm \sqrt{(8)^2 - 4(-2)(7)}}{2(-2)}$

$x = \dfrac{-8 \pm \sqrt{120}}{-4}$

$x \approx -0.7 \quad x \approx 4.7$

Reflected in x-axis, right 2, up 15, stretched vertically
x-intercepts: $(-0.7, 0)$, $(4.7, 0)$
y-intercept: $(0, 7)$
Vertex: $(2, 15)$

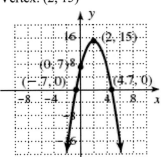

15. $p(x) = 2x^2 - 7x + 3$

$p(x) = 2\left(x^2 - \dfrac{7}{2}x\right) + 3$

$p(x) = 2\left(x^2 - \dfrac{7}{2}x + \dfrac{49}{16}\right) + 3 - \dfrac{49}{8}$

$p(x) = 2\left(x - \dfrac{7}{4}\right)^2 - \dfrac{25}{8};$

$x = \dfrac{7 \pm \sqrt{(-7)^2 - 4(2)(3)}}{2(2)}$

$x = \dfrac{7 \pm \sqrt{25}}{4}$

$x = \dfrac{7 \pm 5}{4}$

$x = 3; \quad x = \dfrac{1}{2}$

Right $\dfrac{7}{4}$, down $\dfrac{25}{8}$, stretched vertically

x-intercepts: $(3, 0)$, $\left(\dfrac{1}{2}, 0\right)$

y-intercept: $(0, 3)$

Vertex: $\left(\dfrac{7}{4}, -\dfrac{25}{8}\right)$

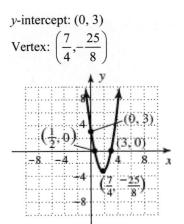

17. $f(x) = -3x^2 - 7x + 6$

$f(x) = -3\left(x^2 + \dfrac{7}{3}x\right) + 6$

$f(x) = -3\left(x^2 + \dfrac{7}{3}x + \dfrac{49}{36}\right) + 6 + \dfrac{49}{12}$

$f(x) = -3\left(x + \dfrac{7}{6}\right)^2 + \dfrac{121}{12};$

$x = \dfrac{7 \pm \sqrt{(-7)^2 - 4(-3)(6)}}{2(-3)}$

$x = \dfrac{7 \pm \sqrt{121}}{-6}$

$x = \dfrac{7 \pm 11}{-6}$

$x = -3; \quad x = \dfrac{2}{3}$

Reflected in x-axis, left $\dfrac{7}{6}$, up $\dfrac{121}{12}$, stretched vertically

x-intercepts: $(-3, 0)$, $\left(\dfrac{2}{3}, 0\right)$

y-intercept: $(0, 6)$

Vertex: $\left(-\dfrac{7}{6}, \dfrac{121}{12}\right)$

19. $p(x) = x^2 - 5x + 2$

$$p(x) = \left(x^2 - 5x + \frac{25}{4}\right) + 2 - \frac{25}{4}$$

$$p(x) = \left(x - \frac{5}{2}\right)^2 - \frac{17}{4};$$

$$x = \frac{5 \pm \sqrt{(-5)^2 - 4(1)(2)}}{2(1)}$$

$$x = \frac{5 \pm \sqrt{17}}{2}$$

$x \approx 4.56; \quad x \approx 0.44$

Right $\dfrac{5}{2}$, down $\dfrac{17}{4}$

x-intercepts: (4.6, 0), (0.4, 0)

y-intercept: (0, 2)

Vertex: $\left(\dfrac{5}{2}, -\dfrac{17}{4}\right)$

21. $f(x) = x^2 + 2x - 6$

$$f(x) = \left(x^2 + 2x\right) - 6$$

$$f(x) = \left(x^2 + 2x + 1\right) - 6 - 1$$

$$f(x) = (x + 1)^2 - 7;$$

$$(x + 1)^2 - 7 = 0$$

$$(x + 1)^2 = 7$$

$$x + 1 = \pm\sqrt{7}$$

$$x = -1 \pm \sqrt{7}$$

$x \approx 1.6; \quad x \approx -3.6$

Left 1, down 7

x-intercepts: (1.6, 0), (-3.6, 0)

y-intercept: (0, -6)

Vertex: $(-1, -7)$

23. $h(x) = -x^2 + 4x + 2$

$$h(x) = -\left(x^2 - 4x\right) + 2$$

$$h(x) = -\left(x^2 - 4x + 4\right) + 2 + 4$$

$$h(x) = -(x - 2)^2 + 6;$$

$$-(x - 2)^2 + 6 = 0$$

$$-(x - 2)^2 = -6$$

$$(x - 2)^2 = 6$$

$$x - 2 = \pm\sqrt{6}$$

$$x = 2 \pm \sqrt{6}$$

$x \approx 4.4; \quad x \approx -0.4$

Reflected across x-axis, right 2, up 6,

x-intercepts: (4.4, 0), (-0.4, 0)

y-intercept: (0, 2)

Vertex: (2, 6)

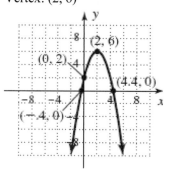

25. $Y_1 = 0.5x^2 + 3x + 7$

$Y_1 = 0.5(x^2 + 6x) + 7$

$Y_1 = 0.5(x^2 + 6x + 9) + 7 - 4.5$

$Y_1 = 0.5(x + 3)^2 + 2.5;$

$0.5(x + 3)^2 + 2.5 = 0$

$0.5(x + 3)^2 = -2.5$

$(x + 3)^2 = -5$

Left 3, up 2.5, compressed vertically
No x-intercepts
y-intercept: $(0, 7)$

Vertex: $\left(-3, \dfrac{5}{2}\right)$

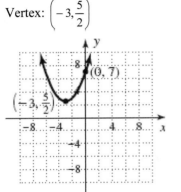

27. $Y_1 = -2x^2 + 10x - 7$

$Y_1 = -2(x^2 - 5x) - 7$

$Y_1 = -2\left(x^2 - 5x + \dfrac{25}{4}\right) - 7 + \dfrac{50}{4}$

$Y_1 = -2\left(x - \dfrac{5}{2}\right)^2 + \dfrac{11}{2};$

$-2\left(x - \dfrac{5}{2}\right)^2 + \dfrac{11}{2} = 0$

$-2\left(x - \dfrac{5}{2}\right)^2 = -\dfrac{11}{2}$

$\left(x - \dfrac{5}{2}\right)^2 = \dfrac{11}{4}$

$x - \dfrac{5}{2} = \pm\dfrac{\sqrt{11}}{2}$

$x = \dfrac{5}{2} \pm \dfrac{\sqrt{11}}{2}$

$x \approx 4.2; \quad x \approx 0.8$

Reflected across x-axis, right $\dfrac{5}{2}$, up $\dfrac{11}{2}$,
stretched vertically
x-intercepts: $(4.2, 0)$, $(0.8, 0)$
y-intercept: $(0, -7)$

Vertex: $\left(\dfrac{5}{2}, \dfrac{11}{2}\right)$

29. $f(x) = 4x^2 - 12x + 3$

$f(x) = 4(x^2 - 3x) + 3$

$f(x) = 4\left(x^2 - 3x + \dfrac{9}{4}\right) + 3 - 9$

$f(x) = 4\left(x - \dfrac{3}{2}\right)^2 - 6;$

$4\left(x - \dfrac{3}{2}\right)^2 - 6 = 0$

$4\left(x - \dfrac{3}{2}\right)^2 = 6$

$\left(x - \dfrac{3}{2}\right)^2 = \dfrac{3}{2}$

$x - \dfrac{3}{2} = \pm\dfrac{\sqrt{3}}{\sqrt{2}}$

$x = \dfrac{3}{2} \pm \dfrac{\sqrt{6}}{2}$

$x \approx 2.7; \quad x \approx 0.3$

Right $\dfrac{3}{2}$, down 6, stretched vertically

x-intercepts: (2.7, 0), (0.3, 0)

y-intercept: (0, 3)

Vertex: $\left(\dfrac{3}{2}, -6\right)$

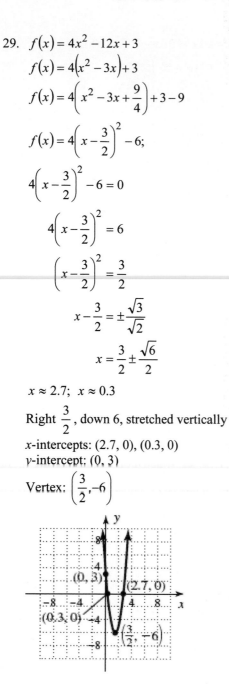

31. $p(x) = \dfrac{1}{2}x^2 + 3x - 5$

$p(x) = \dfrac{1}{2}(x^2 + 6x) - 5$

$p(x) = \dfrac{1}{2}(x^2 + 6x + 9) - 5 - \dfrac{9}{2}$

$p(x) = \dfrac{1}{2}(x + 3)^2 - \dfrac{19}{2};$

$\dfrac{1}{2}(x + 3)^2 - \dfrac{19}{2} = 0$

$\dfrac{1}{2}(x + 3)^2 = \dfrac{19}{2}$

$(x + 3)^2 = 19$

$x + 3 = \pm\sqrt{19}$

$x = -3 \pm \sqrt{19}$

$x \approx 1.4 \quad x \approx -7.4$

Left 3, down $\dfrac{19}{2}$, compressed vertically

x-intercepts: (1.4, 0), (−7.4, 0)

y-intercept: (0, −5)

Vertex: $\left(-3, \dfrac{-19}{2}\right)$

33. Compare to the graph of $y = x^2$, this graph is shifted right 2 and down 1: $y = a(x - 2)^2 - 1$ where a is positive. Choose a point on the graph, using $(3, 0)$

$y = a(x - 2)^2 - 1$

$0 = a(3 - 2)^2 - 1$

$0 = a - 1$

$1 = a$

Equation: $y = 1(x - 2)^2 - 1$

Chapter 3: Polynomial and Rational Functions

35. Compare to the graph of $y = x^2$, this graph is reflected across the x-axis, shifted to the left 2 and up 4: $y = a(x+2)^2 - 4$.

Choose a point on the graph, using $(0,0)$

$0 = a(0+2)^2 + 4$

$0 = a(4) + 4$

$-4 = 4a$

$a = -1$

Equation: $y = -1(x+2)^2 + 4$

37. Compare to the graph of $y = x^2$, this graph is reflected across the x-axis, shifted left 2 and up 3:

$y = a(x+2)^2 + 3$.

Choose a point on the graph, using $(0,-3)$

$-3 = a(0+2)^2 + 3$

$-3 = 4a + 3$

$-6 = 4a$

$\dfrac{-3}{2} = a$

Equation: $y = -\dfrac{3}{2}(x+2)^2 + 3$

39. i. a. $y = (x+3)^2 - 5$

$(x+3)^2 - 5 = 0$

$(x+3)^2 = 5$

$x+3 = \pm\sqrt{5}$

$x = -3 \pm \sqrt{5}$

b. $x = -3 \pm \sqrt{-\dfrac{-5}{1}}$

$x = -3 \pm \sqrt{5}$

ii. a. $y = -(x-4)^2 + 3$

$-(x-4)^2 + 3 = 0$

$-(x-4)^2 = -3$

$(x-4)^2 = 3$

$x-4 = \pm\sqrt{3}$

$x = 4 \pm \sqrt{3}$

b. $x = 4 \pm \sqrt{-\dfrac{3}{-1}}$

$x = 4 \pm \sqrt{3}$

iii. a. $y = 2(x+4)^2 - 7$

$2(x+4)^2 - 7 = 0$

$2(x+4)^2 = 7$

$(x+4)^2 = \dfrac{7}{2}$

$x+4 = \pm\sqrt{\dfrac{7}{2}}$

$x = -4 \pm \dfrac{\sqrt{14}}{2}$

b. $x = -4 \pm \sqrt{-\dfrac{-7}{2}}$

$x = -4 \pm \sqrt{\dfrac{7}{2}}$

$x = -4 \pm \dfrac{\sqrt{14}}{2}$

iv. a. $y = -3(x-2)^2 + 6$

$-3(x-2)^2 + 6 = 0$

$-3(x-2)^2 = -6$

$(x-2)^2 = 2$

$x-2 = \pm\sqrt{2}$

$x = 2 \pm \sqrt{2}$

b. $x = 2 \pm \sqrt{-\dfrac{6}{-3}}$

$x = 2 \pm \sqrt{2}$

v. a. $s(t) = 0.2(t+0.7)^2 - 0.8$

$0.2(t+0.7)^2 - 0.8 = 0$

$0.2(t+0.7)^2 = 0.8$

$(t+0.7)^2 = 4$

$t+0.7 = \pm 2$

$t = -0.7 \pm 2$

$t = -2.7; \ t = 1.3$

b. $t = -0.7 \pm \sqrt{-\dfrac{-0.8}{0.2}}$

$t = -0.7 \pm \sqrt{4}$

$t = -0.7 \pm 2$

$t = -2.7; t = 1.3$

3.1 Exercises

vi. a. $r(t) = -0.5(t - 0.6)^2 + 2$

$$-0.5(t - 0.6)^2 + 2 = 0$$
$$-0.5(t - 0.6)^2 = -2$$
$$(t - 0.6)^2 = 4$$
$$t - 0.6 = \pm 2$$
$$t = 0.6 \pm 2$$
$$t = -1.4; \ t = 2.6$$

b. $t = 0.6 \pm \sqrt{-\dfrac{2}{-0.5}}$

$$t = 0.6 \pm \sqrt{4}$$
$$t = 0.6 \pm 2$$
$$t = -1.4; t = 2.6$$

41. $P(x) = -10x^2 + 3500x - 66000$

a. $(0, {}^-66000)$ When no cars are produced, there is a profit loss of \$66,000.

b. $P(x) = -10x^2 + 3500x - 66000$

$$-10x^2 + 3500x - 66000 = 0$$
$$-10\left(x^2 - 350 + 6600\right) = 0$$
$$10(x \quad 20)(x \quad 330) = 0$$

$x - 20; \ x - 330$
(20, 0) and (330, 0)
No profit will be made if less than 20 cars or more than 330 cars are produced.

c. $x = \dfrac{-b}{2a} = \dfrac{-3500}{-20} = 175$ cars

d. $P(175) = -10(175)^2 + 3500(175) - 66000$
$P(175) = \$240,250$

43. $d(x) = x^2 - 12x$

a. $x = \dfrac{-b}{2a} = \dfrac{12}{2} = 6$ miles

b. $d(6) = (6)^2 - 12(6) = -36$
3600 feet

c. $d(4) = (4)^2 - 12(4) = -32$
3200 feet

d. $x^2 - 12x = 0$
$x(x - 12) = 0$

$x = 0; \ x = 12$
12 miles

45. $P(x) = -0.5x^2 + 175x - 3300$

a. $(0, -3300)$ If no appliances are sold, the loss will be \$3300.

b. $-0.5x^2 + 175x - 3300 = 0$
$$-0.5\left(x^2 - 350x + 6600\right) = 0$$
$$-0.5(x - 20)(x - 330) = 0$$

$x = 20; \ x = 330$
(20, 0) and (330, 0)
If less than 20 or more than 330 appliances are made and sold, there will be no profit.

c. $0 \le x \le 200$ Greatest number of appliances to be produced each day is 200.

d. $x = \dfrac{-b}{2a} = \dfrac{175}{1} = 175$;

$P(175) = -0.5(175)^2 + 175(175) - 3300$
$= \$12,012.50$

47. $h(t) = -16t^2 + 176t$

a.

$$h(2) = -16(2)^2 + 176(2) = 288 \text{ feet}$$

b.

c. $x = \dfrac{-b}{2a} = \dfrac{-176}{-32} = 5.5 \text{ sec}$

$h(5.5) = -16(5.5)^2 + 176(5.5) = 484$ feet

d. $16t^2 - 176t = 0$
$16t(t - 11) = 0$

$t = 11$ sec

49. a. $h(t) = -16t^2 + 32t = 5$

b. $t = 0.5$

$h(0.5) = -16(0.5)^2 + 32(0.5) + 5$

$= 17$ ft

$t = 1.5$

$h(1.5) = -16(1.5)^2 + 32(1.5) + 5$

$= 17$ ft

c. The person is 17 ft height at 0.5 sec and is 17 feet height at 1.5 sec, so max height must occur between $t = 0.5$ and $t = 1.5$.

d. $h(t) = -16t^2 + 32t + 5$

$t = -\dfrac{32}{2(-16)}$

$= 1$ sec

e. $h(1) = -16(1)^2 + 32(1) + 5$

$= 21$ ft

f. $h(t) = -16t^2 + 32t + 5$

$t = \dfrac{-32 \pm \sqrt{32^2 - 4(-16)(5)}}{2(-16)}$

$= \dfrac{-32 \pm \sqrt{1344}}{-32}$

$= \dfrac{-32 \pm (36.66)}{-32}$

$= \dfrac{-32 - (36.66)}{-32}$

≈ 2.2 sec

51. $C(x) = 16x - 63$

$R(x) = -x^2 + 326x - 7463$

$P(x) = R(x) - C(x)$

$P(x) = -x^2 + 326x - 7463 - 16x + 63$

$P(x) = -x^2 + 310x - 7400;$

$x = \dfrac{-b}{2a} = \dfrac{-310}{-2} = 155$

155,000 bottles;

$P(155) = -(155)^2 + 310(155) - 7400 = 16625$

$16,625

53. Let x represent the width.

Let $\dfrac{384 - 4x}{2}$ represent the length

$x\left(\dfrac{384 - 4x}{2}\right) = 192x - 2x^2$

a. Opens downward, x-coordinate of

vertex: $x = \dfrac{-192}{2(-2)} = 48$ ft;

$\dfrac{384 - 4(48)}{2} = 96$ ft

b. $\dfrac{96}{3} = 32;$

32 ft x 48 ft

55. $x = 2 \pm 3i$

$f(x) = (x - (2 + 3i))(x - (2 - 3i))$

$f(x) = (x - 2 - 3i)(x - 2 + 3i)$

$f(x) = ((x - 2)^2 - (3i)^2)$

$f(x) = x^2 - 4x + 4 - 9i^2$

$f(x) = x^2 - 4x + 4 + 9$

$f(x) = x^2 - 4x + 13$

57. a. radicand will be negative – two complex zeroes

b. radicand will be positive – two real zeroes

c. radicand is zero – one real zero

d. two real, rational zeroes

e. two real, irrational zeroes

59. $\dfrac{x^2 - 4x + 4}{x^2 + 3x - 10} \cdot \dfrac{x^2 - 25}{x^2 - 10x + 25}$

$x = \dfrac{(x - 2)^2}{(x + 5)(x - 2)} \cdot \dfrac{(x - 5)(x + 5)}{(x - 5)^2}$

$= \dfrac{x - 2}{x - 5}$

61. $f(x) = 3x^2 + 7x - 6; \ f(x) \le 0$

$3x^2 + 7x - 6 = 0$

$(3x - 2)(x + 3) = 0$

$x = \dfrac{2}{3}; \ x = -3$

Concave up

$x \in \left[-3, \dfrac{2}{3}\right]$

3.2 Exercises

1. synthetic; zero

3. $P(c)$, remainder

5. If polynomial P(x) is divided by the linear factor x-c, the remainder is identical to P(c).

7. $\dfrac{x^3 - 5x^2 - 4x + 23}{x - 2}$

$$\begin{array}{r} x^2 - 3x - 10 \\ x-2\overline{)x^3 - 5x^2 - 4x + 23} \\ \underline{x^3 - 2x^2} \\ -3x^2 - 4x \\ \underline{-3x^2 + 6x} \\ -10x + 23 \\ \underline{-10x + 20} \\ 3 \end{array}$$

$x^3 - 5x^2 - 4x + 23 = (x-2)(x^2 - 3x - 10) + 3$

9. $\left(2x^3 + 5x^2 + 4x + 17\right) \div \left(x + 3\right)$

$$\begin{array}{r} 2x^2 - x + 7 \\ x+3\overline{)2x^3 + 5x^2 + 4x + 17} \\ \underline{2x^3 + 6x^2} \\ -1x^2 + 4x \\ \underline{-x^2 - 3x} \\ 7x + 17 \\ \underline{7x + 21} \\ -4 \end{array}$$

$2x^3 + 5x^2 + 4x + 17 = (x+3)(2x^2 - x + 7) - 4$

11. $(x^3 - 8x^2 + 11x + 20) \div (x - 5)$

$$\begin{array}{r} x^2 - 3x - 4 \\ x-5\overline{)x^3 - 8x^2 + 11x + 20} \\ \underline{x^3 - 5x^2} \\ -3x^2 + 11x \\ \underline{-3x^2 + 15x} \\ -4x + 20 \\ \underline{-4x + 20} \\ \end{array}$$

$(x^3 - 8x^2 + 11x + 20) = (x-5)(x^2 - 3x - 4) + 0$

13. $\dfrac{2x^2 - 5x - 3}{x - 3}$

$$\begin{array}{r} 3\underline{|2\quad -5\quad -3} \\ \underline{\quad\quad 6\quad\;\; 3} \\ 2\quad\;\; 1\quad\;\; 0 \end{array}$$

a) $\dfrac{2x^2 - 5x - 3}{x - 3} = (2x + 1) + \dfrac{0}{x - 3}$

b) $2x^2 - 5x - 3 = (x-3)(2x+1) + 0$

15. $(x^3 - 7x^2 + 6x + 8) \div (x - 2)$

$$\begin{array}{r} 2\underline{|1\quad -7\quad 6\quad\;\; 8} \\ \underline{\quad\quad 2\quad -10\; -8} \\ 1\quad -5\quad -4\quad\;\; 0 \end{array}$$

a) $\dfrac{x^3 - 7x^2 + 16x + 8}{x - 2} = (x^2 - 5x - 4) + \dfrac{0}{x - 2}$

b) $x^3 - 7x^2 + 6x + 8 = (x-2)(x^2 - 5x - 4) + 0$

17. $\dfrac{x^3 - 5x^2 - 4x + 23}{x - 2}$

$$\begin{array}{r} 2\underline{|1\quad -5\quad -4\quad\;\; 23} \\ \underline{\quad\quad 2\quad -6\; -20} \\ 1\quad -3\quad -10\quad\;\; 3 \end{array}$$

a) $\dfrac{x^3 - 5x^2 - 4x + 23}{x - 2} = (x^2 - 3x - 10) + \dfrac{3}{x - 2}$

b) $x^3 - 5x^2 - 4x + 23 = (x-2)(x^2 - 3x - 10) + 3$

19. $(2x^3 - 5x^2 - 11x - 17) \div (x - 4)$

$$\begin{array}{r} 4\underline{|\;2\quad -5\quad -11\quad -17} \\ \underline{\quad\quad 8\quad\;\; 12\quad\;\; 4} \\ 2\quad\;\; 3\quad\;\; 1\quad -13 \end{array}$$

a) $\dfrac{2x^3 - 5x^2 - 11x - 17}{x - 4} = \left(2x^2 + 3x + 1\right) - \dfrac{13}{x - 4}$

b) $2x^3 - 5x^2 - 11x - 17 = (x-4)(2x^2 + 3x + 1) - 13$

21. $(x^3 + 5x^2 + 7) \div (x+1)$

$(x^3 + 5x^2 + 0x + 7) \div (x+1)$

$$\underline{-1|}\ \begin{array}{cccc} 1 & 5 & 0 & 7 \end{array}$$
$$\underline{\ \ -1\ -4\ \ \ 4}$$
$$\begin{array}{cccc} 1 & 4 & -4 & 11 \end{array}$$

$x^3 + 5x^2 + 7 = (x+1)(x^2 + 4x - 4) + 11$

23. $(x^3 - 13x - 12) \div (x-4)$

$(x^3 - 0x^2 - 13x - 12) \div (x-4)$

$$\underline{4|}\ \begin{array}{cccc} 1 & 0 & -13 & -12 \end{array}$$
$$\underline{\ \ \ 4\ \ \ 16\ \ \ 12}$$
$$\begin{array}{cccc} 1 & 4 & 3 & 0 \end{array}$$

$x^3 - 13x - 12 = (x-4)(x^2 + 4x + 3) + 0$

25. $\dfrac{3x^3 - 8x + 12}{x-1}$

$(3x^3 + 0x^2 - 8x + 12) \div (x-1)$

$$\underline{1|}\ \begin{array}{cccc} 3 & 0 & -8 & 12 \end{array}$$
$$\underline{\ \ \ \ 3\ \ \ \ 3\ \ -5}$$
$$\begin{array}{cccc} 3 & 3 & -5 & 7 \end{array}$$

$3x^3 - 8x + 12 = (x-1)(3x^2 + 3x - 5) + 7$

27. $(n^3 + 27) \div (n+3)$

$(n^3 + 0n^2 + 0n + 27) \div (n+3)$

$$\underline{-3|}\ \begin{array}{cccc} 1 & 0 & 0 & 27 \end{array}$$
$$\underline{\ \ -3\ \ \ 9\ -27}$$
$$\begin{array}{cccc} 1 & -3 & 9 & 0 \end{array}$$

$n^3 + 27 = (n+3)(n^2 - 3n + 9) + 0$

29. $(x^4 + 3x^3 - 16x - 8) \div (x-2)$

$(x^4 + 3x^3 + 0x^2 - 16x - 8) \div (x-2)$

$$\underline{2|}\ \begin{array}{ccccc} 1 & 3 & 0 & -16 & -8 \end{array}$$
$$\underline{\ \ \ \ 2\ \ \ 10\ \ \ 20\ \ \ \ 8}$$
$$\begin{array}{ccccc} 1 & 5 & 10 & 4 & 0 \end{array}$$

$x^4 + 3x^3 - 16x - 8$
$= (x-2)(x^3 + 5x^2 + 10x + 4) + 0$

31. $\dfrac{2x^3 + 7x^2 - x + 26}{x^2 + 0x + 3}$

$$\require{enclose}\begin{array}{r} 2x+7 \\ x^2+0x+3 \enclose{longdiv}{2x^3 + 7x^2 - x + 26} \\ \underline{2x^3 + 0x^2 + 6x} \\ 7x^2 - 7x + 26 \\ \underline{7x^2 + 0x + 21} \\ -7x + 5 \end{array}$$

$\dfrac{2x^3 + 7x^2 - x + 26}{x^2 + 3} = (2x+7) + \dfrac{-7x+5}{x^2+3}$

33. $\dfrac{x^4 - 5x^2 - 4x + 7}{x^2 - 1}$

$$\require{enclose}\begin{array}{r} x^2 - 4 \\ x^2+0x-1 \enclose{longdiv}{x^4 + 0x^3 - 5x^2 - 4x + 7} \\ \underline{x^4 + 0x^3 - x^2} \\ -4x^2 - 4x + 7 \\ \underline{-4x^2 + 0x + 4} \\ -4x + 3 \end{array}$$

$\dfrac{x^4 - 5x^2 - 4x + 7}{x^2 - 1} = (x^2 - 4) + \dfrac{-4x+3}{x^2-1}$

35. $P(x) = x^3 - 6x^2 + 5x + 12$

a. $P(-2) = -30$

$$\underline{-2|}\ \begin{array}{cccc} 1 & -6 & 5 & 12 \end{array}$$
$$\underline{\ \ \ -2\ \ \ 16\ \ -42}$$
$$\begin{array}{cccc} 1 & -8 & 21 & \underline{|-30} \end{array}$$

b. $P(5) = 12$

$$\underline{5|}\ \begin{array}{cccc} 1 & -6 & 5 & 12 \end{array}$$
$$\underline{\ \ \ \ \ 5\ \ -5\ \ \ \ 0}$$
$$\begin{array}{cccc} 1 & -1 & 0 & \underline{|12} \end{array}$$

37. $P(x) = 2x^3 - x^2 - 19x + 4$

a. $P(-3) = -2$

$$\underline{-3|}\ \begin{array}{cccc} 2 & -1 & -19 & 4 \end{array}$$
$$\underline{\ \ \ \ -6\ \ \ 21\ \ -6}$$
$$\begin{array}{cccc} 2 & -7 & 2 & \underline{|-2} \end{array}$$

b. $P(2) = -22$

$$\underline{2|}\ \begin{array}{cccc} 2 & -1 & -19 & 4 \end{array}$$
$$\underline{\ \ \ \ \ 4\ \ \ \ \ 6\ \ -26}$$
$$\begin{array}{cccc} 2 & 3 & -13 & \underline{|-22} \end{array}$$

3.2 Exercises

39. $P(x) = x^4 - 4x^2 + x + 1$

   a. $P(-2) = -1$

$$-2 \underline{| \quad 1 \quad\quad 0 \quad -4 \quad\quad 1 \quad\quad 1}$$
$$\underline{\quad\quad\quad -2 \quad\quad 4 \quad\quad 0 \quad -2}$$
$$\quad\quad 1 \quad -2 \quad\quad 0 \quad\quad 1 \quad \boxed{-1}$$

   b. $P(2) = 3$

$$2 \underline{| \quad 1 \quad\quad 0 \quad -4 \quad\quad 1 \quad\quad 1}$$
$$\underline{\quad\quad\quad 2 \quad\quad 4 \quad\quad 0 \quad\quad 2}$$
$$\quad\quad 1 \quad\quad 2 \quad\quad 0 \quad\quad 1 \quad \boxed{3}$$

41. $P(x) = 2x^3 - 7x + 33$

   a. $P(-2)$

$$-2 \underline{| \quad 2 \quad\quad 0 \quad -7 \quad\quad 33}$$
$$\underline{\quad\quad\quad -4 \quad\quad 8 \quad -2}$$
$$\quad\quad 2 \quad -4 \quad\quad 1 \quad\quad 31$$

   b. $P(-3) = 0$

$$-3 \underline{| \quad 2 \quad\quad 0 \quad -7 \quad\quad 33}$$
$$\underline{\quad\quad\quad -6 \quad\quad 18 \quad -33}$$
$$\quad\quad 2 \quad -6 \quad\quad 11 \quad\quad 0$$

43. $Px = 2x^3 + 3x^2 - 9x - 10$

   a. $P\left(\dfrac{3}{2}\right) = -10$

$$\dfrac{3}{2} \underline{| \quad 2 \quad\quad 3 \quad -9 \quad -10}$$
$$\underline{\quad\quad\quad 3 \quad\quad 9 \quad\quad 0}$$
$$\quad\quad 2 \quad\quad 6 \quad\quad 0 \quad \boxed{-10}$$

   b. $P\left(-\dfrac{5}{2}\right) = 0$

$$-\dfrac{5}{2} \underline{| \quad 2 \quad\quad 3 \quad -9 \quad -10}$$
$$\underline{\quad\quad\quad -5 \quad\quad 5 \quad\quad 10}$$
$$\quad\quad 2 \quad -2 \quad -4 \quad \boxed{0}$$

45. $f(x) = x^3 - 3x^2 - 13x + 15$

   a.

$$-3 \underline{| \quad 1 \quad -3 \quad -13 \quad\quad 15}$$
$$\underline{\quad\quad\quad -3 \quad\quad 18 \quad -15}$$
$$\quad\quad 1 \quad -6 \quad\quad 5 \quad\quad 0$$

   yes, $(x + 3)$ is a factor since remainder is 0.

   b.

$$5 \underline{| \quad 1 \quad -3 \quad -13 \quad\quad 15}$$
$$\underline{\quad\quad\quad 5 \quad\quad 10 \quad -15}$$
$$\quad\quad 1 \quad\quad 2 \quad -3 \quad\quad 0$$

   yes, $(x - 5)$ is a factor since remainder is 0.

47. $f(x) = x^3 - 6x^2 + 3x + 10$

$$-2 \underline{| \quad 1 \quad -6 \quad\quad 3 \quad\quad 10}$$

   a.
$$\underline{\quad\quad\quad -2 \quad\quad 16 \quad -38}$$
$$\quad\quad 1 \quad -8 \quad\quad 19 \quad -28$$

   no, $(x + 2)$ is not a factor since remainder is not 0.

$$5 \underline{| \quad 1 \quad -6 \quad\quad 3 \quad\quad 10}$$

   b.
$$\underline{\quad\quad\quad 5 \quad -5 \quad -10}$$
$$\quad\quad 1 \quad -1 \quad -2 \quad\quad 0$$

   yes, $(x - 5)$ is a factor since remainder is 0.

49. $f(x) = -x^3 + 7x - 6$

$$-3 \underline{| \quad -1 \quad\quad 0 \quad\quad 7 \quad -6}$$

   a.
$$\underline{\quad\quad\quad 3 \quad -9 \quad\quad 6}$$
$$\quad\quad -1 \quad\quad 3 \quad -2 \quad\quad 0$$

   yes, $(x + 3)$ is a factor since remainder is 0.

$$2 \underline{| \quad -1 \quad\quad 0 \quad\quad 7 \quad -6}$$

   b.
$$\underline{\quad\quad\quad -2 \quad -4 \quad\quad 6}$$
$$\quad\quad -1 \quad -2 \quad\quad 3 \quad\quad 0$$

   yes, $(x - 2)$ is a factor since remainder is 0.

51. $P(x) = x^3 + 2x^2 - 5x - 6; \quad x = -3$

   verified

$$-3 \underline{| \quad 1 \quad\quad 2 \quad -5 \quad -6}$$
$$\underline{\quad\quad\quad -3 \quad\quad 3 \quad\quad 6}$$
$$\quad\quad 1 \quad -1 \quad -2 \quad \boxed{0}$$

   $P(-3) = 0$

53. $P(x) = x^3 - 7x + 6; \quad x = 2$

   verified

$$2 \underline{| \quad 1 \quad\quad 0 \quad -7 \quad\quad 6}$$
$$\underline{\quad\quad\quad 2 \quad\quad 4 \quad -6}$$
$$\quad\quad 1 \quad\quad 2 \quad -3 \quad \boxed{0}$$

   $P(2) = 0$

55. $P(x) = 9x^3 + 18x^2 - 4x - 8$

$$\dfrac{2}{3} \underline{| \quad 9 \quad\quad 18 \quad -4 \quad -8}$$
$$\underline{\quad\quad\quad 6 \quad\quad 16 \quad\quad 8}$$
$$\quad\quad 9 \quad\quad 24 \quad\quad 12 \quad\quad 0$$

   $P\left(\dfrac{2}{3}\right) = 0$

57. $-2,\ 3,\ -5$; degree 3

$P(x) = (x+2)(x-3)(x+5);$

$P(x) = (x^2 - x - 6)(x+5)$

$P(x) = x^3 + 5x^2 - x^2 - 5x - 6x - 30$

$P(x) = x^3 + 4x^2 - 11x - 30$

59. $-2,\ \sqrt{3},\ -\sqrt{3}$; degree 3

$P(x) = (x+2)(x - \sqrt{3})(x+\sqrt{3});$

$P(x) = (x+2)(x^2 + \sqrt{3}x - \sqrt{3}x - 3)$

$P(x) = (x+2)(x^2 - 3)$

$P(x) = x^3 - 3x + 2x^2 - 6$

$P(x) = x^3 + 2x^2 - 3x - 6$

61. $-5,\ 2\sqrt{3},\ -2\sqrt{3}$; degree 3

$P(x) = (x+5)(x - 2\sqrt{3})(x+2\sqrt{3});$

$P(x) = (x+5)(x^2 + 2\sqrt{3}x - 2\sqrt{3}x - 12)$

$P(x) = (x+5)(x^2 - 12)$

$P(x) = x^3 - 12x + 5x^2 - 60$

$P(x) = x^3 + 5x^2 - 12x - 60$

63. $1,\ -2,\ \sqrt{10},\ -\sqrt{10}$; degree 4

$P(x) = (x-1)(x+2)(x - \sqrt{10})(x+\sqrt{10});$

$P(x) = (x^2 + x - 2)(x^2 + \sqrt{10}x - \sqrt{10}x - 10)$

$P(x) = (x^2 + x - 2)(x^2 - 10)$

$P(x) = x^4 - 10x^2 + x^3 - 10x - 2x^2 + 20$

$P(x) = x^4 + x^3 - 12x^2 - 10x + 20$

65. $P(x) = x^3 - 5x^2 - 2x + 24$

$\underline{-2|1\quad -5\quad -2\quad 24}$

$\underline{\quad\quad -2\quad 14\ -24}$

$\quad 1\ -7\quad 12\quad\ 0$

$P(x) = (x+2)(x^2 - 7x + 12)$

$P(x) = (x+2)(x-3)(x-4)$

67. $p(x) = x^4 + 2x^3 - 12x^2 - 18x + 27$

$\underline{-3|1\quad 2\quad -12\quad -18\quad\ 27}$

$\underline{\quad\quad -3\quad 3\quad\ 27\ -27}$

$\quad 1\ -1\ -9\quad\ 9\quad\ 0$

$p(x) = (x+3)(x^3 - x^2 - 9x + 9)$

$p(x) = (x+3)(x^2(x-1) - 9(x-1))$

$p(x) = (x+3)(x-1)(x^2 - 9)$

$p(x) = (x+3)(x-1)(x+3)(x-3)$

$p(x) = (x+3)^2(x-1)(x-3)$

69. $f(x) = 2x^3 + 11x^2 - x - 30$

$\dfrac{3}{2}\Big|2\quad 11\quad -1\quad -30$

$\underline{\quad\quad\ 3\quad 21\quad 30}$

$\quad 2\quad 14\quad 20\quad\ 0$

$f(x) = \left(x - \dfrac{3}{2}\right)(2x^2 + 14x + 20)$

$f(x) = \left(x - \dfrac{3}{2}\right)2(x^2 + 7x + 10)$

$f(x) = \left(x - \dfrac{3}{2}\right)2(x+2)(x+5)$

$f(x) = 2\left(x - \dfrac{3}{2}\right)(x+2)(x+5)$

71. $p(x) = x^3 - 3x^2 - 9x + 27$

$p(x) = x^2(x-3) - 9(x-3)$

$p(x) = (x-3)(x^2 - 9)$

$p(x) = (x-3)(x+3)(x-3)$

$p(x) = (x+3)(x-3)^2$

73. $p(x) = x^3 - 6x^2 + 12x - 8$

Possible Factors of 8, $\pm 1, \pm 2, \pm 4, \pm 8$

$p(x) = (x - 2)(x^2 - 4x + 4)$

$p(x) = (x - 2)(x - 2)(x - 2)$

$p(x) = (x - 2)^3$

75. $p(x) = (x^2 - 6x + 9)(x^2 - 9)$

$p(x) = (x - 3)(x - 3)(x + 3)(x - 3)$

$p(x) = (x + 3)(x - 3)^3$

77. $p(x) = (x^3 + 4x^2 - 9x - 36)(x^2 + x - 12)$

$p(x) = (x^2(x + 4) - 9(x + 4))(x + 4)(x - 3)$

$p(x) = (x + 4)(x^2 - 9)(x + 4)(x - 3)$

$p(x) = (x + 4)(x + 3)(x - 3)(x + 4)(x - 3)$

$p(x) = (x + 3)(x - 3)^2(x + 4)^2$

79. $640 = 4x^3 - 84x^2 + 432x$

$160 = x^3 - 21x^2 + 108x$

$0 = x^3 - 21x^2 + 108x - 160$

$$\underline{4}\begin{array}{|rrrr} 1 & -21 & 108 & -160 \\ & 4 & -68 & 160 \\ \hline 1 & -17 & 40 & \boxed{0} \end{array}$$

$0 = (x - 4)(x^2 - 17x + 40)$

4 inch squares,

$24 - 8 = 16$ in.

$18 - 8 = 10$ in.

Dimensions of box: 16 in. x 10 in. x 4 in.

81. $P(w) = -0.1w^4 + 2w^3 - 14w^2 + 52w + 5$

a. week 10, 22.5 thousand

$P(5) = -0.1(5)^4 + 2(5)^3 - 14(5)^2 + 52(5) + 5 = 102.5$

$P(10) = -0.1(10)^4 + 2(10)^3 - 14(10)^2 + 52(10) + 5 = 125$

b. one week before closing, 36 thousand

$P(1) = -0.1(1)^4 + 2(1)^3 - 14(1)^2 + 52(1) + 5 = 44.9$

$P(11) = -0.1(11)^4 + 2(11)^3 - 14(11)^2 + 52(11) + 5 = 80.9$

c. week 9

$P(7) = -0.1(7)^4 + 2(7)^3 - 14(7)^2 + 52(7) + 5 = 128.9$

$P(8) = -0.1(8)^4 + 2(8)^3 - 14(8)^2 + 52(8) + 5 = 139.4$

$P(9) = -0.1(9)^4 + 2(9)^3 - 14(9)^2 + 52(9) + 5 = 140.9$

$P(10) = -0.1(10)^4 + 2(10)^3 - 14(10)^2 + 52(10) + 5 = 125$

83. $v(x) = x^3 + 11x^2 + 24x$

a. $v(3) = 198$

$$\underline{3}\begin{array}{|rrrr} 1 & 11 & 24 & 0 \\ & 3 & 42 & 198 \\ \hline 1 & 14 & 66 & 198 \end{array}$$

Volume: 198 ft^3

b. $100 = x^3 + 11x^2 + 24x$

$0 = x^3 + 11x^2 + 24x - 100$

$$\underline{2}\begin{array}{|rrrr} 1 & 11 & 24 & -100 \\ & 2 & 26 & 100 \\ \hline 1 & 13 & 50 & 0 \end{array}$$

Height: 2 ft

c. $y = x^3 + 11x^2 + 24x - 1000$

Use graphing calculator to find the zero: (6.8437621, 0) Depth: about 7 ft

84. a.

$$\underline{4}\begin{array}{|rrrrr} -\dfrac{1}{4} & 6 & -52 & 196 & -260 \\ & -1 & 20 & -128 & 272 \\ \hline -\dfrac{1}{4} & 5 & -32 & 68 & 12 \end{array}$$

$A(4) = 12$: 12,000 people

$$\underline{6}\begin{array}{|rrrrr} -\dfrac{1}{4} & 6 & -52 & 196 & -260 \\ & -\dfrac{3}{2} & 27 & -150 & 276 \\ \hline -\dfrac{1}{4} & \dfrac{9}{2} & -25 & 46 & 16 \end{array}$$

$A(6) = 16$: 16,000 people;

June; $16,000 - 12,000 = 4,000$ more

b.

$$7\rfloor -\frac{1}{4} \quad 6 \quad -52 \quad 196 \quad -260$$

$$-\frac{7}{4} \quad \frac{119}{4} \quad -\frac{623}{4} \quad \frac{1127}{4}$$

$$-\frac{1}{4} \quad \frac{17}{4} \quad -\frac{89}{4} \quad \frac{161}{4} \quad \frac{87}{4}$$

$$8\rfloor -\frac{1}{4} \quad 6 \quad -52 \quad 196 \quad -260$$

$$-2 \quad 32 \quad -160 \quad 288$$

$$-\frac{1}{4} \quad 4 \quad -20 \quad 36 \quad 28$$

$$9\rfloor -\frac{1}{4} \quad 6 \quad -52 \quad 196 \quad -260$$

$$2.25 \quad 33.75 \quad 164.25 \quad 285.75$$

$$-0.25 \quad 3.75 \quad -18.25 \quad 31.75 \quad 25.75$$

$$10\rfloor -\frac{1}{4} \quad 6 \quad -52 \quad 196 \quad -260$$

$$-2.5 \quad 35 \quad -170 \quad 260$$

$$-0.25 \quad 3.75 \quad -18.25 \quad 31.75 \quad 25.75$$

Maximum: 8 month (August)

Using calculator, find max (8.38, 28.83)

Maximum ≈ 29000

c.   October, since $A(10) = 0$

85. $f(x) = x^3 - 3x^2 - 5x + k$

$$-2\rfloor 1 \quad -3 \quad -5 \quad k$$

$$-2 \quad 10 \quad -10$$

$$1 \quad -5 \quad 5 \quad 0$$

$k - 10 = 0$

$k = 10$

87. $p(x) = x^3 - 3x^2 + k + 10$

$$2\rfloor 1 \quad -3 \quad k \quad 10$$

$$2 \quad -2 \quad 2k-4$$

$$1 \quad -1 \quad k-2 \quad 0$$

$2k - 4 = -10$

$2k = -6$

$k = -3$

89. $f(x) = (x - 2i)(x + 2i)(x - 3)$

$f(x) = (x^2 + 4)(x - 3)$

$f(x) = x^3 - 3x^2 + 4x - 12$

$$3\rfloor 1 \quad -3 \quad 4 \quad -12$$

$$3 \quad 0 \quad 12$$

$$1 \quad 0 \quad 4 \quad 0$$

$x^2 + 4 = 0$

$x^2 = -4$

$x = \pm 2i$

The theorems also apply to complex zeroes of polynomials.

91. (a) $1^3 + 2^3 + 3^3 = 1 + 8 + 27 = 36$

$S_3 = 36$

$$3\rfloor 1 \quad 2 \quad 1 \quad 0 \quad 0$$

$$3 \quad 15 \quad 48 \quad 144$$

$$1 \quad 5 \quad 16 \quad 48 \quad \lfloor 144$$

$144 \div 4 = 36$

(b) $1^3 + 2^3 + 3^3 + 4^3 + 5^3$

$= 1 + 8 + 27 + 64 + 125 = 225$

$S_5 = 225$

$$5\rfloor 1 \quad 2 \quad 1 \quad 0 \quad 0$$

$$5 \quad 35 \quad 180 \quad 900$$

$$1 \quad 7 \quad 36 \quad 180 \quad \lfloor 900$$

$900 \div 4 = 225$

93.

	D	r	t
John	1275	5	$\dfrac{1275}{5} = 255$
Rick	1025	4	$\dfrac{1025}{4} = 256.25$

John reaches the finish line in 25 secs while Rick reaches the finish line in 256.25 sec.
Yes, John wins.

95. $(0, 5000), (5, 12000)$

$$m = \frac{12000 - 5000}{5 - 0} = \frac{7000}{5} = 1400$$

$G(t) = 1400t + 5000$ where t is the number of years since 2005.

3.3 Technology Highlight

1. They give an approximate location for each zero.

3. Outputs change sign at each zero.

3.3 Exercises

1. Coefficients

3. $a - bi$

5. b; 4 is not a factor of 6.

7. $P(x) = x^4 + 5x^2 - 36$
$P(x) = (x^2 - 4)(x^2 + 9)$
$P(x) = (x - 2)(x + 2)(x + 3i)(x - 3i)$
Zeroes: $x = 2, x = -2, x = 3i, x = -3i$

9. $Q(x) = x^4 - 16$
$Q(x) = (x^2 + 4)(x^2 - 4)$
$Q(x) = (x + 2i)(x - 2i)(x - 2)(x + 2)$
Zeroes: $x = -2, x = 2, x = 2i, x = -2i$

11. $P(x) = x^3 + x^2 - x - 1$
$P(x) = x^2(x + 1) - (x + 1)$
$P(x) = (x + 1)(x^2 - 1)$
$P(x) = (x + 1)(x + 1)(x - 1)$
Zeroes: $x = -1, x = -1, x = 1$

13. $Q(x) = x^3 - 5x^2 - 25x + 125$
$Q(x) = x^2(x - 5) - 25(x - 5)$
$Q(x) = (x - 5)(x^2 - 25)$
$Q(x) = (x - 5)(x + 5)(x - 5)$
Zeroes: $x = 5, x = -5, x = 5$

15.
$p(x) = (x^2 - 10x + 25)(x^2 + 4x - 45)(x + 9)$
$p(x) = (x - 5)^2(x + 9)(x - 5)(x + 9)$
$p(x) = (x - 5)^3(x + 9)^2$
Zeroes: $x = 5$, multiplicity 3,
$x = -9$, multiplicity 2

17. $P(x) = (x^2 - 5x - 14)(x^2 - 49)(x + 2)$
$P(x) = (x - 7)(x + 2)(x - 7)(x + 7)(x + 2)$
$P(x) = (x - 7)^2(x + 2)^2(x + 7)$
Zeroes: $x = 7$, multiplicity 2,
$x = -2$, multiplicity 2,
$x = -7$, multiplicity 1

19. Degree 3, $x = 3, x = 2i$, $(x = -2i)$
$P(x) = (x - 3)(x - 2i)(x + 2i)$
$P(x) = (x - 3)(x^2 + 4)$
$P(x) = x^3 - 3x^2 + 4x - 12$

21. Degree 4, $x = -1, x = 2, x = i$, $(x = -i)$
$P(x) = (x + 1)(x - 2)(x - i)(x + i)$
$P(x) = (x^2 - x - 2)(x^2 + 1)$
$P(x) = x^4 - x^3 - 2x^2 + x^2 - x - 2$
$P(x) = x^4 - x^3 - x^2 - x - 2$

23. Degree 4, $x = 3, x = 2i$, $(x = 3, x = -2i)$
$P(x) = (x - 3)(x - 3)(x - 2i)(x + 2i)$
$P(x) = (x^2 - 6x + 9)(x^2 + 4)$
$P(x) = x^4 + 4x^2 - 6x^3 - 24x + 9x^2 + 36$
$P(x) = x^4 - 6x^3 + 13x^2 - 24x + 36$

25. Degree 4, $x = -1, x = 1 + 2i$,
$(x = -1, x = 1 - 2i)$
$P(x) = (x + 1)(x + 1)(x - (1 + 2i))(x - (1 - 2i))$
$P(x) = (x^2 + 2x + 1)((x - 1) - 2i)((x - 1) + 2i)$
$P(x) = (x^2 + 2x + 1)((x - 1)^2 - 4i^2)$
$P(x) = (x^2 + 2x + 1)((x - 1)^2 + 4)$
$P(x) = (x^2 + 2x + 1)(x^2 - 2x + 1 + 4)$
$P(x) = (x^2 + 2x + 1)(x^2 - 2x + 5)$
$P(x) = x^4 - 2x^3 + 5x^2 + 2x^3 - 4x^2$
$\qquad + 10x + x^2 - 2x + 5$
$P(x) = x^4 + 2x^2 + 8x + 5$

27. Degree 4, $x = -3, x = 1 + i\sqrt{2}$,

$(x = -3, x = 1 - i\sqrt{2})$

$P(x) = (x+3)(x+3)\left(x - \left(1 + i\sqrt{2}\right)\right)\left(x - \left(1 - i\sqrt{2}\right)\right)$

$P(x) = (x^2 + 6x + 9)\left(\left(x-1\right) - i\sqrt{2}\right)\left(\left(x-1\right) + i\sqrt{2}\right)$

$\quad P(x) = (x^2 + 6x + 9)\left(\left(x-1\right)^2 - i^2\sqrt{4}\right)$

$\quad P(x) = (x^2 + 6x + 9)\left(\left(x-1\right)^2 + 2\right)$

$\quad P(x) = (x^2 + 6x + 9)\left(x^2 - 2x + 1 + 2\right)$

$\quad P(x) = (x^2 + 6x + 9)\left(x^2 - 2x + 3\right)$

$\quad P(x) = x^4 - 2x^3 + 3x^2 + 6x^3 - 12x^2$

$\qquad + 18x + 9x^2 - 18x + 27$

$\quad P(x) = x^4 + 4x^3 + 27$

29. $f(x) = x^3 + 2x^2 - 8x - 5$

a. $[-4, -3]$, yes

$f(-4)$

$= (-4)^3 + 2(-4)^2 - 8(-4) - 5 = -5;$

$f(-3)$

$= (-3)^3 + 2(-3)^2 - 8(-3) - 5 = 10$

b. $[2, 3]$, yes

$f(2) = (2)^3 + 2(2)^2 - 8(2) - 5 = -5;$

$f(3) = (3)^3 + 2(3)^2 - 8(3) - 5 = 16$

31. $h(x) = 2x^3 + 13x^2 + 3x - 36$

a. $[1, 2]$, yes

$h(1) = 2(1)^3 + 13(1)^2 + 3(1) - 36 = -18$

$h(2) = 2(2)^3 + 13(2)^2 + 3(2) - 36 = 38$

b. $[-3, -2]$, yes

$h(-3)$

$= 2(-3)^3 + 13(-3)^2 + 3(-3) - 36 = 18$

$h(-2)$

$= 2(-2)^3 + 13(-2)^2 + 3(-2) - 36 = -6$

33. $f(x) = 4x^3 - 19x - 15$

$\dfrac{\{\pm 1, \pm 15, \pm 3, \pm 5\}}{\{\pm 1, \pm 4, \pm 2\}};$

$\left\{\pm 1, \pm 15, \pm 3, \pm 5, \pm\dfrac{1}{4}, \pm\dfrac{15}{4},\right.$

$\left.\pm\dfrac{3}{4}, \pm\dfrac{5}{4}, \pm\dfrac{1}{2}, \pm\dfrac{15}{2}, \pm\dfrac{3}{2}, \pm\dfrac{5}{2}\right\}$

35. $h(x) = 2x^3 - 5x^2 - 28x + 15$

$\dfrac{\{\pm 1, \pm 15, \pm 3, \pm 5\}}{\{\pm 1, \pm 2\}};$

$\left\{\pm 1, \pm 15, \pm 3, \pm 5, \pm\dfrac{1}{2}, \pm\dfrac{15}{2}, \pm\dfrac{3}{2}, \pm\dfrac{5}{2}\right\}$

37. $p(x) = 6x^4 - 2x^3 + 5x^2 - 28$

$\dfrac{\{\pm 1, \pm 28, \pm 2, \pm 14, \pm 4, \pm 7\}}{\{\pm 1, \pm 6, \pm 2, \pm 3\}};$

$\left\{\pm 1, \pm 28, \pm 2, \pm 14, \pm 4, \pm 7, \pm\dfrac{1}{6}, \pm\dfrac{14}{3}, \pm\dfrac{1}{3}, \pm\dfrac{7}{3}, \pm\dfrac{2}{3},\right.$

$\left.\pm\dfrac{7}{6}, \pm\dfrac{1}{2}, \pm\dfrac{7}{2}, \pm\dfrac{28}{3}, \pm\dfrac{4}{3}\right\}$

39. $Y_1 = 32t^3 - 52t^2 + 17t + 3$

$\dfrac{\{\pm 1, \pm 3\}}{\{\pm 1, \pm 32, \pm 2, \pm 16, \pm 4, \pm 8\}};$

$\left\{\pm 1, \pm\dfrac{1}{32}, \pm\dfrac{1}{2}, \pm\dfrac{1}{16}, \pm\dfrac{1}{4}, \pm\dfrac{1}{8},\right.$

$\left.\pm 3, \pm\dfrac{3}{32}, \pm\dfrac{3}{2}, \pm\dfrac{3}{16}, \pm\dfrac{3}{4}, \pm\dfrac{3}{8}\right\}$

41. $f(x) = x^3 - 13x + 12$

Possible rational zeroes:

$\dfrac{\{\pm 1, \pm 12, \pm 2, \pm 6, \pm 3, \pm 4\}}{\{\pm 1\}};$

$\{\pm 1, \pm 12, \pm 2, \pm 6, \pm 3, \pm 4\}$

$$\begin{array}{r|rrrr} -4 & 1 & 0 & -13 & 12 \\ & & -4 & 16 & -12 \\ \hline & 1 & -4 & 3 & 0 \end{array}$$

$f(x) = (x + 4)(x^2 - 4x + 3)$

$f(x) = (x + 4)(x - 1)(x - 3)$

$x = -4, 1, 3$

3.3 Exercises

43. $h(x) = x^3 - 19x - 30$

Possible rational zeroes:
$$\frac{\{\pm 1, \pm 30, \pm 2, \pm 15, \pm 3, \pm 10, \pm 5, \pm 6\}}{\{\pm 1\}};$$
$$\{\pm 1, \pm 30, \pm 2, \pm 15, \pm 3, \pm 10, \pm 5, \pm 6\}$$

```
-3 | 1    0   -19   -30
   |     -3    9     30
     1   -3   -10    | 0
```

$h(x) = (x + 3)(x^2 - 3x - 10)$
$h(x) = (x + 3)(x + 2)(x - 5)$
$x = -3, -2, 5$

45. $p(x) = x^3 - 2x^2 - 11x + 12$

Possible rational zeroes:
$$\frac{\{\pm 1, \pm 12, \pm 2, \pm 6, \pm 3, \pm 4\}}{\{\pm 1\}};$$
$$\{\pm 1, \pm 12, \pm 2, \pm 6, \pm 3, \pm 4\}$$

```
-3 | 1   -2   -11    12
   |     -3    15   -12
     1   -5    4    | 0
```

$p(x) = (x + 3)(x^2 - 5x + 4)$
$p(x) = (x + 3)(x - 1)(x - 4)$
$x = -3, 1, 4$

47. $Y_1 = x^3 - 6x^2 - x + 30$

Possible rational zeroes:
$$\frac{\{\pm 1, \pm 30, \pm 2, \pm 15, \pm 3, \pm 10, \pm 5, \pm 6\}}{\{\pm 1\}};$$
$$\{\pm 1, \pm 30, \pm 2, \pm 15, \pm 3, \pm 10, \pm 5, \pm 6\}$$

```
-2 | 1   -6   -1    30
   |     -2   16   -30
     1   -8   15   | 0
```

$Y_1 = (x + 2)(x^2 - 8x + 15)$
$Y_1 = (x + 2)(x - 3)(x - 5)$
$x = -2, 3, 5$

49. $Y_3 = x^4 - 15x^2 + 10x + 24$

Possible rational zeroes:
$$\frac{\{\pm 1, \pm 24, \pm 2, \pm 12, \pm 3, \pm 8, \pm 4, \pm 6\}}{\{\pm 1\}};$$
$$\{\pm 1, \pm 24, \pm 2, \pm 12, \pm 3, \pm 8, \pm 4, \pm 6\}$$

```
-4 | 1    0   -15   10    24
   |     -4   16   -4   -24
     1   -4    1    6    | 0
```

```
-1 | 1   -4    1    6
   |     -1    5   -6
     1   -5    6   | 0
```

$Y_3 = (x + 4)(x + 1)(x^2 - 5x + 6)$
$Y_3 = (x + 4)(x + 1)(x - 2)(x - 3)$
$x = -4, -1, 2, 3$

51. $f(x) = x^4 + 7x^3 - 7x^2 - 55x - 42$

Possible rational zeroes:
$$\frac{\{\pm 1, \pm 42, \pm 2, \pm 21, \pm 3, \pm 14, \pm 6, \pm 7\}}{\{\pm 1\}};$$
$$\{\pm 1, \pm 42, \pm 2, \pm 21, \pm 3, \pm 14, \pm 6, \pm 7\}$$

```
-7 | 1    7   -7   -55   -42
   |     -7    0    49    42
     1    0   -7   -6    | 0
```

```
-2 | 1    0   -7   -6
   |     -2    4    6
     1   -2   -3   | 0
```

$f(x) = (x + 7)(x + 2)(x^2 - 2x - 3)$
$f(x) = (x + 7)(x + 2)(x + 1)(x - 3)$
$x = -7, -2, -1, 3$

53. $f(x) = 4x^3 - 7x + 3$

Possible rational zeroes: $\dfrac{\{\pm 1, \pm 3\}}{\{\pm 1, \pm 4, \pm 2\}}$

```
1 | 4    0   -7    3
  |      4    4   -3
    4    4   -3   | 0
```

$f(x) = (x - 1)(4x^2 + 4x - 3)$
$f(x) = (x - 1)(2x + 3)(2x - 1)$
$x = \dfrac{-3}{2}, \dfrac{1}{2}, 1$

162

55. $h(x) = 4x^3 + 8x^2 - 3x - 9$

Possible rational zeroes: $\dfrac{\{\pm 1, \pm 9, \pm 3\}}{\{\pm 1, \pm 4, \pm 2\}}$

$\left\{\pm 1, \pm 9, \pm 3, \pm \dfrac{1}{4}, \pm \dfrac{9}{4}, \pm \dfrac{3}{4}, \pm \dfrac{1}{2}, \pm \dfrac{9}{2}, \pm \dfrac{3}{2}\right\}$

$$\begin{array}{r|rrrr} 1 & 4 & 8 & -3 & -9 \\ & & 4 & 12 & 9 \\ \hline & 4 & 12 & 9 & \underline{|0} \end{array}$$

$h(x) = (x-1)(4x^2 + 12x + 9)$

$h(x) = (x-1)(2x+3)^2$

$x = \dfrac{-3}{2}, 1$

57. $Y_1 = 2x^3 - 3x^2 - 9x + 10$

Possible rational zeroes: $\dfrac{\{\pm 1, \pm 10, \pm 2, \pm 5\}}{\{\pm 1, \pm 2\}}$

$\left\{\pm 1, \pm 10, \pm 2, \pm 5, \pm \dfrac{1}{2}, \pm \dfrac{5}{2}\right\}$

$$\begin{array}{r|rrrr} 1 & 2 & -3 & -9 & 10 \\ & & 2 & -1 & -10 \\ \hline & 2 & -1 & -10 & \underline{|0} \end{array}$$

$Y_1 = (x-1)(2x^2 - x - 10)$

$Y_1 = (x-1)(x+2)(2x-5)$

$x = -2, 1, \dfrac{5}{2}$

59. $p(x) = 2x^4 + 3x^3 - 9x^2 - 15x - 5$

Possible rational zeroes: $\dfrac{\{\pm 1, \pm 5\}}{\{\pm 1, \pm 2\}}$

$\left\{\pm 1, \pm 5, \pm \dfrac{1}{2}, \pm \dfrac{5}{2}\right\}$

$$\begin{array}{r|rrrrr} -1 & 2 & 3 & -9 & -15 & -5 \\ & & -2 & -1 & 10 & 5 \\ \hline & 2 & 1 & -10 & -5 & \underline{|0} \end{array}$$

$p(x) = (x+1)(2x^3 + x^2 - 10x - 5)$

$p(x) = (x+1)(x^2(2x+1) - 5(2x+1))$

$p(x) = (x+1)(2x+1)(x^2 - 5)$

$p(x) = (x+1)(2x+1)(x - \sqrt{5})(x + \sqrt{5})$

$x = -1, \dfrac{-1}{2}, \sqrt{5}, -\sqrt{5}$

61. $r(x) = 3x^4 - 5x^3 + 14x^2 - 20x + 8$

Possible rational zeroes: $\dfrac{\{\pm 1, \pm 8, \pm 2, \pm 4\}}{\{\pm 1, \pm 3\}}$

$\left\{\pm 1, \pm 8, \pm 2, \pm 4, \pm \dfrac{1}{3}, \pm \dfrac{8}{3}, \pm \dfrac{2}{3}, \pm \dfrac{4}{3}\right\}$

$$\begin{array}{r|rrrrr} 1 & 3 & -5 & 14 & -20 & 8 \\ & & 3 & -2 & 12 & -8 \\ \hline & 3 & -2 & 12 & -8 & \underline{|0} \end{array}$$

$r(x) = (x-1)(3x^3 - 2x^2 + 12x - 8)$

$r(x) = (x-1)(x^2(3x-2) + 4(3x-2))$

$r(x) = (x-1)(3x-2)(x^2 + 4)$

$r(x) = (x-1)(3x-2)(x - 2i)(x + 2i)$

$x = 1, \dfrac{2}{3}, \pm 2i$

63. $f(x) = 2x^4 - 9x^3 + 4x^2 + 21x - 18$

Possible rational zeroes:

$\dfrac{\{\pm 1, \pm 18, \pm 2, \pm 9, \pm 3, \pm 6\}}{\{\pm 1, \pm 2\}}$

$\left\{\pm 1, \pm 18, \pm 2, \pm 9, \pm 3, \pm 6, \pm \dfrac{1}{2}, \pm \dfrac{9}{2}, \pm \dfrac{3}{2}\right\}$

$$\begin{array}{r|rrrrr} 1 & 2 & -9 & 4 & 21 & -18 \\ & & 2 & -7 & -3 & 18 \\ \hline & 2 & -7 & -3 & 18 & \underline{|0} \end{array}$$

$$\begin{array}{r|rrrr} 2 & 2 & -7 & -3 & 18 \\ & & 4 & -6 & -18 \\ \hline & 2 & -3 & -9 & \underline{|0} \end{array}$$

$f(x) = (x-1)(x-2)(2x^2 - 3x - 9)$

$f(x) = (x-1)(x-2)(x-3)(2x+3)$

$x = 1, 2, 3, \dfrac{-3}{2}$

65. $h(x) = 3x^4 + 2x^3 - 9x^2 + 4$

Possible rational zeroes: $\dfrac{\{\pm 1, \pm 4, \pm 2\}}{\{\pm 1, \pm 3\}}$

$\left\{\pm 1, \pm 4, \pm 2, \pm \dfrac{1}{3}, \pm \dfrac{4}{3}, \pm \dfrac{2}{3}\right\}$

$$\begin{array}{r|rrrrr} 1 & 3 & 2 & -9 & 0 & 4 \\ & & 3 & 5 & -4 & -4 \\ \hline & 3 & 5 & -4 & -4 & \boxed{0} \end{array}$$

$$\begin{array}{r|rrrr} -2 & 3 & 5 & -4 & -4 \\ & & -6 & 2 & 4 \\ \hline & 3 & -1 & -2 & \boxed{0} \end{array}$$

$h(x) = (x-1)(x+2)(3x^2 - x - 2)$

$h(x) = (x-1)(x+2)(3x+2)(x-1)$

$x = -2, 1, \dfrac{-2}{3}$

67. $P(x) = 2x^4 + 3x^3 - 24x^2 - 68x - 48$

Possible rational zeroes:

$\dfrac{\{\pm 1, \pm 48, \pm 2, \pm 24, \pm 3, \pm 16, \pm 4, \pm 12, \pm 6, \pm 8\}}{\{\pm 1, \pm 2\}}$

$\left\{\pm 1, \pm 48, \pm 2, \pm 24, \pm 3, \pm 16, \pm 4, \pm 12, \pm 6, \pm 8\right.$

$\left. \pm \dfrac{1}{2}, \pm \dfrac{3}{2}\right\}$

$$\begin{array}{r|rrrrr} -2 & 2 & 3 & -24 & -68 & -48 \\ & & -4 & 12 & 44 & 48 \\ \hline & 2 & -1 & -22 & -24 & 0 \end{array}$$

$$\begin{array}{r|rrrr} 4 & 2 & -1 & -22 & -24 \\ & & 8 & 28 & 24 \\ \hline & 2 & 7 & 6 & 0 \end{array}$$

$2x^2 + 7x + 6 = (2x+3)(x+2)$

$P(x) = (x+2)^2 (x-4)(2x+3)$

Zeroes: $x = -2$, multiplicity 2

$x = 4$, multiplicity 1

$x = -\dfrac{3}{2}$, multiplicity 1

69. $r(x) = 3x^4 - 20x^3 + 34x^2 + 12x - 45$

Possible rational zeroes:

$\dfrac{\{\pm 1, \pm 45, \pm 3, \pm 15, \pm 5, \pm 9\}}{\{\pm 1, \pm 3\}}$

$\left\{\pm 1, \pm 45, \pm 3, \pm 15, \pm 5, \pm 9\right.$

$\left. \pm \dfrac{1}{3}, \pm \dfrac{5}{3}\right\}$

$$\begin{array}{r|rrrrr} -1 & 3 & -20 & 34 & 12 & -45 \\ & & -3 & 23 & -57 & 45 \\ \hline & 3 & -23 & 57 & -45 & 0 \end{array}$$

$$\begin{array}{r|rrrr} 3 & 3 & -23 & 57 & -45 \\ & & 9 & -42 & 45 \\ \hline & 3 & -14 & 15 & 0 \end{array}$$

$3x^2 - 14x + 15 = (3x-5)(x-3)$

$r(x) = (x+1)(x-3)^2(3x-5)$

Zeroes: $x = 3$, multiplicity 2

$x = -1$, multiplicity 1

$x = \dfrac{5}{3}$, multiplicity 1

71. $Y_1 = x^5 + 6x^2 - 49x + 42$

Possible rational zeroes:

$\dfrac{\{\pm 1, \pm 42, \pm 2, \pm 21, \pm 3, \pm 14, \pm 6, \pm 7\}}{\{\pm 1\}}$

$\left\{\pm 1, \pm 42, \pm 2, \pm 21, \pm 3, \pm 14, \pm 6, \pm 7\right\}$

$$\begin{array}{r|rrrrrr} 1 & 1 & 0 & 0 & 6 & -49 & 42 \\ & & 1 & 1 & 1 & 7 & -42 \\ \hline & 1 & 1 & 1 & 7 & -42 & \boxed{0} \end{array}$$

$$\begin{array}{r|rrrrr} 2 & 1 & 1 & 1 & 7 & -42 \\ & & 2 & 6 & 14 & 42 \\ \hline & 1 & 3 & 7 & 21 & \boxed{0} \end{array}$$

$Y_1 = (x-1)(x-2)(x^3 + 3x^2 + 7x + 21)$

$Y_1 = (x-1)(x-2)(x^2(x+3) + 7(x+3))$

$Y_1 = (x-1)(x-2)(x+3)(x^2 + 7)$

$Y_1 = (x-1)(x-2)(x+3)(x+\sqrt{7}\,i)(x-\sqrt{7}\,i)$

$x = 1, 2, -3, \pm\sqrt{7}\,i$

73. $P(x) = 3x^5 + x^4 + x^3 + 7x^2 - 24x + 12$

Possible rational zeroes:

$$\frac{\{\pm 1, \pm 12, \pm 2, \pm 6, \pm 3, \pm 4\}}{\{\pm 1, \pm 3\}}$$

$$\left\{\pm 1, \pm 12, \pm 2, \pm 6, \pm 3, \pm 4, \pm \frac{1}{3}, \pm \frac{2}{3}, \pm \frac{4}{3}\right\}$$

$$
\begin{array}{r|rrrrrr}
1 & 3 & 1 & 1 & 7 & -24 & 12 \\
  &   & 3 & 4 & 5 & 12 & -12 \\
\hline
  & 3 & 4 & 5 & 12 & -12 & \boxed{0}
\end{array}
$$

$$
\begin{array}{r|rrrrr}
-2 & 3 & 4 & 5 & 12 & -12 \\
   &   & -6 & 4 & -18 & 12 \\
\hline
   & 3 & -2 & 9 & -6 & \boxed{0}
\end{array}
$$

$P(x) = (x-1)(x+2)(3x^3 - 2x^2 + 9x - 6)$

$P(x) = (x-1)(x+2)(x^2(3x-2) + 3(3x-2))$

$P(x) = (x-1)(x+2)(3x-2)(x^2+3)$

$P(x) = (x-1)(x+2)(3x-2)(x+\sqrt{3}\,i)(x-\sqrt{3}\,i)$

$x = -2, \dfrac{2}{3}, 1, \pm\sqrt{3}\,i$

75. $Y_1 = x^4 - 5x^3 + 20x - 16$

Possible rational zeroes: $\dfrac{\{\pm 1, \pm 16, \pm 2, \pm 8, \pm 4\}}{\{\pm 1\}}$

$\{\pm 1, \pm 16, \pm 2, \pm 8, \pm 4\}$

$$
\begin{array}{r|rrrrr}
1 & 1 & -5 & 0 & 20 & -16 \\
  &   & 1 & -4 & -4 & 16 \\
\hline
  & 1 & -4 & -4 & 16 & \boxed{0}
\end{array}
$$

$Y_1 = (x-1)(x^3 - 4x^2 - 4x + 16)$

$Y_1 = (x-1)(x^2(x-4) - 4(x-4))$

$Y_1 = (x-1)(x-4)(x^2-4)$

$Y_1 = (x-1)(x-4)(x+2)(x-2)$

$x = 1, 2, 4, -2$

77. $r(x) = x^4 + 2x^3 - 5x^2 - 4x + 6$

Possible rational zeroes: $\dfrac{\{\pm 1, \pm 6, \pm 2, \pm 3\}}{\{\pm 1\}}$

$\{\pm 1, \pm 6, \pm 2, \pm 3\}$

$$
\begin{array}{r|rrrrr}
1 & 1 & 2 & -5 & -4 & 6 \\
  &   & 1 & 3 & -2 & -6 \\
\hline
  & 1 & 3 & -2 & -6 & \boxed{0}
\end{array}
$$

$r(x) = (x-1)(x^3 + 3x^2 - 2x - 6)$

$r(x) = (x-1)(x^2(x+3) - 2(x+3))$

$r(x) = (x-1)(x+3)(x^2-2)$

$r(x) = (x+3)(x-1)(x+\sqrt{2})(x-\sqrt{2})$

$x = -3, 1, \pm\sqrt{2}$

79. $p(x) = 2x^4 - x^3 + 3x^2 - 3x - 9$

Possible rational zeroes: $\dfrac{\{\pm 1, \pm 9, \pm 3\}}{\{\pm 1, \pm 2\}}$

$\left\{\pm 1, \pm 9, \pm 3, \pm \frac{1}{2}, \pm \frac{9}{2}, \pm \frac{3}{2}\right\}$

$$
\begin{array}{r|rrrrr}
-1 & 2 & -1 & 3 & -3 & -9 \\
   &   & -2 & 3 & -6 & 9 \\
\hline
   & 2 & -3 & 6 & -9 & \boxed{0}
\end{array}
$$

$p(x) = (x+1)(2x^3 - 3x^2 + 6x - 9)$

$p(x) = (x+1)(x^2(2x-3) + 3(2x-3))$

$p(x) = (x+1)(2x-3)(x^2+3)$

$p(x) = (x+1)(2x-3)(x+\sqrt{3}\,i)(x-\sqrt{3}\,i)$

$x = -1, \dfrac{3}{2}, \pm\sqrt{3}\,i$

81. $f(x) = 2x^5 - 7x^4 + 13x^3 - 23x^2 + 21x - 6$

Possible rational zeroes: $\dfrac{\{\pm 1, \pm 6, \pm 2, \pm 3\}}{\{\pm 1, \pm 2\}}$

$$\left\{ \pm 1, \pm 6, \pm 2, \pm 3, \pm \frac{1}{2}, \pm \frac{3}{2} \right\}$$

```
1 | 2  -7   13   -23   21   -6
  |      2   -5    8   -15    6
  ---------------------------------
    2  -5    8   -15    6  | 0
```

```
2 | 2  -5    8   -15    6
  |      4   -2   12    -6
  -----------------------------
    2  -1    6    -3  | 0
```

$f(x) = (x-1)(x-2)(2x^3 - x^2 + 6x - 3)$

$f(x) = (x-1)(x-2)(x^2(2x-1) + 3(2x-1))$

$f(x) = (x-1)(x-2)(2x-1)(x^2 + 3)$

$f(x) = (x-1)(x-2)(2x-1)(x + \sqrt{3}\,i)(x - \sqrt{3}\,i)$

$x = \dfrac{1}{2}, 1, 2, \pm\sqrt{3}\,i$

83. $f(x) = x^4 - 2x^3 + 4x - 8$

   a. Possible rational zeroes:

   $$\dfrac{\{\pm 1, \pm 8, \pm 2, \pm 4\}}{\{\pm 1\}}$$

   $$\{\pm 1, \pm 8, \pm 2, \pm 4\}$$

   b. Zeroes of unity: none, neither 1 nor -1 is a zero
   $(1 - 2 + 4 - 8 \neq 0), (1 + 2 - 4 - 8 \neq 0)$

   c. # of positive zeroes: 3 or 1 zeroes
   # of negative zeroes: 1 root

   d. Bounds: zeroes must lie between -2 and 2.

```
-2 | 1  -2    0    4   -8
   |     -2    8  -16   24
   ---------------------------
     1  -4    8  -12  | 16
```

```
2 | 1  -2    0    4   -8
  |      2    0    0    8
  ---------------------------
    1   0    0    4  | 0
```

85. $h(x) = x^5 + x^4 - 3x^3 + 5x + 2$

   a. Possible rational zeroes: $\dfrac{\{\pm 1, \pm 2\}}{\{\pm 1\}}$

   $\{\pm 1, \pm 2\}$

   b. Zeroes of unity: -1 is a root
   $(1 + 1 - 3 + 5 + 2 \neq 0), (-1 + 1 + 3 - 5 + 2 = 0)$

   c. # of positive zeroes: 2 or 0 zeroes
   # of negative zeroes: 3 or 1 zeroes

   d. Bounds: zeroes must lie between -3 and 2.

```
-3 | 1   1   -3    0    5    2
   |     -3    6   -9   27  -96
   ----------------------------------
     1  -2    3   -9   32  | -94
```

```
2 | 1   1    3    0    5    2
  |      2    6    6   12   34
  ----------------------------------
    1   3    3    6   17  | 36
```

87. $p(x) = x^5 - 3x^4 + 3x^3 - 9x^2 - 4x + 12$

   a. Possible rational zeroes:
   $\{\pm 1, \pm 12, \pm 2, \pm 6, \pm 3, \pm 4\}$

   b. Zeroes of unity: $x = 1$ and $x = -1$ are zeroes.
   $(1 - 3 + 3 - 9 - 4 + 12 = 0),$
   $(-1 - 3 - 3 - 9 + 4 + 12 = 0)$

   c. # of positive zeroes: 4, 2, or 0 zeroes
   # of negative zeroes: 1 root

   d. Bounds: zeroes must lie between -1 and 4.

```
-1 | 1  -3    3   -9   -4   12
   |     -1    4   -7   16  -12
   ----------------------------------
     1  -4    7  -16   12  | 0
```

```
4 | 1  -3    3   -9   -4   12
  |      4    4   28   76  288
  ----------------------------------
    1   1    7   19   72  | 300
```

89. $r(x) = 2x^4 + 7x^2 + 11x - 20$

    a. Possible rational zeroes:

$$\frac{\{\pm 1, \pm 20, \pm 2, \pm 10, \pm 4, \pm 5\}}{\{\pm 1, \pm 2\}}$$

$$\left\{\pm 1, \pm 20, \pm 2, \pm 10, \pm 4, \pm 5, \pm\frac{1}{2}, \pm\frac{5}{2}\right\}$$

    b. Zeroes of unity: $x = 1$ is a root.

$$(2 + 7 + 11 - 20 = 0), \ (2 + 7 - 11 - 20 \neq 0)$$

    c. # of positive zeroes: 1 root

        # of negative zeroes: 1 root

    d. Bounds: zeroes must lie between -2 and 1.

```
-2| 2   0   7   11  -20
   |    -4   8  -30   38
   ─────────────────────
     2  -4  15  -19  |18
```

```
 1| 2   0   7   11  -20
  |     2   2   9   20
  ─────────────────────
    2   2   9   20  | 0
```

91. $f(x) = 4x^3 - 16x^2 - 9x + 36$

Possible Positive zeroes	Possible Negative zeroes	Possible Complex zeroes	Total number of zeroes
2	1	0	3
0	1	2	3

Possible rational zeroes:

$$\frac{\{\pm 1, \pm 36, \pm 2, \pm 18, \pm 3, \pm 12, \pm 4, \pm 9, \pm 6\}}{\{\pm 1, \pm 4, \pm 2\}} ;$$

$$\left\{\pm 1, \pm 36, \pm 2, \pm 18, \pm 3, \pm 12, \pm 4, \pm 9, \pm 6, \pm\frac{1}{4}, \pm\frac{1}{2}, \pm\frac{9}{2}, \pm\frac{3}{4}, \pm\frac{9}{4}, \pm\frac{3}{2}\right\}$$

```
 4| 4  -16  -9   36
  |     16   0  -36
  ─────────────────
    4    0  -9  | 0
```

$$f(x) = (x - 4)(4x^2 - 9)$$
$$f(x) = (x - 4)(2x - 3)(2x + 3)$$

$$x = \frac{-3}{2}, \frac{3}{2}, 4$$

93. $h(x) = 6x^3 - 73x^2 + 10x + 24$

Possible Positive zeroes	Possible Negative zeroes	Possible Complex zeroes	Total number of zeroes
2	1	0	3
0	1	2	3

Possible rational zeroes:

$$\frac{\{\pm 1, \pm 24, \pm 2, \pm 12, \pm 3, \pm 8, \pm 4, \pm 6\}}{\{\pm 1, \pm 6, \pm 2, \pm 3\}} ;$$

$$\left\{\pm 1, \pm 24, \pm 2, \pm 12, \pm 3, \pm 8, \pm 4, \pm 6, \pm\frac{1}{6}, \right.$$
$$\left. \pm\frac{1}{3}, \pm\frac{1}{2}, \pm\frac{4}{3}, \pm\frac{2}{3}, \pm\frac{3}{2}, \pm\frac{8}{3}\right\}$$

```
12| 6  -73  10   24
  |     72  -12  -24
  ──────────────────
    6   -1  -2  | 0
```

$$h(x) = (x - 12)(6x^2 - x - 2)$$
$$h(x) = (x - 12)(3x - 2)(2x + 1)$$

$$x = \frac{-1}{2}, \frac{2}{3}, 12$$

95. $p(x) = 4x^4 + 40x^3 - 97x^2 - 10x + 24$

Possible Positive zeroes	Possible Negative zeroes	Possible Complex zeroes	Total number of zeroes
2	2	0	4
0	2	2	4
2	0	2	4
0	0	4	4

Possible rational zeroes:

$$\frac{\{\pm 1, \pm 24, \pm 2, \pm 12, \pm 3, \pm 8, \pm 4, \pm 6\}}{\{\pm 1, \pm 4, \pm 2\}}$$

$$\left\{\pm 1, \pm 24, \pm 2, \pm 12, \pm 3, \pm 8, \pm 4, \pm 6, \pm\frac{1}{4}, \pm\frac{1}{2}, \pm\frac{3}{4}, \pm\frac{3}{2}\right\}$$

```
 2| 4   40  -97  -10   24
  |      8   96   -2  -24
  ───────────────────────
    4   48   -1  -12  | 0
```

$$p(x) = (x - 2)(4x^3 + 48x^2 - 1x - 12)$$
$$p(x) = (x - 2)(4x^2(x + 12) - 1(x + 12))$$
$$p(x) = (x - 2)(x + 12)(4x^2 - 1)$$
$$p(x) = (x - 2)(x + 12)(2x - 1)(2x + 1)$$

$$x = -12, \frac{-1}{2}, \frac{1}{2}, 2$$

97. $z = a + bi : |z| = \sqrt{a^2 + b^2}$

    (a) $|3 + 4i| := \sqrt{(3)^2 + (4)^2} = 5$

    (b) $|-5 + 12i| = \sqrt{(-5)^2 + (12)^2} = 13$

    (c) $|1 + \sqrt{3}\,i| = \sqrt{(1)^2 + (\sqrt{3})^2} = 2$

99. $f(x) = 4x^3 - 12x^2 - 24x + 32$

    Possible rational zeroes:

$$\frac{\{\pm 1, \pm 32, \pm 2, \pm 16, \pm 4, \pm 8\}}{\{\pm 1, \pm 4, \pm 2\}} ;$$

$$\left\{\pm 1, \pm 32, \pm 2, \pm 16, \pm 4, \pm 8, \pm \frac{1}{4}, \pm \frac{1}{2}\right\}$$

$$\begin{array}{r|rrrr}
-2 & 4 & -12 & -24 & 32 \\
   &   & -8 & 40 & -32 \\
\hline
   & 4 & -20 & 16 & \boxed{0}
\end{array}$$

$f(x) = (x + 2)(4x^2 - 20x + 16)$

$f(x) = 4(x + 2)(x^2 - 5x + 4)$

$f(x) = 4(x + 2)(x - 4)(x - 1)$

$x = -2, 1, 4$

Yes; grapher shows maximum and minimum values occur at the zeroes of f.

101. $g(x) = 4x^3 - 18x^2 + 2x + 24$

    Possible rational zeroes:

$$\frac{\{\pm 1, \pm 24, \pm 2, \pm 12, \pm 3, \pm 8, \pm 4, \pm 6\}}{\{\pm 1, \pm 4, \pm 2\}} ;$$

$$\left\{\pm 1, \pm 24, \pm 2, \pm 12, \pm 3, \pm 8, \pm 4, \pm 6, \pm \frac{1}{4}, \pm \frac{1}{2}, \pm \frac{3}{4}, \pm \frac{3}{2}\right\}$$

$$\begin{array}{r|rrrr}
-1 & 4 & -18 & 2 & 24 \\
   &   & -4 & 22 & -24 \\
\hline
   & 4 & -22 & 24 & \boxed{0}
\end{array}$$

$g(x) = (x + 1)(4x^2 - 22x + 24)$

$g(x) = 2(x + 1)(2x^2 - 11x + 12)$

$g(x) = 2(x + 1)(2x - 3)(x - 4)$

$x = -1, \dfrac{3}{2}, 4$

Yes; grapher shows maximum and minimum values occur at the zeroes of g.

103. $v = x \cdot x \cdot (x - 1) = x^3 - x^2$

    (a) $x^3 - x^2 = 48$

    $x^3 - x^2 - 48 = 0$

    Possible rational zeroes:

$$\frac{\{\pm 1, \pm 48, \pm 2, \pm 24, \pm 3, \pm 16, \pm 4, \pm 12, \pm 6, \pm 8\}}{\{\pm 1\}} ;$$

$$\{\pm 1, \pm 48, \pm 2, \pm 24, \pm 3, \pm 16, \pm 4, \pm 12, \pm 6, \pm 8\}$$

$$\begin{array}{r|rrrr}
4 & 1 & -1 & 0 & -48 \\
  &   & 4 & 12 & 48 \\
\hline
  & 1 & 3 & 12 & 0
\end{array}$$

    $x = 4$

    $4 \text{ cm} \times 4 \text{ cm} \times 4 \text{ cm}$

    (b) $x^3 - x^2 = 100$

    $x^3 - x^2 - 100 = 0$

    Possible rational zeroes:

    $\{\pm 1, \pm 100, \pm 2, \pm 50, \pm 4, \pm 25, \pm 5, \pm 20, \pm 10\}$

$$\begin{array}{r|rrrr}
5 & 1 & -1 & 0 & -100 \\
  &   & 5 & 20 & 100 \\
\hline
  & 1 & 4 & 20 & \boxed{0}
\end{array}$$

$$\begin{array}{r|rrrr}
5 & 2 & -4 & 0 & -150 \\
  &   & 10 & 30 & 150 \\
\hline
  & 2 & 6 & 30 & 0
\end{array}$$

    $x = 5$

    $5 \text{cm} \times 5 \text{ cm} \times 5 \text{ cm}$

105. V= LWH

    $2w(w)(w - 2) = 150$

    $2w^3 - 4w^2 - 150 = 0$

    $2(w^3 - 2w^2 - 75) = 0$

    Possible rational zeroes:

    $\{\pm 1, \pm 75, \pm 3, \pm 25, \pm 5, \pm 15\}$

$$\begin{array}{r|rrrr}
5 & 2 & -4 & 0 & -150 \\
  &   & 10 & 30 & 150 \\
\hline
  & 2 & 6 & 30 & \boxed{0}
\end{array}$$

    $w = 5; 2w = 10; w - 2 = 3$

    length 10 in., width 5 in., height 3 in.

107. $f(x) = \frac{1}{4}x^4 - 6x^3 + 42x^2 - 72x - 64$

$0 = \frac{1}{4}x^4 - 6x^3 + 42x^2 - 72x - 64$

$0 = x^4 - 24x^3 + 168x^2 - 288x - 256$

Using a grapher:

$x = 4, 8$, between 12 and 13

1994, 1998, 2002; 5 years

109.

$f(x) = -0.4192x^4 + 18.9663x^3 - 319.9714x^2$
$\qquad + 2384.2x - 6615.8$

To solve $f(x) = 0$, find zeros using the graphing calculator.

a. $x = 8.97$ m, $x = 11.29$ m,
   $x = 12.05$ m, $x = 12.94$ m

b. Find maximum $(9.70, 3.71)$
   9.7 m will maximize the efficiency of
   the boat with rating 3.7.

111. $P(x) = 0.2x^3 - 0.24x^2 - 1.04x + 2.68$

a.

$$\begin{array}{r|rrrr} -10 & 0.02 & -0.24 & -1.04 & 2.68 \\ & & 0.2 & 4.4 & -33.6 \\ \hline & 0.02 & -0.44 & 3.36 & -30.92 \end{array}$$

alternate signs, -10 is a lower bound

b.

$$\begin{array}{r|rrrr} 10 & 0.02 & -0.24 & -1.04 & 2.68 \\ & & 0.2 & -0.4 & -14.4 \\ \hline & 0.02 & -0.04 & -1.44 & -11.72 \end{array}$$

no

c. about 14.88

113.A. a. $p(x) = x^2 + 25$
         $p(x) = (x + 5i)(x - 5i)$

     b. $q(x) = x^2 + 9$
        $q(x) = (x + 3i)(x - 3i)$

     c. $r(x) = x^2 + 7$
        $r(x) = (x + i\sqrt{7})(x - i\sqrt{7})$

   B. a. $x^2 - 7 = 0$
         $(x + \sqrt{7})(x - \sqrt{7}) = 0$
         $x + \sqrt{7} = 0$ or $x - \sqrt{7} = 0$
         $x = -\sqrt{7}$ or $x = \sqrt{7}$

b. $x^2 - 12 = 0$
   $(x + \sqrt{12})(x - \sqrt{12}) = 0$
   $x + \sqrt{12} = 0$ or $x - \sqrt{12} = 0$
   $x = -\sqrt{12}$ or $x = \sqrt{12}$
   $x = -2\sqrt{3}$ or $x = 2\sqrt{3}$

c. $x^2 - 18 = 0$
   $(x + \sqrt{18})(x - \sqrt{18}) = 0$
   $x + \sqrt{18} = 0$ or $x - \sqrt{18} = 0$
   $x = -\sqrt{18}$ or $x = \sqrt{18}$
   $x = -3\sqrt{2}$ or $x = 3\sqrt{2}$

115.a. $C(z) = z^3 + (1 - 4i)z^2 + (-6 - 4i)z + 24i$

$z = 4i$;

$$\begin{array}{r|rrrr} 4i & 1 & 1 - 4i & -6 - 4i & 24i \\ & & 4i & 4i & -24i \\ \hline & 1 & 1 & -6 & 0 \end{array}$$

$C(z) = (z - 4i)(z^2 + z - 6)$

$C(z) = (z - 4i)(z + 3)(z - 2)$

b. $C(x) = x^3 + (5 - 9i)x^2 + (4 - 45i)x - 36i$;

$z = 9i$;

$$\begin{array}{r|rrrr} 9i & 1 & 5 - 9i & 4 - 45i & -36i \\ & & 9i & 45i & 36i \\ \hline & 1 & 5 & 4 & 0 \end{array}$$

$C(z) = (z - 9i)(z^2 + 5z + 4)$

$C(z) = (z - 9i)(z + 4)(z + 1)$

c. $C(z) = z^3 + (-2 - 3i)z^2 + (5 + 6i)z - 15i$;

$z = 3i$;

$$\begin{array}{r|rrrr} 3i & 1 & -2 - 3i & 5 + 6i & -15i \\ & & 3i & -6i & 15i \\ \hline & 1 & -2 & 5 & 0 \end{array}$$

$C(z) = (z - 3i)(z^2 - 2z + 5)$;

$z = \dfrac{-(-2) \pm \sqrt{(-2)^2 - 4(1)(5)}}{2(1)}$

$= \dfrac{2 \pm \sqrt{-16}}{2} = 1 \pm 2i$;

$C(z) = (z - 3i)(z - 1 - 2i)(z - 1 + 2i)$

d. $C(z) = z^3 + (-4-i)z^2 + (29+4i)z - 29i$

$z = i$;

$$\begin{array}{r|rrrr} i & 1 & -4-i & 29+4i & -29i \\ & & i & -4i & 29i \\ \hline & 1 & -4 & 29 & \boxed{0} \end{array}$$

$C(z) = (z-i)(z^2 - 4z + 29)$;

$z = \dfrac{-(-4) \pm \sqrt{(-4)^2 - 4(1)(29)}}{2(1)}$

$= \dfrac{4 \pm \sqrt{-100}}{2} = 2 \pm 5i$;

$C(z) = (z-i)(z-2-5i)(z-2+5i)$

e. $C(z) = z^3 + (-2-6i)z^2 + (4+12i)z - 24i$

$z = 6i$;

$$\begin{array}{r|rrrr} 6i & 1 & -2-6i & 4+12i & -24i \\ & & 6i & -12i & 24i \\ \hline & 1 & -2 & 4 & \boxed{0} \end{array}$$

$C(z) = (z-6i)(z^2 - 2z + 4)$;

$z = \dfrac{-(-2) \pm \sqrt{(-2)^2 - 4(1)(4)}}{2(1)}$

$= \dfrac{2 \pm \sqrt{-12}}{?} = 1 \pm \sqrt{3}\, i$;

$C(z) = (z-6i)\left(z-1-\sqrt{3}\, i\right)\left(z-1+\sqrt{3}\, i\right)$

f. $C(z) = z^3 + (-6+4i)z^2 + (11-24i)z + 44i$

$z = -4i$;

$$\begin{array}{r|rrrr} -4i & 1 & -6+4i & 11-24i & 44i \\ & & -4i & 24i & -44i \\ \hline & 1 & -6 & 11 & \boxed{0} \end{array}$$

$C(z) = (z+4i)(z^2 - 6z + 11)$;

$z = \dfrac{-(-6) \pm \sqrt{(-6)^2 - 4(1)(11)}}{2(1)}$

$= \dfrac{6 \pm \sqrt{-8}}{2} = 3 \pm \sqrt{2}\, i$;

$C(z) = (z+4i)\left(z-3-\sqrt{2}\, i\right)\left(z-3+\sqrt{2}\, i\right)$

g.

$C(z) = z^3 + (-2-i)z^2 + (5+4i)z + (-6+3i)$;

$z = 2 - i$;

$$\begin{array}{r|rrrr} 2-i & 1 & -2-i & 5+4i & -6+3i \\ & & 2-i & -2-4i & 6-3i \\ \hline & 1 & -2i & 3 & \boxed{0} \end{array}$$

$C(z) = (z-2+i)(z^2 - 2iz + 3)$;

$z = \dfrac{-(-2i) \pm \sqrt{(-2i)^2 - 4(1)(3)}}{2(1)}$

$= \dfrac{2i \pm \sqrt{-16}}{2} = i \pm 2i = 3i \text{ or } -i$;

$C(z) = (z-2+i)(z-3i)(z+i)$

h.

$C(z) = z^3 - 2z^2 + (19+6i)z + (-20+30i)$;

$z = 2 - 3i$;

$$\begin{array}{r|rrrr} 2-3i & 1 & -2 & 19+6i & -20+30i \\ & & 2-3i & -9-6i & 20-30i \\ \hline & 1 & -3i & 10 & \boxed{0} \end{array}$$

$C(z) = (z-2+3i)(z^2 - 3iz + 10)$;

$z = \dfrac{-(-3i) \pm \sqrt{(-3i)^2 - 4(1)(10)}}{2(1)}$

$= \dfrac{3i \pm \sqrt{49}}{2} = \dfrac{3i \pm 7i}{2} = 5i \text{ or } -2i$;

$C(z) = (z-2+3i)(z-5i)(z+2i)$

117. Let x represent the width.

Let $\dfrac{1200-4x}{2}$ represent the length

$x\left(\dfrac{1200-4x}{2}\right) = 600x - 2x^2$

a. Opens downward, x-coordinate of vertex: $x = \dfrac{-600}{2(-2)} = 150$ ft;

$\dfrac{1200 - 4(150)}{2} = 300$ ft

b. $\dfrac{300}{3} = 100$;

$100(150) = 15{,}000 \text{ ft}^2$

119. Node $(-4, -2)$

$r(x) = a\sqrt{x+4} - 2$

Passing through $(0, 2)$

$2 = a\sqrt{0+4} - 2$

$4 = 2a$

$2 = a$;

$r(x) = 2\sqrt{x+4} - 2$

3.4 Exercises

1. zero, m

3. Bounce, flatter

5. Answers will vary.

7. polynomial, degree 3

9. not a polynomial, sharp turns

11. polynomial, degree 2

13. up/down

15. down/down

17. down/up, $(0, -2)$

19. down/down, $(0, -6)$

21. up/down, $(0, 6)$

23. (a) even
    (b) -3, odd; -1, even; 3, odd
    (c) $f(x) = (x+3)(x+1)^2(x-3)$
    (d) D: $x \in ¡$, R: $y \in [-9, \infty)$

25. (a) even
    (b) -3, odd; -1, odd; 2, odd; 4, odd
    (c) deg 4; $f(x) = -(x+3)(x+1)(x-2)(x-4)$
    (d) D: $x \in ¡$, R: $y \in (-\infty, 25]$

27. (a) odd
    (b) -1, even; 3, odd
    (c) deg 3; $f(x) = -(x+1)^2(x-3)$
    (d) D: $x \in ¡$, R: $y \in ¡$

29. degree 6, up/up, $(0, -12)$

31. degree 5, up/down, $(0, -24)$

33. degree 6, up/up, $(0, -192)$

35. degree 5, up/down, $(0, 2)$

37. b

39. e

41. c

43. $f(x) = (x+3)(x+1)(x-2)$
    end behavior: down/up
    x-intercepts: $(-3,0), (-1,0),$ and $(2,0)$;
    crosses at all x-intercepts
    $f(0) = (0+3)(0+1)(0-2) = -6$
    y-intercept: $(0,-6)$

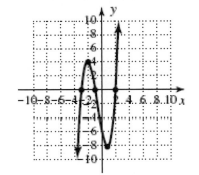

45. $p(x) = -(x+1)^2(x-3)$
    end behavior: up/down
    x-intercepts: $(-1,0)$ and $(3,0)$;
    crosses at $(3,0)$, bounces at $(-1,0)$
    $p(0) = -(0+1)^2(0-3) = 3$
    y-intercept: $(0,3)$

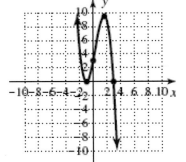

3.4 Exercises

47. $Y_1 = (x+1)^2(3x-2)(x+3)$

end behavior: up/up

x-intercepts: $(-1,0), \left(\dfrac{2}{3},0\right)$ and $(-3,0)$;

crosses at $(-3,0)$ and $\left(\dfrac{2}{3},0\right)$,

bounces at $(-1,0)$

$Y_1 = (0+1)^2(3(0)-2)(0+3) = -6$

y-intercept: $(0,-6)$

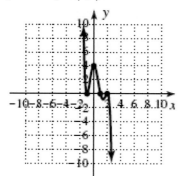

49. $r(x) = -(x+1)^2(x-2)^2(x-1)$

end behavior: up/down

x-intercepts: $(-1,0), (2,0)$ and $(1,0)$;

crosses at $(1,0)$, bounces at $(-1,0)$ and $(2,0)$

$r(0) = -(0+1)^2(0-2)^2(0-1) = 4$

y-intercept: $(0,4)$

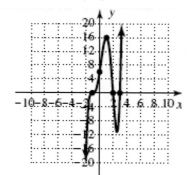

51. $f(x) = (2x+3)(x-1)^3$

end behavior: up/up

x-intercepts: $\left(-\dfrac{3}{2},0\right)$ and $(1,0)$;

crosses at all x-intercepts

$f(0) = (2(0)+3)(0-1)^3 = -3$

y-intercept: $(0,-3)$

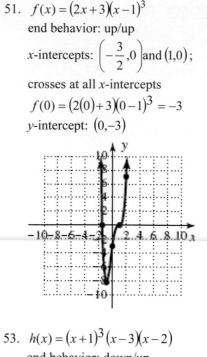

53. $h(x) = (x+1)^3(x-3)(x-2)$

end behavior: down/up

x-intercepts: $(-1,0), (3,0)$ and $(2,0)$;

crosses at all x-intercepts

$h(0) = (0+1)^3(0-3)(0-2) = 6$

y-intercept: $(0,6)$

55. $Y_3 = (x+1)^3(x-1)^2(x-2)$

    end behavior: up/up

    x-intercepts: $(-1,0), (1,0)$ and $(2,0)$;

    crosses at $(-1,0)$ and $(2,0)$, bounces at $(1,0)$;

    $Y_3 = (0+1)^3(0-1)^2(0-2) = -2$

    y-intercept: $(0,-2)$

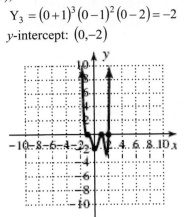

57. $y = x^3 + 3x^2 - 4$

    end behavior: down/up

    Possible rational roots: $\{\pm1, \pm4, \pm2\}$

    $y = (x+2)^2(x-1)$

    x-intercepts: $(-2,0)$ and $(1,0)$

    crosses at $(1,0)$, bounces at $(-2,0)$;

    $y = 0^3 + 3(0)^2 - 4 = -4$

    y-intercept: $(0,-4)$

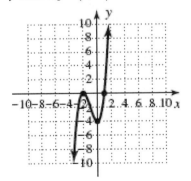

59. $f(x) = x^3 - 3x^2 - 6x + 8$

    end behavior: down/up

    Possible rational roots: $\{\pm1, \pm8, \pm2, \pm4\}$

    $f(x) = (x+2)(x-1)(x-4)$

    x-intercepts: $(-2,0), (1,0)$ and $(4,0)$

    crosses at all x-intercepts;

    $f(0) = 0^3 - 3(0)^2 - 6(0) + 8 = 8$

    y-intercept: $(0,8)$

61. $h(x) = -x^3 - x^2 + 5x - 3$

    end behavior: up/down

    Possible rational roots: $\{\pm1, \pm3\}$

    $h(x) = -1(x+3)(x-1)^2$

    x-intercepts: $(-3,0)$ and $(1,0)$

    crosses at $(-3,0)$, bounces at $(1,0)$;

    $h(0) = -0^3 - (0)^2 + 5(0) - 3 = -3$

    y-intercept: $(0,-3)$

63. $p(x) = -x^4 + 10x^2 - 9$

  end behavior: down/down

  $$p(x) = -1(x^2 - 9)(x^2 - 1)$$

  $$p(x) = -1(x+3)(x-3)(x+1)(x-1)$$

  x-intercepts:

  $(-3,0),\ (3,0),\ (-1,0)$ and $(1,0)$

  crosses at all x-intercepts;

  $$p(0) = -0^4 + 10(0)^2 - 9 = -9$$

  y-intercept: $(0,-9)$

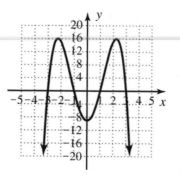

65. $r(x) = x^4 - 9x^2 - 4x + 12$

  end behavior: up/up

  Possible rational roots:

  $\{+1, +12, +2, +6, +3, +4\}$

  $$r(x) = (x+2)^2(x-1)(x-3)$$

  x-intercepts: $(-2,0),\ (1,0)$ and $(3,0)$

  crosses at $(1,0)$ and $(3,0)$, bounces at $(-2,0)$;

  $$r(0) = 0^4 - 9(0)^2 - 4(0) + 12 = 12$$

  y-intercept: $(0,12)$

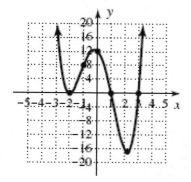

67. $Y_1 = x^4 - 6x^3 + 8x^2 + 6x - 9$

  end behavior: up/up

  Possible rational roots: $\{\pm 1, \pm 9, \pm 3\}$

  $$Y_1 = (x+1)(x-1)(x-3)^2$$

  x-intercepts: $(-1,0),\ (1,0),$ and $(3,0)$

  crosses at $(-1,0)$ and $(1,0)$, bounces at $(3,0)$;

  $$Y_1 = 0^4 - 6(0)^3 + 8(0)^2 + 6(0) - 9 = -9$$

  y-intercept: $(0,-9)$

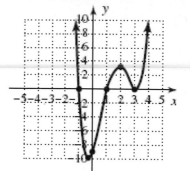

69. $Y_3 = 3x^4 + 2x^3 - 36x^2 + 24x + 32$

  end behavior: up/up

  Possible rational roots:

  $\{+1, +32, +2, +16, +4, +8\}$

  $$Y_3 = (x+4)(3x+2)(x-2)^2$$

  x-intercepts:

  $$(-4,0),\ \left(-\frac{2}{3}, 0\right),\ \text{and } (2,0)$$

  crosses at $(-4,0)$ and $\left(-\frac{2}{3}, 0\right)$,

  bounces at $(2,0)$;

  $$Y_3 = 3(0)^4 + 2(0)^3 - 36(0)^2 + 24(0) + 32 = 32$$

  y-intercept: $(0,32)$

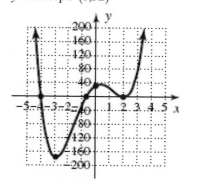

71. $F(x) = 2x^4 + 3x^3 - 9x^2$

   $F(x) = x^2(2x^2 + 3x - 9)$

   end behavior: up/up

   Possible rational roots: $\dfrac{\{\pm 1, \pm 9, \pm 3\}}{\{\pm 1, \pm 2\}}$;

   $\left\{\pm 1, \pm 9, \pm 3, \pm\dfrac{1}{2}, \pm\dfrac{9}{2}, \pm\dfrac{3}{2}\right\}$

   $F(x) = x^2(x+3)(2x-3)$

   x-intercepts:

   $(0,0),\ (-3,0),\ \text{and}\ \left(\dfrac{3}{2},0\right)$

   crosses at (3,0) and $\left(\dfrac{3}{2},0\right)$ bounces at

   $(0,0)$;

   $F(0) = 2(0)^4 + 3(0)^3 - 9(0)^2 = 0$

   y-intercept: $(0,0)$

   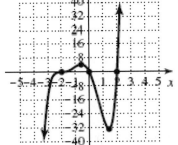

73. $f(x) = x^5 + 4x^4 - 16x^2 - 16x$

   $f(x) = x(x^4 + 4x^3 - 16x - 16)$

   end behavior: down/up

   Possible rational roots: $\{\pm 1, \pm 16, \pm 2, \pm 8, \pm 4\}$

   $f(x) = x(x+2)^3(x-2)$

   x-intercepts: $(0,0),\ (-2,0),\ \text{and}\ (2,0)$

   crosses at all x-intercepts;

   $f(0) = (0)^5 + 4(0)^4 - 16(0)^2 - 16(0) = 0$

   y-intercept: $(0,0)$

   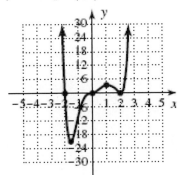

75. $h(x) = x^6 - 2x^5 - 4x^4 + 8x^3$

   $h(x) = x^3(x^3 - 2x^2 - 4x + 8)$

   $h(x) = x^3(x^2(x-2) - 4(x-2))$

   $h(x) = x^3(x-2)(x^2 - 4)$

   $h(x) = x^3(x-2)(x+2)(x-2)$

   end behavior: up/up

   x-intercepts: $(0,0),\ (-2,0)\ \text{and}\ (2,0)$

   crosses at $(-2,0)$ and $(0,0)$, bounces at $(2,0)$;

   $h(0) = (0)^6 - 2(0)^5 - 4(0)^4 + 8(0)^3 = 0$

   y-intercept: $(0,0)$

77. $h(x) = x^5 + 4x^4 - 9x - 36$

$h(x) = (x+4)(x-\sqrt{3})(x+\sqrt{3})(x^2+3)$

$h(x) = (x+4)(x-\sqrt{3})(x+\sqrt{3})(x-\sqrt{3}\,i)(x+\sqrt{3}\,i)$

y-intercept: $(0, -36)$

79. $f(x) = 2x^5 + 5x^4 - 10x^3 - 25x^2 + 12x + 30$

$f(x) = 2\left(x + \dfrac{5}{2}\right)(x-\sqrt{2})(x+\sqrt{2})(x-\sqrt{3})(x+\sqrt{3})$

y-intercept: $(0, 30)$

81. $P(x) = a(x+4)(x-1)(x-3)$;

y-intercept; $(0,2)$

$2 = a(0+4)(0-1)(0-3)$

$2 = 12a$

$\dfrac{1}{6} = a$;

$P(x) = \dfrac{1}{6}(x+4)(x-1)(x-3)$

$P(x) = \dfrac{1}{6}(x^3 - 13x + 12)$

83. $P(x) = (x+3)(x+1)(x-2)(x-4)$

$P(x) = (x^2 + 4x + 3)(x^2 - 6x + 8)$

$P(x) = x^4 - 6x^3 + 8x^2 + 4x^3 - 24x^2 + 32x$

$+ 3x^2 - 18x + 24$

$P(x) = x^4 - 2x^3 - 13x^2 + 14x + 24$

85. $v(t) = -t^4 + 25t^3 - 192t^2 + 432t$

a.

$v(2) = -(2)^4 + 25(2)^3 - 192(2)^2 + 432(2) = 280$

280 vehicles above average;

$v(6) = -(6)^4 + 25(6)^3 - 192(6)^2 + 432(6) = -216$

216 vehicles below average;

$v(11) = -(11)^4 + 25(11)^3 - 192(11)^2 + 432(11) = 154$

154 vehicles below average

b.　$0 = -t^4 + 25t^3 - 192t^2 + 432t$

$0 = -t(t^3 - 25t^2 + 192t - 432)$

Possible rational zeroes:

$\{\pm1, \pm432, \pm2, \pm216, \pm3, \pm144, \pm4, \pm108, \pm6, \pm72,$

$\pm8, \pm54, \pm9, \pm48, \pm12, \pm36, \pm16, \pm27, \pm18, \pm24\}$

$0 = t(t-4)(t-9)(t-12)$

$t = 0, t = 4, t = 9, t = 12$

6 am, 10 am, 3 pm, 6 pm

c.　$x \in [-2, 13, 1]$, $y \in [-300, 300, 30]$

87. a.　3

b.　$9 - 4 = 5$

c.　$B(x) = a(x-4)(x-9)$;

y-intercept: $(1,6)$;

$6 = a(1-4)(1-9)$

$\dfrac{1}{4} = a$;

$B(x) = \dfrac{1}{4}x(x-4)(x-9)$

$B(8) = \dfrac{1}{4}(8)(8-4)(8-9) = -\$80,000$

89. a. $f(x) \to \infty, f(x) \to -\infty$

   b. $g(x) \to \infty, g(x) \to -\infty$,

   $x^4 \geq 0$ for all x.

91. $x^5 - x^4 - x^3 + x^2 - 2x + 3 = 0$

   Possible rational roots: $\{\pm 1, \pm 3\}$

   Testing these four roots by synthetic division shows there are no rational roots.
   Verified

93. $h(x) = (f \circ g)(x) = \left(\dfrac{1}{x}\right)^2 - 2\left(\dfrac{1}{x}\right)$

   $= \dfrac{1}{x^2} - \dfrac{2}{x} = \dfrac{1-2x}{x^2}$;

   $D: x \in \{x \mid x \neq 0\}$;

   $H(x) = (g \circ f)(x) = \dfrac{1}{x^2 - 2x}$;

   $D: x \in \{x \mid x \neq 0, x \neq 2\}$

95. a. $-(2x+5) - (6-x) + 3 = x - 3(x+2)$

   $-2x - 5 - 6 + x + 3 = x - 3x - 6$

   $-x - 8 = -2x - 6$

   $x = 2$

   b. $\sqrt{x+1} + 3 = \sqrt{2x} + 2$

   $\sqrt{x+1} = \sqrt{2x} - 1$

   $\left(\sqrt{x+1}\right)^2 = \left(\sqrt{2x} - 1\right)^2$

   $x + 1 = 2x - 2\sqrt{2x} + 1$

   $-x = -2\sqrt{2x}$

   $(-x)^2 = \left(-2\sqrt{2x}\right)^2$

   $x^2 = 4(2x)$

   $x^2 - 8x = 0$

   $x(x-8) = 0$

   $x = 0 \quad \text{or} \quad x - 8 = 0$

   $x = 8$

   $x = 8 \ (x = 0 \text{ does not check})$

   c. $\dfrac{2}{x-3} + 5 = \dfrac{21}{x^2 - 9} + 4$

   $\dfrac{2}{x-3} + 5 = \dfrac{21}{(x+3)(x-3)} + 4$

   $(x-3)(x+3)\left[\dfrac{2}{x-3} + 5 = \dfrac{21}{(x+3)(x-3)} + 4\right]$

   $2(x+3) + 5(x^2 - 9) = 21 + 4(x^2 - 9)$

   $2x + 6 + 5x^2 - 45 = 21 + 4x^2 - 36$

   $2x + 5x^2 - 39 = 4x^2 - 15$

   $x^2 + 2x - 24 = 0$

   $(x+6)(x-4) = 0$

   $x + 6 = 0 \quad \text{or} \quad x - 4 = 0$

   $x = -6 \qquad \text{or} \quad x = 4$

Mid-Chapter Check

1. a.
$$x^3 + 8x^2 + 7x - 14 = \left(x^2 + 6x - 5\right)(x+2) - 4$$

$$
\begin{array}{r}
x^2 + 6x - 5 \\
x+2\overline{)x^3 + 8x^2 + 7x - 14} \\
-\underline{\left(x^3 + 2x^2\right)} \\
6x^2 + 7x \\
-\underline{\left(6x^2 + 12x\right)} \\
-5x - 14 \\
-\underline{\left(-5x - 10\right)} \\
-4
\end{array}
$$

b. $\dfrac{x^3 + 8x^2 + 7x - 14}{x+2} = x^2 + 6x - 5 - \dfrac{4}{x+2}$;

3. $f(-2) = 7$

$$
\begin{array}{r|rrrrr}
-2 & -3 & 0 & 7 & -8 & 11 \\
   &   & 6 & -12 & 10 & -4 \\
\hline
   & -3 & 6 & -5 & 2 & \boxed{7}
\end{array}
$$

5. $g(2) = (2)^3 - 6(2) - 4 = -8;$

$g(3) = (3)^3 - 6(3) - 4 = 5;$

They have opposite signs.

7. $h(x) = x^4 + 3x^3 + 10x^2 + 6x - 20$

Possible Rational Roots:
$\{\pm 1, \pm 20, \pm 2, \pm 10, \pm 4, \pm 5\}$;

$h(x) = (x+2)(x-1)\left(x^2 + 2x + 10\right)$

$x = -2, x = 1, x = -1 \pm 3i$

9. $q(x) = x^3 + 5x^2 + 2x - 8$

end behavior: down/up

Possible rational roots: $\{\pm 1, \pm 8, \pm 2, \pm 4\}$

$q(x) = (x+4)(x+2)(x-1)$

x-intercepts: $(-4, 0), (-2, 0), (1, 0)$

crosses at all x-intercepts;

$q(0) = 0^3 + 5(0)^2 + 2(0) - 8 = -8$

y-intercept: $(0, -8)$

Reinforcing Basic Concepts

1. 1.532

3.5 Technology Highlight

1. $3,420,000; undefinded since cost becomes negative.

3. very closely

3.5 Exercises

1. as $x \to -\infty, y \to 2$

3. denominator, numerator

5. about $x = 98$

7. $V(x) = \dfrac{1}{(x-1)} + 2$

   a. as $x \to \infty, y \to 2$;
       as $x \to -\infty, y \to 2$;

   b. as $x \to -1^-, y \to -\infty$;
       as $x \to -1^+, y \to \infty$;

9. $Q(x) = \dfrac{1}{(x+2)^2} + 1$

   a. as $x \to \infty, y \to 1$;
       as $x \to -\infty, y \to 1$;

   b. as $x \to -2^-, y \to \infty$;
       as $x \to -2^+, y \to \infty$;

11. reciprocal quadratic,

$$S(x) = \dfrac{1}{(x+1)^2} - 2$$

13. reciprocal function,

$$Q(x) = \dfrac{1}{(x+1)} - 2$$

15. reciprocal quadratic,

$$f(x) = \dfrac{1}{(x+2)^2} - 5$$

17. $y \to -2$

19. $y \to -\infty$

21. $x \to -1, \ y \to \pm\infty$

23. $x - 3 = 0$
$x = 3$
$D: x \in (-\infty, 3) \cup (3, \infty)$

25. $x^2 - 9 = 0$
$x^2 = 9$
$x = 3, x = -3$
$D: x \in (-\infty, -3) \cup (-3, 3) \cup (3, \infty)$

27. $2x^2 + 3x - 5 = 0$
$(2x + 5)(x - 1) = 0$
$2x + 5 = 0$ or $x - 1 = 0$
$2x = -5$ or $x = 1$
$x = -\dfrac{5}{2}$ or $x = 1$
$D: x \in \left(-\infty, -\dfrac{5}{2}\right) \cup \left(-\dfrac{5}{2}, 1\right) \cup (1, \infty)$

29. $x^2 + x + 1 = 0, b^2 - 4ac < 0$
no vertical asymptotes
$D: x \in (-\infty, \infty)$

31. $x^2 - x - 6 = 0$
$(x - 3)(x + 2) = 0$
$x - 3 = 0$ or $x + 2 = 0$
$x = 3$ or $x = -2$
yes   yes

33. $x^2 - 6x + 9 = 0$
$(x - 3)(x - 3) = 0$
$x - 3 = 0$
$x = 3$
no

35. $x^3 + 2x^2 - 4x - 8 = 0$
$x^2(x + 2) - 4(x + 2) = 0$
$(x + 2)(x^2 - 4) = 0$
$(x + 2)(x + 2)(x - 2) = 0$
$x + 2 = 0$ or $x - 2 = 0$
$x = -2$ or $x = 2$
no   yes

37. $Y_1 = \dfrac{2x-3}{x^2+1}$

   (a) $HA: y = 0$

   (b) $2x - 3 = 0$

   $x = \dfrac{3}{2}$

   crosses at $\left(\dfrac{3}{2}, 0\right)$

39. $r(x) = \dfrac{4x^2 - 9}{x^2 - 3x - 18}$

   (a) $HA: y = 4$

   (b) $4x^2 - 9 = 4\left(x^2 - 3x - 18\right)$

   $4x^2 - 9 = 4x^2 - 12x - 72$

   $12x = -63$

   $12x = -63$

   $x = -\dfrac{63}{12} = -\dfrac{21}{4}$

   crosses at $\left(-\dfrac{21}{4}, 4\right)$

41. $p(x) = \dfrac{3x^2 - 5}{x^2 - 1}$

   (a) $HA: y = 3$

   (b) $3x^2 - 5 = 3\left(x^2 - 1\right)$

   $3x^2 - 5 = 3x^2 - 3$

   does not cross

43. $f(x) = \dfrac{x^2 - 3x}{x^2 - 5}$

   $f(x) = \dfrac{x(x-3)}{x^2 - 5}$

   x-intercepts: $(0, 0)$ cross, $(3, 0)$ cross;
   y-intercept: $(0, 0)$

45. $g(x) = \dfrac{x^2 + 3x - 4}{x^2 - 1}$

   $g(x) = \dfrac{(x+4)(x-1)}{(x+1)(x-1)}$

   x-intercept: $(-4, 0)$ cross;
   y-intercept: $(0, 4)$

47. $h(x) = \dfrac{x^3 - 6x^2 + 9x}{4 - x^2}$

   $h(x) = \dfrac{x\left(x^2 - 6x + 9\right)}{4 - x^2}$

   $h(x) = \dfrac{x(x-3)(x-3)}{(2+x)(2-x)}$

   x-intercepts: $(0, 0)$ cross, $(3, 0)$ bounce;
   y-intercept: $(0, 0)$

49. $f(x) = \dfrac{x+3}{x-1}$

   $f(0) = \dfrac{(0)+3}{(0)-1} = -3;$

   y-intercept: $(0, -3)$;

   $x \quad 1 = 0$

   vertical asymptote: $x = 1$

   x-intercept: $(-3, 0)$

   horizontal asymptote: $y = 1$

   deg num = deg den

51. $F(x) = \dfrac{8x}{x^2 + 4}$

$F(0) = \dfrac{8(0)}{(0)^2 + 4} = 0;$

y-intercept: $(0,0)$;

$x^2 + 4 \neq 0$

vertical asymptote: none

x-intercept: $(0,0)$

horizontal asymptote: $y = 0$

deg num < deg den

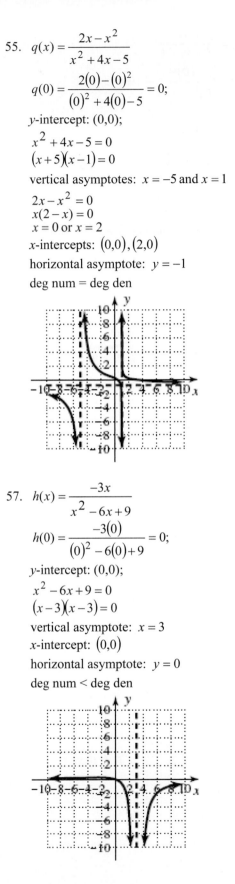

53. $p(x) = \dfrac{-2x^2}{x^2 - 4}$

$p(0) = \dfrac{-2(0)^2}{(0)^2 - 4} = 0;$

y-intercept: $(0,0)$;

$x^2 - 4 = 0$

$(x+2)(x-2) = 0$

vertical asymptote: $x = -2$ and $x = 2$

x-intercept: $(0,0)$

horizontal asymptote: $y = -2$

deg num = deg den

55. $q(x) = \dfrac{2x - x^2}{x^2 + 4x - 5}$

$q(0) = \dfrac{2(0) - (0)^2}{(0)^2 + 4(0) - 5} = 0;$

y-intercept: $(0,0)$;

$x^2 + 4x - 5 = 0$

$(x+5)(x-1) = 0$

vertical asymptotes: $x = -5$ and $x = 1$

$2x - x^2 = 0$

$x(2 - x) = 0$

$x = 0$ or $x = 2$

x-intercepts: $(0,0), (2,0)$

horizontal asymptote: $y = -1$

deg num = deg den

57. $h(x) = \dfrac{-3x}{x^2 - 6x + 9}$

$h(0) = \dfrac{-3(0)}{(0)^2 - 6(0) + 9} = 0;$

y-intercept: $(0,0)$;

$x^2 - 6x + 9 = 0$

$(x-3)(x-3) = 0$

vertical asymptote: $x = 3$

x-intercept: $(0,0)$

horizontal asymptote: $y = 0$

deg num < deg den

3.5 Exercises

59. $Y_1 = \dfrac{x-1}{x^2-3x-4}$

$Y_1 = \dfrac{(0)-1}{(0)^2-3(0)-4} = \dfrac{1}{4}$;

y-intercept: $\left(0, \dfrac{1}{4}\right)$;

$x^2-3x-4=0$
$(x-4)(x+1)=0$

vertical asymptotes: $x=4$ and $x=-1$
x-intercept: $(1,0)$
horizontal asymptote: $y=0$
deg num < deg den

61. $s(x) = \dfrac{4x^2}{2x^2+4}$

$s(0) = \dfrac{4(0)^2}{2(0)^2+4} = 0$;

y-intercept: $(0,0)$;

$2x^2+4 \neq 0$
vertical asymptotes: none
x-intercept: $(0,0)$
horizontal asymptote: $y=2$
deg num = deg den

63. $Y_1 = \dfrac{x^2-4}{x^2-1}$

$Y_1 = \dfrac{(0)^2-4}{(0)^2-1} = 4$;

y-intercept: $(0,4)$;

$Y_1 = \dfrac{(x+2)(x-2)}{(x+1)(x-1)}$

vertical asymptotes: $x=-1$ and $x=1$
x-intercepts: $(-2,0)$ and $(2,0)$
horizontal asymptote: $y=1$
deg num = deg den

65. $v(t) = \dfrac{-2x}{x^3+2x^2-4x-8}$

$v(0) = \dfrac{-2(0)}{(0)^3+2(0)^2-4(0)-8} = 0$

y-intercept: $(0,0)$;

$v(t) = \dfrac{-2x}{x^2(x+2)-4(x+2)}$

$v(t) = \dfrac{-2x}{(x+2)(x^2-4)} = \dfrac{-2x}{(x+2)^2(x-2)}$

vertical asymptotes: $x=-2$ and $x=2$
x-intercepts: $(0,0)$
horizontal asymptote: $y=0$
deg num < deg den

67. VA: $x = -2, x = 3$
    HA: $y = 1$
    $$f(x) = \frac{(x-4)(x+1)}{(x+2)(x-3)}$$

69. VA: $x = -3, x = 3$
    HA: $y = -1$
    $$f(x) = \frac{x^2 - 4}{9 - x^2}$$

71. $D(x) = \dfrac{63x}{x^2 + 20}$

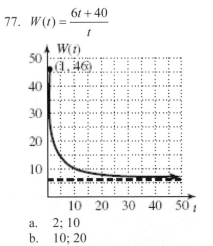

D(x) graph

   a. Population density approaches zero far from town.
   b. 10 miles, 20 miles
   c. 4.5 miles, 704

73. $C(p) = \dfrac{80p}{100 - p}$

   a. $C(20) = \dfrac{80(20)}{100 - 20} = 20$;$20,000

   $C(50) = \dfrac{80(50)}{100 - 50} = 80$;$80,000

   $C(80) = \dfrac{80(80)}{100 - 80} = 320$; $320,000

   Cost increases dramatically

b.

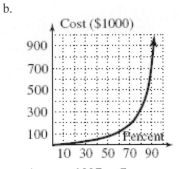

   c. As $p \to 100^-$, $C \to \infty$

75. $C(h) = \dfrac{2h^2 + h}{h^3 + 70}$

   a. According to the graph, 5 hours, about 0.28

   b. $\dfrac{\Delta C}{\Delta h} = \dfrac{C(10) - C(8)}{10 - 8} = \dfrac{0.196 - 0.234}{2}$

   $\dfrac{\Delta C}{\Delta h} = -0.019$;

   $\dfrac{\Delta C}{\Delta h} = \dfrac{C(22) - C(20)}{22 - 20} = \dfrac{0.0924 - 0.102}{2}$

   $\dfrac{\Delta C}{\Delta h} = -0.005$

   As number of hours increases, the rate of change decreases.

   c. Horizontal asymptote:

   As $h \to \infty$, $C \to 0^+$

77. $W(t) = \dfrac{6t + 40}{t}$

W(t) graph

   a. 2; 10
   b. 10; 20
   c. On the average, the number of words remembered for life is 6.

79. a. $C(x) = \dfrac{40+3x}{160+4x}$

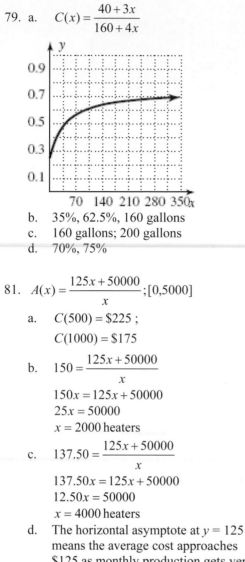

b. 35%, 62.5%, 160 gallons

c. 160 gallons; 200 gallons

d. 70%, 75%

81. $A(x) = \dfrac{125x+50000}{x}$; $[0,5000]$

a. $C(500) = \$225$;
$C(1000) = \$175$

b. $150 = \dfrac{125x+50000}{x}$
$150x = 125x+50000$
$25x = 50000$
$x = 2000$ heaters

c. $137.50 = \dfrac{125x+50000}{x}$
$137.50x = 125x+50000$
$12.50x = 50000$
$x = 4000$ heaters

d. The horizontal asymptote at $y = 125$ means the average cost approaches $\$125$ as monthly production gets very large. Due to the limitations on production (maximum of 5000 heaters) the average cost will never fall below A(5000)=135.

83. $G(n) = \dfrac{336+n(95)}{4+n}$

a. $90 = \dfrac{336+n(95)}{4+n}$
$90(4+n) = 336+n(95)$
$360+90n = 336+95n$
$24 = 5n$
$\dfrac{24}{5} = n$
5 tests

b. $93 = \dfrac{336+n(95)}{4+n}$
$93(4+n) = 336+n(95)$
$372+93n = 336+95n$
$36 = 2n$
$18 = n$

c. HA: $y = 95$
$95 = \dfrac{336+n(95)}{4+n}$
$95(4+n) = 336+n(95)$
$380+95n = 336+95n$
$380 \neq 336$
The horizontal asymptote at $y = 95$ means her average grade will approach 95 as the number of tests taken increases; no.

d. $93 = \dfrac{336+n(100)}{4+n}$
$93(4+n) = 336+n(100)$
$372+93n = 336+100n$
$36 = 7n$
$n \approx 6$

85. a.
$$\frac{\Delta C}{\Delta x} = \frac{\dfrac{250(61)}{100-61} - \dfrac{250(60)}{100-60}}{61-60} = 16.0;$$

$$\frac{\Delta C}{\Delta x} = \frac{\dfrac{250(71)}{100-71} - \dfrac{250(70)}{100-70}}{71-70} = 28.7;$$

$$\frac{\Delta C}{\Delta x} = \frac{\dfrac{250(81)}{100-81} - \dfrac{250(80)}{100-80}}{81-80} = 65.8;$$

$$\frac{\Delta C}{\Delta x} = \frac{\dfrac{250(91)}{100-91} - \dfrac{250(90)}{100-90}}{91-90} = 277.8;$$

b. 12.7, 37.1, 212.0

c.
$$\frac{\Delta C}{\Delta x} = \frac{\dfrac{350(61)}{100-61} - \dfrac{350(60)}{100-60}}{61-60} = 22.4;$$

$$\frac{\Delta C}{\Delta x} = \frac{\dfrac{350(71)}{100-71} - \dfrac{350(70)}{100-70}}{71-70} = 40.2;$$

$$\frac{\Delta C}{\Delta x} = \frac{\dfrac{350(81)}{100-81} - \dfrac{350(80)}{100-80}}{81-80} = 92.1;$$

$$\frac{\Delta C}{\Delta x} = \frac{\dfrac{350(91)}{100-91} - \dfrac{350(90)}{100-90}}{91-90} = 388.9;$$

17.8, 51.9, 296.8
Answers will vary.

87. a.
$$V(x) = \frac{3x^2 - 16x - 20}{x^2 - 3x - 10}$$

$$
\begin{array}{r}
3 \\
x^2 - 3x - 10 \overline{)3x^2 - 16x - 20} \\
\underline{3x^2 - 9x - 30} \\
-7x + 10
\end{array}
$$

$q(x) = 3$, horizontal asymptote at $y = 3$.
$r(x) = -7x + 10$, graph crosses HA at
$$x = \frac{10}{7}$$

b.
$$v(x) = \frac{-2x^2 + 4x + 13}{x^2 - 2x - 3}$$

$$
\begin{array}{r}
-2 \\
x^2 - 2x - 3 \overline{)-2x^2 + 4x + 13} \\
\underline{-2x^2 + 4x + 6} \\
7
\end{array}
$$

$q(x) = -2$, horizontal asymptote at $y = -2$.
$r(x) = 7$, no zeroes-graph will not cross.

89. $3x - 4y = 12$
$-4y = -3x + 12$
$y = \dfrac{3}{4}x - 3$; slope is $\dfrac{3}{4}$;

Slope of perpendicular is $-\dfrac{4}{3}$;

$$y - (-3) = -\frac{4}{3}(x - 2)$$

$$y + 3 = -\frac{4}{3}x + \frac{8}{3}$$

$$y = -\frac{4}{3}x - \frac{1}{3}$$

91. $f(4) = 39$;

$$
\begin{array}{r|rrrr}
4 & 2 & -7 & 5 & 3 \\
  &   & 8 & 4 & 36 \\
\hline
  & 2 & 1 & 9 & \underline{39}
\end{array}
$$

$$f\left(\frac{3}{2}\right) = \frac{3}{2};$$

$$
\begin{array}{r|rrrr}
\frac{3}{2} & 2 & -7 & 5 & 3 \\
  &   & 3 & -6 & -\frac{3}{2} \\
\hline
  & 2 & -4 & -1 & \underline{\frac{3}{2}}
\end{array}
$$

$f(2) = 1$

$$
\begin{array}{r|rrrr}
2 & 2 & -7 & 5 & 3 \\
  &   & 4 & -6 & -2 \\
\hline
  & 2 & -3 & -1 & \underline{1}
\end{array}
$$

3.6 Technology Highlight

1. $(-2, -4)$

3. $(-1, 3)$

3.6 Exercises

1. Nonremovable

3. Two

5. Answers will vary.

7. $f(x) = \dfrac{(x+2)(x-2)}{x+2}$;

$x + 2 \neq 0$
$x \neq -2$;

$f(x) = \begin{cases} \dfrac{x^2 - 4}{x+2}; & x \neq -2 \\ -4 & ; \ x = -2 \end{cases}$

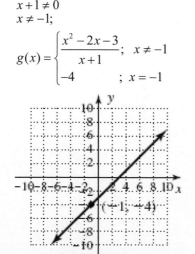

9. $g(x) = \dfrac{(x-3)(x+1)}{x+1}$;

$x + 1 \neq 0$
$x \neq -1$;

$g(x) = \begin{cases} \dfrac{x^2 - 2x - 3}{x+1}; & x \neq -1 \\ -4 & ; \ x = -1 \end{cases}$

11. $h(x) = \dfrac{x(3 - 2x)}{2x - 3} = \dfrac{-x(2x - 3)}{2x - 3}$

$2x - 3 \neq 0$
$x \neq \dfrac{3}{2}$;

$h(x) = \begin{cases} \dfrac{3x - 2x^2}{2x - 3} ; & x \neq \dfrac{3}{2} \\ -\dfrac{3}{2} & ; \ x = \dfrac{3}{2} \end{cases}$

13. $p(x) = \dfrac{(x-2)(x^2 + 2x + 4)}{x - 2}$;

$x - 2 \neq 0$
$x \neq 2$;

$p(x) = \begin{cases} \dfrac{x^3 - 8}{x - 2}; & x \neq 2 \\ 12 & ; \ x = 2 \end{cases}$

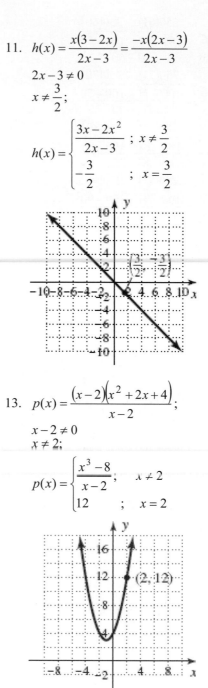

15. $q(x) = \dfrac{x^3 - 7x - 6}{x + 1}$;

$x + 1 \neq 0$
$x \neq -1$;

$$q(x) = \begin{cases} \dfrac{x^3 - 7x - 6}{x + 1} & x \neq -1 \\ \\ -4 & x = -1 \end{cases}$$

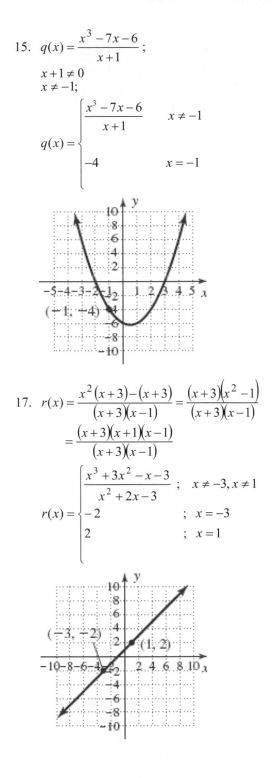

17. $r(x) = \dfrac{x^2(x+3) - (x+3)}{(x+3)(x-1)} = \dfrac{(x+3)(x^2-1)}{(x+3)(x-1)}$

$= \dfrac{(x+3)(x+1)(x-1)}{(x+3)(x-1)}$

$$r(x) = \begin{cases} \dfrac{x^3 + 3x^2 - x - 3}{x^2 + 2x - 3} & ; \quad x \neq -3, x \neq 1 \\ -2 & ; \quad x = -3 \\ 2 & ; \quad x = 1 \end{cases}$$

19. $Y_1 = \dfrac{x^2 - 4}{x}$

$x^2 - 4 = 0$

$x^2 = 4$

$x = \pm 2$

x-intercepts: $(-2, 0)$ and $(2, 0)$;

y-intercept: none;

$Y_1 = \dfrac{x^2}{x} - \dfrac{4}{x} = x - \dfrac{4}{x}$

$q(x) = x$

Oblique Asymptote: $y = x$

Vertical Asymptote: $x = 0$

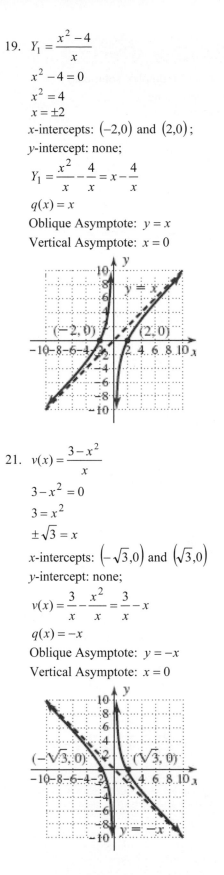

21. $v(x) = \dfrac{3 - x^2}{x}$

$3 - x^2 = 0$

$3 = x^2$

$\pm\sqrt{3} = x$

x-intercepts: $\left(-\sqrt{3}, 0\right)$ and $\left(\sqrt{3}, 0\right)$

y-intercept: none;

$v(x) = \dfrac{3}{x} - \dfrac{x^2}{x} = \dfrac{3}{x} - x$

$q(x) = -x$

Oblique Asymptote: $y = -x$

Vertical Asymptote: $x = 0$

23. $w(x) = \dfrac{x^2 + 1}{x}$

$x^2 + 1 \neq 0$ (complex solutions)

x-intercepts: none
y-intercept: none;

$w(x) = \dfrac{x^2}{x} + \dfrac{1}{x} = x + \dfrac{1}{x}$

$q(x) = x$

Oblique Asymptote: $y = x$
Vertical Asymptote: $x = 0$

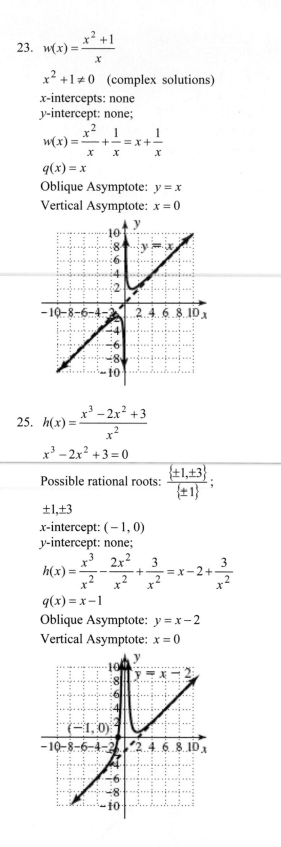

25. $h(x) = \dfrac{x^3 - 2x^2 + 3}{x^2}$

$x^3 - 2x^2 + 3 = 0$

Possible rational roots: $\dfrac{\{\pm1, \pm3\}}{\{\pm1\}}$;

$\pm1, \pm3$

x-intercept: $(-1, 0)$
y-intercept: none;

$h(x) = \dfrac{x^3}{x^2} - \dfrac{2x^2}{x^2} + \dfrac{3}{x^2} = x - 2 + \dfrac{3}{x^2}$

$q(x) = x - 1$

Oblique Asymptote: $y = x - 2$
Vertical Asymptote: $x = 0$

27. $Y_1 = \dfrac{x^3 + 3x^2 - 4}{x^2}$

$x^3 + 3x^2 - 4 = 0$

Possible rational roots: $\dfrac{\{\pm1, \pm4, \pm2\}}{\{\pm1\}}$;

$\pm1, \pm4, \pm2$

x-intercepts: $(1, 0)$; $(-2, 0)$
y-intercept: none;

$Y_1 = \dfrac{x^3}{x^2} + \dfrac{3x^2}{x^2} - \dfrac{4}{x^2} = x + 3 - \dfrac{4}{x^2}$

$q(x) = x + 3$

Oblique Asymptote: $y = x + 3$
Vertical Asymptote: $x = 0$

29. $f(x) = \dfrac{x^3 - 3x + 2}{x^2}$

$x^3 - 3x + 2 = 0$

Possible rational roots: $\dfrac{\{\pm1, \pm2\}}{\{\pm1\}}$;

$\pm1, \pm2$

x-intercepts: $(-2, 0)$ and $(1, 0)$
y-intercept: none;

$f(x) = \dfrac{x^3}{x^2} - \dfrac{3x}{x^2} + \dfrac{2}{x^2} = x - \dfrac{3}{x} + \dfrac{2}{x^2}$

$q(x) = x$

Oblique Asymptote: $y = x$
Vertical Asymptote: $x = 0$

31. $Y_3 = \dfrac{x^3 - 5x^2 + 4}{x^2}$

$x^3 - 5x^2 + 4 = 0$

Possible rational roots: $\dfrac{\{\pm 1, \pm 4, \pm 2\}}{\{\pm 1\}}$;

$\pm 1, \pm 4, \pm 2$

$Y_3 = (x-1)(x^2 - 4x - 4)$;

$x = \dfrac{-(-4) \pm \sqrt{(-4)^2 - 4(1)(-4)}}{2(1)}$

$x = \dfrac{4 \pm \sqrt{32}}{2}$

$x = \dfrac{4 \pm 4\sqrt{2}}{2}$

$x = 2 \pm 2\sqrt{2}$

x-intercepts: $(1,0)$, $(2 + \sqrt{2}, 0)$ and $(2 - \sqrt{2}, 0)$

y-intercept: none;

$Y_3 = \dfrac{x^3}{x^2} - \dfrac{5x^2}{x^2} + \dfrac{4}{x^2} = x - 5 + \dfrac{4}{x^2}$

$q(x) = x - 5$

Oblique Asymptote: $y = x - 5$

Vertical Asymptote: $x = 0$

33. $r(x) = \dfrac{x^3 - x^2 - 4x + 4}{x^2}$

$x^3 - x^2 - 4x + 4 = 0$

$x^2(x - 1) - 4(x - 1) = 0$

$(x - 1)(x^2 - 4) = 0$

$(x - 1)(x + 2)(x - 2) = 0$

$x - 1 = 0$ or $x + 2 = 0$ or $x - 2 = 0$

$x = 1$ or $x = -2$ or $x = 2$

x-intercepts: $(-2, 0)$ and $(1, 0)$ and $(2, 0)$

y-intercept: none;

$r(x) = \dfrac{x^3}{x^2} - \dfrac{x^2}{x^2} - \dfrac{4x}{x^2} + \dfrac{4}{x^2} = x - 1 - \dfrac{4}{x} + \dfrac{4}{x^2}$

$q(x) = x - 1$

Oblique Asymptote: $y = x - 1$

Vertical Asymptote: $x = 0$

35. $g(x) = \dfrac{x^2 + 4x + 4}{x + 3}$

$x^2 + 4x + 4 = 0$

$(x + 2)(x + 2) = 0$

$x + 2 = 0$

$x = -2$

x-intercept: $(-2, 0)$

y-intercept: none;

$$
\begin{array}{r}
x + 1 \\
x + 3 \overline{\smash{)} x^2 + 4x + 4} \\
-\underline{(x^2 + 3x)} \\
x + 4 \\
-\underline{(x + 3)} \\
1
\end{array}
$$

Oblique Asymptote: $y = x + 1$

$x + 3 = 0$

Vertical Asymptote: $x = -3$

37. $f(x) = \dfrac{x^2+1}{x+1}$

$x^2 + 1 \neq 0$ (complex solutions)

x-intercepts: none

y-intercept: $(0,1)$;

$$\begin{array}{r} x-1 \\ x+1\overline{)x^2+0x+1} \\ -\left(x^2+x\right) \\ \hline -x+1 \\ -(-x-1) \\ \hline 2 \end{array}$$

Oblique Asymptote: $y = x - 1$

$x + 1 = 0$

Vertical Asymptote: $x = -1$

Oblique Asymptote: $y = x - 1$

$x + 1 = 0$

Vertical Asymptote: $x = -1$

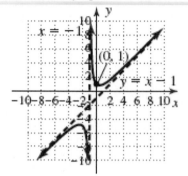

39. $Y_3 = \dfrac{x^2-4}{x+1}$

$x^2 - 4 = 0$

$x^2 = 4$

$x = \pm 2$

x-intercepts: $(-2,0)$ and $(2,0)$;

y-intercept: $(0,-4)$;

$$\begin{array}{r} x-1 \\ x+1\overline{)x^2+0x-4} \\ -\left(x^2+x\right) \\ \hline -x-4 \\ -(-x-1) \\ \hline -3 \end{array}$$

41. $v(x) = \dfrac{x^3-4x}{x^2-1}$

$x\left(x^2-4\right) = 0$

$x(x+2)(x-2) = 0$

$x = 0$ or $x + 2 = 0$ or $x - 2 = 0$

$x = 0$ or $x = -2$ or $x = 2$

x-intercepts: $(-2,0), (2,0)$ and $(0,0)$

y-intercept: $(0,0)$;

$$\begin{array}{r} x \\ x^2-1\overline{)x^3-4x} \\ -\left(x^3-x\right) \\ \hline -3x \end{array}$$

Oblique Asymptote: $y = x$

$x^2 - 1 = 0$

$(x+1)(x-1) = 0$

$x + 1 = 0$ or $x - 1 = 0$

$x = -1$ or $x = 1$

Vertical Asymptote: $x = -1$ or $x = 1$

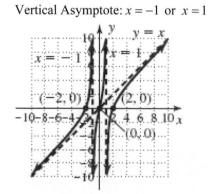

43. $w(x) = \dfrac{16x - x^3}{x^2 + 4}$

$x(16 - x^2) = 0$

$x(4 + x)(4 - x) = 0$

$x = 0$ or $4 + x = 0$ or $4 - x = 0$

$x = 0$ or $x = -4$ or $x = 4$

x-intercepts: $(-4,0),\ (4,0)$ and $(0,0)$

y-intercept: $(0,0)$;

$$
\begin{array}{r}
-x \\
x^2 + 4 \overline{\smash{)}-x^3 + 16x} \\
\underline{-\left(-x^3 - 4x\right)} \\
20x
\end{array}
$$

Oblique Asymptote: $y = -x$

$x^2 + 4 \ne 0$ (complex solutions)

Vertical Asymptote: none

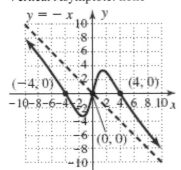

45. $W(x) = \dfrac{x^3 - 3x + 2}{x^2 - 9}$

$x^3 - 3x + 2 = 0$

Possible rational roots: $\dfrac{\pm 1, \pm 2}{\pm 1}$;

$\pm 1, \pm 2$

x-intercept: $(-2, 0)$ and $(1,0)$

y-intercept: $\left(0, -\dfrac{2}{9}\right)$;

$$
\begin{array}{r}
x \\
x^2 - 9 \overline{\smash{)}x^3 - 3x + 2} \\
\underline{-\left(x^3 - 9x\right)} \\
6x + 2
\end{array}
$$

Oblique Asymptote: $y = x$

$x^2 - 9 = 0$

$(x + 3)(x - 3) = 0$

$x + 3 = 0$ or $x - 3 = 0$

$x = -3$ or $x = 3$

Vertical Asymptote: $x = -3$ or $x = 3$

47. $p(x) = \dfrac{x^4 + 4}{x^2 + 1}$

$x^4 + 4 = 0$

Possible rational roots: $\dfrac{\pm 1, \pm 4, \pm 2}{\pm 1}$;

$\pm 1, \pm 4, \pm 2$

$x^4 + 4 \ne 0$ (complex solutions)

x-intercept: none

y-intercept: $(0,4)$;

$$
\begin{array}{r}
x^2 - 1 \\
x^2 + 1 \overline{\smash{)}x^4 + 0x^2 + 4} \\
\underline{-\left(x^4 + x^2\right)} \\
-x^2 + 4 \\
\underline{-\left(-x^2 - 1\right)} \\
5
\end{array}
$$

Oblique Asymptote: $y = x^2 - 1$

$x^2 + 1 \ne 0$ (complex solutions)

Vertical Asymptote: none

49. $q(x) = \dfrac{10 + 9x^2 - x^4}{x^2 + 5}$

$10 + 9x^2 - x^4 = 0$

$(10 - x^2)(1 + x^2) = 0$

$10 - x^2 = 0$ or $1 + x^2 \neq 0$

$10 = x^2$

$\pm\sqrt{10} = x$

x-intercepts: $(-\sqrt{10}, 0)$ and $(\sqrt{10}, 0)$

y-intercept: $(0, 2)$;

$$x^2 + 5 \overline{\smash{\big)}\ {-x^4 + 9x^2 + 10}} \quad \overset{\displaystyle -x^2 + 14}{}$$
$$\underline{-(-x^4 - 5x^2)}$$
$$14x^2 + 10$$
$$\underline{-(14x^2 + 70)}$$
$$-60$$

Oblique Asymptote: $y = -x^2 + 14$

$x^2 + 5 \neq 0$ (complex solutions)

Vertical Asymptote: none

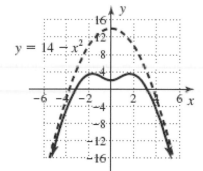

51. $f(x) = \dfrac{x^3}{x} + \dfrac{500}{x} = x^2 + \dfrac{500}{x}$

Oblique Asymptote: $y = x^2$

Minimum: 119.1

53. $A(a) = \dfrac{1}{2}\left(\dfrac{ka^2}{a - h}\right)$

$A(a) = \dfrac{1}{2}\left(\dfrac{6a^2}{a - 5}\right)$

$A(a) = \dfrac{3a^2}{a - 5}$

a. Oblique Asymptote: $y = 3a + 15$

$a - 5 = 0$

Vertical Asymptote: $a = 5$

b. $A(11) = \dfrac{3(11)^2}{11 - 5} = 60.5$

c. $(10, 0)$

55. a. $A(x) = \dfrac{4x^2 + 53x + 250}{x}$

Vertical Asymptote: $x = 0$

Oblique Asymptote: $q(x) = 4x + 53$

b. Cost: \$307, \$372, \$445

Avg Cost: \$307, \$186, \$148.33

c. 8, \$116.25

d. verified

57. a. $S(x, y) = 2x^2 + 4xy$;

$V(x, y) = x^2 y$

b. $12 = x^2 y$

$\dfrac{12}{x^2} = y$;

$S(x) = 2x^2 + 4x\left(\dfrac{12}{x^2}\right)$

$= 2x^2 + \dfrac{48}{x} = \dfrac{2x^3 + 48}{x}$

c. $S(x)$ is asymptotic to $y = 2x^2$

d. $x = 2$ ft 3.5 in;

$y = 2$ ft 3.5 in

59. a. $A(x, y) = xy$;

$R(x, y) = (x - 2.5)(y - 2)$

b. $60 = (x - 2.5)(y - 2)$

$$\frac{60}{x - 2.5} = y - 2$$

$$\frac{60}{x - 2.5} + 2 = y$$

$$\frac{2x + 55}{x - 2.5} = y;$$

$$A(x) = x\left(\frac{60}{x - 2.5} + 2\right)$$

$$A(x) = \frac{60x}{x - 2.5} + 2x$$

$$A(x) = \frac{60x}{x - 2.5} + 2x\left(\frac{x - 2.5}{x - 2.5}\right)$$

$$A(x) = \frac{60x + 2x^2 - 5x}{x - 2.5} = \frac{2x^2 + 55x}{x - 2.5}$$

c. $A(x)$ is asymptotic to $y = 2x + 60$

d. $x \approx 11.16$ in.;

$y \approx 8.93$ in.

61. a. $V = \pi r^2 h$;

$$\frac{V}{\pi r^2} = h$$

b. $S = 2\pi r^2 + 2\pi r\left(\frac{V}{\pi r^2}\right) = 2\pi r^2 + \frac{2V}{r}$

c. $S = 2\pi r^2 + \frac{2V}{r} = \frac{2\pi r^3 + 2V}{r}$

d. $\frac{1200}{\pi r^2} = h$

$r \approx 5.76$ cm, $h \approx 11.51$ cm;

$S \approx 625.13$ cm^3

63. Answers will vary.

65. $S = \frac{\pi r^3 + 2V}{r}$;

$S = \frac{\pi r^3 + 180}{r}$;

$90 = \frac{\pi r^3 + 180}{r}$

$90r = \pi r^3 + 180$

$0 = \pi r^3 - 90r + 180$

Using grapher, $r \approx 3.1$ in., $h \approx 3.0$ in.

67. $-3x + 4y = -16$

$4y - 3x \quad 16$

$y = \frac{3}{4}x - 4$; $m = \frac{3}{4}$; $(0, -4)$

69. a. $\left(\overline{AB}\right)^2 = 12^2 + 5^2$

$\overline{AB} = \sqrt{169} = 13$;

Perimeter $= 12 + 5 + 13 = 30$ cm

b. $\left(\overline{CB}\right)^2 = \overline{AB} \cdot \overline{DB}$

$5^2 = 13 \cdot \overline{DB}$

$\frac{25}{13} = \overline{DB}$;

$\left(\overline{CD}\right)^2 + \left(\overline{DB}\right)^2 = 5^2$

$\left(\overline{CD}\right)^2 + \left(\frac{25}{13}\right)^2 = 5^2$

$\left(\overline{CD}\right)^2 = \frac{3600}{169}$

$\overline{CD} = \frac{60}{13}$ cm

c. $A = \frac{1}{2}(13)\left(\frac{60}{13}\right) = 30$ cm^2

d. $A_{BDC} = \frac{1}{2}\left(\frac{60}{13}\right)\left(\frac{25}{13}\right) = \frac{750}{169} \approx 4.4$ cm^2;

$A_{ADC} = 30 - 4.4 = \frac{4320}{169} \approx 25.6$ cm^2

3.7 Technology Highlight

1. $P(x) < 0 : x \in (-3.1, -1.7) \cup (1.3, 2.4)$

3.7 Exercises

1. Vertical, multiplicity

3. Empty

5. Answers will vary.

7. $f(x) = -x^2 + 4x; \quad f(x) > 0$
$-x^2 + 4x = 0$
$-x(x - 4) = 0$
$x = 0; \quad x = 4$
Concave down
$x \in (0, 4)$

9. $h(x) = x^2 + 4x - 5; \quad h(x) \geq 0$
$x^2 + 4x - 5 = 0$
$(x + 5)(x - 1) = 0$
$x = -5; \quad x = 1$
Concave up
$x \in (-\infty, -5] \cup [1, \infty)$

11. $q(x) = 2x^2 - 5x - 7; \quad q(x) < 0$
$2x^2 - 5x - 7 = 0$
$(2x - 7)(x + 1) = 0$
$x = \dfrac{7}{2}; \quad x = -1$
Concave up
$x \in \left(-1, \dfrac{7}{2}\right)$

13. $7 \geq x^2$
$7 = x^2$
$\pm\sqrt{7} = x$
$x \in \left[-\sqrt{7}, \sqrt{7}\right]$

15. $x^2 + 3x \leq 6$
$x^2 + 3x - 6 = 0$
$x = \dfrac{-3 \pm \sqrt{3^2 - 4(1)(-6)}}{2(1)}$
$x = \dfrac{-3 \pm \sqrt{9 + 24}}{2}$
$x = \dfrac{-3 \pm \sqrt{33}}{2}$
Concave up
$x \in \left(-\infty, \dfrac{-3 - \sqrt{33}}{2}\right] \cup \left[\dfrac{-3 + \sqrt{33}}{2}, \infty\right)$

17. $3x^2 \geq -2x + 5$
$3x^2 + 2x - 5 = 0$
$(3x + 5)(x - 1) = 0$
$x = \dfrac{-5}{3}; \quad x = 1$
Concave up
$x \in \left(-\infty, \dfrac{-5}{3}\right] \cup [1, \infty)$

19. $s(x) = x^2 - 8x + 16; \quad s(x) \geq 0$
$x^2 - 8x + 16 = 0$
$(x - 4)(x - 4) = 0$
$x = 4$
Concave up
$x \in (-\infty, \infty)$

21. $r(x) = 4x^2 + 12x + 9; \quad r(x) < 0$
$4x^2 + 12x + 9 = 0$
$(2x + 3)(2x + 3) = 0$
$x = -\dfrac{3}{2}$
Concave up
No solution

23. $g(x) = -x^2 + 10x - 25; \quad g(x) < 0$
$-x^2 + 10x - 25 = 0$
$-(x^2 - 10x + 25) = 0$
$-(x - 5)(x - 5) = 0$
$x = 5$
Concave down
$x \in (-\infty, 5) \cup (5, \infty)$

25. $-x^2 > 2$

$-x^2 = 2$

$x^2 = -2$

$x = \sqrt{-2}$

No x-intercepts
Concave down
No solution

27. $x^2 - 2x > -5$

$x^2 - 2x + 5 = 0$

$x = \dfrac{2 \pm \sqrt{(-2)^2 - 4(1)(5)}}{2(1)}$

$x = \dfrac{2 \pm \sqrt{4 - 20}}{2}$

$x = \dfrac{2 \pm \sqrt{-16}}{2}$

No x-intercepts
Concave up
$x \in (-\infty, \infty)$

29. $p(x) = 2x^2 - 6x + 9; \quad p(x) \geq 0$

$2x^2 - 6x + 9 = 0$

$x = \dfrac{6 \pm \sqrt{(-6)^2 - 4(2)(9)}}{2(2)}$

$x = \dfrac{6 \pm \sqrt{36 - 72}}{4}$

$x = \dfrac{6 \pm \sqrt{-36}}{4}$

No x-intercepts
Concave up
$x \in (-\infty, \infty)$

31. $h(x) = \sqrt{x^2 - 25}$

$x^2 - 25 \geq 0$

To find zeroes, solve

$x^2 = 25$

$x = \pm 5$

Use a number line diagram, plot (-5, 0) and (5, 0). Sketch a parabola opening upward. The graph is above the x-axis when domain is $\quad x \in (-\infty, -5] \cup [5, \infty)$

33. $q(x) = \sqrt{x^2 - 5x}$

To find zeroes, solve

$x^2 - 5x = 0$

$x(x - 5) = 0$

$x = 0$ or $x = 5$

Use a number line diagram, plot (0, 0) and (5, 0). Sketch a parabola opening upward. The graph is above the x-axis when domain is: $\quad x \in (-\infty, 0] \cup [5, \infty)$

35. $t(x) = \sqrt{-x^2 + 3x - 4}$

To find the zeroes, solve

$-x^2 + 3x - 4 = 0$

$a = -1, b = 3, c = -4$

$x = \dfrac{-3 \pm \sqrt{(-3)^2 - 4(-1)(-4)}}{2(-1)}$

$x = \dfrac{-3 \pm \sqrt{-7}}{-1}$

No solution

37. $(x + 3)(x - 5) < 0$

$x \in (-3, 5)$

39. $(x + 1)^2 (x - 4) \geq 0$

$x \in [4, \infty) \cup \{-1\}$

41. $(x + 2)^3 (x - 2)^2 (x - 4) \geq 0$

$x \in (-\infty, -2] \cup \{2\} \cup [4, \infty)$

43. $x^2 + 4x + 1 < 0$;

$x = \dfrac{-(4) \pm \sqrt{(4)^2 - 4(1)(1)}}{2(1)} = \dfrac{-4 \pm \sqrt{12}}{2}$

$= \dfrac{-4 \pm 2\sqrt{3}}{2} = -2 \pm \sqrt{3}$;

$x \in \left(-2 - \sqrt{3}, -2 + \sqrt{3}\right)$

45. $x^3 + x^2 - 5x + 3 \le 0$

Possible rational roots: $\dfrac{\{\pm 1, \pm 3\}}{\{\pm 1\}}$;

$\{\pm 1, \pm 3\}$

$$\begin{array}{r|rrrr} 1 & 1 & 1 & -5 & 3 \\ & & 1 & 2 & -3 \\ \hline & 1 & 2 & -3 & \underline{|0} \end{array}$$

$(x-1)(x^2 + 2x - 3) \le 0$

$(x-1)(x+3)(x-1) \le 0$

$(x+3)(x-1)^2 \le 0$

$x \in (-\infty, -3] \cup \{1\}$

47. $x^3 - 7x + 6 > 0$

Possible rational roots: $\dfrac{\{\pm 1, \pm 6, \pm 2, \pm 3\}}{\{\pm 1\}}$;

$\{\pm 1, \pm 6, \pm 2, \pm 3\}$

$$\begin{array}{r|rrrr} 1 & 1 & 0 & -7 & 6 \\ & & 1 & 1 & -6 \\ \hline & 1 & 1 & -6 & \underline{|0} \end{array}$$

$(x-1)(x^2 + x - 6) > 0$

$(x-1)(x+3)(x-2) > 0$

$x \in (-3, 1) \cup (2, \infty)$

49. $x^4 - 10x^2 > -9$

$x^4 - 10x^2 + 9 > 0$

$(x^2 - 1)(x^2 - 9) > 0$

$(x+1)(x-1)(x+3)(x-3) > 0$

$x \in (-\infty, -3) \cup (-1, 1) \cup (3, \infty)$

51. $x^4 - 9x^2 > 4x - 12$

$x^4 - 9x^2 - 4x + 12 > 0$

Possible rational roots:

$\dfrac{\{\pm 1, \pm 12, \pm 2, \pm 6, \pm 3, \pm 4\}}{\{\pm 1\}}$;

$\{\pm 1, \pm 12, \pm 2, \pm 6, \pm 3, \pm 4\}$

$$\begin{array}{r|rrrrr} 1 & 1 & 0 & -9 & -4 & 12 \\ & & 1 & 1 & -8 & -12 \\ \hline & 1 & 1 & -8 & -12 & \underline{|0} \end{array}$$

$$\begin{array}{r|rrrr} 3 & 1 & 1 & -8 & -12 \\ & & 3 & 12 & 12 \\ \hline & 1 & 4 & 4 & \underline{|0} \end{array}$$

$(x-1)(x-3)(x^2 + 4x + 4) > 0$

$(x-1)(x-3)(x+2)^2 > 0$

$x \in (-\infty, -2) \cup (-2, 1) \cup (3, \infty)$

53. $x^4 - 6x^3 \le -8x^2 - 6x + 9$

$x^4 - 6x^3 + 8x^2 + 6x - 9 \le 0$

Possible rational roots: $\dfrac{\{\pm 1, \pm 9, \pm 3\}}{\{\pm 1\}}$;

$\{\pm 1, \pm 9, \pm 3\}$

$$\begin{array}{r|rrrrr} 1 & 1 & 6 & 8 & 6 & 9 \\ & & -1 & 7 & -15 & 9 \\ \hline & 1 & -7 & 15 & -9 & \underline{|0} \end{array}$$

$$\begin{array}{r|rrrr} 1 & 1 & -7 & 15 & -9 \\ & & 1 & -6 & 9 \\ \hline & 1 & -6 & 9 & \underline{|0} \end{array}$$

$(x+1)(x-1)(x^2 - 6x + 9) \le 0$

$(x+1)(x-1)(x-3)^2 \le 0$

$x \in [-1, 1] \cup \{3\}$

55. $\dfrac{x+3}{x-2} \le 0$

$x \in [-3, 2)$

57. $\dfrac{x+1}{x^2+4x+4}<0$

$\dfrac{x+1}{(x+2)^2}<0$

neg   neg   pos

   -2   -1

$x\in(-\infty,-2)\cup(-2,-1)$

59. $\dfrac{2-x}{x^2-x-6}\geq 0$

$\dfrac{2-x}{(x-3)(x+2)}\geq 0$

pos   neg   pos neg

   -2   2   3

$x\in(-\infty,-2)\cup[2,3)$

61. $\dfrac{2x-x^2}{x^2+4x-5}<0$

$\dfrac{x(2-x)}{(x+5)(x-1)}<0$

neg pos neg pos neg

  -5   0   1   2

$x\in(-\infty,-5)\cup(0,1)\cup(2,\infty)$

63. $\dfrac{x^2-4}{x^3-13x+12}\geq 0$

Possible rational roots of denominator:

$\dfrac{\{\pm 1,\pm 12,\pm 2,\pm 6,\pm 3,\pm 4\}}{\{\pm 1\}}$;

$\{\pm 1,\pm 12,\pm 2,\pm 6,\pm 3,\pm 4\}$

$\underline{1|}\ \ 1\quad 0\quad -13\quad 12$
$\qquad\quad 1\quad\ \ 1\quad -12$
$\overline{\qquad 1\quad 1\quad -12\ \ \underline{|0}}$

$x^3-13x+12=(x-1)(x^2+x-12)$
$\qquad\qquad\qquad =(x-1)(x+4)(x-3)$;

$\dfrac{(x+2)(x-2)}{(x-1)(x+4)(x-3)}\geq 0$

neg pos neg pos neg pos

  -4 -2 1   2   3

$x\in(-4,-2]\cup(1,2]\cup(3,\infty)$

65. $\dfrac{x^2+5x-14}{x^3+x^2-5x+3}>0$

Possible rational roots of denominator:

$\dfrac{\{\pm 1,\pm 3\}}{\{\pm 1\}}$; $\{\pm 1,\pm 3\}$;

$\underline{1|}\ \ 1\quad 1\quad -5\quad 3$
$\qquad\quad\ \ 1\quad 2\quad -3$
$\overline{\qquad 1\quad 2\quad -3\ \ \underline{|0}}$

$x^3+x^2-5x+3=(x-1)(x^2+2x-3)$
$\qquad\qquad\qquad\quad =(x-1)(x+3)(x-1)$;

$\dfrac{(x+7)(x-2)}{(x-1)^2(x+3)}>0$

neg pos neg neg pos

 -7   -3   1 2

$x\in(-7,-3)\cup(2,\infty)$

67. $\dfrac{2}{x-2}\leq\dfrac{1}{x}$

$\dfrac{2}{x-2}-\dfrac{1}{x}\leq 0$

$\dfrac{2x-x+2}{x(x-2)}\leq 0$

$\dfrac{x+2}{x(x-2)}\leq 0$

neg pos neg   pos

 -2   0   2

$x\in(-\infty,-2]\cup(0,2)$

69. $\dfrac{x-3}{x+17}>\dfrac{1}{x-1}$

$\dfrac{x-3}{x+17}-\dfrac{1}{x-1}>0$

$\dfrac{(x-3)(x-1)-1(x+17)}{(x+17)(x-1)}>0$

$\dfrac{x^2-4x+3-x-17}{(x+17)(x-1)}>0$

$\dfrac{x^2-5x-14}{(x+17)(x-1)}>0$

$\dfrac{(x-7)(x+2)}{(x+17)(x-1)}>0$

pos neg pos neg pos

 -17   -2   1   7

$x\in(-\infty,-17)\cup(-2,1)\cup(7,\infty)$

71. $\dfrac{x+1}{x-2} \ge \dfrac{x+2}{x+3}$

$\dfrac{x+1}{x-2} - \dfrac{x+2}{x+3} \ge 0$

$\dfrac{(x+1)(x+3)-(x+2)(x-2)}{(x-2)(x+3)} \ge 0$

$\dfrac{x^2+4x+3-x^2+4}{(x-2)(x+3)} \ge 0$

$\dfrac{4x+7}{(x-2)(x+3)} \ge 0$

neg   pos   neg   pos
   ○—————●——————————○——————▶
  -3   -$\frac{7}{4}$ 2

$x \in \left(-3, -\dfrac{7}{4}\right] \cup (2, \infty)$

73. $\dfrac{x+2}{x^2+9} > 0$

$x^2 + 9$ has no real roots

       neg   pos
————————————————○—————————▶
               -2

$x \in (-2, \infty)$

75. $\dfrac{x^3+1}{x^2+1} > 0$

$\dfrac{(x+1)(x^2-x+1)}{x^2+1} > 0$

$x^2 - x + 1, \; x^2 + 1$ have no real roots

        neg   pos
————————————————○—————————▶
               -1

$x \in (-1, \infty)$

77. $\dfrac{x^4-5x^2-36}{x^2-2x+1} > 0$

$\dfrac{(x^2-9)(x^2+4)}{(x-1)^2} > 0$

$\dfrac{(x+3)(x-3)(x^2+4)}{(x-1)^2} > 0$

$x^2 + 4$ has no real roots

   pos   neg neg   pos
————————○——————○——○————————▶
       -3   1 3

$x \in (-\infty, -3) \cup (3, \infty)$

79. $x^2 - 2x \ge 15$

$x^2 - 2x - 15 \ge 0$

$(x-5)(x+3) \ge 0$

$x \in (-\infty, -3] \cup [5, \infty)$

81. $x^3 \ge 9x$

$x^3 - 9x \ge 0$

$x(x^2-9) \ge 0$

$x(x+3)(x-3) \ge 0$

$x \in [-3, 0] \cup [3, \infty)$

83. $-4x + 12 < -x^3 + 3x^2$

$x^3 - 3x^2 - 4x + 12 < 0$

$x^2(x-3) - 4(x-3) < 0$

$(x-3)(x^2-4) < 0$

$(x-3)(x+2)(x-2) < 0$

$x \in (-\infty, -2) \cup (2, 3)$

85. $\dfrac{x^2-x-6}{x^2-1} \ge 0$

$\dfrac{(x+2)(x-3)}{(x+1)(x-1)} \ge 0$

$x \in (-\infty, -2] \cup (-1, 1) \cup [3, \infty)$

87. b

89. b

91. a. $D = -\left(4p^3 + 27(p+1)^2\right)$

    $D = -\left(4p^3 + 27\left(p^2 + 2p + 1\right)\right)$

    $D = -\left(4p^3 + 27p^2 + 54p + 27\right)$

    verified

  b. $-\left(4p^3 + 27p^2 + 54p + 27\right) = 0$

    Possible rational roots: $\dfrac{\{\pm 1, \pm 3, \pm 9, \pm 27\}}{\{\pm 1, \pm 2, \pm 4\}}$

    $D = -(p+3)^2\left(p + \dfrac{3}{4}\right)$

    $p = -3, q = -3 + 1 = -2$

    $p = -\dfrac{3}{4}, q = -\dfrac{3}{4} + 1 = \dfrac{1}{4}$

  c. $-(p+3)^2\left(p + \dfrac{3}{4}\right) > 0$

    $(-\infty, -3) \cup \left(-3, -\dfrac{3}{4}\right)$

  d. Verified

93. $d(x) = k\left(x^3 - 192x + 1024\right)$

  a. $\dfrac{k\left(x^3 - 3(8)^2 x + 2(8)^3\right)}{k} < 189$

    $x^3 - 192x + 1024 < 189$

    $x^3 - 192x + 835 < 0$

    Possible rational roots:
    $\pm 1, \pm 835, \pm 5, \pm 167$

    $(x-5)\left(x^2 + 5x - 167\right) < 0$

    $x \in (5, 8]$

  b. $(4)^3 - 192(4) + 1024 = 320$ units

  c. $\dfrac{k\left(x^3 - 3(8)^2 x + 2(8)^3\right)}{k} > 475$

    $x^3 - 192x + 1024 > 475$

    $x^3 - 192x + 549 > 0$

    Possible rational roots:
    $\dfrac{\{\pm 1, \pm 3, \pm 9, \pm 61, \pm 183, \pm 549\}}{\{\pm 1\}}$

    $(x-3)\left(x^2 + 3x - 183\right) > 0$

    $x \in [0, 3)$

  d. $\dfrac{k\left(x^3 - 3(8)^2 x + 2(8)^3\right)}{k} \le 648$

    $x^3 - 192x + 1024 \le 648$

    $x^3 - 192x + 376 \le 0$

    Possible rational roots:
    $\dfrac{\{\pm 1, \pm 2, \pm 4, \pm 8, \pm 47, \pm 94, \pm 188, \pm 376\}}{\{\pm 1\}}$

    $(x-2)\left(x^2 + 2x - 188\right) \le 0$

    2 feet

95. a. $R = \dfrac{2D}{t_1 + t_2}$

    $40 = \dfrac{2(80)}{t_1 + t_2}$

    $1 = \dfrac{4}{t_1 + t_2}$

    $1 = \dfrac{4}{\dfrac{80}{r_1} + \dfrac{80}{r_2}}$

    $1 = \dfrac{4r_1 r_2}{80r_1 + 80r_2}$

    $80r_1 + 80r_2 = 4r_1 r_2$

    $20r_1 + 20r_2 = r_1 r_2$

    $20r_2 - r_1 r_2 = -20r_1$

    $r_2(20 - r_1) = -20r_1$

    $r_2 = \dfrac{-20r_1}{20 - r_1}$

    $r_2 = \dfrac{20r_1}{r_1 - 20}$

    Verified

  b. Horizontal: $r_2 = 20$, as r_1 increases, r_2 decreases to maintain $R = 40$.

    Vertical: $r_1 = 20$, as r_1 decreases, r_2 increases to maintain $R = 40$.

c. $\dfrac{20r_1}{r_1 - 20} > r_1$

$\dfrac{20r_1}{r_1 - 20} - r_1 > 0$

$\dfrac{20r_1}{r_1 - 20} - \dfrac{r_1(r_1 - 20)}{r_1 - 20} > 0$

$\dfrac{20r_1 - r_1^2 + 20r_1}{r_1 - 20} > 0$

$\dfrac{40r_1 - r_1^2}{r_1 - 20} > 0$

$\dfrac{r_1(40 - r_1)}{r_1 - 20} > 0$

Critical points: 0, 20, 40

$r_1 \in (20, 40)$

97. $R(t) = 0.01t^2 + 0.1t + 30$

a. $0.01t^2 + 0.1t + 30 < 42$

$0.01t^2 + 0.1t - 12 < 0$

$t^2 + 10t - 1200 < 0$

$(t + 40)(t - 30) < 0$

$[0°, 30°)$

b. $R(t) = 0.01t^2 + 0.1t + 20$

$0.01t^2 + 0.1t + 30 > 36$

$0.01t^2 + 0.1t - 6 > 0$

$t^2 + 10t - 600 > 0$

$(t - 20)(t + 30) > 0$

$(20°, \infty)$

c. $0.01t^2 + 0.1t + 30 > 60$

$0.01t^2 + 0.1t - 30 > 0$

$t^2 + 10t - 3000 > 0$

$(t + 60)(t - 50) > 0$

$(50°, \infty)$

99. a. $\dfrac{2n^3 + 3n^2 + n}{6} \geq 30$

$2n^3 + 3n^2 + n \geq 180$

$2n^3 + 3n^2 + n - 180 \geq 0$

Possible rational roots:

$\{\pm 1, \pm 180, \pm 2, \pm 90, \pm 3, \pm 60, \pm 4, \pm 45, \pm 5, \pm 36,$

$\pm 6, \pm 30, \pm 9, \pm 20, \pm 10, \pm 18, \pm 12, \pm 15, \pm \dfrac{1}{2},$

$\pm \dfrac{3}{2}, \pm \dfrac{45}{2}, \pm \dfrac{5}{2}, \pm \dfrac{9}{2}, \pm \dfrac{15}{2}\}$

$(n - 4)(2n^2 + 11n + 45) \geq 0$

$n \geq 4$

b. $\dfrac{2n^3 + 3n^2 + n}{6} \leq 285$

$2n^3 + 3n^2 + n \leq 1710$

$2n^3 + 3n^2 + n - 1710 \leq 0$

Possible rational roots:

$\{\pm 1, \pm 1710, \pm 2, \pm 855, \pm 3, \pm 570, \pm 5, \pm 342,$

$\pm 6, \pm 285, \pm 9, \pm 190, \pm 10, \pm 171, \pm 15,$

$\pm 114, \pm 18, \pm 95, \pm 19, \pm 90, \pm 30, \pm 57,$

$\pm 38, \pm 45, \pm \dfrac{1}{2}, \pm \dfrac{855}{2}, \pm \dfrac{3}{2}, \pm \dfrac{5}{2},$

$\pm \dfrac{285}{2}, \pm \dfrac{9}{2}, \pm \dfrac{171}{2}, \pm \dfrac{15}{2}, \pm \dfrac{95}{2},$

$\pm \dfrac{19}{2}, \pm \dfrac{57}{2}, \pm \dfrac{45}{2}\}$

$(n - 9)(2n^2 + 21n + 190) \leq 0$

$n \leq 9$

c. $\dfrac{2n^3+3n^2+n}{6}\le 999$

$2n^3+3n^2+n\le 5994$

$2n^3+3n^2+n-5994\le 0$

Possible rational roots:

$\{\pm1,\pm5994,\pm2,\pm2997,\pm3,\pm1998,\pm6,\pm999,\pm9,$

$\pm666,\pm18,\pm333,\pm27,\pm222,\pm37,\pm162,\pm54,$

$\pm111,\pm74,\pm81,\pm\dfrac{1}{2},\pm\dfrac{3}{2},\pm\dfrac{999}{2},\pm\dfrac{9}{2},\pm\dfrac{333}{2},$

$\pm\dfrac{27}{2},\pm\dfrac{37}{2},\pm\dfrac{111}{2},\pm\dfrac{81}{2};\}$

Not factorable

$\dfrac{2(13)^3+3(13)^2+(13)}{6}\le 999$

$819\le 999$

$n=13$

101.a. yes, $x^2\ge 0$

b. yes, $\dfrac{x^2}{x^2+1}\ge 0$

103. $x(x+2)(x-1)^2>0;\quad \dfrac{x(x+2)}{(x-1)^2}>0$

105. $R(x)=\dfrac{x^2-16x+28}{(x-8)^2}$

$R(x)=\dfrac{(x-14)(x-2)}{(x-8)^2}$

Solve: $\dfrac{(x-14)(x-2)}{(x-8)^2}<0$

$R(x)<0$ for $x\in(2,8)\cup(8,14)$

107. $f(x)=\dfrac{x^2+2x-8}{x+4}$

$f(x)=\dfrac{(x+4)(x-2)}{x+4}=x-2$

$f(x)=-6$ when $x=-4$

$F(x)=\begin{cases} f(x) & x\ne -4 \\ -6 & x=-4 \end{cases}$

109. $3x+1<10$ and $x^2-3<1$

$3x<9\quad$ and $x^2-4<0$

$x<3\quad$ and $(x+2)(x-2)<0$

3.8 Exercises

1. Constant

3. $y = \dfrac{k}{x^2}$

5. Answers will vary.

7. $d = kr$

9. $F = ka$

11. $y = kx$

$0.6 = k(24)$

$0.025 = k$

$y = 0.025x$

x	$f(x) = 0.025x$
500	$f(500) = 0.025(500) = 12.5$
650	$16.25 = 0.025x$ $650 = x$
750	$f(750) = 0.025(750) = 18.75$

13. $w = kh$

$344.25 = k(37.5)$

$9.18 = k$;

$w = 9.18h$

$w = 9.18(35)$

$w = \$321.30$

k represents the hourly wage.

15. a. $s = kh$

$192 = k(47)$

$\dfrac{192}{47} = k$;

$s = \dfrac{192}{47} h$

b.

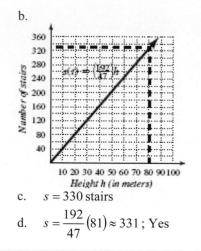

c. $s = 330$ stairs

d. $s = \dfrac{192}{47}(81) \approx 331$; Yes

17. $A = kS^2$

19. $P = kc^2$

21. $p = kq^2$

$280 = k(50)^2$

$\dfrac{280}{(50)^2} = k$

$0.112 = k$;

$p = 0.112q^2$

q	$p(q) = 0.112q^2$
45	$p(45) = 0.112(45)^2 = 226.8$
55	$338.8 = 0.112q^2$ $3025 = q^2$ $55 = q$
70	$p(70) = 0.112(70)^2 = 548.8$

23. $A = ks^2$

$3528 = k(14\sqrt{3})^2$

$3528 = 588k$

$\dfrac{3528}{588} = k$

$6 = k$;

$A = 6s^2$;

$A = 6(303,600)^2$;

$A = 553,037,760,000 \text{ cm}^2$

$A = 55,303,776 \text{ m}^2$

25. a. $d = kt^2$

$169 = k(3.25)^2$

$169 = 10.5625k$

$16 = k;$

$d = 16t^2$

b.

c.   According to the graph, about 3.5 seconds

d. $196 = 16t^2$

$12.25 = t^2$

$3.5\,\text{sec} = t$

Yes, it was close.

e. $121 = 16t^2$

$7.5625 = t^2$

$2.75 = t$

2.75 seconds

27. $F = \dfrac{k}{d^2}$

29. $S = \dfrac{k}{L}$

31. $Y = \dfrac{k}{Z^2}$

$1369 = \dfrac{k}{3^2}$

$12321 = k;$

$Y = \dfrac{12321}{Z^2}$

Z	Y
37	$Y(37) = \dfrac{12321}{37^2} = 9$
74	$2.25 = \dfrac{12321}{Z^2}$ $2.25Z^2 = 12321$ $Z = 74$
111	$Y(111) = \dfrac{12321}{111^2} = 1$

33. $w = \dfrac{k}{r^2}$

$75 = \dfrac{k}{(6400)^2}$

$3072000000 = k;$

$w = \dfrac{3072000000}{r^2}$

$w = \dfrac{3072000000}{(8000)^2}$

$w = 48$ kg

35. $I = krt$

37. $A = kh(B + b)$

39. $V = ktr^2$

41. $C = \dfrac{kR}{S^2}$

    $21 = \dfrac{k(7)}{(1.5)^2}$

    $47.25 = 7k$

    $6.75 = k$;

    $C = \dfrac{6.75R}{S^2}$

R	S	C
120	6	22.5
200	12.5	8.64
350	15	10.5

    $22.5 = \dfrac{6.75(120)}{S^2}$

    $22.5S^2 = 810$

    $S = 6$;

    $C = \dfrac{6.75(200)}{(12.5)^2} = \dfrac{1350}{156.25} = 8.64$;

    $10.5 = \dfrac{6.75R}{(15)^2}$

    $2362.5 = 6.75R$

    $350 = R$

43. $E - kmv^2$

    $200 = k(1)(20)^2$

    $0.5 = k$;

    $E = 0.5mv^2$;

    $E = 0.5(1)(35)^2$

    $E = 612.5$ joules

45. $R(A) = \sqrt[3]{A} - 1$

    $f(x) = \sqrt[3]{x}$; Cube root family

Amount A	Rate r
1.0	0.0
1.05	0.016
1.10	0.032
1.15	0.048
1.20	0.063
1.25	0.077

    $R(A) = \sqrt[3]{1.17} - 1 = 0.054 = 5.4\%$

    Interest Rate: 5.4%

47. $T = \dfrac{k}{V}$

    $4 = \dfrac{k}{12}$

    $48 = k$;

    $T = \dfrac{48}{V}$;

    $T = \dfrac{48}{1.5}$

    $T = 32$ volunteers

49. $M = kE$

    $16 = k(96)$

    $\dfrac{16}{96} = k$

    $\dfrac{1}{6} = k$;

    $M = \dfrac{1}{6}E$;

    $M = \dfrac{1}{6}(250)$

    $M \approx 41.7$ kg

51. $D = k\sqrt{S}$

$108 = k\sqrt{25}$

$21.6 = k;$

$D = 21.6\sqrt{S};$

$D = 21.6\sqrt{45}$

$D \approx 144.9 \text{ ft}$

53. $C = kLD$

$76.50 = k(36)\left(\dfrac{1}{4}\right)$

$76.50 = 9k$

$8.5 = k;$

$C = 8.5LD;$

$C = 8.5(24)\left(\dfrac{3}{8}\right)$

$C = \$76.50$

55. $C = \dfrac{kp_1p_2}{d^2}$

$300 = \dfrac{k(300000)(420000)}{430^2}$

$55470000 = 1.26 \times 10^{11} k$

$4.4 \times 10^{-4} = k;$

$C = \dfrac{\left(4.4 \times 10^{-4}\right)p_1p_2}{d^2}$

$C = \dfrac{\left(4.4 \times 10^{-4}\right)(170000)(550000)}{430^2} \approx 222.5$

about 223 calls

57. $V = k \cdot l \cdot w^2$

$12.27 = k \cdot (3.75) \cdot (2.50)^2$

$\dfrac{12.27}{(3.75) \cdot (2.50)^2} = k;$

a. $V = \dfrac{12.27}{3.75(2.50)^2} \cdot (4.65) \cdot (3.10)^2$

$V \approx 23.39 \text{ cm}^3$

b. $\dfrac{23.39}{12.27} \approx 1.91 \text{ or } 191\%$

59. a. $M = k(w)h^2\left(\dfrac{1}{L}\right)$

b. $270 = k(18)(2)^2\left(\dfrac{1}{8}\right)$

$270 = 9k$

$30 = k;$

$M = 30(18)(2)^2\left(\dfrac{1}{12}\right) = 180 \text{ lb}$

61. $f(x) = k\dfrac{1}{x}$

$\dfrac{\Delta y}{\Delta x} = \dfrac{f(0.6) - f(0.5)}{0.6 - 0.5} = -\dfrac{10}{3}$

$g(x) = k\dfrac{1}{x^2}$

$\dfrac{\Delta y}{\Delta x} = \dfrac{g(0.6) - g(0.5)}{0.6 - 0.5} = -\dfrac{110}{9}$

From $x = 0.7$ to $x = 0.8$, the rate of decrease will be less because as x approaches infinity, y approaches 0.

$x \to \infty, \quad y \to 0^+$

63. $I = \dfrac{k}{d^2}$

   a. $I = \dfrac{k}{5^2}$

        $25I = k$;

        $2I = \dfrac{25I}{d^2}$

        $d^2 = \dfrac{25I}{2I}$

        $d^2 = \dfrac{25}{2}$

        $d = \sqrt{\dfrac{25}{2}} \approx 3.5\,\text{ft}$

   b. $I = \dfrac{k}{12^2}$

        $144I = k$;

        $3I = \dfrac{144I}{d^2}$

        $d^2 = \dfrac{144I}{3I}$

        $d^2 = \dfrac{144}{3}$

        $d = \sqrt{\dfrac{144}{3}} \approx 6.9\,\text{ft}$

65. $x^3 + 4x^2 + 8x = 0$

     $x\left(x^2 + 4x + 8\right) = 0$;

     $a = 1, \quad b = 4, \quad c = 8$

     $x = \dfrac{-4 \pm \sqrt{(4)^2 - 4(1)(8)}}{2(1)}$

     $x = \dfrac{-4 \pm \sqrt{16 - 32}}{2}$

     $x = \dfrac{-4 \pm \sqrt{-16}}{2}$

     $x = \dfrac{-4 \pm 4i}{2}$;

     $x = -2 \pm 2i; \quad x = 0$

67. $f(x) = -2|x - 3| + 5$

     Right 3, up 5, reflected across the x-axis.

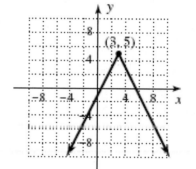

Chapter 3: Polynomial and Rational Functions

Chapter 3 Summary and Concept Review

1. $f(x) = x^2 + 8x + 15$

$$0 = x^2 + 8x + 15$$
$$0 = \left(x^2 + 8x + 16\right) + 15 - 16$$
$$0 = (x+4)^2 - 1$$
$$0 = (x+4)^2 - 1$$
$$1 = (x+4)^2$$
$$\pm 1 = x + 4$$
$$x = -4 \pm 1$$

x-intercepts: $(-5, 0)$ and $(-3, 0)$
Vertex: $(-4, -1)$

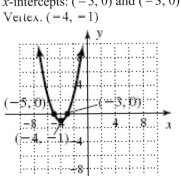

3. $f(x) = 4x^2 - 12x + 3$

$$0 = 4x^2 - 12x + 3$$
$$0 = 4\left(x^2 - 3x\right) + 3$$
$$0 = 4\left(x^2 - 3x + \frac{9}{4}\right) + 3 - 9$$
$$0 = 4\left(x - \frac{3}{2}\right)^2 - 6$$
$$0 = 4\left(x - \frac{3}{2}\right)^2 - 6$$
$$6 = 4\left(x - \frac{3}{2}\right)^2$$
$$\frac{3}{2} = \left(x - \frac{3}{2}\right)^2$$
$$x - \frac{3}{2} = \sqrt{\frac{3}{2}}$$
$$x - \frac{3}{2} = \pm\frac{\sqrt{6}}{2}$$
$$x = \frac{3}{2} \pm \frac{\sqrt{6}}{2}$$

x-intercepts: $(2.7, 0)$ and $(0.3, 0)$;
y-intercept: $(0, 3)$

$$f(0) = 4(0)^2 - 12(0) + 3 = 3$$

Vertex: $\left(\frac{3}{2}, -6\right)$

5. $\dfrac{x^3 + 4x^2 - 5x - 6}{x - 2}$

$$
\begin{array}{r}
x^2 + 6x + 7 \\
x-2 \overline{\smash{)}\ x^3 + 4x^2 - 5x - 6} \\
\underline{-\left(x^3 - 2x^2\right)} \\
6x^2 - 5x \\
\underline{-\left(6x^2 - 12x\right)} \\
7x - 6 \\
\underline{-\left(7x - 14\right)} \\
8
\end{array}
$$

$q(x) = x^2 + 6x + 7$
$R = 8$

7. Since $R = 0$, -7 is a root and $x + 7$ is a factor.

$$
\begin{array}{r|rrrrr}
-7 & 2 & 13 & -6 & 9 & 14 \\
   &   & -14 & 7 & -7 & -14 \\
\hline
   & 2 & -1 & 1 & 2 & \underline{|\,0}
\end{array}
$$

9. $p(x) = x^3 + 2x^2 - 11x - 12$

Possible rational roots: $\pm1, \pm12, \pm2, \pm6, \pm3, \pm4$

$$
\begin{array}{r|rrrr}
-4 & 1 & 2 & -11 & -12 \\
   &   & -4 & 8 & 12 \\
\hline
   & 1 & -2 & -3 & \underline{|\,0}
\end{array}
$$

$p(x) = (x+4)\left(x^2 - 2x - 3\right)$
$p(x) = (x+4)(x+1)(x-3)$

11. $P(x) = 4x^3 + 8x^2 - 3x - 1$

$$\frac{1}{2}\begin{array}{|rrrr}4 & 8 & -3 & -1 \\ & 2 & 5 & 1 \\ \hline 4 & 10 & 2 & \underline{|\,0}\end{array}$$

Since $R = 0$, $\dfrac{1}{2}$ is a root and

$\left(x - \dfrac{1}{2}\right)$ is a factor.

13. $h(x) = x^3 + 9x^2 + 13x - 10$

$$-7\,]\begin{array}{rrrr}1 & 9 & 13 & -10 \\ & -7 & -14 & 7 \\ \hline 1 & 2 & -1 & \underline{|-3}\end{array}$$

$h(-7) = -3$

15. $C(x) = (x-1)^2(x+2i)(x-2i)$

$C(x) = (x^2 - 2x + 1)(x^2 - 4i^2)$

$C(x) = (x^2 - 2x + 1)(x^2 + 4)$

$C(x) = x^4 + 4x^2 - 2x^3 - 8x + x^2 + 4$

$C(x) = x^4 - 2x^3 + 5x^2 - 8x + 4$

17. $p(x) = 4x^3 - 16x^2 + 11x + 10$

Possible rational roots:

$\dfrac{\{\pm 1, \pm 10, \pm 2, \pm 5\}}{\{\pm 1, \pm 2, \pm 4\}}$;

$\left\{\pm 1, \pm 10, \pm 2, \pm 5, \pm\dfrac{1}{2}, \pm\dfrac{5}{2}, \pm\dfrac{1}{4}, \pm\dfrac{5}{4}\right\}$

19. $P(x) = 2x^3 - 3x^2 - 17x - 12$

Possible rational roots:

$\dfrac{\{\pm 1, \pm 12, \pm 2, \pm 6, \pm 3, \pm 4\}}{\{\pm 1\}}$

$$4\,]\begin{array}{rrrr}2 & -3 & -17 & -12 \\ & 8 & 20 & 12 \\ \hline 2 & 5 & 3 & 0\end{array}$$

$P(x) = (x - 4)(2x^2 + 5x + 3)$

$P(x) = (x - 4)(x + 1)(2x + 3)$

21. $P(x) = x^4 - 3x^3 - 8x^2 + 12x + 6$

$[-2, -1]$

$P(-2) = (-2)^4 - 3(-2)^3 - 8(-2)^2 + 12(-2) + 6 = -10$;

$P(-1) = (-1)^4 - 3(-1)^3 - 8(-1)^2 + 12(-1) + 6 = -10$;

$[1, 2]$

$P(1) = (1)^4 - 3(1)^3 - 8(1)^2 + 12(1) + 6 = 8$;

$P(2) = (2)^4 - 3(2)^3 - 8(2)^2 + 12(2) + 6 = -10$;

$[2, 3]$

$P(3) = (3)^4 - 3(3)^3 - 8(3)^2 + 12(3) + 6 = -30$;

$[4, 5]$

$P(4) = (4)^4 - 3(4)^3 - 8(4)^2 + 12(4) + 6 = -10$;

$P(5) = (5)^4 - 3(5)^3 - 8(5)^2 + 12(5) + 6 = 116$

Sign changes in intervals: $[1, 2]$, $[4, 5]$, verified

23. $f(x) = -3x^5 + 2x^4 + 9x - 4$

$f(0) = -3(0)^5 + 2(0)^4 + 9(0) - 4 = -4$

degree 5; up/down; $(0, -4)$

25. $p(x) = (x + 1)^3(x - 2)^2$

end behavior: down/up

bounce at $(2, 0)$; cross at $(-1, 0)$

$p(0) = (0 + 1)^3(0 - 2)^2 = 4$

y-intercept: $(0, 4)$

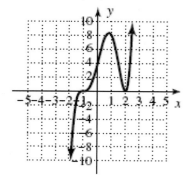

27. $h(x) = x^4 - 6x^3 + 8x^2 + 6x - 9$

end behavior: up/up

Possible rational roots: $\dfrac{\{\pm 1, \pm 9, \pm 3\}}{\{\pm 1\}}$

$h(x) = (x+1)(x-1)(x-3)^2$

bounce at $(3,0)$; cross at $(-1,0)$ and $(1,0)$

y-intercept: $(0,-9)$

29. $V(x) = \dfrac{x^2 - 9}{x^2 - 3x - 4}$

$V(x) = \dfrac{(x+3)(x-3)}{(x-4)(x+1)}$

a. $\{x \mid x \in R, x \neq -1, 4\}$

b. HA: $y = 1$
(deg num = deg den)
VA: $x = -1, x = 4$

c. $V(0) = \dfrac{0^2 - 9}{0^2 - 3(0) - 4} = \dfrac{9}{4}$

y-intercept $\left(0, \dfrac{9}{4}\right)$;

x-intercepts : $(-3, 0)$ and $(3, 0)$

d. $V(1) = \dfrac{1^2 - 9}{1^2 - 3(1) - 4} = \dfrac{4}{3}$

31. $v(x) = \dfrac{x^2 - 4x}{x^2 - 4}$

$v(0) = \dfrac{(0)^2 - 4(0)}{(0)^2 - 4} = 0$;

y-intercept: $(0,0)$

$v(x) = \dfrac{x(x-4)}{(x+2)(x-2)}$;

vertical asymptotes: $x = -2$ and $x = 2$

x-intercepts: $(0,0)$ and $(4,0)$

horizontal asymptote: $y = 1$

(deg num = deg den)

33. $V(x) = \dfrac{(x+3)(x-4)}{(x+2)(x-3)}$

$V(x) = \dfrac{x^2 - x - 12}{x^2 - x - 6}$;

$V(0) = \dfrac{(0)^2 - (0) - 12}{(0)^2 - (0) - 6} = 2$

35. $h(x) = \dfrac{x^3 - 2x^2 - 9x + 18}{x - 2}$

$h(x) = \dfrac{x^2(x-2) - 9(x-2)}{x-2}$

$h(x) = \dfrac{(x-2)(x^2-9)}{x-2}$

$h(x) = \dfrac{(x-2)(x+3)(x-3)}{x-2}$

If $x = 2$, $x^2 - 9 = (2)^2 - 9 = -5$

Removable discontinuity at $(2, -5)$.

37. $h(x) = \dfrac{x^2 - 2x}{x - 3}$

$h(0) = \dfrac{(0)^2 - 2(0)}{(0) - 3} = 0;$

y-intercept: $(0,0)$

$h(x) = \dfrac{x(x-2)}{x-3};$

vertical asymptote: $x = 3$

x-intercepts: $(0,0)$ and $(2,0)$

horizontal asymptote: none
(deg num > deg den)

oblique asymptote: $y = x + 1$

39. $A(x) = \dfrac{x^2 - 2x + 6}{x}$

a.

b. about 2450 favors

c. $A(2.45) \approx 2.90$
   about \$2.90 each

41. $\dfrac{x^2 - 3x - 10}{x - 2} \geq 0$

$\dfrac{(x-5)(x+2)}{x-2} \geq 0$

Outputs are positive for $x \in [-2, 2) \cup [5, \infty)$

43. $y = k\sqrt[3]{x};$

$52.5 = k\sqrt[3]{27}$

$\dfrac{52.5}{3} = k$

$17.5 = k;$

$y = 17.5\sqrt[3]{x};$

x	y
216	105
0.343	12.25
729	157.5

45. $t = \dfrac{kuv}{w};$

$30 = \dfrac{k(2)(3)}{5}$

$t = \dfrac{25uv}{w};$

$t = \dfrac{25(8)(12)}{15} = 160;$

Chapter 3 Mixed Review

1. Vertex $\left(\dfrac{1}{2}, \dfrac{9}{2}\right)$

$$y = a\left(x - \dfrac{1}{2}\right)^2 + \dfrac{9}{2}, \ (2,0)$$

$$0 = a\left(2 - \dfrac{1}{2}\right)^2 + \dfrac{9}{2}$$

$$-\dfrac{9}{2} = a\left(\dfrac{3}{2}\right)^2$$

$$-2 = a;$$

$$y = -2\left(x - \dfrac{1}{2}\right)^2 + \dfrac{9}{2}$$

3. $C(s) = \dfrac{1}{180}s^2 - \dfrac{8}{9}s + \dfrac{680}{9}$

The s value of the vertex: $\dfrac{-\left(-\dfrac{8}{9}\right)}{2\left(\dfrac{1}{180}\right)} = 80$

$$C(80) = \dfrac{1}{180}(80)^2 - \dfrac{8}{9}(80) + \dfrac{680}{9} = 40$$

80 GB, $40.00

5. $\dfrac{x^4 - 3x^2 + 5x - 1}{x + 2}$

$$
\begin{array}{r|rrrrr}
-2 & 1 & 0 & -3 & 5 & -1 \\
   &   & -2 & 4 & -2 & -6 \\
\hline
   & 1 & -2 & 1 & 3 & \underline{-7}
\end{array}
$$

$q(x) = x^3 - 2x^2 + x + 3$
$R = -7$

7. $P(x) = 6x^3 - 23x^2 - 40x + 31$

(a) $P(-1) = 42$

$$
\begin{array}{r|rrrr}
-1 & 6 & -23 & -40 & 31 \\
   &   & -6 & 29 & 11 \\
\hline
   & 6 & -29 & -11 & \underline{42}
\end{array}
$$

(b) $P(1) = -26$

$$
\begin{array}{r|rrrr}
1 & 6 & -23 & -40 & 31 \\
  &   & 6 & -17 & -57 \\
\hline
  & 6 & -17 & -57 & \underline{-26}
\end{array}
$$

(c) $P(5) = 6$

$$
\begin{array}{r|rrrr}
5 & 6 & -23 & -40 & 31 \\
  &   & 30 & 35 & -25 \\
\hline
  & 6 & 7 & -5 & \underline{6}
\end{array}
$$

9. a. $6x^3 + x^2 - 20x - 12 = 0$
Possible rational roots:
$\left\{\pm 1, \pm 12, \pm 2, \pm 6, \pm 3, \pm 4,\right.$
$\left. \pm\dfrac{1}{6}, \pm\dfrac{1}{3}, \pm\dfrac{1}{2}, \pm\dfrac{2}{3}, \pm\dfrac{3}{2}, \pm\dfrac{4}{3}\right\}$

$x = 9$ and $x = \dfrac{8}{3}$ CANNOT be roots.

b. $P(x) = x^4 - x^3 + 7x^2 - 9x - 18$
Possible rational roots: $\left\{\pm 1, \pm 18, \pm 2, \pm 9, \pm 3, \pm 6\right\}$

$$
\begin{array}{r|rrrrr}
2 & 1 & -1 & 7 & -9 & -18 \\
  &   & 2 & 2 & 18 & 18 \\
\hline
  & 1 & 1 & 9 & 9 & \underline{0}
\end{array}
$$

$$
\begin{array}{r|rrrr}
-1 & 1 & 1 & 9 & 9 \\
   &   & -1 & 0 & -9 \\
\hline
   & 1 & 0 & 9 & \underline{0}
\end{array}
$$

$P(x) = (x - 2)(x + 1)(x^2 + 9)$
$P(x) = (x - 2)(x + 1)(x + 3i)(x - 3i)$
$x = 2, x = -1, x = -3i, x = 3i$

11. $p(x) = \dfrac{x^2 - 2x}{x^2 - 2x + 1}$

$$p(0) = \dfrac{(0)^2 - 2(0)}{(0)^2 - 2(0) + 1} = 0$$

y-intercept: $(0,0)$

$$p(x) = \dfrac{x(x - 2)}{(x - 1)^2};$$

vertical asymptote: $x = 1$
x-intercepts: $(0,0)$ and $(2,0)$
horizontal asymptote: $y = 1$

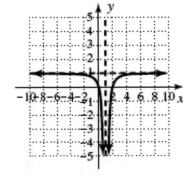

13. $r(x) = \dfrac{x^3 - 13x + 12}{x^2}$

$r(0) = \dfrac{(0)^3 - 13(0) + 12}{(0)^2} = $ undefined

y-intercept: none

Possible rational roots:
$\{\pm 1, \pm 12, \pm 2, \pm 6, \pm 3, \pm 4\}$

$r(x) = \dfrac{(x+4)(x-1)(x-3)}{x^2}$;

vertical asymptote: $x = 0$

x-intercepts: $(-4, 0), (1, 0)$ and $(3, 0)$

horizontal asymptote: none

$r(x) = x - \dfrac{13}{x} + \dfrac{12}{x^2}$

oblique asymptote: $y = x$

15. $x^3 - 4x < 12 - 3x^2$

$x^3 + 3x^2 - 4x - 12 < 0$

Possible rational roots:
$\{\pm 1, \pm 12, \pm 2, \pm 6, \pm 3, \pm 4\}$

$(x+2)(x-2)(x+3) < 0$

$x \in (-\infty, -3) \cup (-2, 2)$

17. a. $V(s) = (24 - 2x)(16 - 2x)(x)$

$= (384 - 80x + 4x^2)(x)$

$= 384x - 80x^2 + 4x^3$

$= 4x^3 - 80x^2 + 384x$

b. $512 = 4x^3 - 80x^2 + 384x$

$0 = 4x^3 - 80x^2 + 384x - 512$

$0 = x^3 - 20x^2 + 96x - 128$

c. For $0 < x < 8$, possible rational zeroes are: 1, 2, and 4

d. $x = 4$ inches

$$\begin{array}{r|rrr} 4 & 1 & -20 & 96 & -128 \\ & & 4 & -64 & 128 \\ \hline & 1 & -16 & 32 & 0 \end{array}$$

e. $(x - 4)(x^2 - 16x + 32)$

$x = \dfrac{-(-16) \pm \sqrt{(-16)^2 - 4(1)(32)}}{2(1)}$

$x = \dfrac{16 \pm \sqrt{128}}{2} = 8 - 4\sqrt{2} \approx 2.34$ inches

19. $R = kL\left(\dfrac{1}{A}\right)$

Chapter 3 Practice Test

1. a. $f(x) = -x^2 + 10x - 16$

    $f(x) = -(x^2 - 10x) - 16$

    $f(x) = -(x^2 - 10x + 25) - 16 + 25$

    $f(x) = -(x-5)^2 + 9$;

    Vertex: (5, 9), opens downward.

    $f(0) = -0^2 + 10(0) - 16 = -16$

    y-intercept $(0, \,^-16)$

    $0 = -x^2 + 10x - 16$

    $x^2 - 10x + 16 = 0$

    $(x-8)(x-2) = 0$

    $x = 8$ or $x = 2$

    x-intercepts $(8,0)$ and $(2,0)$

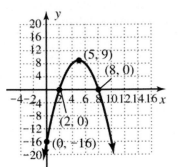

    b. $g(x) = \dfrac{1}{2}x^2 + 4x + 16$

    $g(x) = \dfrac{1}{2}(x^2 + 8x) + 16$

    $g(x) = \dfrac{1}{2}(x^2 + 8x + 16) + 16 - 8$

    $g(x) = \dfrac{1}{2}(x+4)^2 + 8$;

    Vertex: $(-4, 8)$, opens upward.

    $g(0) = \dfrac{1}{2}(0)^2 + 4(0) + 16 = 16$

y-intercept (0, 16)

$0 = \dfrac{1}{2}x^2 + 4x + 16$

$0 = x^2 + 8x + 32$

$b^2 - 4ac = 64 - 4(1)(32) < 0$

No x-intercepts

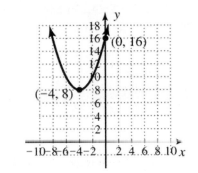

3. $d(t) = t^2 - 14t$

    a. $d(4) = (4)^2 - 14(4) = -40$

        40 ft

        $d(6) = (6)^2 - 14(6) = -48$0

        48 ft

    b. $d(t) = t^2 - 14t + 49 - 49$

        $d(t) = (t-7)^2 - 49$

        49 ft

    c. $2(7) = 14$ seconds

5. $\dfrac{x^3 + 4x^2 - 5x - 20}{x+2} = x^2 + 2x - 9 + \dfrac{-2}{x+2}$

$$\begin{array}{r|rrrr} -2 & 1 & 4 & -5 & -20 \\ & & -2 & -4 & 18 \\ \hline & 1 & 2 & -9 & \underline{|-2} \end{array}$$

7. $f(x) = 2x^3 + 4x^2 - 5x + 2$

    $f(-3) = -1$

$$\begin{array}{r|rrrr} -3 & 2 & 4 & -5 & 2 \\ & & -6 & 6 & -3 \\ \hline & 2 & -2 & 1 & \underline{|-1} \end{array}$$

9. $Q(x) = (x^2 - 3x + 2)(x^3 - 2x^2 - x + 2)$

    $Q(x) = (x-2)(x-1)(x^2(x-2) - (x-2))$

    $Q(x) = (x-2)(x-1)(x-2)(x^2 - 1)$

    $Q(x) = (x-2)(x-1)(x-2)(x+1)(x-1)$

    $Q(x) = (x-2)^2(x-1)^2(x+1)$

    2 multiplicity 2

    1 multiplicity 2, -1 multiplicity 1

11. $f(x) = \frac{1}{2}x^3 - 7x^2 + 28x - 32$

  (a) $0 = \frac{1}{2}x^3 - 7x^2 + 28x - 32$

  $0 = x^3 - 14x^2 + 56x - 64$
  Possible rational roots:
  $\{\pm 1, \pm 64, \pm 2, \pm 32, \pm 4, \pm 16, \pm 8\}$

  $\underline{2\rfloor\ 1\ -14\quad 56\quad -64}$
  $\qquad\quad 2\ -24\quad 64$
  $\overline{\quad 1\ -12\quad 32\quad\ \ 0}$

  $0 = (x-2)(x^2 - 12x + 32)$
  $0 = (x-2)(x-4)(x-8)$
  $x = 2,\ x = 4,\ x = 8$
  1992, 1994, 1998

  (b) 4 years (1992-1994, 1998-2000)
  (c) surplus of $2.5 million

13. $g(x) = x^4 - 9x^2 - 4x + 12$

  end behavior: up/up
  Possible rational roots:
  $\dfrac{\{\pm 1, \pm 12, \pm 2, \pm 6, \pm 3, \pm 4\}}{\{\pm 1\}}$

  $\underline{-2\rfloor\ 1\quad 0\quad -9\quad -4\quad 12}$
  $\qquad\qquad\ -2\quad 4\quad 10\ -12$
  $\overline{\quad 1\ -2\ -5\quad 6\ \rfloor\underline{0}}$

  $\underline{-2\rfloor\ 1\ -2\ -5\quad 6}$
  $\qquad\qquad\ -2\quad 8\ -6$
  $\overline{\quad 1\ -4\quad 3\ \rfloor\underline{0}}$

  $g(x) = (x+2)^2(x^2 - 4x + 3)$
  $g(x) = (x+2)^2(x-1)(x-3)$

  -2 multiplicity 2, 1 multiplicity 1, 3 multiplicity 1
  bounce at $(-2,0)$; cross at $(1,0)$ and $(3,0)$
  $g(0) = 0^4 - 9(0)^2 - 4(0) + 12 = 12$
  y-intercept: $(0,12)$

15. $C(x) = \dfrac{300x}{100 - x}$

  a. VA: $x = 100$; removal of 100% of the contaminants

  b. From 80% to 85%:
  $C(85) = \dfrac{300(85)}{100 - (85)} = 1700$;
  $1,700,000$;
  $C(80) = \dfrac{300(80)}{100 - (80)} = 1200$;
  $1700 - 1200 = 500, \$500,000$;
  From 90% to 95 %:
  $C(95) = \dfrac{300(95)}{100 - (95)} = 5700$;
  $5,700,000$
  $C(90) = \dfrac{300(90)}{100 - (90)} = 2700$;
  $5700 - 2700 = 3000; \$3,000,000$;
  It becomes cost prohibitive to remove all the contaminants.

  c. $2200 = \dfrac{300x}{100 - x}$
  $2200(100 - x) = 300x$
  $220000 - 2200x = 300x$
  $220000 = 2500x$
  $88 = x$
  $x = 88\%$

17. $\overline{C}(x) = \dfrac{2x^2 + 25x + 128}{x}$

  Using grapher: $x = 8$; 800 items
  Minimizes costs

19. $C(h) = \dfrac{2h^2 + 5h}{h^3 + 55}$

a.

b. $h^3 + 55 = 0$

$h^3 = -55$

$h = -\sqrt[3]{55}$, no

c. $C(2) = \dfrac{2(2)^2 + 5(2)}{(2)^3 + 55} \approx 0.286 - 28.6\%$;

$C(8) = \dfrac{2(8)^2 + 5(8)}{(8)^3 + 55} \approx 0.296 = 29.6\%$

d. $\dfrac{2h^2 + 5h}{h^3 + 55} < 0.2$

Using grapher: ≈ 11.7 hours

e. Using grapher: 4 hours, 43.7%

f. A trace amount of the chemical will remain in the bloodstream.

Chapter 3 Calculator Exploration

1. $Y_1 = (x^3 - 6x^2 + 32)(x^2 + 1)$

$= (x-4)^2 (x+2)(x^2 + 1)$

$Y_2 = x^3 - 6x^2 + 32 = (x-4)^2 (x+2)$

$Y_3 = x + 2$

3. They do not affect the solution.

Chapter 3 Strengthening Core Skills

1. $x^3 - 3x - 18 \le 0$

$\dfrac{\{\pm 1, \pm 18, \pm 2, \pm 9, \pm 3, \pm 6\}}{\{\pm 1\}}$

$$\begin{array}{r|rrrr} 3 & 1 & 0 & -3 & -18 \\ & & 3 & 9 & 18 \\ \hline & 1 & 3 & 6 & \underline{|\,0} \end{array}$$

$(x-3)(x^2 + 3x + 6) \le 0$

$x \in (-\infty, 3]$

3. $x^3 - 13x + 12 < 0$

$\dfrac{\{\pm 1, \pm 12, \pm 2, \pm 6, \pm 3, \pm 4\}}{\{\pm 1\}}$

$$\begin{array}{r|rrrr} -4 & 1 & 0 & -13 & 12 \\ & & -4 & 16 & -12 \\ \hline & 1 & -4 & 3 & \underline{|\,0} \end{array}$$

$(x+4)(x^2 - 4x + 3) < 0$

$(x-3)(x-1)(x+4) < 0$

$x \in (-\infty, -4) \cup (1, 3)$

5. $x^4 - x^2 - 12 > 0$

$(x^2 - 4)(x^2 + 3) > 0$

$(x-2)(x+2)(x^2 + 3) > 0$

$(x^2 + 3)$ does not affect the solution set.

$x \in (-\infty, -2) \cup (2, \infty)$

Cumulative Review Chapters R-3

1. $\dfrac{1}{R} = \dfrac{1}{R_1} + \dfrac{1}{R_2}$

$RR_1 R_2 \left[\dfrac{1}{R} \right] = \left[\dfrac{1}{R_1} + \dfrac{1}{R_2} \right] RR_1 R_2$

$R_1 R_2 = RR_2 + RR_1$

$R_1 R_2 = R(R_2 + R_1)$

$\dfrac{R_1 R_2}{R_1 + R_2} = R$

3. a. $x^3 - 1$

$= (x-1)(x^2 + x + 1)$

b. $x^3 - 3x^2 - 4x + 12$

$= x^2 (x-3) - 4(x-3)$

$= (x-3)(x^2 - 4)$

$= (x-3)(x+2)(x-2)$

5. $x + 3 < 5$ or $5 - x < 4$
   $x < 2$ or $-x < -1$
   $x < 2$ or $x > 1$
   $x \in (-\infty, \infty)$

7. $(2 - 3i)^2 - 4(2 - 3i) + 13 = 0$
   $4 - 12i + 9i^2 - 8 + 12i + 13 = 0$
   $4 - 12i - 9 - 8 + 12i + 13 = 0$
   $0 = 0$
   Verified

9. $(1,\ 17), (61,\ 28)$
   $m = \dfrac{28 - 17}{61 - 1} = \dfrac{11}{60}$;

   $y - 17 = \dfrac{11}{60}(x - 1)$

   $y - 17 = \dfrac{11}{60}x - \dfrac{11}{60}$

   $y = \dfrac{11}{60}x + \dfrac{1009}{60}$;

   $y = \dfrac{11}{60}(121) + \dfrac{1009}{60} = 39$ minutes;

   Driving time increases 11 minutes every 60 days.

11. $y = 1.18x^2 - 10.99x + 4.6$;
    Using grapher, the profit is first earned in the 9th month.

13. $f(x) = \sqrt[3]{2x - 3}$;
    $x = \sqrt[3]{2y - 3}$
    $x^3 = 2y - 3$
    $x^3 + 3 = 2y$
    $\dfrac{x^3 + 3}{2} = y$
    $f^{-1}(x) = \dfrac{x^3 + 3}{2}$

    $(f \circ f^{-1})(x) = \sqrt[3]{2\left(\dfrac{x^3 + 3}{2}\right) - 3}$

    $= \sqrt[3]{x^3 + 3 - 3} = \sqrt[3]{x^3} = x$;

    $(f^{-1} \circ f)(x) = \dfrac{(\sqrt[3]{2x - 3})^3 + 3}{2}$

    $= \dfrac{2x - 3 + 3}{2} = \dfrac{2x}{2} = x$;

    Verified

15. $F(x) = -f(x + 1) + 2$
    Reflected in x-axis, left 1, up 2

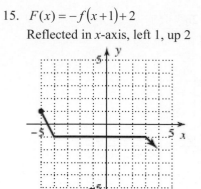

17. $Y = \dfrac{kX}{Z^2}$;

    $10 = \dfrac{k(32)}{(4)^2}$

    $5 = k$;

    $Y = \dfrac{5X}{Z^2}$;

    $1.4 = \dfrac{5x}{(15)^2}$

    $x = 63$

19. $f(x) = x^3 - 3x^2 - 6x + 8$
    Possible rational roots: $\{\pm 1, \pm 8, \pm 2, \pm 4\}$

    $\begin{array}{r|rrrr} 1 & 1 & -3 & -6 & 8 \\ & & 1 & -2 & -8 \\ \hline & 1 & -2 & -8 & \boxed{0} \end{array}$

    $f(x) = (x - 1)(x^2 - 2x - 8)$
    $f(x) = (x - 1)(x + 2)(x - 4)$
    $f(x) = (x - 1)(x + 2)(x - 4)$

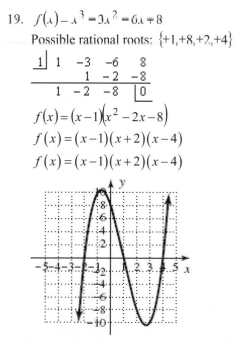

4.1 Technology Highlight

1. a. $f(x) = 2x + 1$;

    $x = 2y + 1$

    $x - 1 = 2y$

    $\dfrac{x-1}{2} = y$

    $f^{-1}(x) = \dfrac{x-1}{2}$

    b, c verified using a graphing calculator

    d.

    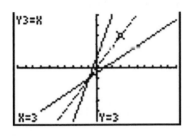

3. a. $h(x) = \dfrac{x}{x+1}$;

    $x = \dfrac{y}{y+1}$

    $x(y+1) = y$

    $xy + x = y$

    $xy - y = -x$

    $y(x-1) = -x$

    $y = \dfrac{-x}{x-1}$

    $f^{-1}(x) = \dfrac{-x}{x-1} = \dfrac{x}{1-x}$

    b, c verified using a graphing calculator

    d.

    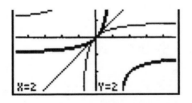

4.1 Exercises

1. Second; one

3. $(-11, -2), (-5, 0), (1, 2), (19, 4)$

5. False, answers will vary.

7. One-to-one

9. One-to-one

11. Not a function

13. One-to-one

15. Not one-to-one

17. Not one-to-one; $y = 7$ is paired with $x = -2$ and $x = 2$.

19. One-to-one

21. One-to-one

23. Not one-to-one; for $p(t) > 5$, one y corresponds to two x-values.

25. One-to-one

27. One-to-one

29. $f(x) = \{(-2, 1), (-1, 4), (0, 5), (2, 9), (5, 15)\}$
    $f^{-1}(x) = \{(1, -2), (4, -1), (5, 0), (9, 2), (15, 5)\}$

31. $v(x) = \{(-4, 3), (-3, 2), (0, 1), (5, 0), (12, -1), (21, -2), (32, -3)\}$
    $v^{-1}(x) = \{(3, -4), (2, -3), (1, 0), (0, 5), (-1, 12), (-2, 21), (-3, 32)\}$

33. $f(x) = x + 5$
    $y = x + 5$;
    $x = y + 5$
    $x - 5 = y$
    $f^{-1}(x) = x - 5$

35. $p(x) = -\dfrac{4}{5}x$

$y = -\dfrac{4}{5}x;$

$x = -\dfrac{4}{5}y$

$-\dfrac{5}{4}x = y$

$p^{-1}(x) = -\dfrac{5}{4}x$

37. $f(x) = 4x + 3$

Multiply by 4, add 3
Inverse:
Subtract 3, divide by 4

$f^{-1}(x) = \dfrac{x-3}{4}$

39. $Y_1 = \sqrt[3]{x-4}$

Subtract 4, take cube root
Inverse:
Cube x, add 4

$Y_1^{-1} = x^3 + 4$

41. $f(x) = \sqrt[3]{x-2}$

a. $f(10) - \sqrt[3]{10-2} - \sqrt[3]{8} - 2,$

$f(-6) = \sqrt[3]{-6-2} = \sqrt[3]{-8} = -2;$

$f(1) = \sqrt[3]{1-2} = \sqrt[3]{-1} = -1;$

$(-6, -2), (1, -1), (10, 2)$

b. $f(x) = \sqrt[3]{x-2};$

Interchange x and y to find the inverse.

$x = \sqrt[3]{y-2}$

$x^3 = y-2$

$x^3 + 2 = y$

$f^{-1}(x) = x^3 + 2$

c. $f^{-1}(-2) = (-2)^3 + 2 = -6;$

$f^{-1}(-1) = (-1)^3 + 2 = 1;$

$f^{-1}(2) = (2)^3 + 2 = 10$

43. $f(x) = x^3 + 1$

a. $f(0) = (0)^3 + 1 = 1;$

$f(1) = (1)^3 + 1 = 2;$

$f(-1) = (-1)^3 + 1 = 0;$

$(0,1), (1,2), (-1,0)$

b. $f(x) = x^3 + 1;$

Interchange x and y to find the inverse.

$x = y^3 + 1$

$x - 1 = y^3$

$\sqrt[3]{x-1} = y$

$f^{-1}(x) = \sqrt[3]{x-1}$

c. $f^{-1}(1) = \sqrt[3]{1-1} = 0;$

$f^{-1}(2) = \sqrt[3]{2-1} = 1;$

$f^{-1}(0) = \sqrt[3]{0-1} = -1$

45. $f(x) = \dfrac{8}{x+2}$

a. $f(0) = \dfrac{8}{0+2} = 4;$

$f(2) = \dfrac{8}{2+2} = 2;$

$f(6) = \dfrac{8}{6+2} = 1;$

$(0,4), (2,2), (6,1)$

b. $f(x) = \dfrac{8}{x+2};$

Interchange x and y to find the inverse.

$x = \dfrac{8}{y+2}$

$x(y+2) = 8$

$xy + 2x = 8$

$xy = 8 - 2x$

$y = \dfrac{8}{x} - 2$

$f^{-1}(x) = \dfrac{8}{x} - 2$

c. $f^{-1}(4) = \dfrac{8}{4} - 2 = 0;$

$f^{-1}(2) = \dfrac{8}{2} - 2 = 2;$

$f^{-1}(1) = \dfrac{8}{1} - 2 = 6$

47. $f(x) = \dfrac{x}{x+1}$

a. $f(0) = \dfrac{0}{0+1} = 0$;

$f(1) = \dfrac{1}{1+1} = \dfrac{1}{2}$;

$f(-2) = \dfrac{-2}{-2+1} = 2$;

$(0,0), \left(1, \dfrac{1}{2}\right), (-2,2)$

b. $f(x) = \dfrac{x}{x+1}$;

Interchange x and y to find the inverse.

$x = \dfrac{y}{y+1}$

$x(y+1) = y$

$xy + x = y$

$xy - y = -x$

$y(x-1) = -x$

$y = \dfrac{-x}{x-1}$

$y = \dfrac{x}{1-x}$

$f^{-1}(x) = \dfrac{x}{1-x}$

c. $f^{-1}(0) = \dfrac{0}{1-0} = 0$;

$f^{-1}\left(\dfrac{1}{2}\right) = \dfrac{\dfrac{1}{2}}{1-\dfrac{1}{2}} = 1$;

$f^{-1}(2) = \dfrac{2}{1-2} = -2$

49. $f(x) = (x+5)^2$

a. Parabola with vertex (-5,0)
Restricting domain to $x \geq -5$ leaves
right branch of $f(x) = (x+5)^2$ with
range $y \geq 0$

b. For $x \geq -5$

$f(x) = (x+5)^2$

$y = (x+5)^2$

Interchange x and y to find the inverse.

$x = (y+5)^2$

$\pm\sqrt{x} = \sqrt{(y+5)^2}$

$\pm\sqrt{x} = y+5$

use $\sqrt{x}$ since $x \geq -5$

$\sqrt{x} - 5 = y$

$f^{-1}(x) = \sqrt{x} - 5$,

Domain $x \in [0, \infty)$, Range $y \in [-5, \infty)$

51. $v(x) = \dfrac{8}{(x-3)^2}$

a. Restricting domain to $x > 3$, range
$y > 0$

b. $v(x) = \dfrac{8}{(x-3)^2}$

Interchange x and y to find the inverse.

$x = \dfrac{8}{(y-3)^2}$

$x(y-3)^2 = 8$

$(y-3)^2 = \dfrac{8}{x}$

$y - 3 = \pm\sqrt{\dfrac{8}{x}}$

use $+\sqrt{\dfrac{8}{x}}$ since $x > 3$

$y = 3 + \sqrt{\dfrac{8}{x}}$

$v^{-1}(x) = 3 + \sqrt{\dfrac{8}{x}}$,

Domain $x \in (0, \infty)$, Range $y \in (3, \infty)$

53. $p(x) = (x+4)^2 - 2$

    a. Restricting domain to $x \geq -4$, range

        $y \geq -2$

    b. $p(x) = (x+4)^2 - 2$

        Interchange x and y to find the inverse.

        $x = (y+4)^2 - 2$

        $x + 2 = (y+4)^2$

        $\pm\sqrt{x+2} = y + 4$

        use $\sqrt{x+2}$ since $x \geq -4$

        $\sqrt{x+2} - 4 = y$

        $p^{-1}(x) = \sqrt{x+2} - 4$

        Domain $x \in [-2, \infty)$,, Range $y \in [-4, \infty)$

55. $f(x) = -2x + 5$; $g(x) = \dfrac{x-5}{-2}$

    $(f \circ g)(x) = f[g(x)]$

    $= -2(g(x)) + 5$

    $= -2\left(\dfrac{x-5}{-2}\right) + 5 = x - 5 + 5 = x;$

    $(g \circ f)(x) = g[f(x)]$

    $= \dfrac{f(x) - 5}{-2}$

    $= \dfrac{-2x + 5 - 5}{-2}$

    $= \dfrac{-2x}{-2}$

    $= x$

57. $f(x) = \sqrt[3]{x+5}$; $g(x) = x^3 - 5$

    $(f \circ g)(x) = f[g(x)]$

    $= \sqrt[3]{g(x) + 5}$

    $= \sqrt[3]{x^3 - 5 + 5}$

    $= \sqrt[3]{x^3}$

    $= x;$

    $(g \circ f)(x) = g[f(x)]$

    $= (f(x))^3 - 5$

    $= \left(\sqrt[3]{x+5}\right)^3 - 5$

    $= x + 5 - 5$

    $= x$

59. $f(x) = \dfrac{2}{3}x - 6$; $g(x) = \dfrac{3}{2}x + 9$

    $(f \circ g)(x) = f[g(x)]$

    $= \dfrac{2}{3}g(x) - 6$

    $= \dfrac{2}{3}\left(\dfrac{3}{2}x + 9\right) - 6$

    $= x + 6 - 6$

    $= x;$

    $(g \circ f)(x) = g[f(x)]$

    $= \dfrac{3}{2}f(x) + 9$

    $= \dfrac{3}{2}\left(\dfrac{2}{3}x - 6\right) + 9$

    $= x - 9 + 9$

    $= x$

61. $f(x) = x^2 - 3$; $x \geq 0$; $g(x) = \sqrt{x+3}$

    $(f \circ g)(x) = f[g(x)]$

    $= (g(x))^2 - 3$

    $= \left(\sqrt{x+3}\right)^2 - 3$

    $= x + 3 - 3$

    $= x;$

    $(g \circ f)(x) = g[f(x)]$

    $= \sqrt{f(x) + 3}$

    $= \sqrt{x^2 - 3 + 3}$

    $= \sqrt{x^2}$

    $= x$

63. $f(x) = 3x - 5$

$y = 3x - 5$

$x = 3y - 5$

$x + 5 = 3y$

$\dfrac{x+5}{3} = y$

$f^{-1}(x) = \dfrac{x+5}{3}$

$(f \circ f^{-1})(x) = f\left[f^{-1}(x)\right]$

$= 3(f^{-1}(x)) - 5$

$= 3\left(\dfrac{x+5}{3}\right) - 5$

$= x + 5 - 5$

$= x;$

$(f^{-1} \circ f)(x) = f^{-1}[f(x)]$

$= \dfrac{f(x)+5}{3}$

$= \dfrac{3x - 5 + 5}{3}$

$= \dfrac{3x}{3}$

$= x$

65. $f(x) = \dfrac{x-5}{2}$

$y = \dfrac{x-5}{2}$

$x = \dfrac{y-5}{2}$

$2x = y - 5$

$2x + 5 = y$

$f^{-1}(x) = 2x + 5$

$(f \circ f^{-1})(x) = f\left[f^{-1}(x)\right]$

$= \dfrac{f^{-1}(x) - 5}{2}$

$= \dfrac{2x + 5 - 5}{2}$

$= \dfrac{2x}{2}$

$= x;$

$(f^{-1} \circ f)(x) = f^{-1}[f(x)]$

$= 2(f(x)) + 5$

$= 2\left(\dfrac{x-5}{2}\right) + 5$

$= x - 5 + 5$

$= x$

67. $f(x) = \dfrac{1}{2}x - 3$

$y = \dfrac{1}{2}x - 3$

$x = \dfrac{1}{2}y - 3$

$x + 3 = \dfrac{1}{2}y$

$2x + 6 = y$

$f^{-1}(x) = 2x + 6$

$(f \circ f^{-1})(x) = f\left[f^{-1}(x)\right]$

$= \dfrac{1}{2}(f^{-1}(x)) - 3$

$= \dfrac{1}{2}(2x + 6) - 3$

$= x + 3 - 3$

$= x;$

$(f^{-1} \circ f)(x) = f^{-1}[f(x)]$

$= 2(f(x)) + 6$

$= 2\left(\dfrac{1}{2}x - 3\right) + 6$

$= x - 6 + 6$

$= x$

69. $f(x) = x^3 + 3$

$y = x^3 + 3$

$x = y^3 + 3$

$x - 3 = y^3$

$\sqrt[3]{x-3} = y$

$f^{-1}(x) = \sqrt[3]{x-3}$

$\left(f \circ f^{-1}\right)(x) = f\left[f^{-1}(x)\right]$

$= \left(f^{-1}(x)\right)^3 + 3$

$= \left(\sqrt[3]{x-3}\right)^3 + 3$

$= x - 3 + 3$

$= x;$

$\left(f^{-1} \circ f\right)(x) = f^{-1}[f(x)]$

$= \sqrt[3]{f(x) - 3}$

$= \sqrt[3]{x^3 + 3 - 3}$

$= \sqrt[3]{x^3}$

$= x$

71. $f(x) = \sqrt[3]{2x+1}$

$y = \sqrt[3]{2x+1}$

$x = \sqrt[3]{2y+1}$

$x^3 = 2y + 1$

$x^3 - 1 = 2y$

$\dfrac{x^3 - 1}{2} = y$

$f^{-1}(x) = \dfrac{x^3 - 1}{2}$

$\left(f \circ f^{-1}\right)(x) = f\left[f^{-1}(x)\right]$

$= \sqrt[3]{2\left(f^{-1}(x)\right) + 1}$

$= \sqrt[3]{2\left(\dfrac{x^3 - 1}{2}\right) + 1}$

$= \sqrt[3]{x^3 - 1 + 1}$

$= \sqrt[3]{x^3}$

$= x;$

$\left(f^{-1} \circ f\right)(x) = f^{-1}[f(x)]$

$= \dfrac{(f(x))^3 - 1}{2}$

$= \dfrac{\left(\sqrt[3]{2x+1}\right)^3 - 1}{2}$

$= \dfrac{2x + 1 - 1}{2}$

$= \dfrac{2x}{2}$

$= x$

73. $f(x) = \dfrac{(x-1)^3}{8}$

$y = \dfrac{(x-1)^3}{8}$

$x = \dfrac{(y-1)^3}{8}$

$8x = (y-1)^3$

$\sqrt[3]{8x} = y - 1$

$2\sqrt[3]{x} + 1 = y$

$f^{-1}(x) = 2\sqrt[3]{x} + 1$

$\left(f \circ f^{-1}\right)(x) = f\left[f^{-1}(x)\right]$

$= \dfrac{\left(f^{-1}(x) - 1\right)^3}{8}$

$= \dfrac{\left(2\sqrt[3]{x} + 1 - 1\right)^3}{8}$

$= \dfrac{\left(2\sqrt[3]{x}\right)^3}{8}$

$= \dfrac{8x}{8}$

$= x;$

$\left(f^{-1} \circ f\right)(x) = f^{-1}[f(x)]$

$= 2\sqrt[3]{f(x)} + 1$

$= 2\sqrt[3]{\dfrac{(x-1)^3}{8}} + 1$

$= 2\left(\dfrac{x-1}{2}\right) + 1$

$= x - 1 + 1$

$= x$

75. $f(x) = \sqrt{3x+2}$,

$x \in \left[-\dfrac{2}{3}, \infty\right), y \in [0, \infty)$

$y = \sqrt{3x+2}$

$x = \sqrt{3y+2}$

$x^2 = 3y + 2$

$x^2 - 2 = 3y$

$\dfrac{x^2 - 2}{3} = y$

$f^{-1}(x) = \dfrac{x^2 - 2}{3}; x \geq 0; \ y \in \left[-\dfrac{2}{3}, \infty\right)$

$(f \circ f^{-1})(x) = f(f^{-1}(x))$

$= \sqrt{3(f^{-1}(x)) + 2}$

$= \sqrt{3\left(\dfrac{x^2 - 2}{3}\right) + 2}$

$= \sqrt{x^2 - 2 + 2}$

$= \sqrt{x^2}$

$= |x|$

$= x; \ \text{since } x \geq 0$

$(f^{-1} \circ f)(x) = f^{-1}[f(x)]$

$= \dfrac{(f(x))^2 - 2}{3}$

$= \dfrac{\left(\sqrt{3x+2}\right)^2 - 2}{3}$

$= \dfrac{3x + 2 - 2}{3}$

$= \dfrac{3x}{3}$

$= x$

77. $p(x) = 2\sqrt{x-3}$

$x \in [3, \infty), y \in [0, \infty)$

$y = 2\sqrt{x-3}$

$x = 2\sqrt{y-3}$

$\dfrac{x}{2} = \sqrt{y-3}$

$\dfrac{x^2}{4} = y - 3$

$\dfrac{x^2}{4} + 3 = y$

$p^{-1}(x) = \dfrac{x^2}{4} + 3; x \geq 0 \ ; y \in [3, \infty)$

$(p \circ p^{-1})(x) = p(p^{-1}(x))$

$= 2\sqrt{p^{-1}(x) - 3}$

$= 2\sqrt{\dfrac{x^2}{4} + 3 - 3}$

$= 2\sqrt{\dfrac{x^2}{4}}$

$= 2\left(\dfrac{x}{2}\right)$

$= x;$

$(p^{-1} \circ p)(x) = p^{-1}[p(x)]$

$= \dfrac{(p(x))^2}{4} + 3$

$= \dfrac{\left(2\sqrt{x-3}\right)^2}{4} + 3$

$= \dfrac{4(x-3)}{4} + 3$

$= x - 3 + 3$

$= x$

79. $v(x) = x^2 + 3; x \geq 0, y \in [3, \infty)$

$y = x^2 + 3$

$x = y^2 + 3$

$x - 3 = y^2$

$\sqrt{x - 3} = y$

$v^{-1}(x) = \sqrt{x - 3}$; $x \geq 3, y \in [0, \infty)$

$\left(v \circ v^{-1}\right)(x) = v\left[v^{-1}(x)\right]$

$= \left(v^{-1}(x)\right)^2 + 3$

$= \left(\sqrt{x - 3}\right)^2 + 3$

$= x - 3 + 3$

$= x;$

$\left(v^{-1} \circ v\right)(x) = v^{-1}[v(x)]$

$= \sqrt{v(x) - 3}$

$= \sqrt{x^2 + 3 - 3}$

$= \sqrt{x^2}$

$= |x|$

$= x;$ since $x \geq 0$

81. $f(x) = 4x + 1$, $f^{-1}(x) = \dfrac{x - 1}{4}$

$\left(f \circ f^{-1}\right)(x) = f\left[f^{-1}(x)\right]$

$= 4\left(f^{-1}(x)\right) + 1$

$= 4\left(\dfrac{x - 1}{4}\right) + 1$

$= x - 1 + 1$

$= x;$

$\left(f^{-1} \circ f\right)(x) = f^{-1}[f(x)]$

$= \dfrac{f(x) - 1}{4}$

$= \dfrac{4x + 1 - 1}{4}$

$= \dfrac{4x}{4}$

$= x$

83. $f(x) = \sqrt[3]{x + 2}$; $f^{-1}(x) = x^3 - 2$

$\left(f \circ f^{-1}\right)(x) = f\left[f^{-1}(x)\right]$

$= \sqrt[3]{f^{-1}(x) + 2}$

$= \sqrt[3]{x^3 - 2 + 2}$

$= \sqrt[3]{x^3}$

$= x;$

$\left(f^{-1} \circ f\right)(x) = f^{-1}[f(x)]$

$= (f(x))^3 - 2$

$= \left(\sqrt[3]{x + 2}\right)^3 - 2$

$= x + 2 - 2$

$= x$

85. $f(x) = 0.2x + 1$; $f^{-1}(x) = 5x - 5$

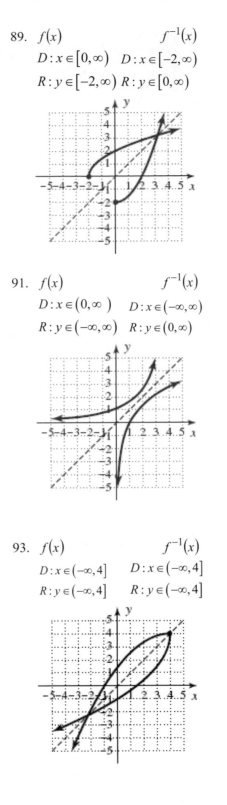

$\left(f \circ f^{-1}\right)(x) = f\left[f^{-1}(x)\right]$

$= 0.2\left(f^{-1}(x)\right) + 1$

$= 0.2(5x - 5) + 1$

$= x - 1 + 1$

$= x;$

$\left(f^{-1} \circ f\right)(x) = f^{-1}[f(x)]$

$= 5(f(x)) - 5$

$= 5(0.2x + 1) - 5$

$= x + 5 - 5$

$= x$

87. $f(x) = (x + 2)^2; x \geq -2$; $f^{-1}(x) = \sqrt{x} - 2$

$\left(f \circ f^{-1}\right)(x) = f\left(f^{-1}(x)\right)$

$= \left(f^{-1}(x) + 2\right)^2$

$= \left(\sqrt{x} - 2 + 2\right)^2$

$= \left(\sqrt{x}\right)^2$

$= x;$

$\left(f^{-1} \circ f\right)(x) = f^{-1}[f(x)]$

$= \sqrt{f(x)} - 2$

$= \sqrt{(x + 2)^2} - 2$

$= x + 2 - 2$

$= x$

89. $f(x)$ $\qquad$ $f^{-1}(x)$

$D: x \in [0, \infty)$ $\quad D: x \in [-2, \infty)$

$R: y \in [-2, \infty)$ $\quad R: y \in [0, \infty)$

91. $f(x)$ $\qquad$ $f^{-1}(x)$

$D: x \in (0, \infty)$ $\quad D: x \in (-\infty, \infty)$

$R: y \in (-\infty, \infty)$ $\quad R: y \in (0, \infty)$

93. $f(x)$ $\qquad$ $f^{-1}(x)$

$D: x \in (-\infty, 4]$ $\quad D: x \in (-\infty, 4]$

$R: y \in (-\infty, 4]$ $\quad R: y \in (-\infty, 4]$

95. $f(x) = \frac{1}{2}x - 8.5$

   a. $f(80) = \frac{1}{2}(80) - 8.5 = 31.5\,\text{cm}$

   b. $y = \frac{1}{2}x - 8.5$

   $x = \frac{1}{2}y - 8.5$

   $x + 8.5 = \frac{1}{2}y$

   $2x + 17 = y$

   $f^{-1}(x) = 2x + 17$

   $f^{-1}(31.5) = 2(31.5) + 17 = 80$

   It gives the distance of the projector from screen.

97. $f(x) = -\frac{7}{2}x + 59$

   a. $f(35) = -\frac{7}{2}(35) + 59 = -63.5°\text{F}$

   b. $y = -\frac{7}{2}x + 59$

   $x = -\frac{7}{2}y + 59$

   $x - 59 = -\frac{7}{2}y$

   $-\frac{2}{7}(x - 59) = y$

   $f^{-1}(x) = -\frac{2}{7}(x - 59)$

   Independent: temperature
   Dependent: altitude

   c. $f^{-1}(-18) = -\frac{2}{7}(-18 - 59)$

   $= -\frac{2}{7}(-77) = 22$

   The approximate altitude is 22000 feet.

99. $f(x) = 16x^2; x \geq 0$

   a. $f(3) = 16(3)^2 = 16(9) = 144\,\text{ft}$

   b. $y = 16x^2$

   $x = 16y^2$

   $\frac{x}{16} = y^2$

   $\frac{\sqrt{x}}{4} = y$

   $f^{-1}(x) = \frac{\sqrt{x}}{4};$

   Independent: distance fallen
   Dependent: time fallen

   c. $f^{-1}(784) = \frac{\sqrt{784}}{4} = \frac{28}{4} = 7\,\text{sec}$

101. $f(x) = \frac{1}{3}\pi x^3$

   a. $f(30) = \frac{1}{3}\pi(30)^3 = 9000\pi \approx 28260\,\text{ft}^3$

   b. $y = \frac{1}{3}\pi x^3$

   $x = \frac{1}{3}\pi y^3$

   $3x = \pi y^3$

   $\frac{3x}{\pi} = y^3$

   $\sqrt[3]{\frac{3x}{\pi}} = y$

   $f^{-1}(x) = \sqrt[3]{\frac{3x}{\pi}};$

   Independent: volume
   Dependent: height

   c. $f^{-1}(763.02) = \sqrt[3]{\frac{3(763.02)}{\pi}} = 9\,\text{ft}$

103. $f(x) = \{(x,y)|\, y = 3x - 6\}$

    a. $f(2) = 3(2) - 6 = 0,\ (2,0);$

        $f(0) = 3(0) - 6 = -6,\ (0,-6);$

        $f(1) = 3(1) - 6 = -3,\ (1,-3);$

        $f(3) = 3(3) - 6 = 3,\ (3,3);$

        $f(-1) = 3(-1) - 6 = -9,\ (-1,-9)$

    b. $(0,2),(-6,0),(-3,1),(3,3),(-9,-1)$

$$f^{-1}(x) = \left\{(x,y)\Big|\, y = \frac{x}{3} + 2\right\}$$

$$f^{-1}(0) = \frac{0}{3} + 2 = 2,\ (0,2);$$

$$f^{-1}(-6) = \frac{-6}{3} + 2 = 0,\ (-6,0);$$

$$f^{-1}(-3) = \frac{-3}{3} + 2 = 1,\ (-3,1);$$

$$f^{-1}(3) = \frac{3}{3} + 2 = 3,\ (3,3);$$

$$f^{-1}(-9) = \frac{-9}{3} + 2 = -1,\ (-9,-1)$$

105. $f(x) = \dfrac{2}{3}\left(x - \dfrac{1}{2}\right)^5 + \dfrac{4}{5}$

$$y = \frac{2}{3}\left(x - \frac{1}{2}\right)^5 + \frac{4}{5}$$

$$x = \frac{2}{3}\left(y - \frac{1}{2}\right)^5 + \frac{4}{5}$$

$$x - \frac{4}{5} = \frac{2}{3}\left(y - \frac{1}{2}\right)^5$$

$$\frac{3}{2}\left(x - \frac{4}{5}\right) = \left(y - \frac{1}{2}\right)^5$$

$$\sqrt[5]{\frac{3}{2}\left(x - \frac{4}{5}\right)} = y - \frac{1}{2}$$

$$\sqrt[5]{\frac{3}{2}\left(x - \frac{4}{5}\right)} + \frac{1}{2} = y$$

    d. $f^{-1}(x) = \sqrt[5]{\dfrac{3}{2}\left(x - \dfrac{4}{5}\right)} + \dfrac{1}{2}$

107. $f(x) = x^2 - x - 2;\ f(x) \le 0$

    $x^2 - x - 2 = 0$

    $(x - 2)(x + 1) = 0$

    $x = 2;\ x = -1$

    Concave up

    $x \in [-1, 2]$

109.a. Perimeter of a rectangle:

        $P = 2l + 2w$

    b. Area of a circle:

        $A = \pi r^2$

    c. Volume of a cylinder:

        $V = \pi r^2 h$

    d. Volume of a cone:

        $V = \dfrac{1}{3}\pi r^2 h$

    e. Circumference of a circle:

        $C = 2\pi r$

    f. Area of a triangle:

        $A = \dfrac{1}{2}bh$

    g. Area of a trapezoid:

        $A = \dfrac{1}{2}(b_1 + b_2)h$

    h. Volume of a sphere:

        $V = \dfrac{4}{3}\pi r^3$

    i. Pythagorean Theorem:

        $a^2 + b^2 = c^2$

4.2 Exercises

4.2 Technology Highlight

1. $3^x = 22$; $x \approx 2.8$

3. $e^{x-1} = 9$; $x \approx 3.2$

4.2 Exercises

1. b^x, b, b, x

3. $a, 1$

5. False; for $|b| < 1$ and $x_2 > x_1, b^{x_2} < b^{x_1}$
   so the function is decreasing.

7. $P(t) = 2500 \cdot 4^t$;

   $P(2) = 2500 \cdot 4^2 = 40000$;

   $P\left(\dfrac{1}{2}\right) = 2500 \cdot 4^{\frac{1}{2}} = 5000$;

   $P\left(\dfrac{3}{2}\right) = 2500 \cdot 4^{\frac{3}{2}} = 20000$;

   $P\left(\sqrt{3}\right) = 2500 \cdot 4^{\sqrt{3}} \approx 27589.162$

9. $f(x) = 0.5 \cdot 10^x$;

   $f(3) = 0.5 \cdot 10^3 = 500$;

   $f\left(\dfrac{1}{2}\right) = 0.5 \cdot 10^{\frac{1}{2}} \approx 1.581$;

   $f\left(\dfrac{2}{3}\right) = 0.5 \cdot 10^{\frac{2}{3}} \approx 2.321$;

   $f\left(\sqrt{7}\right) = 0.5 \cdot 10^{\sqrt{7}} \approx 221.168$

11. $V(n) = 10{,}000\left(\dfrac{2}{3}\right)^n$;

    $V(0) = 10{,}000\left(\dfrac{2}{3}\right)^0 = 10000$;

    $V(4) = 10{,}000\left(\dfrac{2}{3}\right)^4 \approx 1975.309$;

    $V(4.7) = 10{,}000\left(\dfrac{2}{3}\right)^{4.7} \approx 1487.206$;

    $V(5) = 10{,}000\left(\dfrac{2}{3}\right)^5 \approx 1316.872$

13. $y = 3^x$
    y-intercept: (0, 1)

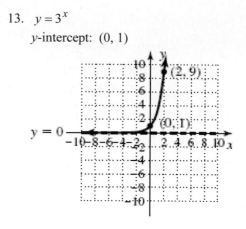

increasing

15. $y = \left(\dfrac{1}{3}\right)^x$
    y-intercept: (0, 1)

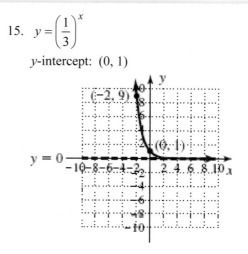

decreasing

17. $y = 3^x + 2$
    up 2

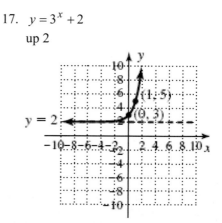

228

19. $y = 3^{x+3}$

left 3

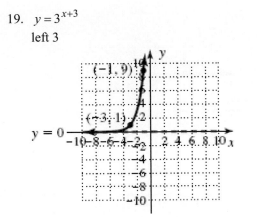

21. $y = 2^{-x}$

reflected in the y-axis

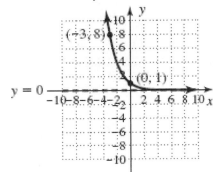

23. $y = 2^{-x} + 3$

reflected in the y-axis, up 3

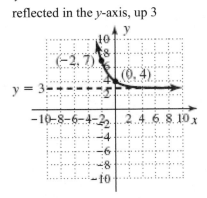

25. $y = 2^{x+1} - 3$

left 1, down 3

27. $y = \left(\dfrac{1}{3}\right)^{x} + 1$

up 1

29. $y = \left(\dfrac{1}{3}\right)^{x-2}$

right 2

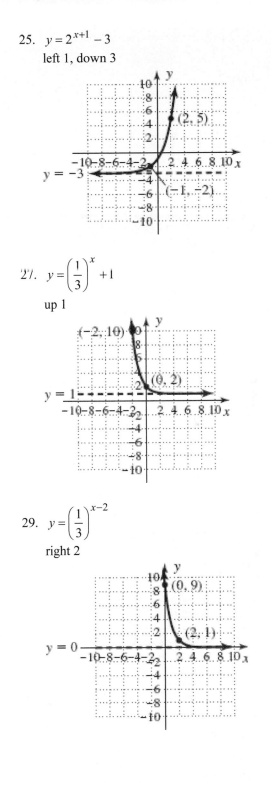

31. $y = \left(\dfrac{1}{3}\right)^x - 2$

    down 2

    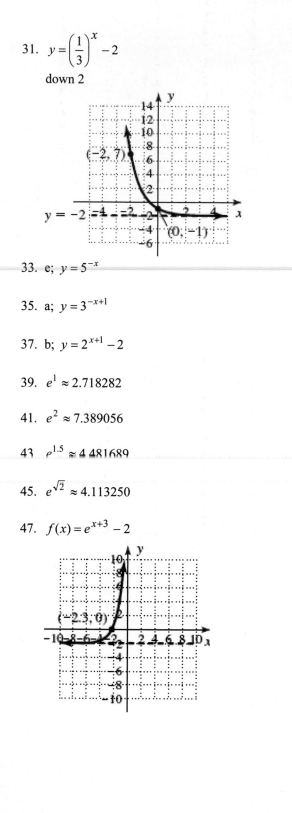

33. e; $y = 5^{-x}$

35. a; $y = 3^{-x+1}$

37. b; $y = 2^{x+1} - 2$

39. $e^1 \approx 2.718282$

41. $e^2 \approx 7.389056$

43. $e^{1.5} \approx 4.481689$

45. $e^{\sqrt{2}} \approx 4.113250$

47. $f(x) = e^{x+3} - 2$

49. $r(t) = -e^t + 2$

    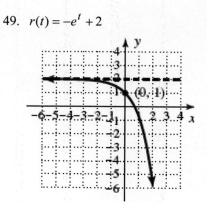

51. $p(x) = e^{-x+2} - 1$

    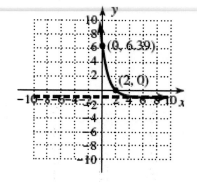

53. $10^x = 1000$

    $10^x = 10^3$

    $x = 3$

55. $25^x = 125$

    $5^{2x} = 5^3$

    $2x = 3$

    $x = \dfrac{3}{2}$

57. $8^{x+2} = 32$

    $2^{3(x+2)} = 2^5$

    $3x + 6 = 5$

    $3x = -1$

    $x = -\dfrac{1}{3}$

59. $32^x = 16^{x+1}$

    $2^{5x} = 2^{4(x+1)}$

    $5x = 4x + 4$

    $x = 4$

61. $\left(\dfrac{1}{5}\right)^x = 125$

$\left(\dfrac{1}{5}\right)^x = \left(\dfrac{1}{5}\right)^{-3}$

$x = -3$

63. $\left(\dfrac{1}{3}\right)^{2x} = 9^{x-6}$

$\left(\dfrac{1}{3}\right)^{2x} = \left(\dfrac{1}{3}\right)^{-2(x-6)}$

$2x = -2x + 12$

$4x = 12$

$x = 3$

65. $\left(\dfrac{1}{9}\right)^{x-5} = 3^{3x}$

$\left(\dfrac{1}{3}\right)^{2(x-5)} = \left(\dfrac{1}{3}\right)^{-1(3x)}$

$2x - 10 = -3x$

$5x = 10$

$x = 2$

67. $25^{3x} = 125^{x-2}$

$5^{6x} = 5^{3(x-2)}$

$6x = 3x - 6$

$3x = -6$

$x = -2$

69. $\dfrac{e^4}{e^{2-x}} = e^3 e^1$

$e^{4-(2-x)} = e^4$

$4 - (2 - x) = 4$

$4 - 2 + x = 4$

$2 + x = 4$

$x = 2$

71. $\left(e^{2x-4}\right)^3 = \dfrac{e^{x+5}}{e^2}$

$e^{6x-12} = e^{x+5-2}$

$6x - 12 = x + 3$

$5x - 12 = 3$

$5x = 15$

$x = 3$

74. $P(x) = \left(\dfrac{1}{4}\right)^x$

(a) $P(1) = \left(\dfrac{1}{4}\right)^1 = \dfrac{1}{4}$

(b) $P(4) = \left(\dfrac{1}{4}\right)^4 = \dfrac{1}{256}$

(c) as $x \to \infty$, $P \to 0$

77. $V(t) = V_0 \cdot \left(\dfrac{4}{5}\right)^t$

(a) $V(1) = 125000 \cdot \left(\dfrac{4}{5}\right)^1 = \$100,000$

(b) $64000 = 125000 \cdot \left(\dfrac{4}{5}\right)^t$

$\dfrac{64}{125} = \left(\dfrac{4}{5}\right)^t$

$\left(\dfrac{4}{5}\right)^3 = \left(\dfrac{4}{5}\right)^t$

$t = 3$ yr

79. $V(t) = V_0 \cdot \left(\dfrac{5}{6}\right)^t$

(a) $V(5) = 216000 \cdot \left(\dfrac{5}{6}\right)^5 \approx \$86,806$

(b) $125000 = 216000 \cdot \left(\dfrac{5}{6}\right)^t$

$\dfrac{125}{216} = \left(\dfrac{5}{6}\right)^t$

$\left(\dfrac{5}{6}\right)^3 = \left(\dfrac{5}{6}\right)^t$

$t = 3$ yr

81. $R(t) = R_0 \cdot 2^t$

(a) $R(4) = 2.5 \cdot 2^4 = \$40$ million

(b) $320 = 2.5 \cdot 2^t$

$128 = 2^t$

$2^7 = 2^t$

$t = 7$ yr

83. $T(x) = 0.85^x$

  $T(7) = 0.85^7 = 0.32058 \approx 32\%$,
  transparent

85. $T(11) = 0.85^{11} = 0.167734 \approx 17\%$,
  transparent

87. $P(t) = P_0(1.05)^t$

  $P(10) = 20000(1.05)^{10} \approx \$32,578$

89. $Q(t) = Q_0\left(\dfrac{1}{2}\right)^{\frac{t}{h}}$

  (a) $Q(24) = 64\left(\dfrac{1}{2}\right)^{\frac{24}{8}} = 8$ grams

  (b) $1 = 64\left(\dfrac{1}{2}\right)^{\frac{t}{8}}$

  $\dfrac{1}{64} = \left(\dfrac{1}{2}\right)^{\frac{t}{8}}$

  $\left(\dfrac{1}{2}\right)^6 = \left(\dfrac{1}{2}\right)^{\frac{t}{8}}$

  $6 = \dfrac{t}{8}$

  $t = 48$ minutes

91. $f(20) = \left(\dfrac{1}{2}\right)^{20} = 9.5 \times 10^{-7}$;

  Answers will vary.

93. $5^{3x} = 27$, find 5^{2x}

  $\left(5^x\right)^3 = 3^3$

  $5^x = 3^1$

  $\left(5^x\right)^2 = 3^2$

  $5^{2x} = 9$

95. $\left(\dfrac{1}{2}\right)^{x+1} = \dfrac{1}{3}$, find $\left(\dfrac{1}{2}\right)^{-x}$

  $\dfrac{1}{2}\left(\dfrac{1}{2}\right)^x = \dfrac{1}{3}$

  $\left(\dfrac{1}{2}\right)^x = \dfrac{2}{3}$

  $\left(\left(\dfrac{1}{2}\right)^x\right)^{-1} = \left(\dfrac{2}{3}\right)^{-1}$

  $\left(\dfrac{1}{2}\right)^{-x} = \dfrac{3}{2}$

97. $\left(1 + \dfrac{1}{x}\right)^x$

  a. $\dfrac{f(x+0.01) - f(x)}{x+0.01-x}$

  $= \dfrac{\left(1 + \dfrac{1}{x+0.01}\right)^{x+0.01} - \left(1 + \dfrac{1}{x}\right)^x}{0.01}$

  TABLE:
  At $x = 1$, 0.3842;
  At $x = 4$, 0.0564;
  At $x = 10$, 0.0114
  At $x = 20$, 0.0031
  The rate of growth seems to be approaching 0.

  b. Using TABLE At $e = 2.718281828...$
  $x = 16,608$

  c. yes, the secant lines are becoming virtually horizontal $(y = e)$

99. $D : x \in [-2, \infty), R : y \in [-1, \infty)$

101. a. Volume of a sphere: $\dfrac{4}{3}\pi r^3$

  b. Area of a triangle: $\dfrac{1}{2}bh$

  c. Volume of a rectangular prism: lwh

  d. Pythagorean theorem: $a^2 + b^2 = c^2$

4.3 Exercises

1. $\log_b x$, b, b, greater

3. $(1,0)$; 0

5. 5; answers will vary

7. $3 = \log_2 8$
   $2^3 = 8$

9. $-1 = \log_7 \frac{1}{7}$
   $7^{-1} = \frac{1}{7}$

11. $0 = \log_9 1$
    $9^0 = 1$

13. $\frac{1}{3} = \log_8 2$
    $8^{\frac{1}{3}} = 2$

15. $1 = \log_2 2$
    $2^1 = 2$

17. $\log_7 49 = 2$
    $7^2 = 49$

19. $\log_{10} 100 = 2$
    $10^2 = 100$

21. $\log_e (54.598) \approx 4$
    $e^4 \approx 54.598$

23. $4^3 = 64$
    $\log_4 64 = 3$

25. $3^{-2} = \frac{1}{9}$
    $\log_3 \left(\frac{1}{9}\right) = -2$

27. $e^0 = 1$
    $0 = \log_e 1$

29. $\left(\frac{1}{3}\right)^{-3} = 27$
    $\log_{\frac{1}{3}} 27 = -3$

31. $10^3 = 1000$
    $\log 1000 = 3$

33. $10^{-2} = \frac{1}{100}$
    $\log \frac{1}{100} = -2$

35. $4^{\frac{3}{2}} = 8$
    $\log_4 8 = \frac{3}{2}$

37. $4^{\frac{-3}{2}} = \frac{1}{8}$
    $\log_4 \frac{1}{8} = \frac{-3}{2}$

39. $\log_4 4$
    $= 1$

41. $\log_{11} 121 = x$
    $11^x = 121$
    $11^x = 11^2$
    $x = 2$

43. $\log_e e = \log_e e^1 = 1$

45. $\log_4 2 = x$
    $4^x = 2$
    $2^{2x} = 2^1$
    $2x = 1$
    $x = \frac{1}{2}$

47. $\log_7 \dfrac{1}{49} = x$

$7^x = \dfrac{1}{49}$

$7^x = 7^{-2}$
$x = -2$

49. $\log_e \dfrac{1}{e^2} = \log_e e^{-2} = -2$

51. $\log 50 = 1.6990$

53. $\ln 1.6 = 0.4700$

55. $\ln 225 = 5.4161$

57. $\log \sqrt{37} = 0.7841$

59. $f(x) = \log_2 x + 3$
Shift up 3

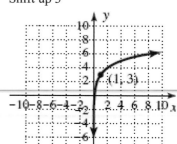

61. $h(x) = \log_2(x-2) + 3$
Shift right 2, up 3

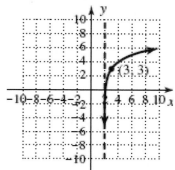

63. $q(x) = \ln(x+1)$
Shift left 1

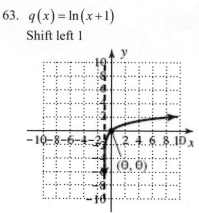

65. $Y_1 = -\ln(X+1)$
Reflected across x–axis, shift left 1

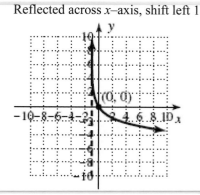

67. $y = \log_b(x+2)$, II

69. $y = 1 - \log_b x$, VI

71. $y = \log_b x + 2$, V

73. $y = \log_6\left(\dfrac{x+1}{x-3}\right)$

$\dfrac{x+1}{x-3} > 0, x \neq 3$

critical values: -1 and 3

pos   neg   pos

-1   3

$x \in (-\infty, -1) \cup (3, \infty)$

75. $y = \log_5 \sqrt{2x-3}$

$2x - 3 > 0$

$2x > 3$

$x > \dfrac{3}{2}$

$x \in \left(\dfrac{3}{2}, \infty \right)$

77. $y = \log(9 - x^2)$

$9 - x^2 > 0$;

$(3 + x)(3 - x) > 0$

critical values: 3 and -3

neg pos neg

-3   3

$x \in (-3, 3)$

79. $f(x) = -\log_{10} x$

$x = 7.94 \times 10^{-5}$

$f(7.94 \times 10^{-5}) = -\log_{10}(7.94 \times 10^{-5})$

$pH \approx 4.1$; acid

81. $M(I) = \log\left(\dfrac{I}{I_0} \right)$

a. $I = 50{,}000 I_0$

$M(50000 I_0) = \log\left(\dfrac{50000 I_0}{I_0} \right)$

$M(50000 I_0) = \log(50000) \approx 4.7$

b. $I = 75{,}000 I_0$

$M(75000 I_0) = \log\left(\dfrac{75000 I_0}{I_0} \right)$

$M(75000 I_0) = \log(75000) \approx 4.9$

83. $M(I) = \log\left(\dfrac{I}{I_0} \right)$

1989: $6.2 = \log\left(\dfrac{I}{I_0} \right)$

$10^{6.2} = \dfrac{I}{I_0}$

$I = 10^{6.2} I_0$;

2006: $6.7 = \log\left(\dfrac{I}{I_0} \right)$

$10^{6.7} = \dfrac{I}{I_0}$

$I = 10^{6.7} I_0$;

Comparing the 1989 earthquakes to the 2006

earthquakes: $\dfrac{10^{6.7} I_0}{10^{6.2} I_0} \approx 3.2$ times

85. $M(I) = 6 - 2.5 \cdot \log\left(\dfrac{I}{I_0} \right)$

a. $I = 27 \cdot I_0$

$M(27 I_0) = 6 - 2.5 \cdot \log\left(\dfrac{27 I_0}{I_0} \right)$

≈ 2.4

b. $I = 85 \cdot I_0$

$M(85 I_0) = 6 - 2.5 \cdot \log\left(\dfrac{85 I_0}{I_0} \right)$

≈ 1.2

87. $D(I) = 10 \cdot \log\left(\dfrac{I}{I_0} \right)$

a. $I = 10^{-14}$

$D(10^{-14}) = 10 \cdot \log\left(\dfrac{10^{-14}}{10^{-16}} \right) = 20$ dB

b. $I = 10^{-4}$

$D(10^{-4}) = 10 \cdot \log\left(\dfrac{10^{-4}}{10^{-16}} \right) = 120$ dB

4.3 Exercises

89. $D(I) = 10 \cdot \log\left(\dfrac{I}{I_0}\right)$;

Aircompressor: $D(I) = 110$

$110 = 10 \cdot \log\left(\dfrac{I}{I_0}\right)$

$\dfrac{110}{10} = \log\left(\dfrac{I}{I_0}\right)$

$11 = \log\left(\dfrac{I}{I_0}\right)$

$10^{11} = \dfrac{I}{I_0}$

$I = 10^{11} I_0$;

Hair Dryer: $D(I) = 75$

$75 = 10 \cdot \log\left(\dfrac{I}{I_0}\right)$

$\dfrac{75}{10} = \log\left(\dfrac{I}{I_0}\right)$

$7.5 = \log\left(\dfrac{I}{I_0}\right)$

$10^{7.5} = \dfrac{I}{I_0}$

$I = 10^{7.5} I_0$;

Comparison: $\dfrac{10^{11} I_0}{10^{7.5} I_0} \approx 3162$ times

91. $H = (30T + 8000) \cdot \ln\left(\dfrac{P_0}{P}\right)$

$T = -10, P = 34, P_0 = 76$

$H = (30(-10) + 8000) \cdot \ln\left(\dfrac{76}{34}\right)$

$H \approx 6194$ meters

93. $H = (30T + 8000) \cdot \ln\left(\dfrac{P_0}{P}\right)$

a. $T = 8, P = 39.3, P_0 = 76$

$H = (30(8) + 8000) \cdot \ln\left(\dfrac{76}{39.3}\right)$

$H \approx 5,434$ meters

b. $T = 12, P = 47.1, P_0 = 76$

$H = (30(12) + 8000) \cdot \ln\left(\dfrac{76}{47.1}\right)$

$H \approx 4,000$ meters

95. $N(A) = 1500 + 315 \cdot \ln(A)$

(a) $N(10) = 1500 + 315 \cdot \ln(10)$

$= 2225$ items

(b) $N(50) = 1500 + 315 \cdot \ln(50)$

$= 2732$ items

(c) $\approx \$117,000$

(d) $\dfrac{\Delta N}{\Delta A} = \dfrac{8}{1}$

$\dfrac{N(39.4) - N(39.3)}{39.4 - 39.3} = \dfrac{8}{1}$

97. $C(x) = 42 \ln x - 270$

(a) $C(2500) = 42 \ln 2500 - 270$

≈ 58.6 cfm

(b) $40 = 42 \ln x - 270$

$310 = 42 \ln x$

$\dfrac{310}{42} = \ln x$

$e^{\frac{310}{42}} = x$

$x \approx 1,605$ ft^2

99. $P(x) = 95 - 14 \cdot \log_2(x)$

a. 1 day

$P(1) = 95 - 14 \cdot \log_2(1) = 95\%$

b. 4 days

$P(4) = 95 - 14 \cdot \log_2(4) = 67\%$

c. 16 days

$P(16) = 95 - 14 \cdot \log_2(16) = 39\%$

101. $f(x) = -\log_{10} x$

$x = 5.1 \times 10^{-5}$

$f(5.1 \times 10^{-5}) = -\log_{10}(5.1 \times 10^{-5})$

pH ≈ 4.3; acid

103. a. Threshold of audibility
0 dB

b. Lawn Mower
90 dB

c. Whisper
15 dB

d. Loud rock concert
120 dB

e. Lively party
100 dB

f. Jet engine
140 dB

Many sources give the threshold of pain as 120dB; answers will vary.

105.a. $\log_{64} \dfrac{1}{16} = x$

Convert to exponential form:

$64^x = \dfrac{1}{16}$

$4^{3x} = 4^{-2}$

$3x = -2$

$x = \dfrac{-2}{3}$

b. $\log_{\frac{4}{9}}\left(\dfrac{27}{8}\right) = x$

Convert to exponential form:

$\left(\dfrac{4}{9}\right)^x - \dfrac{27}{8}$

$\left(\dfrac{2}{3}\right)^{2x} = \left(\dfrac{2}{3}\right)^{-3}$

$2x = -3$

$x = \dfrac{-3}{2}$

c. $\log_{0.25} 32 = x$

$(0.25)^x = 32$

$\left(\dfrac{1}{4}\right)^x = 2^5$

$\left(4^{-1}\right)^x = 2^5$

$\left(2^{-2}\right)^x = 2^5$

$2^{-2x} = 2^5$

$-2x = 5$

$x = \dfrac{-5}{2}$

107. $g(x) = \sqrt[3]{x+2} - 1$

D: $x \in R$

R: $y \in R$

109. $x \in (-\infty, -5)$;

$f(x) = (x+5)(x-4)^2$

$f(x) = (x+5)(x^2 - 8x + 16)$

$f(x) = x^3 - 8x^2 + 16x + 5x^2 - 40x + 80$

$f(x) = x^3 - 3x^2 - 24x + 80$

Chapter 4 Mid-Chapter Check

1. a. $27^{\frac{2}{3}} = 9$

$\dfrac{2}{3} = \log_{27} 9$

b. $81^{\frac{5}{4}} = 243$

$\dfrac{5}{4} = \log_{81} 243$

3. a. $4^{2x} = 32^{x-1}$

$\left(2^2\right)^{2x} = \left(2^5\right)^{x-1}$

$2^{4x} = 2^{5x-5}$

$4x = 5x - 5$

$x = 5$

b. $\left(\dfrac{1}{3}\right)^{4b} = 9^{2b-5}$

$\left(3^{-1}\right)^{4b} = \left(3^2\right)^{2b-5}$

$3^{-4b} = 3^{4b-10}$

$-4b = 4b - 10$

$-8b = -10$

$b = \dfrac{5}{4}$

5. $V(t) = V_0\left(\dfrac{9}{8}\right)^t$

a. $V(3) = 50{,}000\left(\dfrac{9}{8}\right)^3 = \$71{,}191.41$

b. 6 yr

7. $f(x) = \sqrt{x-3} + 1$

$D: x \in [3, \infty); R: y \in [1, \infty);$

$f(x) = \sqrt{x-3} + 1$

Interchange x and y.

$x = \sqrt{y-3} + 1$

$x - 1 = \sqrt{y-3}$

$(x-1)^2 = y - 3$

$(x-1)^2 + 3 = y$

$f^{-1}(x) = (x-1)^2 + 3$

$D: x \in [1, \infty); R: y \in [3, \infty)$

9. (a) $\dfrac{2}{3} = \log_{27} 9$

$27^{\frac{2}{3}} = 9$

$\left(3^3\right)^{\frac{2}{3}} = 9$

$3^2 = 9$, verified

(b) $1.4 \approx \ln 4.0552$

$e^{1.4} \approx 4.0552$,

verified on calculator

Reinforcing Basic Concepts

1. $14 - 11.8 = 2.2; \ 10^{2.2} \approx 158$

About 158 times

2. $11.2 - 8.5 = 2.7; \ 10^{2.7} \approx 501$

About 501 times

3. $7.5 - 3.4 = 4.1; \ 10^{4.1} \approx 12{,}589$

About 12,589 times

4. $9.5 - 6.9 = 2.6; \ 10^{2.6} \approx 398$

About 398 times

5. $9.1 - 4.5 = 4.6; \ 10^{4.6} \approx 39{,}811$

About 39,811 times

4.4 Exercises

1. e

3. Extraneous

5. $\ln(4x+3)+\ln 2 = 3.2$
$\ln 2(4x+3) = 3.2$
$\ln(8x+6) = 3.2$
$e^{3.2} = 8x+6$
$e^{3.2} - 6 = 8x$
$\dfrac{e^{3.2}-6}{8} = x$
$x = 2.316366273$

7. $\ln x = 3.4$
$e^{\ln x} = e^{3.4}$
$x = e^{3.4}$
$x \approx 29.964$

9. $\log x = \dfrac{1}{4}$
$10^{\log x} = 10^{\frac{1}{4}}$
$x = 10^{\frac{1}{4}}$
$x \approx 1.778$

11. $e^x = 9.025$
$\ln e^x = \ln 9.025$
$x = \ln 9.025$
$x \approx 2.200$

13. $10^x = 18.197$
$\log 10^x = \log 18.197$
$x = \log 18.197$
$x \approx 1.260$

15. $4e^{x-2}+5 = 70$
$4e^{x-2} = 65$
$e^{x-2} = 16.25$
$\ln e^{x-2} = \ln 16.25$
$x-2 = \ln 16.25$
$x = 2 + \ln 16.25$
$x \approx 4.7881$

17. $10^{x+5} - 228 = -150$
$10^{x+5} = 78$
$\log 10^{x+5} = \log 78$
$x+5 = \log 78$
$x = -5 + \log 78$
$x \approx -3.1079$

19. $-150 = 290.8 - 190e^{-0.75x}$
$-440.8 = -190e^{-0.75x}$
$\dfrac{-440.8}{-190} = e^{-0.75x}$
$\dfrac{58}{25} = e^{-0.75x}$
$\ln\left(\dfrac{58}{25}\right) = \ln e^{-0.75x}$
$\ln\left(\dfrac{58}{25}\right) = -0.75x$
$\dfrac{\ln\left(\dfrac{58}{25}\right)}{-0.75} = x$
$x \approx -1.1221$

21. $3\ln(x+4)-5 = 3$
$3\ln(x+4) = 8$
$\ln(x+4) = \dfrac{8}{3}$
$x+4 = e^{\frac{8}{3}}$
$x = e^{\frac{8}{3}} - 4$
$x \approx 10.3919$

23. $-1.5 = 2\log(5-x)-4$
$2.5 = 2\log(5-x)$
$1.25 = \log(5-x)$
$10^{1.25} = 10^{\log(5-x)}$
$10^{1.25} = 5-x$
$x = 5 - 10^{1.25}$
$x \approx -12.7828$

4.4 Exercises

25. $\dfrac{1}{2}\ln(2x+5)+3=3.2$

$\dfrac{1}{2}\ln(2x+5)=0.2$

$\ln(2x+5)=0.4$

$e^{\ln(2x+5)}=e^{0.4}$

$2x+5=e^{0.4}$

$2x=e^{0.4}-5$

$x=\dfrac{e^{0.4}-5}{2}$

$x\approx -1.7541$

27. $\ln(2x)+\ln(x-7)$

$=\ln(2x(x-7))$

$=\ln(2x^2-14x)$

29. $\log(x+1)+\log(x-1)$

$=\log((x+1)(x-1))$

$=\log(x^2-1)$

31. $\log_3 28-\log_3 7$

$=\log_3\left(\dfrac{28}{7}\right)$

$=\log_3(4)$

33. $\log x-\log(x+1)$

$=\log\left(\dfrac{x}{x+1}\right)$

35. $\ln(x-5)-\ln x$

$=\ln\left(\dfrac{x-5}{x}\right)$

37. $\ln(x^2-4)-\ln(x+2)$

$=\ln\left(\dfrac{x^2-4}{x+2}\right)$

$=\ln\left(\dfrac{(x+2)(x-2)}{x+2}\right)$

$=\ln(x-2)$

39. $\log_2 7+\log_2 6$

$=\log_2(7\cdot 6)$

$=\log_2 42$

41. $\log_5(x^2-2x)+\log_5 x^{-1}$

$=\log_5\left(x^{-1}(x^2-2x)\right)$

$=\log_5(x-2)$

43. $\log 8^{x+2}=(x+2)\log 8$

45. $\ln 5^{2x-1}=(2x-1)\ln 5$

47. $\log\sqrt{22}=\log 22^{\frac{1}{2}}=\dfrac{1}{2}\log 22$

49. $\log_5 81=\log_5 3^4=4\log_5 3$

51. $\log(a^3 b)=\log a^3+\log b=3\log a+\log b$

53. $\ln\left(x\sqrt[4]{y}\right)=\ln x+\ln y^{\frac{1}{4}}$

$=\ln x+\dfrac{1}{4}\ln y$

55. $\ln\left(\dfrac{x^2}{y}\right)=\ln x^2-\ln y$

$=2\ln x-\ln y$

57. $\log\left(\sqrt{\dfrac{x-2}{x}}\right)=\log\left(\dfrac{x-2}{x}\right)^{\frac{1}{2}}$

$=\dfrac{1}{2}\log\left(\dfrac{x-2}{x}\right)$

$=\dfrac{1}{2}\left[\log(x-2)-\log x\right]$

59. $\ln\left(\dfrac{7x\sqrt{3-4x}}{2(x-1)^3}\right)$

$=\ln\left(7x\sqrt{3-4x}\right)-\ln\left(2(x-1)^3\right)$

$=\ln 7x+\ln\sqrt{3-4x}-\left[\ln 2+\ln(x-1)^3\right]$

$=\ln 7x+\ln(3-4x)^{\frac{1}{2}}-\ln 2-\ln(x-1)^3$

$=\ln 7+\ln x+\dfrac{1}{2}\ln(3-4x)-\ln 2-3\ln(x-1)$

61. $\log_7 60 = \dfrac{\ln 60}{\ln 7} = 2.104076884$

63. $\log_5 152 = \dfrac{\ln 152}{\ln 5} \approx 3.121512475$

65. $\log_3 1.73205 = \dfrac{\log 1.73205}{\log 3}$

    ≈ 0.499999576

67. $\log_{0.5} 0.125 = \dfrac{\log 0.125}{\log 0.5} = 3$

69. $f(x) = \log_3 x = \dfrac{\log x}{\log 3}$;

    $f(5) = \dfrac{\log 5}{\log 3} \approx 1.4650$;

    $f(15) = \dfrac{\log 15}{\log 3} \approx 2.4650$;

    $f(45) = \dfrac{\log 45}{\log 3} \approx 3.4650$;

    Outputs increase by 1; $f\left(3^3 \cdot 5\right) \approx 4.465$

71. $h(x) = \log_9 x = \dfrac{\log x}{\log 9}$;

    $h(2) = \dfrac{\log 2}{\log 9} \approx 0.3155$;

    $h(4) = \dfrac{\log 4}{\log 9} \approx 0.6309$;

    $h(8) = \dfrac{\log 8}{\log 9} \approx 0.9464$;

    Outputs are multiples of 0.3155;

    $h\left(2^4\right) = 4(0.3155) \approx 1.2619$

73. $\log 4 + \log(x - 7) = 2$

    $\log 4(x - 7) = 2$

    $4x - 28 = 10^2$

    $4x = 128$

    $x = 32$;

    Check:

    $\log 4 + \log(32 - 7) = 2$

    $\log 4 + \log 25 = 2$

    $\log 100 = 2$

    $\log 10^2 = 2$

    $2 = 2$

75. $\log(2x - 5) - \log 78 = -1$

    $\log\left(\dfrac{2x - 5}{78}\right) = -1$

    $\dfrac{2x - 5}{78} = 10^{-1}$

    $\dfrac{2x - 5}{78} = \dfrac{1}{10}$

    $2x - 5 = \dfrac{78}{10}$

    $2x = 12.8$

    $x = 6.4$;

    Check:

    $\log(2(6.4) - 5) - \log 78 = -1$

    $\log(7.8) - \log 78 = -1$

    $\log\left(\dfrac{7.8}{78}\right) = -1$

    $\log(0.1) = -1$

    $\log 10^{-1} = -1$

    $-1 = -1$

77. $\log(x - 15) - 2 = -\log x$

    $\log(x - 15) + \log x = 2$

    $\log x(x - 15) = 2$

    $x(x - 15) = 10^2$

    $x^2 - 15x = 100$

    $x^2 - 15x - 100 = 0$

    $(x - 20)(x + 5) = 0$

    $x - 20 = 0$ or $x + 5 = 0$

    $x = 20$ or $x = -5$;

    Check $x = 20$:

    $\log(20 - 15) - 2 = -\log 20$

    $\log(5) - 2 = -\log 20$

    $-1.3010 = -1.3010$;

    Check $x = -5$:

    $\log(-5 - 15) - 2 = -\log(-5)$

    -5 is not in the domain;

    $x = 20$, $x = -5$ is extraneous

4.4 Exercises

79. $\log(2x+1)=1-\log x$

$\log(2x+1)+\log x=1$

$\log x(2x+1)=1$

$x(2x+1)=10^1$

$2x^2+x=10$

$2x^2+x-10=0$

$(2x+5)(x-2)=0$

$2x+5=0$ or $x-2=0$

$x=\dfrac{-5}{2}$ or $x=2$;

Check $x=-\dfrac{5}{2}$:

$\log\left(2\left(-\dfrac{5}{2}\right)+1\right)=1-\log\left(-\dfrac{5}{2}\right)$

$\log(-4)=1-\log\left(-\dfrac{5}{2}\right)$

$-\dfrac{5}{2}$ is not in the domain;

Check $x=2$:

$\log(2(2)+1)=1-\log(2)$

$\log(5)=1-\log(2)$

$0.69897=0.69897$;

$x=2, x=-\dfrac{5}{2}$ is extraneous

81. $\log(5x+2)=\log 2$

$5x+2=2$

$5x=0$

$x=0$

83. $\log_4(x+2)-\log_4 3=\log_4(x-1)$

$\log_4\left(\dfrac{x+2}{3}\right)=\log_4(x-1)$

$\dfrac{x+2}{3}=x-1$

$x+2=3x-3$

$-2x=-5$

$x=\dfrac{5}{2}$

85. $\ln(8x-4)=\ln 2+\ln x$

$\ln(8x-4)=\ln(2x)$

$8x-4=2x$

$6x=4$

$x=\dfrac{2}{3}$

87. $\log(2x-1)+\log 5=1$

Write in exponential form:

$\log(10x-5)=1$

$10^1=10x-5$

$15=10x$

$x=\dfrac{3}{2}$

89. $\log_2 9+\log_2(x+3)=3$

Write in exponential form:

$\log_2(9x+27)=3$

$2^3=9x+27$

$8=9x+27$

$-19=9x$

$x=\dfrac{-19}{9}$

91. $\ln(x+7)+\ln 9=2$

Write in exponential form:

$\ln(9x+63)=2$

$e^2=9x+63$

$e^2-63=9x$

$x=\dfrac{e^2-63}{9}$

93. $\log(x+8)+\log x=\log(x+18)$

Write in exponential form:

$\log(x^2+8x)=\log(x+18)$

$x^2+8x=x+18$

$x^2+7x-18=0$

$(x+9)(x-2)=0$

$x+9=0$ or $x-2=0$

$x=-9$ or $x=2$

$x=2, -9$ is extraneous

Chapter 4: Exponential and Logarithmic Functions

95. $\ln(2x+1) = 3 + \ln 6$

$e^{3+\ln 6} = 2x + 1$

$e^{3+\ln 6} - 1 = 2x$

$\dfrac{e^{3+\ln 6} - 1}{2} = x$

$\dfrac{e^3 e^{\ln 6} - 1}{2} = x$

$\dfrac{e^3(6) - 1}{2} = x$

$\dfrac{6e^3 - 1}{2} = x$

$x = 3e^3 - \dfrac{1}{2}$

$x \approx 59.75661077$

97. $\log(-x-1) = \log(5x) - \log x$

$\log(-x-1) = \log 5$

$-x - 1 = 5$

$-x = 6$

$x = -6$

$x = -\,-6$ is extraneous, No Solution

99. $\ln(2t+7) = \ln(3) - \ln(t+1)$

$\ln(2t+7) = \ln\left(\dfrac{3}{t+1}\right)$

$2t + 7 = \dfrac{3}{t+1}$

$(t+1)(2t+7) = \left(\dfrac{3}{t+1}\right)(t+1)$

$2t^2 + 9t + 7 = 3$

$2t^2 + 9t + 4 = 0$

$(2t+1)(t+4) = 0$

$2t + 1 = 0$ or $t + 4 = 0$

$2t = -1$ or $t = -4$

$t = -\dfrac{1}{2}, -4$ is extraneous

101. $\log(x-1) - \log x = \log(x-3)$

$\log\dfrac{x-1}{x} = \log(x-3)$

$\dfrac{x-1}{x} = x - 3$

$x - 1 = x(x-3)$

$x - 1 = x^2 - 3x$

$0 = x^2 - 4x + 1$

$a = 1, b = -4, c = 1$

$x = \dfrac{-(-4) \pm \sqrt{(-4)^2 - 4(1)(1)}}{2(1)}$

$x = \dfrac{4 \pm \sqrt{12}}{2}$

$x = \dfrac{4 \pm 2\sqrt{3}}{2}$

$x = 2 \pm \sqrt{3}$

$x = 2 + \sqrt{3}, x = 2 - \sqrt{3}$ is extraneous

103. $7^{x+2} = 231$

$\ln 7^{x+2} = \ln 231$

$(x+2)\ln 7 = \ln 231$

$x + 2 = \dfrac{\ln 231}{\ln 7}$

$x = \dfrac{\ln 231}{\ln 7} - 2$

$x \approx 0.7968$

105. $5^{3x-2} = 128{,}965$

$\ln 5^{3x-2} = \ln 128965$

$(3x-2)\ln 5 = \ln 128965$

$3x - 2 = \dfrac{\ln 128695}{\ln 5}$

$3x = \dfrac{\ln 128695}{\ln 5} + 2$

$x = \dfrac{\ln 128965}{3\ln 5} + \dfrac{2}{3}$

$x \approx 3.1038$

107. $2^{x+1} = 3^x$

$\ln 2^{x+1} = \ln 3^x$

$(x+1)\ln 2 = x \ln 3$

$x \ln 2 + \ln 2 = x \ln 3$

$x \ln 2 - x \ln 3 = -\ln 2$

$x(\ln 2 - \ln 3) = -\ln 2$

$x = \dfrac{-\ln 2}{\ln 2 - \ln 3}$

$x = \dfrac{\ln 2}{\ln 3 - \ln 2}$

$x \approx 1.7095$

109. $5^{2x+1} = 9^{x+1}$

$\ln 5^{2x+1} = \ln 9^{x+1}$

$(2x+1)\ln 5 = (x+1)\ln 9$

$2x \ln 5 + \ln 5 = x \ln 9 + \ln 9$

$2x \ln 5 - x \ln 9 = \ln 9 - \ln 5$

$x(2\ln 5 - \ln 9) = \ln 9 - \ln 5$

$x = \dfrac{\ln 9 - \ln 5}{2 \ln 5 - \ln 9}$

$x \approx 0.5753$

111. $\dfrac{250}{1 + 4e^{-0.06x}} = 200$

$250 - 200\left(1 + 4e^{-0.06x}\right)$

$\dfrac{250}{200} = 1 + 4e^{-0.06x}$

$\dfrac{1}{4} = 4e^{-0.06x}$

$\dfrac{1}{16} = e^{-0.06x}$

$\ln \dfrac{1}{16} = \ln e^{-0.06x}$

$\ln \dfrac{1}{16} = -0.06x$

$\dfrac{\ln \dfrac{1}{16}}{-0.06} = x$

$x \approx 46.2$

113. $P = \dfrac{C}{1 + ae^{-kt}}$

$P\left(1 + ae^{-kt}\right) = C$

$1 + ae^{-kt} = \dfrac{C}{P}$

$ae^{-kt} = \dfrac{C}{P} - 1$

$e^{-kt} = \dfrac{\dfrac{C}{P} - 1}{a}$

$\ln e^{-kt} = \ln\left(\dfrac{\dfrac{C}{P} - 1}{a}\right)$

$-kt = \ln\left(\dfrac{\dfrac{C}{P} - 1}{a}\right)$

$t = \dfrac{\ln\left(\dfrac{\dfrac{C}{P} - 1}{a}\right)}{-k};$

$C = 450,\ a = 8,\ P = 400,\ k = 0.075$

$t = \dfrac{\ln\left(\dfrac{\dfrac{450}{400} - 1}{8}\right)}{-0.075} \approx 55.45$

Chapter 4: Exponential and Logarithmic Functions

115. $P(t) = \dfrac{750}{1 + 24e^{-0.075t}}$

a. $P(0) = \dfrac{750}{1 + 24e^{-0.075(0)}}$

$P(0) = 30$ fish

b. $300 = \dfrac{750}{1 + 24e^{-0.075t}}$

$300\left(1 + 24e^{-0.075t}\right) = 750$

$1 + 24e^{-0.075t} = \dfrac{750}{300}$

$24e^{-0.075t} - \dfrac{3}{2}$

$e^{-0.075t} = \dfrac{1}{16}$

$\ln e^{-0.075t} = \ln \dfrac{1}{16}$

$-0.075t = \ln \dfrac{1}{16}$

$t = \dfrac{\ln \dfrac{1}{16}}{-0.075}$

$t \approx 37$ months

117. $H = (30T + 8,000)\ln\left(\dfrac{P_0}{P}\right)$, $P_0 = 76$

a. $H = 18250, T = -75$

$18,250 = (30(-75) + 8,000)\ln\left(\dfrac{76}{P}\right)$

$18,250 = 5,750\ln\left(\dfrac{76}{P}\right)$

$\dfrac{18,250}{5,750} = \ln\left(\dfrac{76}{P}\right)$

$e^{\frac{73}{23}} = \dfrac{76}{P}$

$Pe^{\frac{73}{23}} = 76$

$P = \dfrac{76}{e^{\frac{73}{23}}}$

$P \approx 3.2$ cmHg

119. $T = T_R + (T_0 - T_R)e^{-kh}$

$32 = -20 + (75 - (-20))e^{-0.012h}$

$52 = 95e^{-0.012h}$

$\dfrac{52}{95} = e^{-0.012h}$

$\ln \dfrac{52}{95} = \ln e^{-0.012h}$

$\ln \dfrac{52}{95} = -0.012h$

$\dfrac{\ln \dfrac{52}{95}}{-0.012} = h$

$h \approx 50.2$ min

121. $T = k \ln \dfrac{V_n}{V_f}$

$3 = 5\ln \dfrac{28500}{V_f}$

$\dfrac{3}{5} = \ln \dfrac{28500}{V_f}$

$e^{\frac{3}{5}} = \dfrac{28500}{V_f}$

$V_f e^{\frac{3}{5}} = 28500$

$V_f = \dfrac{28500}{e^{\frac{3}{5}}}$

$V_f = \$15,641$

123. $T(p) = \dfrac{-\ln p}{k}$

a. $k = 0.072$

$T(0.65) = \dfrac{-\ln 0.65}{0.072} \approx 5.98$

About 6 hours

b. $24 = \dfrac{-\ln P}{0.072}$

$24(0.072) = -\ln P$

$1.728 = -\ln P$

$-1.728 = \ln P$

$e^{-1.728} = P$

$P \approx 0.1776$ or 18.0%

4.4 Exercises

125. $V_s = V_e \ln\left(\dfrac{M_s}{M_s - M_f}\right)$

$6 = 8\ln\left(\dfrac{100}{100 - M_f}\right)$

$\dfrac{6}{8} = \ln\left(\dfrac{100}{100 - M_f}\right)$

$\dfrac{3}{4} = \ln\left(\dfrac{100}{100 - M_f}\right)$

$e^{\frac{3}{4}} = \dfrac{100}{100 - M_f}$

$e^{\frac{3}{4}}\left(100 - M_f\right) = 100$

$100e^{\frac{3}{4}} - M_f e^{\frac{3}{4}} = 100$

$100e^{\frac{3}{4}} = 100 + M_f e^{\frac{3}{4}}$

$100e^{\frac{3}{4}} - 100 = M_f e^{\frac{3}{4}}$

$\dfrac{100e^{\frac{3}{4}} - 100}{e^{\frac{3}{4}}} = M_f$

$M_f = 52.76$ tons

127. $P(t) = 5.9 + 12.6\ln t$

a. $P(5) = 5.9 + 12.6\ln 5 = 26$ planes

b. $34 = 5.9 + 12.6\ln t$

$28.1 = 12.6\ln t$

$\dfrac{28.1}{12.6} = \ln t$

$e^{\frac{28.1}{12.6}} = t$

$t \approx 9$ days

129. Answers will vary.

131. a. d
b. e
c. b
d. f
e. a
f. c

133. $3e^{2x} - 4e^x - 7 = -3$

Let $u = e^x$

$3u^2 - 4u - 4 = 0$

$(3u + 2)(u - 2) = 0$

$3u + 2 = 0$ or $u - 2 = 0$

$3u = -2$ or $u = 2$

$u = -\dfrac{2}{3}$ or $u = 2$

$e^x = -\dfrac{2}{3}$ or $e^x = 2$

$\ln e^x = \ln\left(-\dfrac{2}{3}\right)$ or $\ln e^x = \ln 2$

$x \neq \ln\left(-\dfrac{2}{3}\right)$ or $x = \ln 2$

$x \approx 0.69314718$

135.a. $f(x) = 3^{x-2}$; $g(x) = \log_3 x + 2$

$(f \circ g)(x) = 3^{(\log_3 x + 2) - 2} = 3^{\log_3 x} = x$;

$(g \circ f)(x) = \log_3 \left(3^{x-2}\right) + 2 =$

$(x-2)\log_3 3 + 2 = x - 2 + 2 = x$

135.b. $f(x) = e^{x-1}$; $g(x) = \ln x + 1$

$(f \circ g)(x) = e^{(\ln x + 1) - 1} = e^{\ln x} = x$;

$(g \circ f)(x) = \ln\left(e^{x-1}\right) + 1 =$

$(x-1)\ln e + 1 = x - 1 + 1 = x$

137.a. $y = e^{x \ln 2} = e^{\ln 2^x} = 2^x$;

$y = 2^x$

$\ln y = x \ln 2$

$e^{\ln y} = e^{x \ln 2}$

$y = e^{x \ln 2}$

b. $y = b^x$

$\ln y = \ln b^x$

$\ln y = x \ln b$

$e^{\ln y} = e^{x \ln b}$

$y = e^{xr}$ for $r = \ln b$

139. Answers will vary.

141. b

143. $r(x) = \dfrac{x^2 - 4}{x - 1}$

$r(x) = \dfrac{(x+2)(x-2)}{x-1}$

VA: $x = 1$

HA: none (deg num > deg den)

$$
\begin{array}{r}
x + 1 \\
x - 1 \overline{\smash{)}\, x^2 + 0x - 4} \\
-\left(x^2 - x\right) \\
\hline
x - 4 \\
-(x - 1) \\
\hline
-3
\end{array}
$$

Oblique Asymptote: $y = x + 1$

x-intercepts: $(-2, 0)$ and $(2, 0)$;

$r(0) = \dfrac{0^2 - 4}{0 - 1} = 4$

y-intercept: $(0, 4)$

4.5 Technology Highlight

1. $A = P\left(1 + \dfrac{r}{n}\right)^{nt}$

   Doubling time, find x when $y = 2000$

   $Y_1 = 1000\left(1 + \dfrac{0.08}{4}\right)^{4x} \approx 8.75$ yr;

   $Y_1 = 1000\left(1 + \dfrac{0.08}{12}\right)^{12x} \approx 8.69$ yr;

   $Y_1 = 1000\left(1 + \dfrac{0.08}{365}\right)^{365x} \approx 8.665$ yr;

   $Y_1 = 1000\left(1 + \dfrac{0.08}{365 \cdot 24}\right)^{365 \cdot 24x} \approx 8.664$ yr

   8.75 yr compounded quarterly; 8.69 yr compounded monthly; 8.665 yr compounded daily; 8.664 yr compounded hourly

3. No. Examples will vary.

4.5 Exercises

1. Compound

3. $Q_0 e^{-rt}$

5. Answers will vary.

7. $I = prt$;

   9 months $= \dfrac{3}{4}$ year;

   $229.50 = p(0.0625)(0.75)$

   $\dfrac{229.50}{(0.0625)(0.75)} = p$

   $\$4896 = p$

9. $I = prt$

   $297.50 - 260 = 260r\left(\dfrac{3}{52}\right)$

   $37.5 = 15r$

   $\dfrac{37.5}{15} = r$

   $2.50 = r$

   $r = 250\%$

11. $A = p(1 + rt)$

    $2500 = p\left(1 + 0.0625\left(\dfrac{31}{12}\right)\right)$

    $2500 = p\left(\dfrac{223}{192}\right)$

    $p \approx \$2152.47$

13. $A = p(1 + rt)$

    $149925 = 120000(1 + 0.0475t)$

    $\dfrac{1999}{1600} = 1 + 0.0475t$

    $\dfrac{399}{1600} = 0.0475t$

    5.25 years $= t$

15. $I - prt$

    $40 = 200r\left(\dfrac{13}{52}\right)$

    $40 = 50r$

    $0.80 = r$

    $r = 80\%$

17. $A = p(1 + r)^t$

    $48428 = 38000(1 + 0.0625)^t$

    $\dfrac{12107}{9500} = (1 + 0.0625)^t$

    $\ln \dfrac{12107}{9500} = \ln(1 + 0.0625)^t$

    $\ln \dfrac{12107}{9500} = t \ln(1 + 0.0625)$

    $\dfrac{\ln \dfrac{12107}{9500}}{\ln(1 + 0.0625)} = t$

    $t \approx 4$ years

19. $A = p(1+r)^t$

$4575 = 1525(1+0.071)^t$

$3 = (1+0.071)^t$

$\ln 3 = \ln(1+0.071)^t$

$\ln 3 = t \ln(1+0.071)$

$\dfrac{\ln 3}{\ln(1+0.071)} = t$

$t \approx 16$ years

21. $P = \dfrac{A}{(1+r)^t}$

$P = \dfrac{10000}{(1+0.0575)^5}$

$P \approx \$7561.33$

23. $A = p\left(1+\dfrac{r}{n}\right)^{nt}$

$129500 = 90000\left(1+\dfrac{0.07125}{52}\right)^{52t}$

$\dfrac{259}{180} = \left(1+\dfrac{0.07125}{52}\right)^{52t}$

$\ln\left(\dfrac{259}{180}\right) = \ln(1.001370192)^{52t}$

$\ln\left(\dfrac{259}{180}\right) = 52t \ln(1.001370192)$

$\dfrac{\ln\left(\dfrac{259}{180}\right)}{52 \ln(1.001370192)} = t$

$t \approx 5$ years

25. $A = p\left(1+\dfrac{r}{n}\right)^{nt}$

$10000 = 5000\left(1+\dfrac{0.0925}{365}\right)^{365t}$

$2 = (1.000253425)^{365t}$

$\ln 2 = \ln(1.000253425)^{365t}$

$\ln 2 = 365t \ln(1.000253425)$

$\dfrac{\ln 2}{365 \ln(1.000253425)} = t$

$t \approx 7.5$ years

27. $A = p\left(1+\dfrac{r}{n}\right)^{nt}$

$A = 10\left(1+\dfrac{0.10}{10}\right)^{10(10)} \approx \$27.04$, No

29. $A = p\left(1+\dfrac{r}{n}\right)^{nt}$

(a) $A = 175000\left(1+\dfrac{0.0875}{2}\right)^{2(4)}$

$\approx \$246496.05$, No

(b) $r \approx 9.12\%$

31. $A = pe^{rt}$

$2500 = 1750e^{0.045t}$

$\dfrac{10}{7} = e^{0.045t}$

$\ln\left(\dfrac{10}{7}\right) = \ln e^{0.045t}$

$\ln\left(\dfrac{10}{7}\right) = 0.045t \ln e$

$\ln\left(\dfrac{10}{7}\right) = 0.045t$

$\dfrac{\ln\left(\dfrac{10}{7}\right)}{0.045} = t$

$t \approx 7.9$ years

33. $A = pe^{rt}$

$10000 = 5000e^{0.0925t}$

$2 = e^{0.0925t}$

$\ln 2 = \ln e^{0.0925t}$

$\ln 2 = 0.0925t \ln e$

$\dfrac{\ln 2}{0.0925} = t$

$t \approx 7.5$ years

35. $A = pe^{rt}$

(a) $A = 12500e^{0.086(5)} = 19215.72$ euros , No

(b) $20000 = 12500e^{r(5)}$

$\dfrac{8}{5} = e^{5r}$

$\ln\left(\dfrac{8}{5}\right) = \ln e^{5r}$

$\ln\left(\dfrac{8}{5}\right) = 5r \ln e$

$\dfrac{\ln\left(\dfrac{8}{5}\right)}{5} = r$

$r \approx 9.4\%$

37. $A - pe^{rt}$

(a) $A = 12000e^{0.055(7)} \approx 17635.37$ euros ,
No

(b) $20000 = Pe^{0.055(7)}$

$\dfrac{20000}{e^{0.055(7)}} = P$

$P \approx 13{,}609$ euros

39. $T = \dfrac{1}{r} \cdot \ln\left(\dfrac{A}{P}\right)$

$8 = \dfrac{1}{0.05} \cdot \ln\left(\dfrac{A}{200000}\right)$

$0.4 = \ln\left(\dfrac{A}{200000}\right)$

By definition, $\ln x = y$ iff $e^y = x$

$e^{0.4} = \dfrac{A}{200000}$

$200000e^{0.4} = A$

No, $298364.94;

$8 = \dfrac{1}{0.05} \cdot \ln\left(\dfrac{350000}{P}\right)$

$0.4 = \ln\left(\dfrac{350000}{P}\right)$

By definition, $\ln x = y$ iff $e^y = x$

$e^{0.4} = \dfrac{350000}{P}$

$\dfrac{350000}{e^{0.4}} = P$

$P = \$234{,}612.01$

41. $A = \dfrac{p\left[(1+R)^{nt} - 1\right]}{R}$

$$10000 = \dfrac{90\left[\left(1+\dfrac{0.0775}{12}\right)^{12t} - 1\right]}{\dfrac{0.0775}{12}}$$

$$\dfrac{775}{12} = 90\left[\left(1+\dfrac{0.0775}{12}\right)^{12t} - 1\right]$$

$$\dfrac{155}{216} = \left(1+\dfrac{0.0775}{12}\right)^{12t} - 1$$

$$\dfrac{371}{216} = \left(1+\dfrac{0.0775}{12}\right)^{12t}$$

$$\ln\left(\dfrac{371}{216}\right) = \ln\left(1+\dfrac{0.0775}{12}\right)^{12t}$$

$$\ln\left(\dfrac{371}{216}\right) = 12t \ln\left(1+\dfrac{0.0775}{12}\right)$$

$$\dfrac{\ln\left(\dfrac{371}{216}\right)}{12\ln\left(1+\dfrac{0.0775}{12}\right)} = t$$

$$\approx 7 \text{ years}$$

43. $A = \dfrac{p\left[(1+R)^{nt} - 1\right]}{R}$

$$30000 = \dfrac{50\left[\left(1+\dfrac{0.062}{12}\right)^{12t} - 1\right]}{\dfrac{0.062}{12}}$$

$$155 = 50\left[\left(1+\dfrac{0.062}{12}\right)^{12t} - 1\right]$$

$$3.1 = \left(1+\dfrac{0.062}{12}\right)^{12t} - 1$$

$$4.1 = \left(1+\dfrac{0.062}{12}\right)^{12t}$$

$$\ln 4.1 = \ln\left(1+\dfrac{0.062}{12}\right)^{12t}$$

$$\ln(4.1) = 12t \ln\left(1+\dfrac{0.062}{12}\right)$$

$$\dfrac{\ln(4.1)}{12\ln\left(1+\dfrac{0.062}{12}\right)} = t$$

$$\approx 23 \text{ years}$$

45. $A = \dfrac{p\left[(1+R)^{nt} - 1\right]}{R}$

(a) $A = \dfrac{250\left[\left(1+\dfrac{0.085}{12}\right)^{12(5)} - 1\right]}{\dfrac{0.085}{12}}$

$\approx \$18610.61$, No

(b) $22500 = \dfrac{p\left[\left(1+\dfrac{0.085}{12}\right)^{12(5)} - 1\right]}{\dfrac{0.085}{12}}$

$$159.375 = p\left[\left(1+\dfrac{0.085}{12}\right)^{12(5)} - 1\right]$$

$$\dfrac{159.375}{\left[\left(1+\dfrac{0.085}{12}\right)^{12(5)} - 1\right]} = p$$

$$p \approx \$302.25$$

47. $A = p + prt$

  a. $A - p = prt$

  $$\frac{A - p}{pr} = t$$

  b. $A = p(1 + rt)$

  $$\frac{A}{1 + rt} = p$$

49. $A = P\left(1 + \dfrac{r}{n}\right)^{nt}$

  a. $\dfrac{A}{P} = \left(1 + \dfrac{r}{n}\right)^{nt}$

  $$\sqrt[nt]{\frac{A}{P}} = 1 + \frac{r}{n}$$

  $$\sqrt[nt]{\frac{A}{P}} - 1 = \frac{r}{n}$$

  $$n\left(\sqrt[nt]{\frac{A}{P}} - 1\right) = r$$

  b. $\ln\left(\dfrac{A}{P}\right) = \ln\left(1 + \dfrac{r}{n}\right)^{nt}$

  $$\ln\left(\frac{A}{P}\right) = nt \ln\left(1 + \frac{r}{n}\right)$$

  $$\frac{\ln\left(\dfrac{A}{P}\right)}{n \ln\left(1 + \dfrac{r}{n}\right)} = t$$

51. $Q(t) = Q_0 e^{rt}$

  a. $\dfrac{Q(t)}{e^{rt}} = Q_0$

  b. $\dfrac{Q(t)}{Q_0} = e^{rt}$

  $$\ln\left(\frac{Q(t)}{Q_0}\right) = \ln e^{rt}$$

  $$\ln\left(\frac{Q(t)}{Q_0}\right) = rt \ln e$$

  $$\frac{\ln\left(\dfrac{Q(t)}{Q_0}\right)}{r} = t$$

53. $P = \dfrac{AR}{1 - (1 + R)^{-nt}}$

  $$P = \frac{125000\left(\dfrac{0.055}{12}\right)}{1 - \left(1 + \left(\dfrac{0.055}{12}\right)\right)^{-12(30)}}$$

  $P \approx \$709.74$

55. $Q(t) = Q_0 e^{rt}$

  (a) $2000 = 1000 e^{r(12)}$

  $2 = e^{12r}$

  $\ln 2 = \ln e^{12r}$

  $\ln 2 = 12r \ln e$

  $\dfrac{\ln 2}{12} = r$

  $r \approx 5.78\%$

  (b) $200000 = 1000 e^{(0.0578)t}$

  $200 = e^{(0.0578)t}$

  $\ln 200 = \ln e^{(0.0578)t}$

  $\ln 200 = 0.0578t \ln e$

  $\dfrac{\ln 200}{0.0578} = t$

  $t \sim 91.67$ hours

57. $r = \dfrac{\ln 2}{t}$

  $r = \dfrac{\ln 2}{8}$

  $r \approx 0.087$ or $r \approx 8.7\%$;

  $Q(t) = Q_0 e^{-rt}$

  $0.5 = Q_0 e^{-0.087(3)}$

  $\dfrac{0.5}{e^{-0.087(3)}} = Q_0$

  $Q_0 \approx 0.65$ grams

59. $r = \dfrac{\ln 2}{t}$

$r = \dfrac{\ln 2}{432}$;

$Q(t) = Q_0 e^{-rt}$

$2.7 = 10 e^{-\frac{\ln 2}{432} t}$

$0.27 = e^{-\frac{\ln 2}{432} t}$

$\ln 0.27 = \ln e^{-\frac{\ln 2}{432} t}$

$\ln 0.27 = -\dfrac{\ln 2}{432} t \ln e$

$\dfrac{\ln 0.27}{-\dfrac{\ln 2}{432}} = t$

≈ 816 years

61. $T = -8267 \cdot \ln p$

$17255 = -8267 \cdot \ln p$

$\dfrac{17255}{-8267} = \ln p$

$e^{-\frac{17255}{8267}} = e^{\ln p}$

$e^{-\frac{17255}{8267}} = p$

$p \approx 0.124$

About 12.4 %

63. $A = p e^{rt}$

$A = 10000 e^{0.062(120)} = \$17,027,502.21$

Answers will vary.

65. $A = p\left(1 + \dfrac{r}{n}\right)^{nt}$

$25000 = 6000\left(1 + \dfrac{r}{365}\right)^{365(18)}$

$\dfrac{25}{6} = \left(1 + \dfrac{r}{365}\right)^{6570}$

$\sqrt[6570]{\dfrac{25}{6}} = 1 + \dfrac{r}{365}$

$\sqrt[6570]{\dfrac{25}{6}} - 1 = \dfrac{r}{365}$

$365\left[\sqrt[6570]{\dfrac{25}{6}} - 1\right] = r$

$r \approx 7.93\%$

67. $2000^2 + 1580^2 = x^2$

$x \approx 2548.8$ meters

69. $P(x) = (x - 3)(x + 1)(x - (1 + 2i))(x - (1 - 2i))$

$P(x) = \left(x^2 - 2x - 3\right)\left((x - 1)^2 - 4i^2\right)$

$P(x) = \left(x^2 - 2x - 3\right)\left(x^2 - 2x + 1 + 4\right)$

$P(x) = \left(x^2 - 2x - 3\right)\left(x^2 - 2x + 5\right)$

$P(x) = x^4 - 2x^3 + 5x^2 - 2x^3 + 4x^2$

$\qquad\qquad -10x - 3x^2 + 6x - 15$

$P(x) = x^4 - 4x^3 + 6x^2 - 4x - 15$

Chapter 4 Summary and Concept Review

1. $h(x) = -|x-2| + 3$; No

3. $s(x) = \sqrt{x-1} + 5$; Yes

5. $f(x) = x^2 - 2, \ x \ge 0$

$\quad y = x^2 - 2$

$\quad x = y^2 - 2$

$\quad x + 2 = y^2$

$\quad \sqrt{x+2} = y$

$\quad f^{-1}(x) = \sqrt{x+2}$

$\quad \left(f \circ f^{-1}\right)(x) = f\left[f^{-1}(x)\right]$

$\quad = \left(f^{-1}(x)\right)^2 - 2$

$\quad = \left(\sqrt{x+2}\right)^2 - 2$

$\quad = x + 2 - 2$

$\quad = x;$

$\quad \left(f^{-1} \circ f\right)(x) = f^{-1}[f(x)]$

$\quad = \sqrt{f(x) + 2}$

$\quad = \sqrt{x^2 - 2 + 2}$

$\quad = \sqrt{x^2}$

$\quad = x$

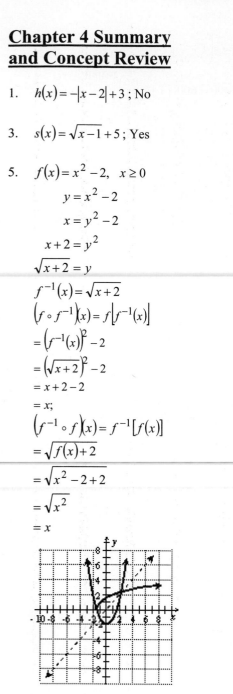

7. $f(x):$

$\quad \begin{cases} D : x \in [-4, \infty) \\ R : y \in [0, \infty) \end{cases}$

$\quad f^{-1}(x):$

$\quad \begin{cases} D : x \in [0, \infty) \\ R : y \in [-4, \infty) \end{cases}$

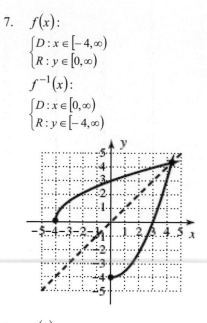

9. $f(x):$

$\quad \begin{cases} D : x \in (-\infty, \infty) \\ R : y \in (0, \infty) \end{cases}$

$\quad f^{-1}(x):$

$\quad \begin{cases} D : x \in (0, \infty) \\ R : y \in (-\infty, \infty) \end{cases}$

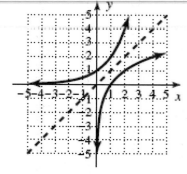

11. $y = 2^x + 3$

Asymptote: $y = 3$

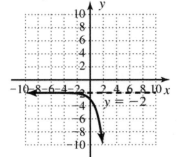

13. $y = -e^{x+1} - 2$

Left 1, reflected across the x-axis, down 2,

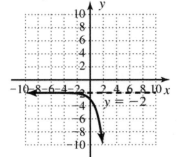

15. $4^x = \dfrac{1}{16}$

$4^x = 4^{-2}$

$x = -2$

17. $20000 = 142000 \cdot (0.85)^t$

$\dfrac{10}{71} = 0.85^t$

$\ln\left(\dfrac{10}{71}\right) = \ln 0.85^t$

$\ln\left(\dfrac{10}{71}\right) = t \ln 0.85$

$\dfrac{\ln\left(\dfrac{10}{71}\right)}{\ln 0.85} = t$

About 12.1 years

19. $\log_5 \dfrac{1}{125} = -3$

$5^{-3} = \dfrac{1}{125}$

21. $5^2 = 25$

$\log_5 25 = 2$

23. $3^4 = 81$

$\log_3 81 = 4$

25. $\ln \dfrac{1}{e} = x$

$e^x = \dfrac{1}{e}$

$e^x = e^{-1}$

$x = -1$

27. $f(x) = \log_2 x$

Asymptote: $x = 0$

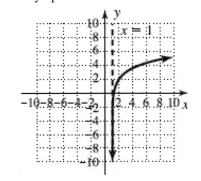

29. $f(x) = 2 + \ln(x - 1)$

Asymptote: $x = 1$

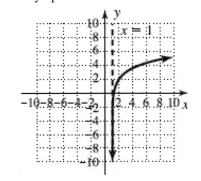

31. $g(x) = \log\sqrt{2x+3}$

    $2x + 3 > 0$

      $2x > -3$

        $x > -\dfrac{3}{2}$

    Domain: $x \in (-\dfrac{3}{2}, \infty)$

33. a. $\ln x = 32$

       $e^{32} = x$

  b. $\log x = 2.38$

       $10^{2.38} = x$

  c. $e^x = 9.8$

       $\ln e^x = \ln 9.8$

       $x = \ln 9.8$

  d. $10^x = \sqrt{7}$

       $\log 10^x = \log\sqrt{7}$

       $x = \log\sqrt{7}$

35. a. $\ln 7 + \ln 6$

       $\ln 42$

  b. $\log_9 2 + \log_9 15$

       $\log_9 30$

  c. $\ln(x+3) - \ln(x-1)$

       $\ln\left(\dfrac{x+3}{x-1}\right)$

  d. $\log x + \log(x+1)$

       $\log\left(x^2 + x\right)$

37. a. $\ln\left(x\sqrt[4]{y}\right)$

      $= \ln x + \ln y^{\frac{1}{4}}$

      $= \ln x + \dfrac{1}{4}\ln y$

  b. $\ln\left(\sqrt[3]{pq}\right)$

      $= \ln p^{\frac{1}{3}} + \ln q$

      $= \dfrac{1}{3}\ln p + \ln q$

  c. $\log\left(\dfrac{\sqrt[3]{x^5 y^4}}{\sqrt{x^5 y^3}}\right)$

      $= \log\left(\sqrt[3]{x^5 y^4}\right) - \log\sqrt{x^5 y^3}$

      $= \log x^{\frac{5}{3}} y^{\frac{4}{3}} - \log x^{\frac{5}{2}} y^{\frac{3}{2}}$

      $= \log x^{\frac{5}{3}} + \log y^{\frac{4}{3}} - \log x^{\frac{5}{2}} - \log y^{\frac{3}{2}}$

      $= \dfrac{5}{3}\log x + \dfrac{4}{3}\log y - \dfrac{5}{2}\log x - \dfrac{3}{2}\log y$

  d. $\log\left(\dfrac{4\sqrt[3]{p^5 q^4}}{\sqrt{p^3 q^2}}\right)$

      $= \log 4\sqrt[3]{p^5 q^4} - \log\sqrt{p^3 q^2}$

      $= \log 4 p^{\frac{5}{3}} q^{\frac{4}{3}} - \log p^{\frac{3}{2}} q$

      $= \log 4 + \log p^{\frac{5}{3}} + \log q^{\frac{4}{3}} - \left(\log p^{\frac{3}{2}} + \log q\right)$

      $= \log 4 + \dfrac{5}{3}\log p + \dfrac{4}{3}\log q - \dfrac{3}{2}\log p - \log q$

39. $2^x = 7$

    $\ln 2^x = \ln 7$

    $x\ln 2 = \ln 7$

    $x = \dfrac{\ln 7}{\ln 2}$

41. $e^{x-2} = 3^x$

$\ln e^{x-2} = \ln 3^x$

$x - 2 = x \ln 3$

$x - x \ln 3 = 2$

$x(1 - \ln 3) = 2$

$x = \dfrac{2}{1 - \ln 3}$

43. $\log x + \log(x - 3) = 1$

$\log x(x - 3) = 1$

$10^1 = x(x - 3)$

$0 = x^2 - 3x - 10$

$0 = (x - 5)(x + 2)$

$x = 5$ or $x = -2$

5, -2 is extraneous

45. $R(h) = \dfrac{\ln(2)}{h}$

a. $R(3.9) = \dfrac{\ln(2)}{3.9}$

$\approx 17.77\%$

b. $0.0289 = \dfrac{\ln(2)}{h}$

$0.0289h = \ln 2$

$h = \dfrac{\ln 2}{0.0289}$

About 23.98 days

47. $I = \mathrm{P}\,rt$

$27.75 = 600r\left(\dfrac{3}{12}\right)$

$4\left(\dfrac{27.75}{600}\right) = r$

$r = 0.185$

18.5%

49. $A = \dfrac{p\left[(1 + R)^{nt} - 1\right]}{R}$

(a) $A = \dfrac{260\left[\left(1 + \dfrac{0.075}{12}\right)^{12(4)} - 1\right]}{\dfrac{0.075}{12}}$

$A \approx \$14501.72$,

No

(b) $15000 = \dfrac{p\left[\left(1 + \dfrac{0.075}{12}\right)^{12(4)} - 1\right]}{\dfrac{0.075}{12}}$

$93.75 = p\left[\left(1 + \dfrac{0.075}{12}\right)^{12(4)} - 1\right]$

$\dfrac{93.75}{\left[\left(1 + \dfrac{0.075}{12}\right)^{12(4)} - 1\right]} = p$

$p \approx \$268.93$

Chapter 4 Mixed Review

1. a. $\log_2 30$

   $\dfrac{\log 30}{\log 2} \approx 4.9069$

   b. $\log_{0.25} 8$

   $\dfrac{\log 8}{\log 0.25} = -1.5$

   c. $\log_8 2$

   $\dfrac{\log 2}{\log 8} = \dfrac{1}{3}$

3. a. $\log_{10} 20^2$

   $= 2\log_{10} 20$

   b. $\log 10^{0.05x}$

   $= 0.05x \log 10$

   $= 0.05x$

   c. $\ln 2^{x-3}$

   $= (x-3)\ln 2$

5. $y = 5 \cdot 2^{-x}$

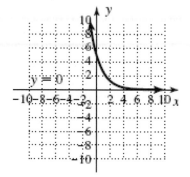

7. $y = \log_2(-x) - 4$

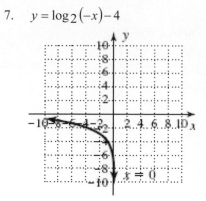

9. a. $\log_5 625 = 4$

   $5^4 = 625$

   b. $\ln 0.15x = 0.45$

   $e^{0.45} = 0.15x$

   c. $\log(0.1 \times 10^8) = 7$

   $10^7 = 0.1 \times 10^8$

11. $g(x) = \sqrt{x-1} + 2$

    a. $D : x \in [1, \infty), R : y \in [2, \infty)$

    b. $g(x) = \sqrt{x-1} + 2$

    Interchange x and y.

    $x = \sqrt{y-1} + 2$

    $x - 2 = \sqrt{y-1}$

    $(x-2)^2 = y - 1$

    $(x-2)^2 + 1 = y$

    $g^{-1}(x) = (x-2)^2 + 1$

    $D : x \in [2, \infty), R : y \in [1, \infty)$

    c. Answers will vary.

13. $10^{x-4} = 200$

$\log 10^{x-4} = \log 200$

$(x-4)\log 10 = \log 2 \cdot 10^2$

$x - 4 = \log 2 + \log 10^2$

$x - 4 = \log 2 + 2\log 10$

$x - 4 = \log 2 + 2$

$x = 6 + \log 2$

15. $\log_2(2x-5) + \log_2(x-2) = 4$

$\log_2(2x-5)(x-2) = 4$

$2^4 = (2x-5)(x-2)$

$16 = 2x^2 - 9x + 10$

$0 = 2x^2 - 9x - 6$

$a = 2, b = -9, c = -6$

$x = \dfrac{-(-9) \pm \sqrt{(-9)^2 - 4(2)(-6)}}{2(2)}$

$x = \dfrac{9 \pm \sqrt{129}}{4}$

$x = \dfrac{9 + \sqrt{129}}{4}$; $x = \dfrac{9 - \sqrt{129}}{4}$ is extraneous

17. $6.5 = \log\left(\dfrac{I}{2\text{x}10^{11}}\right)$

$10^{6.5} = \dfrac{I}{2\text{x}10^{11}}$

$10^{6.5}\left(2\text{x}10^{11}\right) = I$

$2 \cdot 10^{0.5} \cdot 10^{17} = I$

$2 \cdot \sqrt{10} \cdot 10^{17} = I$

$I \approx 6.3\text{x}10^{17}$

19. $r(n) = 2(0.8)^n$

$r(6) = 2(0.8)^6 = 0.524$

0.52 m;

n	$r(n) = 2\left(0.8^n\right)$
1	1.6 m
2	1.28 m
3	1.02 m
4	0.82 m
5	0.66 m
6	0.52 m

Chapter 4 Practice Test

1. $\log_3 81 = 4$

   $3^4 = 81$

3. $\log_b\left(\dfrac{\sqrt{x^5}\,y^3}{z}\right)$

   $= \log_b \sqrt{x^5}\,y^3 - \log_b z$

   $= \log_b x^{\frac{5}{2}} y^3 - \log_b z$

   $= \log_b x^{\frac{5}{2}} + \log_b y^3 - \log_b z$

   $= \dfrac{5}{2}\log_b x + 3\log_b y - \log_b z$

5. $5^{x-7} = 125$

   $5^{x-7} = 5^3$

   $x - 7 = 3$

   $x = 10$

7. $\log_a 45$

   $= \log_a\left(3^2 \cdot 5\right)$

   $= \log_a 3^2 + \log_a 5$

   $= 2\log_a 3 + \log_a 5$

   $= 2(0.48) + 1.72$

   $= 2.68$

9. $g(x) = -2^{x-1} + 3$

   HA: $y = 3$

11. a. $\log_3 100$

    $= \dfrac{\log 100}{\log 3}$

    $= \dfrac{\log 10^2}{\log 3}$

    $= \dfrac{2\log 10}{\log 3}$

    $= \dfrac{2}{\log 3}$

    ≈ 4.19

    b. $\log_6 0.235$

    $= \dfrac{\log 0.235}{\log 6}$

    ≈ -0.81

13. $3^{x-1} = 89$

    $\ln 3^{x-1} = \ln 89$

    $(x-1)\ln 3 = \ln 89$

    $x - 1 = \dfrac{\ln 89}{\ln 3}$

    $x = 1 + \dfrac{\ln 89}{\ln 3}$

15. $3000 = 8000(0.82)^t$

    $\dfrac{3}{8} = (0.82)^t$

    $\ln\left(\dfrac{3}{8}\right) = \ln(0.82)^t$

    $\ln\left(\dfrac{3}{8}\right) = t\ln(0.82)$

    $\dfrac{\ln\left(\dfrac{3}{8}\right)}{\ln 0.82} = t$

    $t \approx 5$ years

17. $Q(t) = -2600 + 1900 \ln t$

$3000 = -2600 + 1900 \ln t$

$5600 = 1900 \ln t$

$\dfrac{56}{19} = \ln t$

$e^{\frac{56}{19}} = e^{\ln t}$

$e^{\frac{56}{19}} = t$

$t \approx 19.1$ months

19. $A = \dfrac{p\left[(1+R)^{nt} - 1\right]}{R}$

(a) $A = \dfrac{50\left[\left(1 + \dfrac{0.0825}{12}\right)^{12(3)} - 1\right]}{\dfrac{0.0825}{12}}$

$A \approx \$3697.88$

No

(b) $4000 = \dfrac{p\left[\left(1 + \dfrac{0.0825}{12}\right)^{12(5)} - 1\right]}{\dfrac{0.0825}{12}}$

$27.5 = p\left[\left(1 + \dfrac{0.0825}{12}\right)^{60} - 1\right]$

$\dfrac{27.5}{\left[\left(1 + \dfrac{0.0825}{12}\right)^{60} - 1\right]} = p$

$p \approx \$54.09$

Chapter 4: Calculator Exploration and Discovery

1. $a = 25,\ b = 0.5,\ c = 2500$

3. b

5. $b = 0.6$

7. Verified

Strengthening Core Skills

1. Answers will vary.

3. Answers will vary.

Cumulative Review Chapters 1 to 4

1. $x^2 - 4x + 53 = 0$
$a = 1, b = -4, c = 53$

$$x = \frac{-(-4) \pm \sqrt{(-4)^2 - 4(1)(53)}}{2(1)}$$

$$x = \frac{4 \pm \sqrt{-196}}{2}$$

$$x = \frac{4 \pm 14i}{2}$$

$x = 2 \pm 7i$

3. $(4 + 5i)^2 - 8(4 + 5i) + 41 = 0$
$-9 + 40i - 32 - 40i + 41 = 0$
$0 = 0$

5. $f(x) = x^3 - 2,\ g(x) = \sqrt[3]{x + 2}$;
$f(g(x)) = \left(\sqrt[3]{x + 2}\right)^3 - 2 = x + 2 + x = x$;
$g(f(x)) = \sqrt[3]{x^3 - 2 + 2} = \sqrt[3]{x^3} = x$
Since $(f \circ g)(x) = (g \circ f)(x)$, they are inverse functions.

7. $1991 \rightarrow$ year 1
(a) $(1, 3100), (9, 6740)$

$$m = \frac{6740 - 3100}{9 - 1} = 455$$

$y - 3100 = 455(x - 1)$
$y - 3100 = 455x - 455$
$y = 455x + 2645$
$T(t) = 455t + 2645$

(b) $\dfrac{\Delta T}{\Delta t} = \dfrac{455}{1}$, triple births increase by 455 each year.

(c) In 1996, $T(6) = 455(6) + 2645 = 5375$ sets of triplets
In 2007,
$t = 17, T(17) = 455(17) + 2645 = 10{,}380$ sets of triplets

9. $h(x) = \begin{cases} -4 & -10 \le x < -2 \\ -x^2 & -2 \le x < 3 \\ 3x - 18 & x \ge 3 \end{cases}$

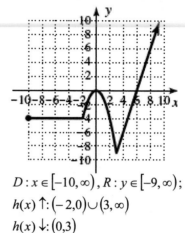

$D : x \in [-10, \infty),\ R : y \in [-9, \infty)$;
$h(x) \uparrow : (-2, 0) \cup (3, \infty)$
$h(x) \downarrow : (0, 3)$

11. $f(x) = x^4 - 3x^3 - 12x^2 + 52x - 48$
Possible rational roots:
$\dfrac{\{\pm 1, \pm 48, \pm 2, \pm 24, \pm 3, \pm 16, \pm 4, \pm 12, \pm 6, \pm 8\}}{\{\pm 1\}}$;
$\{\pm 1, \pm 48, \pm 2, \pm 24, \pm 3, \pm 16, \pm 4, \pm 12, \pm 6, \pm 8\}$

$$\begin{array}{r|rrrrr} 3 & 1 & -3 & -12 & 52 & -48 \\ & & 3 & 0 & -36 & 48 \\ \hline & 1 & 0 & -12 & 16 & \boxed{0} \end{array}$$

$$\begin{array}{r|rrrr} 2 & 1 & 0 & -12 & 16 \\ & & 2 & 4 & -16 \\ \hline & 1 & 2 & -8 & \boxed{0} \end{array}$$

$f(x) = (x - 3)(x - 2)(x^2 + 2x - 8)$
$f(x) = (x - 3)(x - 2)(x - 2)(x + 4)$
$x = 3, x = 2$ (multiplicity 2), $x = -4$

Chapter 4: Practice Test

13. $V = \dfrac{1}{2}\pi b^2 a$

$\dfrac{2V}{\pi a} = b^2$

$\sqrt{\dfrac{2V}{\pi a}} = b$

15. a) $f(x) = \dfrac{2x+3}{5}$

$y = \dfrac{2x+3}{5}$

$x = \dfrac{2y+3}{5}$

$5x - 2y + 3$

$5x - 3 = 2y$

$\dfrac{5x-3}{2} = y$

$f^{-1}(x) = \dfrac{5x-3}{2}$

b)

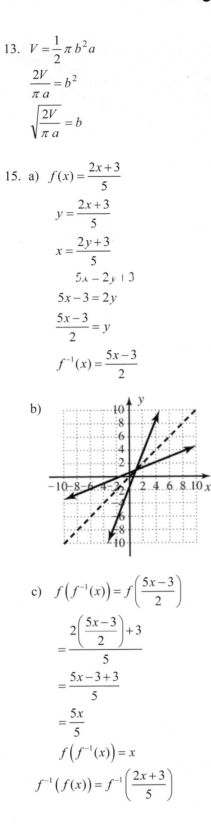

c) $f\left(f^{-1}(x)\right) = f\left(\dfrac{5x-3}{2}\right)$

$= \dfrac{2\left(\dfrac{5x-3}{2}\right)+3}{5}$

$= \dfrac{5x-3+3}{5}$

$= \dfrac{5x}{5}$

$f\left(f^{-1}(x)\right) = x$

$f^{-1}\left(f(x)\right) = f^{-1}\left(\dfrac{2x+3}{5}\right)$

$= \dfrac{5\left(\dfrac{2x+3}{5}\right)-3}{2}$

$= \dfrac{2x+3-3}{2}$

$= \dfrac{2x}{2}$

$f^{-1}\left(f(x)\right) = x$

17. $\ln(x+3) + \ln(x-2) = \ln 24$

$\ln(x+3)(x-2) = \ln 24$

$(x+3)(x-2) = 24$

$x^2 + x - 6 = 24$

$x^2 + x - 30 = 0$

$(x+6)(x-5) = 0$

$x + 6 = 0$ or $x - 5 = 0$

$x = -6$ or $x = 5$

$x = 5, x = -6$ is an extraneous root

19. a) Sportwagon:

$H(3000) = 123\ln(3000) - 897$

≈ 88 hp

Minivan:

$H(3000) = 193\ln(3000) - 1464$

≈ 81 hp

b) $123\ln r - 897 = 193\ln r - 1464$

$123\ln r + 567 = 193\ln r$

$567 = 193\ln r - 123\ln r$

$567 = 70\ln r$

$\dfrac{567}{70} = \ln r$

$e^{\frac{567}{70}} = r$

$r \approx 3294$ rpm

c) Sportwagon:

$H(5600) = 123\ln(5600) - 897$

≈ 164.6 hp

Minivan:

$H(5800) = 193\ln(5800) - 1464$

≈ 208.46 hp

Minivan, 208 hp @ 5800 rpm

MWT 2 Exercises

MWT II

1. e

3. a

5. d

7. Linear

9. Exponential

11. Logistic

13. Exponential

15. As time increases, the amount of radioactivity decreases but it will never truly reach 0 or a negative value. Due to this and the shape, exponential with $b < 1$ and $k > 0$ is the best choice.

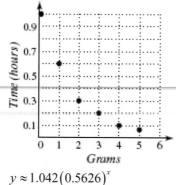

$y \approx 1.042(0.5626)^x$

17. Sales will increase rapidly, then level off as the market is saturated with ads and advertising becomes less effective, possibly modeled by a logarithmic function.
$y \approx 120.4938 + 217.2705 \ln x$

19.a.

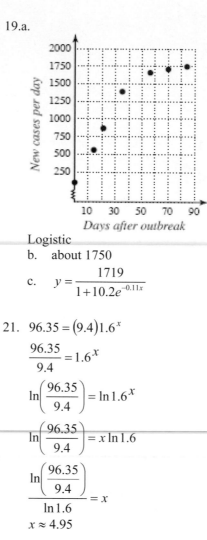

Logistic

b. about 1750

c. $y = \dfrac{1719}{1 + 10.2e^{-0.11x}}$

21. $96.35 = (9.4)1.6^x$

$\dfrac{96.35}{9.4} = 1.6^x$

$\ln\left(\dfrac{96.35}{9.4}\right) = \ln 1.6^x$

$\ln\left(\dfrac{96.35}{9.4}\right) = x \ln 1.6$

$\dfrac{\ln\left(\dfrac{96.35}{9.4}\right)}{\ln 1.6} = x$

$x \approx 4.95$

23. $4.8x^{2.5} = 468.75$

$x^{2.5} = \dfrac{468.75}{4.8}$

$\left(x^{2.5}\right)^{\frac{2}{5}} = \left(\dfrac{468.75}{4.8}\right)^{\frac{2}{5}}$

$x = 6.25$

25. $52 = 63.9 - 6.8 \ln x$

$-11.9 = -6.8 \ln x$

$1.75 = \ln x$

$e^{1.75} = e^{\ln x}$

$e^{1.75} = x$

$x = 5.75$

27. $52 = \dfrac{67}{1 + 20e^{-0.62x}}$

$\left(1 + 20e^{-0.62x}\right)\left[52 = \dfrac{67}{1 + 20e^{-0.62x}}\right]$

$\left(1 + 20e^{-0.62x}\right)52 = 67$

$1 + 20e^{-0.62x} = \dfrac{67}{52}$

$20e^{-0.62x} = \dfrac{15}{52}$

$e^{-0.62x} = \dfrac{3}{208}$

$\ln e^{-0.62x} = \ln\left(\dfrac{3}{208}\right)$

$-0.62x \ln e = \ln\left(\dfrac{3}{208}\right)$

$x = \dfrac{\ln\left(\dfrac{3}{208}\right)}{-0.62}$

$x \approx 6.84$

29. Logarithmic, $y = -27.4 + 13.5 \ln x$

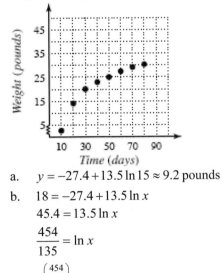

Time (days)

a. $y = -27.4 + 13.5 \ln 15 \approx 9.2$ pounds

b. $18 = -27.4 + 13.5 \ln x$

$45.4 = 13.5 \ln x$

$\dfrac{454}{135} = \ln x$

$e^{\left(\frac{454}{135}\right)} = e^{\ln x}$

$x \approx 29$ days

c. $y = -27.4 + 13.5 \ln 100 \approx 34.8$ pounds

31. Lograthmic, $y = 78.8 - 10.3 \ln x$

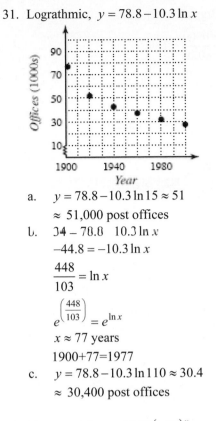

a. $y = 78.8 - 10.3 \ln 15 \approx 51$

$\approx 51,000$ post offices

b. $34 - 78.8 \quad 10.3 \ln x$

$-44.8 = -10.3 \ln x$

$\dfrac{448}{103} = \ln x$

$e^{\left(\frac{448}{103}\right)} = e^{\ln x}$

$x \approx 77$ years

1900+77=1977

c. $y = 78.8 - 10.3 \ln 110 \approx 30.4$

$\approx 30,400$ post offices

33. Exponential, $y = 50.21(1.07)^x$

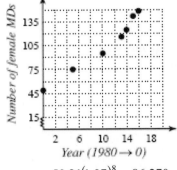

Year (1980 → 0)

a. $y = 50.21(1.07)^8 \approx 86.270$

$\approx 86,270$ female MD's

b. $y = 50.21(1.07)^{25} \approx 272.511$

$\approx 272,511$ female MD's

c. $100 = 50.21(1.07)^x$

$$\frac{10000}{5021} = 1.07^x$$

$$\ln\left(\frac{10000}{5021}\right) = \ln 1.07^x$$

$$\ln\left(\frac{10000}{5021}\right) = x\ln 1.07$$

$$\frac{\ln\left(\dfrac{10000}{5021}\right)}{\ln 1.07} = x$$

$x \approx 10.2$

1980+10.2=1990.2, year 1990

35. Exponential, $y = 346.79(0.94)^x$

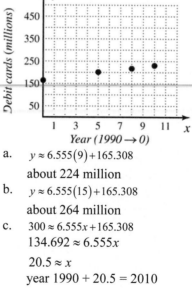

a. $y = 346.79(0.94)^{13} \approx 155.142$
    $\approx 155{,}142$ farms

b. $y = 346.79(0.94)^{24} \approx 78.548$
    $\approx 78{,}548$ farms

c. $150 = 346.79(0.94)^x$

$$\frac{150}{346.79} = 0.94^x$$

$$\ln\left(\frac{150}{346.79}\right) = \ln 0.94^x$$

$$\ln\left(\frac{150}{346.79}\right) = x\ln 0.94$$

$$\frac{\ln\left(\dfrac{150}{346.79}\right)}{\ln 0.94} = x$$

$x \approx 13.5$

1980+13.5=1993.5, year 1993

37. Quadratic, $y \approx 0.576x^2 - 8.879x + 394$

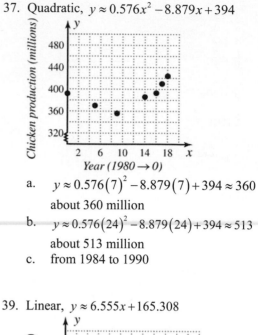

a. $y \approx 0.576(7)^2 - 8.879(7) + 394 \approx 360$
    about 360 million

b. $y \approx 0.576(24)^2 - 8.879(24) + 394 \approx 513$
    about 513 million

c. from 1984 to 1990

39. Linear, $y \approx 6.555x + 165.308$

a. $y \approx 6.555(9) + 165.308$
    about 224 million

b. $y \approx 6.555(15) + 165.308$
    about 264 million

c. $300 \approx 6.555x + 165.308$
    $134.692 \approx 6.555x$

    $20.5 \approx x$
    year 1990 + 20.5 = 2010

Modeling with Technology 2:
Exponential, Logarithmic, and Other Regression Models

41. Linear, $P(t) = 0.51t + 22.51$

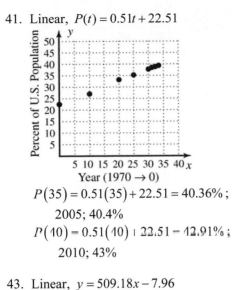

Year (1970 → 0)

$P(35) = 0.51(35) + 22.51 = 40.36\%$;

2005; 40.4%

$P(10) = 0.51(10) + 22.51 = 42.91\%$;

2010; 43%

43. Linear, $y = 509.18x - 7.96$

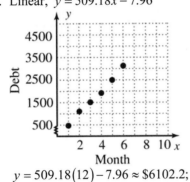

Month

$y = 509.18(12) - 7.96 \approx \$6102.2$;

Debt by the end of December will be about $6100;

$10000 = 509.18x - 7.96$

$10007.96 = 509.18x$

$19.7 \approx x$;

The debt load will exceed $10,000 by the next July.

45. Exponential, $y = (103.83)1.0595^x$

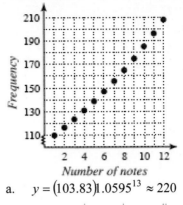

Number of notes

a. $y = (103.83)1.0595^{13} \approx 220$

b. $370.00 = (103.83)1.0595^x$

$3.563517288 = 1.0595^x$

$\ln 3.563517288 = \ln 1.0595^x$

$\ln 3.563517288 = x \ln 1.0595$

$x \approx 22$

The 22nd note, or F#.

c. Frequency doubles, yes.

47.

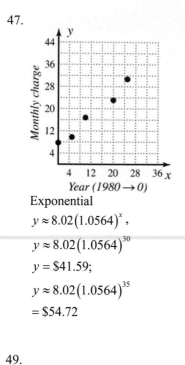

Exponential

$y \approx 8.02(1.0564)^x$,

$y \approx 8.02(1.0564)^{30}$

$y = \$41.59$;

$y \approx 8.02(1.0564)^{35}$

$= \$54.72$

49.

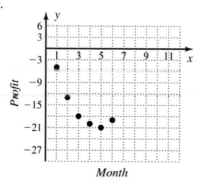

Quadratic, $y \approx 1.18x^2 - 10.99x + 4.60$;

Using grapher, the profit is first earned in the 8th month.

51. Logistic: $y = \dfrac{222.133}{1+32.280e^{-0.336x}}$

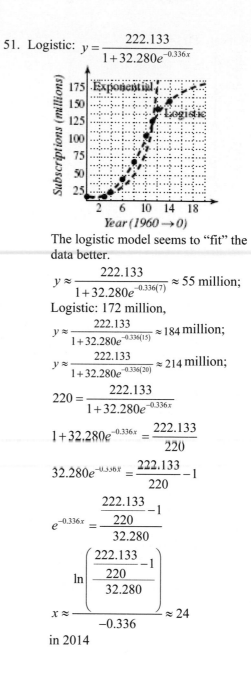

The logistic model seems to "fit" the data better.

$y \approx \dfrac{222.133}{1+32.280e^{-0.336(7)}} \approx 55$ million;

Logistic: 172 million,

$y \approx \dfrac{222.133}{1+32.280e^{-0.336(15)}} \approx 184$ million;

$y \approx \dfrac{222.133}{1+32.280e^{-0.336(20)}} \approx 214$ million;

$220 = \dfrac{222.133}{1+32.280e^{-0.336x}}$

$1+32.280e^{-0.336x} = \dfrac{222.133}{220}$

$32.280e^{-0.336x} = \dfrac{222.133}{220} - 1$

$e^{-0.336x} = \dfrac{\dfrac{222.133}{220}-1}{32.280}$

$x \approx \dfrac{\ln\left(\dfrac{\dfrac{222.133}{220}-1}{32.280}\right)}{-0.336} \approx 24$

in 2014

53.

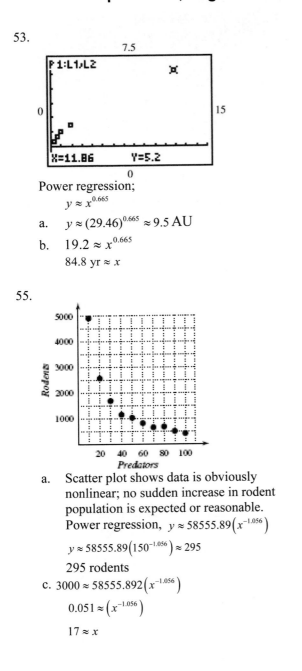

Power regression;

$y \approx x^{0.665}$

a. $y \approx (29.46)^{0.665} \approx 9.5 \, \text{AU}$

b. $19.2 \approx x^{0.665}$

$84.8 \, \text{yr} \approx x$

55.

a. Scatter plot shows data is obviously nonlinear; no sudden increase in rodent population is expected or reasonable. Power regression, $y \approx 58555.89\left(x^{-1.056}\right)$

$y \approx 58555.89\left(150^{-1.056}\right) \approx 295$

295 rodents

c. $3000 \approx 58555.892\left(x^{-1.056}\right)$

$0.051 \approx \left(x^{-1.056}\right)$

$17 \approx x$

57. a. Linear, $w = 1.24L - 15.83$

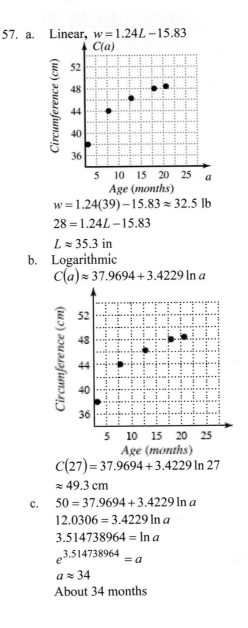

$w = 1.24(39) - 15.83 \approx 32.5 \, \text{lb}$

$28 = 1.24L - 15.83$

$L \approx 35.3 \, \text{in}$

b. Logarithmic

$C(a) \approx 37.9694 + 3.4229 \ln a$

$C(27) = 37.9694 + 3.4229 \ln 27$

$\approx 49.3 \, \text{cm}$

c. $50 = 37.9694 + 3.4229 \ln a$

$12.0306 = 3.4229 \ln a$

$3.514738964 = \ln a$

$e^{3.514738964} = a$

$a \approx 34$

About 34 months

5.1 Technology Highlight

1. $(-1, 4)$

5.1 Exercises

1. Inconsistent

3. Consistent; independent

5. Multiply the 1st equation by 6 and the 2nd equation by 10.

7. $\begin{cases} 7x - 4y = 24 \\ 4x + 3y = 15 \end{cases}$

$\begin{cases} y = \dfrac{7}{4}x - 6 \\ y = -\dfrac{4}{3}x + 5 \end{cases}$

9. $A: \quad y = x + 2$

11. $C: \quad x + 3y = -3$

13. $E: \quad y = x + 2$
$\qquad x + 3y = -3$

15. $\begin{cases} 3x + y = 11 \\ -5x + y = -13 \end{cases}$

$(3, 2)$

$3x + y = 11 \qquad\qquad -5x + y = -13$

$3(3) + 2 = 11 \qquad\quad -5(3) + 2 = -13$

$9 + 2 = 11 \qquad\qquad -15 + 2 = -13$

$\qquad 11 = 11 \qquad\qquad\quad -13 = -13$

Yes

17. $\begin{cases} 8x - 24y = -17 \\ 12x + 30y = 2 \end{cases}$

$\left(-\dfrac{7}{8}, \dfrac{5}{12}\right)$

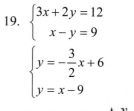

$8x - 24y = -17$

$8\left(-\dfrac{7}{8}\right) - 24\left(\dfrac{5}{12}\right) = -17$

$-7 - 10 = -17$

$-17 = -17;$

$12x + 30y = 2$

$12\left(-\dfrac{7}{8}\right) + 30\left(\dfrac{5}{12}\right) = 2$

$-\dfrac{84}{8} + \dfrac{150}{12} = 2$

$2 = 2$

Yes

19. $\begin{cases} 3x + 2y = 12 \\ x - y = 9 \end{cases}$

$\begin{cases} y = -\dfrac{3}{2}x + 6 \\ y = x - 9 \end{cases}$

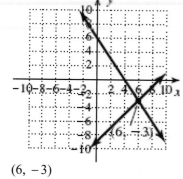

$(6, -3)$

21. $\begin{cases} 5x - 2y = 4 \\ x + 3y = -15 \end{cases}$

$\begin{cases} y = \dfrac{5}{2}x - 2 \\ y = -\dfrac{1}{3}x - 5 \end{cases}$

Estimate: $(-1.1, -4.6)$

23. $\begin{cases} x = 5y - 9 \\ x - 2y = -6 \end{cases}$

$\begin{array}{l} x - 2y = -6 \\ (5y - 9) - 2y = -6 \\ 5y - 9 - 2y = -6 \\ \qquad 3y = 3 \\ \qquad y = 1; \end{array}$

$\begin{array}{l} x - 2y = -6 \\ x - 2(1) = -6 \\ x - 2 = -6 \\ \qquad x = -4 \end{array}$

$(-4, 1)$

25. $\begin{cases} y = \dfrac{2}{3}x - 7 \\ 3x - 2y = 19 \end{cases}$

$3x - 2y = 19$

$3x - 2\left(\dfrac{2}{3}x - 7\right) = 19$

$3x - \dfrac{4}{3}x + 14 = 19$

$\dfrac{5}{3}x = 5$

$x = 3;$

$y = \dfrac{2}{3}x - 7$

$y = \dfrac{2}{3}(3) - 7$

$y = 2 - 7$

$y = -5$

$(3, -5)$

27. $\begin{cases} 3x - 4y = 24 \\ 5x + y = 17 \end{cases}$

Equation 2, variable y

$\begin{cases} 3x - 4y = 24 \\ y = -5x + 17 \end{cases}$

$\begin{array}{l} 3x - 4y = 24 \\ 3x - 4(-5x + 17) = 24 \\ 3x + 20x - 68 = 24 \\ \qquad 23x = 92 \\ \qquad x = 4; \end{array}$

$5x + y = 17$

$5(4) + y = 17$

$20 + y = 17$

$y = -3$

$(4, -3)$

29. $\begin{cases} 0.7x + 2y = 5 \\ x - 1.4y = 11.4 \end{cases}$

Equation 2, variable x

$\begin{cases} 0.7x + 2y = 5 \\ x = 1.4y + 11.4 \end{cases}$

$\begin{array}{l} 0.7x + 2y = 5 \\ 0.7(1.4y + 11.4) + 2y = 5 \\ 0.98y + 7.98 + 2y = 5 \\ \qquad 2.98y = -2.98 \\ \qquad y = -1; \end{array}$

$x - 1.4y = 11.4$

$x - 1.4(-1) = 11.4$

$x + 1.4 = 11.4$

$x = 10$

$(10, -1)$

31. $\begin{cases} 5x - 6y = 2 \\ x + 2y = 6 \end{cases}$

Equation 2, variable x

$\begin{cases} 5x - 6y = 2 \\ x = -2y + 6 \end{cases}$

$5x - 6y = 2$
$5(-2y + 6) - 6y = 2$
$-10y + 30 - 6y = 2$
$\qquad -16y = -28$
$\qquad\qquad y = \dfrac{7}{4};$

$x + 2y = 6$

$x + 2\left(\dfrac{7}{4}\right) = 6$

$x + \dfrac{7}{2} = 6$

$\qquad x = \dfrac{5}{2}$

$\left(\dfrac{5}{2}, \dfrac{7}{4}\right)$

33. $\begin{cases} 2x - 4y = 10 \\ 3x + 4y = 5 \end{cases}$

R1 + R2 = Sum
$2x - 4y = 10$
$3x + 4y = 5$

$5x = 15$
$x = 3;$

$3x + 4y = 5$
$3(3) + 4y = 5$
$9 + 4y = 5$
$\qquad 4y = -4$
$\qquad\quad y = -1$

$(3, -1)$

35. $\begin{cases} 4x - 3y = 1 \\ 3y = -5x - 19 \end{cases}$

$\begin{cases} 4x - 3y = 1 \\ 5x + 3y = -19 \end{cases}$

R1 + R2 = Sum
$4x - 3y = 1$

$5x + 3y = -19$

$9x = -18$
$x = -2;$

$3y = -5x - 19$
$3y = -5(-2) - 19$
$3y = 10 - 19$
$3y = -9$

$y = -3$

$(-2, -3)$

37. $\begin{cases} 2x = -3y + 17 \\ 4x - 5y = 12 \end{cases}$

$\begin{cases} 2x + 3y = 17 \\ 4x - 5y = 12 \end{cases}$

-2R1 + R2 = Sum
$-4x - 6y = -34$

$4x - 5y = 12$

$-11y = -22$
$y - 2;$

$2x = -3y + 17$
$2x = -3(2) + 17$
$2x = -6 + 17$
$2x = 11$

$x = \dfrac{11}{2}$

$\left(\dfrac{11}{2}, 2\right)$

39. $\begin{cases} 0.5x + 0.4y = 0.2 \\ 0.3y = 1.3 + 0.2x \end{cases}$

$\begin{cases} 0.5x + 0.4y = 0.2 \\ -0.2x + 0.3y = 1.3 \end{cases}$

$20R1 + 50\ R2 = \text{Sum}$

$10x + 8y = 4$

$-10x + 15y = 65$

$23y = 69$

$y = 3;$

$0.5x + 0.4y = 0.2$

$0.5x + 0.4(3) = 0.2$

$0.5x + 1.2 = 0.2$

$0.5x = -1$

$x = -2$

$(-2, 3)$

41. $\begin{cases} 0.32m - 0.12n = -1.44 \\ -0.24m + 0.08n = 1.04 \end{cases}$

$200\ R1 + 300\ R2 = \text{Sum}$

$64m - 24n = -288$

$-72m + 24n = 312$

$-8m = 24$

$m = -3;$

$0.32m - 0.12n = -1.44$

$0.32(-3) - 0.12n = -1.44$

$-0.96 - 0.12n = -1.44$

$-0.12n = -0.48$

$n = 4$

$(-3, 4)$

43. $\begin{cases} -\dfrac{1}{6}u + \dfrac{1}{4}v = 4 \\ \dfrac{1}{2}u - \dfrac{2}{3}v = -11 \end{cases}$

$18R1 + 6R2 = \text{Sum}$

$-3u + 4.5v = 72$

$3u - 4v = -66$

$0.5v = 6$

$v = 12;$

$-\dfrac{1}{6}u + \dfrac{1}{4}v = 4$

$-\dfrac{1}{6}u + \dfrac{1}{4}(12) = 4$

$\dfrac{1}{6}u + 3 = 4$

$-\dfrac{1}{6}u = 1$

$u = -6$

$(-6, 12)$

45. $\begin{cases} 4x + \dfrac{3}{4}y = 14 \\ -9x + \dfrac{5}{8}y = -13 \end{cases}$

$9R1 + 4R2 = \text{Sum}$

$36x + \dfrac{27}{4}y = 126$

$-36x + \dfrac{5}{2}y = -52$

$\dfrac{37}{4}y = 74$

$y = 8;$

$4x + \dfrac{3}{4}y = 14$

$4x + \dfrac{3}{4}(8) = 14$

$4x + 6 = 14$

$4x = 8$

$x = 2$

$(2, 8);$ Consistent/independent

47. $\begin{cases} 0.2y = 0.3x + 4 \\ 0.6x - 0.4y = -1 \end{cases}$

$\begin{cases} -0.3x + 0.2y = 4 \\ 0.6x - 0.4y = -1 \end{cases}$

2R1 + R2 = Sum

$-0.6x + 0.4y = 8$

$0.6x - 0.4y = -1$

$0 \neq 7$

No Solution; Inconsistent

49. $\begin{cases} 6x - 22 = -y \\ 3x + \dfrac{1}{2}y = 11 \end{cases}$

$\begin{cases} 6x + y = 22 \\ 3x + \dfrac{1}{2}y = 11 \end{cases}$

R1 − 2R2 = Sum

$6x + y = 22$

$-6x - y = -22$

$0 = 0$

$\{(x, y) \mid 6x + y = 22\}$

Consistent/dependent

51. $\begin{cases} 10x + 35y = 5 \\ y = 0.25x \end{cases}$

$-10x + 35y = -5$

$-10x + 35(0.25x) = -5$

$-10x + 8.75x = -5$

$-1.25x = -5$

$x = 4;$

$y = 0.25x$

$y = 0.25(4)$

$y = 1$

(4, 1); Consistent/Independent

53. $\begin{cases} 7a + b = -25 \\ 2a - 5b = 14 \end{cases}$

5R1 + R2 = Sum

$35a + 5b = -125$

$\underline{2a - 5b = 14}$

$37a = -111$

$a = -3;$

$2a - 5b = 14$

$2(-3) - 5b = 14$

$-6 - 5b = 14$

$-5b = 20$

$b = -4$

(−3, −4); Consistent/Independent

55. $\begin{cases} 4a = 2 - 3b \\ 6b + 2a = 7 \end{cases}$

$\begin{cases} 4a + 3b = 2 \\ 2a + 6b = 7 \end{cases}$

R1 − 2R2 = Sum

$4a + 3b = 2$

$-4a - 12b = -14$

$-9b = -12$

$b = \dfrac{4}{3};$

$4a + 3b = 2$

$4a + 3\left(\dfrac{4}{3}\right) = 2$

$4a + 4 = 2$

$4a = -2$

$a = -\dfrac{1}{2}$

$\left(-\dfrac{1}{2}, \dfrac{4}{3}\right)$; Consistent/Independent

57. $\begin{cases} 2x + 4y = 6 \\ x + 12 = 4y \end{cases}$

$2x + 4y = 6$

$2x + x + 12 = 6$

$3x = -6$

$x = -2;$

$x + 12 = 4y$

$-2 + 12 = 4y$

$10 = 4y$

$\dfrac{5}{2} = y$

$\left(-2, \dfrac{5}{2}\right)$

59. $\begin{cases} 5x - 11y = 21 \\ 11y = 5 - 8x \end{cases}$

$5x - 11y = 21$

$5x - (5 - 8x) = 21$

$5x - 5 + 8x = 21$

$13x = 26$

$x = 2;$

$11y = 5 - 8x$

$11y = 5 - 8(2)$

$11y = 5 - 16$

$11y = -11$

$y = -1$

(2, −1)

Chapter 5: Systems of Equations and Inequalities

61. $\begin{cases} (R+C)T_1 = D_1 \\ (R-C)T_2 = D_2 \end{cases}$

$\begin{cases} (R+C)\,1 = 5 \\ (R-C)\,3 = 9 \end{cases}$

$\begin{cases} R+C = 5 \\ 3R - 3C = 9 \end{cases}$

$3R1 + R2 = \text{Sum}$
$3R + 3C = 15$
$3R - 3C = 9$
$6R = 24$
$\quad R = 4;$
$\quad R + C = 5$
$\quad 4 + C = 5$
$\qquad C = 1$

The current was 1 mph.
He can row 4 mph in still water.

63. Let a represent the number of adult tickets.
Let c represent the number of child tickets.

$\begin{cases} 9a + 6.50c = 30495 \\ a + c = 3800 \end{cases}$

Multiply the second equation by -9.

$\begin{cases} 9a + 6.50c = 30495 \\ -9a - 9c = -34200 \end{cases}$

$\qquad -2.5c = -3705$

$\qquad c = \dfrac{-3705}{-2.5}$

$\qquad c = 1482;$

$a + 1482 = 3800$
$\quad a = 3800 - 1482$
$\quad a = 2318$

2318 adult tickets and 1482 child tickets were sold.

65. Let r represent the price per gallon of regular unleaded gasoline.
Let p represent represent the price per gallon of premium gasoline.

$\begin{cases} 20r + 17p = 144.89 \\ p = 0.10 + r \end{cases}$

$20r + 17(0.10 + r) = 144.89$
$20r + 1.7 + 17r = 144.89$
$\qquad 37r = 144.89 - 1.7$
$\qquad r = \$3.87$
$p = 0.10 + 3.87 = \$3.97$

Premium: \$3.97, Regular: \$3.87.

67. Let s represent the loan made to the science major.
Let n represent the loan made to the nursing student.

$\begin{cases} 0.07s + 0.06n = 635 \\ s + n = 10000 \end{cases}$

$s = 10000 - n$
$0.07(10000 - n) + 0.06n = 635$
$\quad 700 - 0.07n + 0.06n = 635$
$\qquad -0.01n = 635 - 700$
$\qquad -0.01n = -65$
$\qquad n = \dfrac{-65}{-0.01}$
$\qquad n = 6500;$
$\quad s = 10000 - 65000 = 3500$

\$6500 was loaned to the nursing student.
\$3500 was loaned to the science major.

69. Let q represent the number of quarters.
Let d represent represent the number of dimes.

$\begin{cases} q + d = 225 \\ 0.25q + 0.10d = 45 \end{cases}$

$q = 225 - d;$
$\quad 0.25(225 - d) + 0.10d = 45$
$\quad 56.25 - 0.25d + 0.10d = 45$
$\qquad -0.15d = 45 - 56.25$
$\qquad -0.15d = -11.25$
$\qquad d = 75;$
$\quad q = 225 - 75 = 150$

150 quarters, 75 dimes

5.1 Exercises

71. Let x represent the number of lawns serviced each month.

   (a) Total Cost: $C(x) = 75x + 4000$

      Projected Revenue: $R(x) = 115x$

$$\begin{cases} y = 75x + 4000 \\ y = 115x \end{cases}$$

$$75x + 4000 = 115x$$

$$4000 = 40x$$

$$100 = x$$

      100 lawns/month

   (b) 115(100)=$11,500/month.

73. $y = 1.5x + 3$

   (a) $5.40 = 1.5x + 3$

$$2.40 = 1.5x$$

$$1.6 = x$$

      Supply: 1.6 billion bu;

      $y = -2.20x + 12$

$$5.40 = -2.20x + 12$$

$$-6.6 = -2.20x$$

$$3 = x$$

      Demand: 3 billion bu.

      Yes, supply is less than demand.

   (b) $7.05 = 1.5x + 3$

$$4.05 = 1.5x$$

$$2.7 = x$$

      Supply: 2.7 billion bu;

$$7.05 = -2.20x + 12$$

$$-4.95 = -2.20x$$

$$2.25 = x$$

      Demand:2.25 billion bu.

      Yes, demand is less than supply.

   (c) $1.5x + 3 = -2.20x + 12$

$$3.70x = 9$$

$$x \approx 2.43 \text{ billion bu;}$$

$$y = 1.5(2.43) + 3$$

$$y \approx \$6.65$$

75. Let c represent the speed of the current. Let b represent the speed of the boat in still water.

To the drop point: $4 = (b - c)2$

Return to the drop point: $4 = (b + c)\dfrac{1}{2}$

$$\begin{cases} 4 = 2b - 2c \\ 4 = \dfrac{1}{2}b + \dfrac{1}{2}c \end{cases}$$

4R2

$$\begin{cases} 4 = 2b - 2c \\ 4(4) = 4\left(\dfrac{1}{2}b + \dfrac{1}{2}c\right) \end{cases}$$

$$\begin{cases} 4 = 2b - 2c \\ 16 = 2b + 2c \end{cases}$$

$$R1 + R2$$

$$20 = 4b$$

$$5 = b$$

5 mph;

$$4 = (b - c)2$$

$$4 = (5 - c)2$$

$$4 = 10 - 2c$$

$$-6 = -2c$$

$$3 = c$$

3 mph

   (a) Speed of current, 3 mph

   (b) Speed of boat in still water, 5 mph

77. (a) Let w represent the speed of the walkway.

      Let j represent Jason's walking speed.

      With walkway: $256 = (j + w)32$

      Opposite direction: $256 = (j - w)320$

$$\begin{cases} 256 = 32j + 32w \\ 256 = 320j - 320w \end{cases}$$

$$\dfrac{R1}{32}, \dfrac{R2}{32}$$

$$\begin{cases} 8 = j + w \\ 8 = 10j - 10w \end{cases}$$

$$-10R1$$

$$\begin{cases} -80 = -10j - 10w \\ 8 = 10j - 10w \end{cases}$$

$$-72 = -20w$$

$$3.6 = w$$

3.6 ft/sec

   (b) $8 = j + w$

$$8 = j + 3.6$$

$$j = 4.4 \text{ ft/sec}$$

Chapter 5: Systems of Equations and Inequalities

79. Let d represent the year the Declaration was signed.
Let c represent the year the Civil War ended.
$$\begin{cases} c+d = 3641 \\ c-d = 89 \end{cases}$$
$$2c = 3730$$
$$c = \frac{3730}{2}$$
$$c = 1865$$
$$1865 + d = 3641$$
$$d = 3641 - 1865$$
$$d = 1776$$
The Declaration was signed in 1776.
The Civil War ended in 1865.

81. Let x represent Tahiti's land area.
Let y represent Tonga's land area.
$$\begin{cases} x+y = 692 \\ x = 112 + y \end{cases}$$
$$112 + y + y = 692$$
$$112 + 2y = 692$$
$$2y = 580$$
$$y = 290 \text{ mi}^2;$$
$$x+y = 692$$
$$x + 290 = 692$$
$$x = 402 \text{ mi}^2;$$
Tahiti: 402 mi^2
Tonga: 290 mi^2

83. Different slopes so they cannot be the same line or parallel lines. Therefore the system is consistent and the equations are independent.

85. Let x represent the amount invested at 6%. Let y represent the amount invested at 8.5%.
$$\begin{cases} 0.06x + 0.085y = 1250 \\ 0.085x + 0.06y = 1375 \end{cases}$$
Multiply both equations by 1000.
$$\begin{cases} 60x + 85y = 1250000 \\ 85x + 60y = 1375000 \end{cases}$$
Multiply the first equation by -85 and the second equation by 60.

$$\begin{cases} -5100x - 7225y = -106250000 \\ 5100x + 3600y = 82500000 \end{cases}$$
$$-3625y = -23750000$$
$$y = \frac{-23750000}{-3625}$$
$$y \approx 6552$$
$$60x + 85(6552) = 1250000$$
$$60x + 556920 = 1250000$$
$$60x = 1250000 - 556920$$
$$60x = 693080$$
$$x = \frac{693080}{60}$$
$$x \approx 11551$$
$6,552 invested at 8.5%.
$11,551 invested at 6%.

87. Possible factors: $\dfrac{\{\pm 1, \pm 10, \pm 2, \pm 5\}}{\{\pm 1, \pm 3\}}$

```
5 | 3  -19   15   27  -10
  |      15  -20  -25   10
  --------------------------
    3   -4   -5    2  | 0
```

```
2 | 3  -4  -5   2
  |     6   4  -2
  -----------------
    3   2  -1 | 0
```
$(x-5)(x-2)(3x^2 + 2x - 1) = 0$
$(x-5)(x-2)(x+1)(3x-1) = 0$

89. $y = x^2 - 6x - 16$
$y = (x^2 - 6x + 9) - 16 - 9$
$y = (x-3)^2 - 25$
$f(x) \le 0$ on $[-2, 8]$

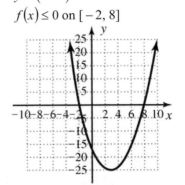

277

Technology Highlight

1. a. Answers will vary.
   b. Answers will vary.
   c. $(-9, -6, -5)$ and $(-2, 1, 2)$ are solutions and $(6, 2, 4)$ is not a solution.

5.2 Exercises

1. Triple

3. Equivalent systems

5. $2(2) + (-5) + z = 4$
   $4 + (-5) + z = 4$
   $-1 + z = 4$
   $z = 5$;
   Substitute and solve for the remaining variable.

7. $x + 2y + z = 9$
   Answers will vary.

9. $-x + y + 2z = -6$
   Answers will vary.

11. $\begin{cases} x + y - 2z = -1 \\ 4x - y + 3z = 3 \\ 3x + 2y - z = 4 \end{cases}$
    $x + y - 2z = -1$
    $0 + 3 - 2(2) = -1$
    $3 - 4 = -1$
    $-1 = -1$;

    $4x - y + 3z = 3$
    $4(0) - 3 + 3(2) = 3$
    $-3 + 6 = 3$
    $3 = 3$;
    $3x + 2y - z = 4$
    $3(0) + 2(3) - 2 = 4$
    $6 - 2 = 4$
    $4 = 4$
    Yes

$x + y - 2z = -1$
$-3 + 4 - 2(1) = -1$
$1 - 2 = -1$
$-1 = -1$;
$4x - y + 3z = 3$
$4(-3) - 4 + 3(1) = 3$
$-12 - 4 + 3 = 3$
$-13 \neq 3$
No

13. $\begin{cases} x - y - 2z = -10 \\ x - z = 1 \\ z = 4 \end{cases}$
    $x - z = 1$
    $x - 4 = 1$
    $x = 5$;
    $x - y - 2z = -10$
    $5 - y - 2(4) = -10$
    $5 - y - 8 = -10$
    $-y - 3 = -10$
    $-y = -7$
    $y = 7$
    $(5, 7, 4)$

15. $\begin{cases} x + 3y + 2z = 16 \\ -2y + 3z = 1 \\ 8y - 13z = -7 \end{cases}$
    $4R2 + R3 = \text{Sum}$
    $-8y + 12z = 4$
    $8y - 13z = -7$
    $-z = -3$
    $z = 3$;
    $-2y + 3z = 1$
    $-2y + 3(3) = 1$
    $-2y + 9 = 1$
    $-2y = -8$
    $y = 4$;
    $x + 3y + 2z = 16$
    $x + 3(4) + 2(3) = 16$
    $x + 12 + 6 = 16$
    $x + 18 = 16$
    $x = -2$
    $(-2, 4, 3)$

17. $\begin{cases} 2x - y + 4z = -7 \\ x + 2y - 5z = 13 \\ y - 4z = 9 \end{cases}$

R1 + R3 = Sum

$2x - y + 4z = -7$

$\underline{y - 4z = 9}$

$2x = 2$

$x = 1$

Substitute $x = 1$ into R2 then new

R2 − 2R3 = Sum

$1 + 2y - 5z = 13$

$2y - 5z = 12$

$\underline{-2y + 8z = -18}$

$3z = -6$

$z = -2;$

$2x - y + 4z = -7$

$2(1) - y + 4(-2) = -7$

$2 - y - 8 = -7$

$-y - 6 = -7$

$-y = -1$

$y = 1$

$(1, 1, -2)$

19. $\begin{cases} -x + y + 2z = -10 \\ x + y - z = 7 \\ 2x + y + z = 5 \end{cases}$

R1 + R2 = Sum

$-x + y + 2z = -10$

$\underline{x + y - z = 7}$

$2y + z = -3$

2R1 + R2 = Sum

$-2x + 2y + 4z = -20$

$\underline{2x + y + z = 5}$

$3y + 5z = -15$

$\begin{cases} 2y + z = -3 \\ 3y + 5z = -15 \end{cases}$

-5R1 + R2 = Sum

$-10y - 5z = 15$

$\underline{3y + 5z = -15}$

$-7y = 0$

$y = 0;$

$2y + z = -3$

$2(0) + z = -3$

$z = -3;$

$x + y - z = 7$

$x + 0 - (-3) = 7$

$x + 3 = 7$

$x = 4$

$(4, 0, -3)$

21. $\begin{cases} 3x + y - 2z = 3 \\ x - 2y + 3z = 10 \\ 4x - 8y + 5z = 5 \end{cases}$

-4R2 + R3 = Sum

$-4x + 8y - 12z = -40$

$\underline{4x - 8y + 5z = 5}$

$-7z = -35$

$z = 5$

Substitute into R1 and R2

$3x + y - 2(5) = 3$

$3x + y - 10 = 3$

$3x + y = 13$

$x - 2y + 3(5) = 10$

$x - 2y + 15 = 10$

$x - 2y = -5$

$\begin{cases} 3x + y = 13 \\ x - 2y = -5 \end{cases}$

2R1 + R2 = Sum

$6x + 2y = 26$

$\underline{x - 2y = -5}$

$7x = 21$

$x = 3;$

$3x + y - 2z = 3$

$3(3) + y - 2(5) = 3$

$9 + y - 10 = 3$

$y - 1 = 3$

$y = 4$

$(3, 4, 5)$

23. $\begin{cases} 3x - y + z = 6 \\ 2x + 2y - z = 5 \\ 2x - y + z = 5 \end{cases}$

R1 + R2 = Sum

$\begin{cases} 3x - y + z = 6 \\ 2x + 2y - z = 5 \end{cases}$

$5x + y = 11;$

R2 + R3 = Sum

$\begin{cases} 2x + 2y - z = 5 \\ 2x - y + z = 5 \end{cases}$

$4x + y = 10$

$\begin{cases} 5x + y = 11 \\ 4x + y = 10 \end{cases}$

$-$R1 + R2 = Sum

$\begin{cases} -5x - y = -11 \\ 4x + y = 10 \end{cases}$

$-x = -1$

$x = 1;$

$5x + y = 11$

$5(1) + y = 11$

$y = 6;$

$3x - y + z = 6$

$3(1) - 6 + z = 6$

$3 - 6 + z = 6$

$-3 + z = 6$

$z = 9$

$(1, 6, 9)$

---

27. $\begin{cases} 4x + y + 3z = 8 \\ x - 2y + 3z = 2 \end{cases}$

2R1

$\begin{cases} 8x + 2y + 6z = 8 \\ x - 2y + 3z = 2 \end{cases}$

R1 + R2 = Sum

$9x + 9z = 18$

$9z = 18 - 9x$

$z = 2 - x;$

$x - 2y + 3z = 2$

$x - 2y + 3(2 - x) = 2$

$x - 2y + 6 - 3x = 2$

$-2x + 6 - 2y = 2$

$-2y = 2x - 4$

$y = -x + 2$

$y = 2 - x;$

$(x, 2 - x, 2 - x)$

Using "p" as our parameter, the solution could be written in $(p, 2 - p, 2 - p)$ parametric form.

---

25. $\begin{cases} 3x + y + 2z = 3 \\ x - 2y + 3z = 1 \\ 4x - 8y + 12z = 7 \end{cases}$

2R1 + R2 = Sum

$6x + 2y + 4z = 6$

$x - 2y + 3z = 1$

$7x + 7z = 7$

8R1 + R3 = Sum

$24x + 8y + 16z = 24$

$4x - 8y + 12z = 7$

$28x + 28z = 31$

$\begin{cases} 7x + 7z = 7 \\ 28x + 28z = 31 \end{cases}$

-4R1 + R2 = Sum

$-28x - 28z = -28$

$28x + 28z = 31$

$0 \neq 3$

No solution; inconsistent

---

29. $\begin{cases} 6x - 3y + 7z = 2 \\ 3x - 4y + z = 6 \end{cases}$

R1 − 2R2 = Sum

$6x - 3y + 7z = 2$

$-6x + 8y - 2z = -12$

$5y + 5z = -10$

$5y = -5z - 10$

$y = -z - 2;$

$3x - 4y + z = 6$

$3x - 4(-z - 2) + z = 6$

$3x + 4z + 8 + z = 6$

$3x + 5z = -2$

$3x = -5z - 2$

$x = -\dfrac{5}{3}z - \dfrac{2}{3}$

$\left(-\dfrac{5}{3}z - \dfrac{2}{3}, -z - 2, z \right)$

Using "p" as our parameter, the solution could be written in

$\left(-\dfrac{5}{3}p - \dfrac{2}{3}, -p - 2, p \right)$ parametric form.

Other solutions are possible.

31. $\begin{cases} 3x - 4y + 5z = 5 \\ -x + 2y - 3z = -3 \\ 3x - 2y + z = 1 \end{cases}$

R2 + R3 = Sum

$-x + 2y - 3z = -3$

$3x - 2y + z = 1$

$2x - 2z = -2$

$-2z = -2x - 2$

$z = x + 1;$

$-x + 2y - 3z = -3$

$-x + 2y - 3(x + 1) = -3$

$-x + 2y - 3x - 3 = -3$

$2y - 4x = 0$

$2y = 4x$

$y = 2x;$

$(x, 2x, x + 1)$

Using "p" as our parameter, the solution could be written in $(p, 2p, p + 1)$ parametric form. Other solutions are possible.

33. $\begin{cases} x + 2y - 3z = 1 \\ 3x + 5y - 8z = 7 \\ x + y - 2z = 5 \end{cases}$

R1 − R3 = Sum

$x + 2y - 3z = 1$

$-x - y + 2z = -5$

$y - z = -4$

$y - z = -4$

$y = z - 4;$

$x + y - 2z = 5$

$x + z - 4 - 2z = 5$

$x - z = 9$

$x = z + 9$

$(z + 9, z - 4, z)$

Using "p" as our parameter, the solution could be written in $(p + 9, p - 4, p)$ parametric form. Other solutions are possible.

35. $\begin{cases} -0.2x + 1.2y - 2.4z = -1 \\ 0.5x - 3y + 6z = 2.5 \\ x - 6y + 12z = 5 \end{cases}$

-2R2 + R3 = Sum

$-1x + 6y - 12z = -5$

$x - 6y + 12z = 5$

$0 = 0$

$\{(x, y, z) | x - 6y + 12z = 5\}$

37. $\begin{cases} x + 2y - z = 1 \\ x + z = 3 \\ 2x - y + z = 3 \end{cases}$

R1 + 2R3 = Sum

$x + 2y - z = 1$

$4x - 2y + 2z = 6$

$5x + z = 7$

$\begin{cases} x + z = 3 \\ 5x + z = 7 \end{cases}$

$-$R1 + R2 = Sum

$-x - z = -3$

$5x + z = 7$

$4x = 4$

$x = 1;$

$x + z = 3$

$1 + z = 3$

$z = 2;$

$x + 2y - z = 1$

$1 + 2y - 2 = 1$

$2y - 1 = 1$

$2y = 2$

$y = 1$

$(1, 1, 2)$

39. $\begin{cases} 2x - 5y - 4z = 6 \\ x - 2.5y - 2z = 3 \\ -3x + 7.5y + 6z = -9 \end{cases}$

R1 − 2R2 = Sum

$2x - 5y - 4z = 6$

$-2x + 5y + 4z = -6$

$0 = 0$

$\left\{(x, y, z) \,\middle|\, x - \dfrac{5}{2}y - 2z = 3\right\}$

5.2 Exercises

41. $\begin{cases} 4x - 5y - 6z = 5 \\ 2x - 3y + 3z = 0 \\ x + 2y - 3z = 5 \end{cases}$

R1 − 2R2 = Sum

$4x - 5y - 6z = 5$

$-4x + 6y - 6z = 0$

$y - 12z = 5$

R1 − 4R3 = Sum

$4x - 5y - 6z = 5$

$-4x - 8y + 12z = -20$

$-13y + 6z = -15$

$\begin{cases} y - 12z = 5 \\ -13y + 6z = -15 \end{cases}$

R1 + 2R2 = Sum

$y - 12z = 5$

$-26y + 12z = -30$

$-25y = -25$

$y = 1;$

$y - 12z = 5$

$1 - 12z = 5$

$-12z = 4$

$z = -\dfrac{1}{3};$

$x + 2y - 3z = 5$

$x + 2(1) - 3\left(-\dfrac{1}{3}\right) = 5$

$x + 2 + 1 = 5$

$x + 3 = 5$

$x = 2$

$\left(2, 1, -\dfrac{1}{3}\right)$

43. $\begin{cases} 2x + 3y - 5z = 4 \\ x + y - 2z = 3 \\ x + 3y - 4z = -1 \end{cases}$

R1 − 2R2 = Sum

$2x + 3y - 5z = 4$

$-2x - 2y + 4z = -6$

$y - z = -2$

R1 − 2R3 = Sum

$2x + 3y - 5z = 4$

$-2x - 6y + 8z = -2$

$-3y + 3z = 6$

$y - z = -2;$

$y - z = -2$

$y = z - 2;$

$x + y - 2z = 3$

$x + z - 2 - 2z = 3$

$x - z = 5$

$x = z + 5$

$\left(z + 5, z - 2, z\right)$

Using "p" as our parameter, the solution could be written in $\left(p + 5, p - 2, p\right)$ parametric form. Other solutions are possible.

45. $\begin{cases} \dfrac{x}{2} + \dfrac{y}{3} - \dfrac{z}{2} = 2 \\ \dfrac{2x}{3} - y - z = 8 \\ \dfrac{x}{6} + 2y + \dfrac{3z}{2} = 6 \end{cases}$

3R1 + R2 = Sum

$\dfrac{3}{2}x + y - \dfrac{3}{2}z = 6$

$\dfrac{2x}{3} - y - z = 8$

$\dfrac{13}{6}x - \dfrac{5}{2}z = 14$

2R2 + R3 = Sum

$\dfrac{4}{3}x - 2y - 2z = 16$

$\dfrac{x}{6} + 2y + \dfrac{3z}{2} = 6$

$\dfrac{3}{2}x - \dfrac{1}{2}z = 22$

$$\begin{cases} \dfrac{13}{6}x - \dfrac{5}{2}z = 14 \\ \dfrac{3}{2}x - \dfrac{1}{2}z = 22 \end{cases}$$

R1 − 5R2 = Sum

$$\dfrac{13}{6}x - \dfrac{5}{2}z = 14$$

$$-\dfrac{15}{2}x + \dfrac{5}{2}z = -110$$

$$-\dfrac{16}{3}x = -96$$

$$x = 18;$$

$$\dfrac{3}{2}x - \dfrac{1}{2}z = 22$$

$$\dfrac{3}{2}(18) - \dfrac{1}{2}z = 22$$

$$27 - \dfrac{1}{2}z = 22$$

$$-\dfrac{1}{2}z = -5$$

$$z = 10;$$

$$\dfrac{2}{3}x - y - z = 8$$

$$\dfrac{2}{3}(18) - y - 10 = 8$$

$$12 - y - 10 = 8$$

$$2 - y = 8$$

$$-y = 6$$

$$y = -6$$

$$(18, -6, 10)$$

47. $$\begin{cases} -A + 3B + 2C = 11 \\ 2B + C = 9 \\ B + 2C = 8 \end{cases}$$

R2 − 2R3

$$2B + C = 9$$

$$-2B - 4C = -16$$

$$-3C = -7$$

$$C = \dfrac{7}{3};$$

$$2B + C = 9$$

$$2B + \dfrac{7}{3} = 9$$

$$2B = \dfrac{20}{3}$$

$$B = \dfrac{10}{3};$$

$$-A + 3B + 2C = 11$$

$$-A + 3\left(\dfrac{10}{3}\right) + 2\left(\dfrac{7}{3}\right) = 11$$

$$-A + 10 + \dfrac{14}{3} = 11$$

$$-A + \dfrac{44}{3} = 11$$

$$-A = -\dfrac{11}{3}$$

$$A = \dfrac{11}{3}$$

$$\left(\dfrac{11}{3}, \dfrac{10}{3}, \dfrac{7}{3}\right)$$

49. $$\begin{cases} A - 2B = 5 \\ B + 3C = 7 \\ 2A - B - C = 1 \end{cases}$$

−2R1 + R3 = Sum

$$-2A + 4B = -10$$

$$-2A - B - C = 1$$

$$3B - C = -9$$

$$\begin{cases} B + 3C = 7 \\ 3B - C = -9 \end{cases}$$

−3R1 + R2 = Sum

$$-3B - 9C = -21$$

$$3B - C = -9$$

$$-10C = -30$$

$$C = 3;$$

$$B + 3C = 7$$

$$B + 3(3) = 7$$

$$B + 9 = 7$$

$$B = -2;$$

$$A - 2B = 5$$

$$A - 2(-2) = 5$$

$$A + 4 = 5$$

$$A = 1$$

$$(1, -2, 3)$$

51.
$$\begin{cases} C = 3 \\ 2A + 3C = 10 \\ 3B - 4C = -11 \end{cases}$$

$2A + 3C = 10$
$2A + 3(3) = 10$
$2A + 9 = 10$
$2A = 1$
$A = \dfrac{1}{2};$

$3B - 4C = -11$
$3B - 4(3) = -11$
$3B - 12 = -11$
$3B = 1$
$B = \dfrac{1}{3}$

$\left(\dfrac{1}{2}, \dfrac{1}{3}, 3\right)$

53.
$$\left| \dfrac{Ax + By + Cz - D}{\sqrt{A^2 + B^2 + C^2}} \right|$$

$A = 1, B = 1, C = 1, D = 6;$
$x = 3, y = 4, z = 5;$

$\left| \dfrac{1(3) + 1(4) + 1(5) - 6}{\sqrt{1^2 + 1^2 + 1^2}} \right| = \dfrac{6}{\sqrt{3}} \approx 3.464 \text{ units}$

55. Let M represent the amount paid for the Monet.
Let P represent the amount paid for the Picasso.
Let V represent the amount paid for the Van Gogh.

$$\begin{cases} M + P + V = 7 \\ M = P + 0.8 \\ V = 2M + 0.2 \end{cases}$$

$$\begin{cases} M + P + V = 7 \\ M - P = 0.8 \\ V - 2M = 0.2 \end{cases}$$

R1 + R2 = Sum
$M + P + V = 7$
$M - P = 0.8$
$2M + V = 7.8$

$$\begin{cases} -2M + V = 0.2 \\ 2M + V = 7.8 \end{cases}$$

R1 + R2 = Sum

$-2M + V = 0.2$
$2M + V = 7.8$
$2V = 8$
$V = 4;$

$V = 2M + 0.2$
$4 = 2M + 0.2$
$3.8 = 2M$
$1.9 = M;$

$M = P + 0.8$
$1.9 = P + 0.8$
$1.1 = P$

Monet: \$1,900,000
Picasso: \$1,100,000
Van Gogh: \$4,000,000

57. Let c represent the gestation period of a camel.
Let e represent the gestation period of an elephant.
Let r represent the gestation period of a rhinoceros.

$$\begin{cases} c + e + r = 1520 \\ r = c + 58 \\ 2c - 162 = e \end{cases}$$

$c + e + r = 1520$
$c + e + c + 58 = 1520$
$2c + e - 1462$

$$\begin{cases} 2c - e = 162 \\ 2c + e = 1462 \end{cases}$$

R1 + R2 = Sum
$2c - e = 162$
$2c + e = 1462$
$4c = 1624$
$c = 406;$

$r = c + 58$
$r = 406 + 58$
$r = 464;$

$2c - 162 = e$
$2(406) - 162 = e$
$812 - 162 = e$
$650 = e$

Camel: 406 days
Elephant: 650 days
Rhinoceros: 464 days

59. Let x represent the wingspan of the California Condor.
Let y represent the wingspan of the Wandering Albatross.
Let z represent the wingspan of the prehistoric Quetzalcoatlus.

$$\begin{cases} x + y + z = 18.6 \\ z = 5y - 2x \\ 6x = 5y \end{cases}$$

$$\begin{cases} x + y + z = 18.6 \\ 2x - 5y + z = 0 \\ 6x - 5y = 0 \end{cases}$$

$-R1$

$$\begin{cases} -x - y - z = -18.6 \\ 2x - 5y + z = 0 \\ 6x - 5y = 0 \end{cases}$$

$R1 + R2$ yields $x - 6y = -18.6$

Sub-system

$$\begin{cases} x - 6y = -18.6 \\ 6x - 5y = 0 \end{cases}$$

$-6R1$

$$\begin{cases} -6x + 36y = 111.6 \\ 6x - 5y = 0 \end{cases}$$

$R1 + R2$

$31y = 111.6$

$y = 3.6;$

Solve for x in R3

$6x = 5y$

$6x = 5(3.6)$

$x = 3;$

$x + y + z = 18.6$

$3.6 + 3 + z = 18.6$

$z = 12;$

Albatross: 3.6 m

Condor: 3.0 m

Quetzalcoatlus: 12.0 m

61. Let f represent the number of $5 gold pieces.
Let t represent the number of $10 gold pieces.
Let w represent the number of $20 gold pieces.

$$\begin{cases} f + t + w = 250 \\ 5f + 10t + 20w = 1875 \\ f = 7w \end{cases}$$

$-10R1 + R2 = \text{Sum}$

$-10f - 10t - 10w = -2500$

$\underline{5f + 10t + 20w = 1875}$

$-5f + 10w = -625$

$$\begin{cases} -5f + 10w = -625 \\ f - 7w = 0 \end{cases}$$

$R1 + 5R2 = \text{Sum}$

$-5f + 10w = -625$

$\underline{5f - 35w = 0}$

$-25w = -625$

$w = 25;$

$f = 7w$

$f = 7(25)$

$f = 175;$

$f + t + w = 250$

$175 + t + 25 = 250$

$200 + t = 250$

$t = 50$

175 $5 gold pieces

50 $10 gold pieces

25 $20 gold pieces

63. $$\begin{cases} A + B = 0 \\ -6A - 3B + C = 1 \\ 9A = -9 \end{cases}$$

Solve for A in R3

$9A = -9$

$A = -1;$

Solve for B in R1

$A + B = 0$

$-1 + B = 0$

$B = 1;$

Solve for C in R2

$-6A - 3B + C = 1$

$-6(-1) - 3(1) + C = 1$

$6 - 3 + C = 1$

$C = -2;$

$A = -1, B = 1, C = -2;$

$$\frac{A}{x} + \frac{B}{x-3} + \frac{C}{(x-3)^2}$$

$$= \frac{-1}{x} + \frac{1}{x-3} - \frac{2}{(x-3)^2}$$

$$= \frac{-1(x-3)^2}{x(x-3)^2} + \frac{1x(x-3)}{x(x-3)^2} - \frac{2x}{x(x-3)^2}$$

$$= \frac{-x^2 + 6x - 9 + x^2 - 3x - 2x}{x(x-3)^2}$$

$$= \frac{x-9}{x(x-3)^2}$$

Verified

65. $x^2 + y^2 + Dx + Ey + F = 0$

$$\begin{cases} (2)^2 + (-1)^2 + D(2) + E(-1) + F = 0 \\ (4)^2 + (-3)^2 + D(4) + E(-3) + F = 0 \\ (2)^2 + (-5)^2 + D(2) + E(-5) + F = 0 \end{cases}$$

$$\begin{cases} 2D - E + F = -5 \\ 4D - 3E + F = -25 \\ 2D - 5E + F = -29 \end{cases}$$

R1 + (−1)R3
$4E = 24$
$E = 6;$

R1 − R2
$-2D + 2E = 20;$
$-2D + 2(6) = 20$
$-2D = 8$
$D = -4;$

$2D - E + F = -5$
$2(-4) - 6 + F = -5$
$F = 9;$

$x^2 + y^2 - 4x + 6y + 9 = 0$

67. $p(x) = 2x^2 - x - 3$; $p(x) \le 0$

$p(x) = (2x - 3)(x + 1)$

$x = \frac{3}{2}; \quad x = -1$

$\left[-1, \frac{3}{2}\right]$

69. $\log(x+2) + \log(x) = \log(3)$

$\log x(x+2) = \log 3$

$x^2 + 2x - 3 = 0$

$(x+3)(x-1) = 0$

$x = -3$ or $x = 1$

$x = 1$ since $x = -3$ will not check.

Chapter 5
Mid-Chapter Check

1. $\begin{cases} x - 3y = -2 \\ 2x + y = 3 \end{cases}$

$x = 3y - 2$

$2x + y = 3$
$2(3y - 2) + y = 3$
$6y - 4 + y = 3$
$7y = 7$
$y = 1$

$x - 3y = -2$
$x - 3(1) = -2$
$x - 3 = -2$
$x = 1$
(1, 1); Consistent

3. Let x represent the amount of 40% acid.
Let y represent the amount of 48% acid.

$\begin{cases} x + 10 = y \\ 0.40x + 0.64(10) = 0.48y \end{cases}$

$\begin{cases} x + 10 = y \\ 40x + 64(10) = 48y \end{cases}$

$40x + 640 = 48(x + 10)$

$40x + 640 = 48x + 480$

$-8x = -160$

$x = 20$

20 ounces

5. $\begin{cases} x + 2y - 3z = 3 \\ 2x + 4y - 6z = 6 \\ x - 2y + 5z = -1 \end{cases}$

The second equation is a multiple of the first equation; 2R1 = R2.

7. $\begin{cases} 2x+3y-4z=-4 \\ x-2y+z=0 \\ -3x-2y+2z=-1 \end{cases}$

R1 + 4R2 = Sum
$2x+3y-4z=-4$
$4x-8y+4z=0$
$6x-5y=-4$

R1 + 2R3 = Sum
$2x+3y-4z=-4$
$-6x-4y+4z=-2$
$-4x-y=-6$

$\begin{cases} 6x-5y=-4 \\ -4x-y=-6 \end{cases}$

R1 − 5R2 = Sum
$6x-5y=-4$
$20x+5y=30$
$26x=26$
$x=1;$

$-4x-y=-6$
$-4(1)-y=-6$
$-4-y=-6$
$-y=-2$
$y=2;$

$x-2y+z=0$
$1-2(2)+z=0$
$1-4+z=0$
$-3+z=0$
$z=3$
$(1, 2, 3)$

9. Let x represent Mozart's age.
   Let y represent Morphy's age.
   Let z represent Pascal's age.

$\begin{cases} x+y+z=37 \\ y=2x-3 \\ z=y+3 \end{cases}$

$\begin{cases} x+y+z=37 \\ -2x+y=-3 \\ -y+z=3 \end{cases}$

−R1 + R2 = Sum
$-x-y-z=-37$
$-2x+y=-3$
$-3x-z=-40$

R1 + R3 = Sum
$x+y+z=37$
$-y+z=3$
$x+2z=40$

$\begin{cases} -3x-z=-40 \\ x+2z=40 \end{cases}$

2R1 + R2 = Sum
$-6x-2z=-80$
$x+2z=40$
$-5x=-40$
$x=8;$

$y=2x-3$
$y=2(8)-3$
$y=16-3$
$y=13;$

$z=y+3$
$z=13+3$
$z=16$

Mozart: 8 years
Morphy: 13 years
Pascal: 16 years

Reinforcing Basic Concepts

1. $\begin{cases} 15.3R+35.7P=211.14 \\ P=R+0.10 \end{cases}$

   Premium: \$4.17/gal
   Regular: \$4.07/gal

5.3 Exercises

1. a. Circle and Line
        3 or 4 not possible

    c. Circle and Parabola

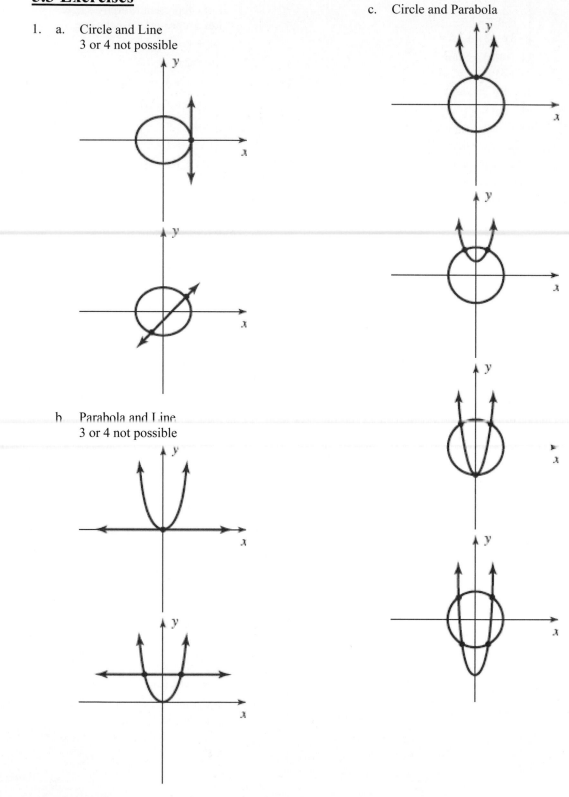

    b. Parabola and Line
        3 or 4 not possible

d. Circle and Absolute Value
Function

e. Absolute Value Function and Line

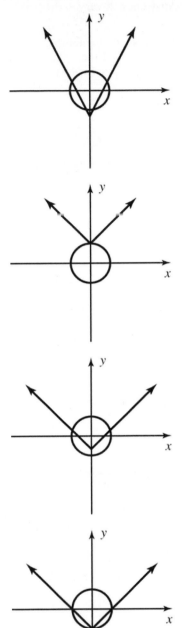

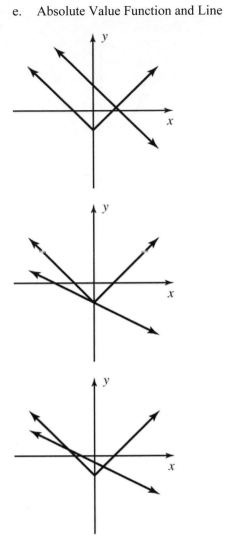

f.　Absolute Value Function and Parabola

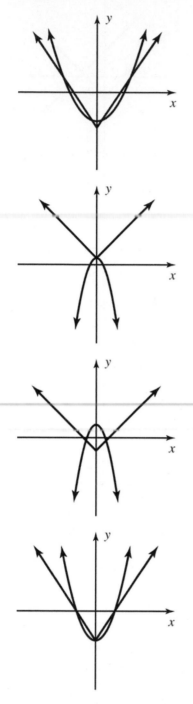

3.　Region, solutions

5.　Answers will vary.

7.　$\begin{cases} x^2 + y = 6 & \text{Parabola} \\ x + y = 4 & \text{Line} \end{cases}$

Solutions: $(-1,5),(2,2)$

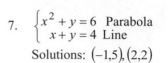
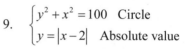
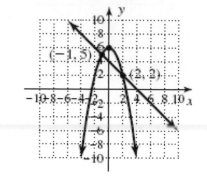

9.　$\begin{cases} y^2 + x^2 = 100 & \text{Circle} \\ y = |x - 2| & \text{Absolute value} \end{cases}$

Solutions: $(-6,8),(8,6)$

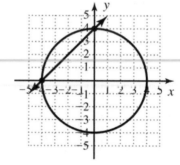

11.　$\begin{cases} -(x-1)^2 + 2 = y & \text{Parabola} \\ y - x^2 = -3 & \text{Parabola} \end{cases}$

Solutions: $(-1,-2),(2,1)$

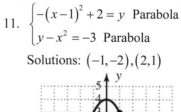

13. $\begin{cases} x^2 + y^2 = 25 \\ y - x = 1 \end{cases}$

2nd equation : $y = x + 1$

$$x^2 + (x+1)^2 = 25$$
$$x^2 + x^2 + 2x + 1 = 25$$
$$2x^2 + 2x - 24 = 0$$
$$x^2 + x - 12 = 0$$
$$(x+4)(x-3) = 0$$

$x = -4, x = 3$;
$y = -4 + 1 = -3$;
$y = 3 + 1 = 4$

Solutions: $(-4, -3), (3, 4)$

15. $\begin{cases} x^2 + y = 9 \\ -2x + y = 1 \end{cases}$

$-2x + y = 1$

$y = 2x + 1$;

Substitute in Equation 2.
$$x^2 + 2x + 1 = 9$$
$$x^2 + 2x - 8 = 0$$
$$(x+4)(x-2) = 0$$
$$x + 4 = 0 \text{ or } x - 2 = 0$$
$$x = -4 \text{ or } x = 2$$

If $x = -4, -2x + y = 1$
$$-2(-4) + y = 1$$
$$y = -7;$$

If $x = 2, -2x + y = 1$
$$-2(2) + y = 1$$
$$y = 5;$$

Solutions: $(2, 5), (-4, -7)$

17. $\begin{cases} x^2 + y = 13 \\ x^2 + y^2 = 25 \end{cases}$

$x^2 + y = 13$

$x^2 = 13 - y;$

$$13 - y + y^2 = 25$$
$$y^2 - y - 12 = 0$$
$$(y-4)(y+3) = 0$$

$y = 4$ or $y = -3$;

If $y=4, x^2 + y = 13$

$$x^2 + 4 = 13$$
$$x^2 = 9$$
$$x = \pm 3,$$

If $y = -3, x^2 + y = 13$

$$x^2 - 3 = 13$$
$$x^2 = 16$$
$$x = \pm 4;$$

Solutions: $(-3, 4), (-4, -3),$
$\qquad (3, 4), (4, -3)$

19. $\begin{cases} x^2 + y^2 = 25 \\ \dfrac{1}{4}x^2 + y = 1 \end{cases}$

$-4R2$

$\begin{cases} x^2 + y^2 = 25 \\ -x^2 - 4y = -4 \end{cases}$

R1+R2
$$y^2 - 4y = 21$$
$$y^2 - 4y - 21 = 0$$
$$(y-7)(y+3) = 0$$
$$y - 7 = 0 \text{ or } y + 3 = 0$$
$$y = 7 \quad \text{ or } y = -3;$$

If $y = 7, x^2 + y^2 = 25$

$$x^2 + 7^2 = 25$$
$$x^2 = -24$$

Not real;

If $y = -3, x^2 + y^2 = 25$

$$x^2 + (-3)^2 = 25$$
$$x^2 = 16$$
$$x = \pm 4;$$

Solutions: $(4, -3), (-4, -3)$

21. $\begin{cases} x^2 + y^2 = 4 \\ y + x^2 = 5 \end{cases}$

$-1R1$

$\begin{cases} -x^2 - y^2 = -4 \\ x^2 + y = 5 \end{cases}$

$R1 + R2$

$-y^2 + y = 1$

$0 = y^2 - y + 1$

$a = 1, b = -1, c = 1$

$b^2 - 4ac = (-1)^2 - 4(1)(1) = -3;$

$y = \dfrac{1 \pm \sqrt{-3}}{2}$

No real solutions.
No solution.

23. $\begin{cases} x^2 + y^2 = 65 \\ y = 3x + 25 \end{cases}$

$x^2 + (3x + 25)^2 = 65$

$x^2 + 9x^2 + 150x + 625 = 65$

$10x^2 + 150x + 560 = 0$

$x^2 + 15x + 56 = 0$

$(x + 8)(x + 7) = 0$

$x = -8$ or $x = -7;$

If $x = -8, y = 3(-8) + 25 = 1;$

If $x = -7, y = 3(-7) + 25 = 4;$

$(-8, 1), (-7, 4)$

25. $\begin{cases} y - 5 = \log x \\ y = 6 - \log(x - 3) \end{cases}$

$\begin{cases} y = \log x + 5 \\ y = 6 - \log(x - 3) \end{cases}$

$\log x + 5 = 6 - \log(x - 3)$

$\log x + \log(x - 3) = 1$

$\log x(x - 3) = 1$

$10 = x^2 - 3x$

$0 = x^2 - 3x - 10$

$0 = (x - 5)(x + 2)$

$x = 5$ or $x = -2$

$x = -2$ is extraneous

$y = \log 5 + 5;$

Solution: $(5, \log 5 + 5)$

27. $\begin{cases} y = \ln(x^2) + 1 \\ y - 1 = \ln(x + 12) \end{cases}$

Substitute in Equation 2

$y = \ln(x^2) + 1$

$\ln(x^2) + 1 - 1 = \ln(x + 12)$

$\ln(x^2) = \ln(x + 12)$

$x^2 = x + 12$

$x^2 - x - 12 = 0$

$(x - 4)(x + 3) = 0$

$x = 4$ or $x = -3;$

If $x = 4, y = \ln(4^2) + 1$

$y = \ln 16 + 1;$

If $x = -3, y = \ln\left((-3)^2\right) + 1$

$y = \ln 9 + 1;$

Solutions: $(-3, \ln 9 + 1), (4, \ln 16 + 1)$

29. $\begin{cases} y - 9 = e^{2x} \\ 3 = y - 7e^x \end{cases}$

$\begin{cases} y - 9 = e^{2x} \\ -y + 3 = -7e^x \end{cases}$

$R1 + R2$

$-6 = e^{2x} - 7e^x$

$0 = e^{2x} - 7e^x + 6$

$(e^x - 6)(e^x - 1) = 0$

$e^x = 6$ or $e^x = 1$

$\ln e^x = \ln 6$ or $\ln e^x = \ln 1$

$x = \ln 6$ or $x = 0$

If $x = \ln 6, y - 9 = e^{2x}$

$y - 9 = e^{2\ln 6}$

$y - 9 = e^{\ln 36}$

$y - 9 = 36$

$y = 45;$

If $x = 0, y - 9 = e^{2(0)}$

$y - 9 = e^0$

$y - 9 = 1$

$y = 10;$

Solutions: $(0, 10), (\ln 6, 45)$

31. $\begin{cases} y = 4^{x+3} \\ y - 2^{x^2+3x} = 0 \end{cases}$

$$4^{x+3} = 2^{x^2+3x}$$
$$2^{2x+6} = 2^{x^2+3x}$$
$$2x + 6 = x^2 + 3x$$
$$0 = x^2 + x - 6$$
$$(x+3)(x-2) = 0$$
$$x = -3 \text{ or } x = 2;$$
$$y = 4^{-3+3} = 1;$$
$$y = 4^{2+3} = 1024;$$
Solutions: $(-3,1),(2,1024)$

33. $\begin{cases} x^3 - y = 2x \\ y - 5x = -6 \end{cases}$

$$y = 5x - 6$$
$$x^3 - 5x + 6 = 2x$$
$$x^3 - 7x + 6 = 0$$

Possible rational roots: $\dfrac{\pm 1, \pm 6, \pm 2, \pm 3}{\pm 1}$

$$\{\pm 1, \pm 6, \pm 2, \pm 3\};$$
$$(x+3)(x-2)(x-1) = 0$$
$$x = -3 \text{ or } x = 2 \text{ or } x = 1;$$
$$y = 5(-3) - 6 = -21;$$
$$y = 5(2) - 6 = 4;$$
$$y = 5(1) - 6 = -1;$$
Solutions: $(-3,-21),(2,4),(1,-1)$

35. $\begin{cases} x^2 - 6x = y - 4 \\ y - 2x = -8 \end{cases}$

Solve for y in Equation 2.
$y = 2x - 8$, Substitute in Equation 1.
$$x^2 - 6x = 2x - 8 - 4$$
$$x^2 - 8x + 12 = 0$$
$$(x-6)(x-2) = 0$$
$$x - 6 = 0 \text{ or } x - 2 = 0$$
$$x = 6 \quad \text{ or } x = 2;$$
If $x = 6, y - 2x = -8$
$$y - 2(6) = -8$$
$$y = 4;$$
If $x = 2, y - 2x = -8$
$$y - 2(2) = -8$$
$$y = -4;$$
Solutions: $(2,-4),(6,4)$

37. $\begin{cases} x^2 + y^2 = 34 \\ y^2 + (x-3)^2 = 25 \end{cases}$

Solve for y in Equation 1.
$$y^2 = -x^2 + 34$$
$$y = \pm\sqrt{-x^2 + 34};$$
Solve for y in Equation 2.
$$y^2 + (x-3)^2 = 25$$
$$y^2 = -(x-3)^2 + 25$$
$$y = \pm\sqrt{-(x-3)^2 + 25};$$
Using a graphing calculator:
Solutions: $(3,5),(3,-5)$

39. $\begin{cases} y = 2^x - 3 \\ y + 2x^2 = 9 \end{cases}$

$$y1 = 2^x - 3;$$
$$y2 = -2x^2 + 9;$$
Solutions: $(-2.43, -2.81),(2,1)$

41. $\begin{cases} y = \dfrac{1}{(x-3)^2} + 2 \\ (x-3)^2 + y^2 = 10 \end{cases}$

$$y1 = \dfrac{1}{(x-3)^2} + 2;$$
$$y^2 = -(x-3)^2 + 10$$
$$y2 = \pm\sqrt{-(x-3)^2 + 10};$$
Solutions:
$(0.72, 2.19),(2, 3),(4, 3),(5.28, 2.19)$

43. $\begin{cases} y - x^2 \geq 1 & \text{parabola} \\ x + y \leq 3 & \text{line} \end{cases}$

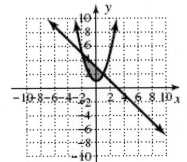

45. $\begin{cases} x^2 + y^2 > 16 & \text{circle} \\ x^2 + y^2 \le 64 & \text{circle} \end{cases}$

Inequality 1, circle with center (0,0), radius 4.
Inequality 2, circle with center (0,0), radius 8.

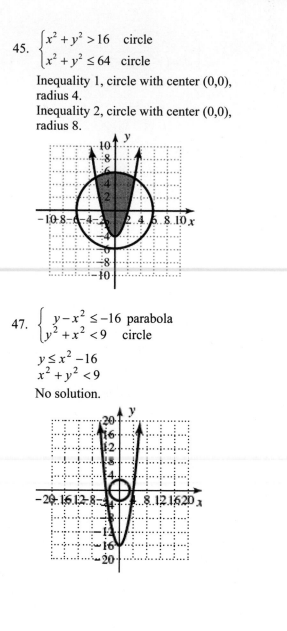

47. $\begin{cases} y - x^2 \le -16 & \text{parabola} \\ y^2 + x^2 < 9 & \text{circle} \end{cases}$

$y \le x^2 - 16$
$x^2 + y^2 < 9$

No solution.

49. $\begin{cases} y^2 + x^2 \le 25 & \text{circle} \\ |x| - 1 > -y & \text{absolute value} \end{cases}$

Inequality 1, circle with center (0,0), radius 5.
Inequality 2, absolute value with vertex (0,1).

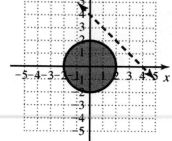

51. $h = \sqrt{r^2 - d^2}$;
If $r = 50, d = 20$
$h = \sqrt{50^2 - 20^2}$
$h \approx 45.8$ ft.;
If $r = 50, d = 30$
$h = \sqrt{50^2 - 30^2}$
$h = 40$ ft.;
If $r = 50, d = 40$
$h = \sqrt{50^2 - 40^2}$
$h = 30$ ft

53. $C(x) = 2.5x^2 - 120x + 3500$
$R(x) = -2x^2 + 180x - 500$
$C(x) = R(x)$

$2.5x^2 - 120x + 3500 = -2x^2 + 180x - 500$
$4.5x^2 - 300x + 4000 = 0$
Using a graphing calculator,
$x \approx 18.426$ or $x \approx 48.241$
The company breaks even if either 18,400 or 48,200 cars are sold.

55. a. $8P^2 - 8P - 4D = 12$

$-4D = -8P^2 + 8P + 12$

$D = 2P^2 - 2P - 3$

minimum: \$1.83 (when $D = 0$)

 b. $\begin{cases} 10P^2 + 6D = 144 \\ 8P^2 - 8P - 4D = 12 \end{cases}$

$10P^2 + 6D = 144$

$\dfrac{144 - 10P^2}{6} = D$;

$8P^2 - 8P - 4D = 12$

$8P^2 - 8P - 4\left(\dfrac{144 - 10P^2}{6}\right) = 12$

$2P^2 - 2P - \left(\dfrac{144 - 10P^2}{6}\right) = 3$

$12P^2 - 12P - 144 + 10P^2 = 18$

$22P^2 - 12P - 162 = 0$

$11P^2 - 6P - 81 = 0$

$(11P + 27)(P - 3) = 0$

$P = -\dfrac{27}{11}$ or $P = \$3$;

$10(3)^2 + 6D = 144$

$6D = 54$

$D = 9$

90,000 gallons

57. $85 = lw$

$37 = 2l + 2w$;

$\dfrac{85}{l} = w$

$37 = 2l + 2\left(\dfrac{85}{l}\right)$

$37l = 2l^2 + 170$

$0 = 2l^2 - 37l + 170$

$0 = (2l - 17)(l - 10)$

$l = \dfrac{17}{2}, l = 10$;

$w = \dfrac{85}{\left(\dfrac{17}{2}\right)} = 10$;

$w = \dfrac{85}{(10)} = 8.5$;

$8.5 \text{ m} \times 10 \text{ m}$

59. Area : 45km^2

Diagonal : $\sqrt{106}$ km

$45 = lw$

$l = \dfrac{45}{w}$

$l^2 + w^2 = 106$

$\left(\dfrac{45}{w}\right)^2 + w^2 = 106$

$2025 + w^4 = 106w^2$

$w^4 - 106w^2 + 2025 = 0$

$\left(w^2 - 25\right)\left(w^2 - 81\right) = 0$

$w = \pm 5, \pm 9$

$5 \text{ km}, 9 \text{ km}$

61. Surface Area $= 928\,\text{ft}^2$

    Edges $= 164\,\text{ft}$

    $$4w + 4l + 4w = 164$$
    $$4l + 8w = 164$$
    $$4l = 164 - 8w$$
    $$l = 41 - 2w;$$

    $$928 = 4lw + 2w^2$$
    $$928 = 4(41 - 2w)w + 2w^2$$
    $$928 = 164w - 8w^2 + 2w^2$$
    $$6w^2 - 164w + 928 = 0$$
    $$3w^2 - 82w + 464 = 0$$
    $$(3w - 58)(w - 8) = 0$$

    $$w = \frac{58}{3} \ \text{ or } \ w = 8;$$

    w=8, $l = 41 - 2(8) = 25$

    8 ft x 8 ft x 25 ft

63. Answers will vary.

65. Height : 18 inches

    Surface Area : $4806\,\text{in}^2$

    $$4806 = lw + 2(18l) + 2(18w)$$
    $$4806 = lw + 36l + 36w$$
    $$4806 - 36l = lw + 36w$$
    $$4806 - 36l = w(l + 36)$$
    $$\frac{4806 - 36l}{l + 36} = w;$$

    $$108(231) = 18(l)w$$
    $$108(231) = 18l\left(\frac{4806 - 36l}{l + 36}\right)$$

    $$24948(l + 36) = 18l(4806 - 36l)$$

    $$24948l + 898128 = 86508l - 648l^2$$

    $$648l^2 - 61560l + 898128 = 0$$

    $$l^2 - 95l + 1386 = 0$$
    $$(l - 18)(l - 77) = 0$$

    18 in. x 18 in. x 77 in.

67. a. $3x^2 + 4x - 12 = 0$

       $a = 3, b = 4, c = -12$

       $$x = \frac{-4 \pm \sqrt{16 - 4(3)(-12)}}{6}$$

       $$x = \frac{-4 \pm \sqrt{160}}{6}$$

       $$x = \frac{-2 \pm 2\sqrt{10}}{3}$$

    b. $\sqrt{3x + 1} - \sqrt{2x} = 1$

       $\sqrt{3x + 1} = 1 + \sqrt{2x}$

       $\left(\sqrt{3x + 1}\right)^2 = \left(1 + \sqrt{2x}\right)^2$

       $3x + 1 = 1 + 2\sqrt{2x} + 2x$

       $x = 2\sqrt{2x}$

       $(x)^2 = \left(2\sqrt{2x}\right)^2$

       $x^2 = 4(2x)$

       $x^2 - 8x = 0$

       $x(x - 8) = 0$

       $x = 0, x = 8$

    c. $\dfrac{1}{x + 2} + \dfrac{3}{x^2 + 5x + 6} = \dfrac{2}{x + 3}$

       $\dfrac{1}{x + 2} + \dfrac{3}{(x + 2)(x + 3)} = \dfrac{2}{x + 3}$

       $(x + 2)(x + 3)\left[\dfrac{1}{x + 2} + \dfrac{3}{(x + 2)(x + 3)} = \dfrac{2}{x + 3}\right]$

       $x + 3 + 3 = 2(x + 2)$

       $x + 6 = 2x + 4$

       $2 = x$

       $x = 2$

69. a. Let 2001 be year 0.

       $$m = \frac{\Delta \text{value}}{\Delta \text{time}} = \frac{4500 - 3300}{0 - 3} = -400$$

       The computer depreciates by $400 a year.

    b. $y = -400x + 4500$

    c. $y = -400(7) + 4500 = \$1700$

    d. $700 = -400x + 4500$

       $-3800 = -400x$

       $9.5 = x$

       9.5 years

5.4 Technology Highlight

Exercise 1:

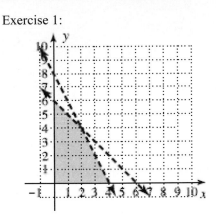

Exercise 3:

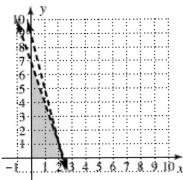

5.4 Exercises

1. Half planes

3. Solution

5. The feasible region may be bordered by three or more oblique lines, with two of them intersecting outside and away from the feasible region.

7. $2x + y > 3$

(0, 0)   No
$$2x + y > 3$$
$$2(0) + 0 > 3$$
$$0 + 0 > 3$$
$$0 > 3;$$

(3, −5)   No
$$2x + y > 3$$
$$2(3) + (−5) > 3$$
$$6 − 5 > 3$$
$$1 > 3;$$

(−3, −4)   No
$$2x + y > 3$$
$$2(−3) + (−4) > 3$$
$$−6 − 4 > 3$$
$$−10 > 3;$$

(−3, 9)   No
$$2x + y > 3$$
$$2(−3) + 9 > 3$$
$$−6 + 9 > 3$$
$$3 > 3$$

9. $4x − 2y \le −8$

(0, 0)   No
$$4x − 2y \le −8$$
$$4(0) − 2(0) \le −8$$
$$0 − 0 \le −8$$
$$0 \le 0,$$

(−3, 5)   Yes
$$4x − 2y \le −8$$
$$4(−3) − 2(5) \le −8$$
$$−12 − 10 \le −8$$
$$−22 \le −8;$$

(−3, −2)   Yes
$$4x − 2y \le −8$$
$$4(−3) − 2(−2) \le −8$$
$$−12 + 4 \le −8$$
$$−8 \le −8;$$

(−1, 1)   No
$$4x − 2y \le −8$$
$$4(−1) − 2(1) \le −8$$
$$−4 − 2 \le −8$$
$$−6 \le −8$$

11. $x + 2y < 8$
$$2y < −x + 8$$
$$y < −\frac{1}{2}x + 4$$

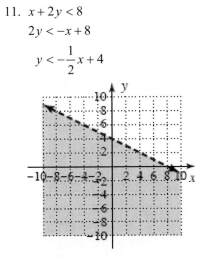

13. $2x - 3y \geq 9$

$-3y \geq -2x + 9$

$y \leq \dfrac{2}{3}x - 3$

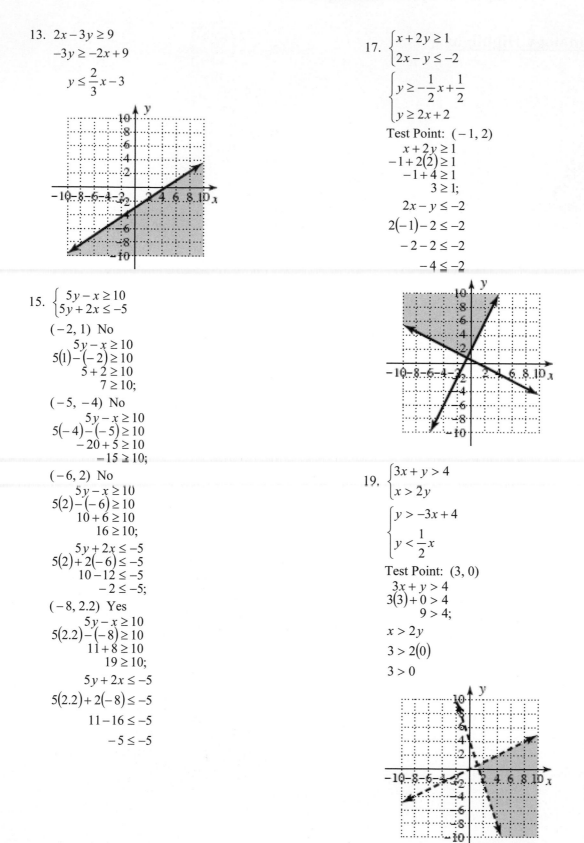

15. $\begin{cases} 5y - x \geq 10 \\ 5y + 2x \leq -5 \end{cases}$

$(-2, 1)$ No

$5y - x \geq 10$
$5(1) - (-2) \geq 10$
$5 + 2 \geq 10$
$7 \geq 10;$

$(-5, -4)$ No

$5y - x \geq 10$
$5(-4) - (-5) \geq 10$
$-20 + 5 \geq 10$
$-15 \geq 10;$

$(-6, 2)$ No

$5y - x \geq 10$
$5(2) - (-6) \geq 10$
$10 + 6 \geq 10$
$16 \geq 10;$

$5y + 2x \leq -5$
$5(2) + 2(-6) \leq -5$
$10 - 12 \leq -5$
$-2 \leq -5;$

$(-8, 2.2)$ Yes

$5y - x \geq 10$
$5(2.2) - (-8) \geq 10$
$11 + 8 \geq 10$
$19 \geq 10;$

$5y + 2x \leq -5$
$5(2.2) + 2(-8) \leq -5$
$11 - 16 \leq -5$
$-5 \leq -5$

17. $\begin{cases} x + 2y \geq 1 \\ 2x - y \leq -2 \end{cases}$

$\begin{cases} y \geq -\dfrac{1}{2}x + \dfrac{1}{2} \\ y \geq 2x + 2 \end{cases}$

Test Point: $(-1, 2)$

$x + 2y \geq 1$
$-1 + 2(2) \geq 1$
$-1 + 4 \geq 1$
$3 \geq 1;$

$2x - y \leq -2$
$2(-1) - 2 \leq -2$
$-2 - 2 \leq -2$
$-4 \leq -2$

19. $\begin{cases} 3x + y > 4 \\ x > 2y \end{cases}$

$\begin{cases} y > -3x + 4 \\ y < \dfrac{1}{2}x \end{cases}$

Test Point: $(3, 0)$

$3x + y > 4$
$3(3) + 0 > 4$
$9 > 4;$

$x > 2y$
$3 > 2(0)$
$3 > 0$

21. $\begin{cases} 2x+y<4 \\ 2y>3x+6 \end{cases}$

$\begin{cases} y<-2x+4 \\ y>\dfrac{3}{2}x+3 \end{cases}$

Test Point: $(-3,3)$

$2x+y<4$
$2(-3)+3<4$
$-6+3<4$
$-3<4;$

$2y>3x+6$
$2(3)>3(-3)+6$
$6>-9+6$
$6>-3$

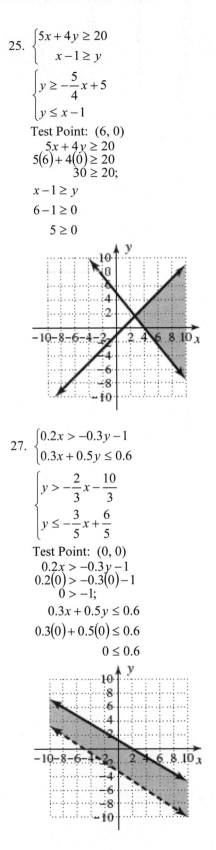

23. $\begin{cases} x>-3y-2 \\ x+3y\le6 \end{cases}$

$\begin{cases} y>-\dfrac{1}{3}x-\dfrac{2}{3} \\ y\le-\dfrac{1}{3}x+2 \end{cases}$

Test Point: $(0,0)$

$x>-3y-2$
$0>-3(0)-2$
$0>-2;$

$x+3y\le6$
$0+3(0)\le6$
$0\le6$

25. $\begin{cases} 5x+4y\ge20 \\ x-1\ge y \end{cases}$

$\begin{cases} y\ge-\dfrac{5}{4}x+5 \\ y\le x-1 \end{cases}$

Test Point: $(6,0)$

$5x+4y\ge20$
$5(6)+4(0)\ge20$
$30\ge20;$

$x-1\ge y$
$6-1\ge0$
$5\ge0$

27. $\begin{cases} 0.2x>-0.3y-1 \\ 0.3x+0.5y\le0.6 \end{cases}$

$\begin{cases} y>-\dfrac{2}{3}x-\dfrac{10}{3} \\ y\le-\dfrac{3}{5}x+\dfrac{6}{5} \end{cases}$

Test Point: $(0,0)$

$0.2x>-0.3y-1$
$0.2(0)>-0.3(0)-1$
$0>-1;$

$0.3x+0.5y\le0.6$
$0.3(0)+0.5(0)\le0.6$
$0\le0.6$

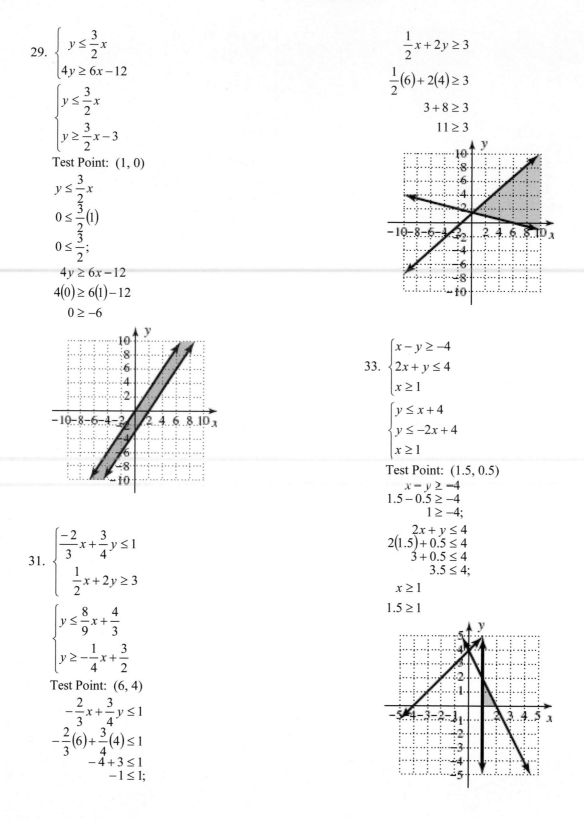

29. $\begin{cases} y \le \dfrac{3}{2}x \\ 4y \ge 6x - 12 \end{cases}$

$\begin{cases} y \le \dfrac{3}{2}x \\ y \ge \dfrac{3}{2}x - 3 \end{cases}$

Test Point: $(1, 0)$

$y \le \dfrac{3}{2}x$

$0 \le \dfrac{3}{2}(1)$

$0 \le \dfrac{3}{2}$;

$4y \ge 6x - 12$

$4(0) \ge 6(1) - 12$

$0 \ge -6$

$\dfrac{1}{2}x + 2y \ge 3$

$\dfrac{1}{2}(6) + 2(4) \ge 3$

$3 + 8 \ge 3$

$11 \ge 3$

31. $\begin{cases} \dfrac{-2}{3}x + \dfrac{3}{4}y \le 1 \\ \dfrac{1}{2}x + 2y \ge 3 \end{cases}$

$\begin{cases} y \le \dfrac{8}{9}x + \dfrac{4}{3} \\ y \ge -\dfrac{1}{4}x + \dfrac{3}{2} \end{cases}$

Test Point: $(6, 4)$

$-\dfrac{2}{3}x + \dfrac{3}{4}y \le 1$

$-\dfrac{2}{3}(6) + \dfrac{3}{4}(4) \le 1$

$-4 + 3 \le 1$

$-1 \le 1$;

33. $\begin{cases} x - y \ge -4 \\ 2x + y \le 4 \\ x \ge 1 \end{cases}$

$\begin{cases} y \le x + 4 \\ y \le -2x + 4 \\ x \ge 1 \end{cases}$

Test Point: $(1.5, 0.5)$

$x - y \ge -4$

$1.5 - 0.5 \ge -4$

$1 \ge -4$;

$2x + y \le 4$

$2(1.5) + 0.5 \le 4$

$3 + 0.5 \le 4$

$3.5 \le 4$;

$x \ge 1$

$1.5 \ge 1$

35. $\begin{cases} y \le x+3 \\ x+2y \le 4 \\ y \ge 0 \end{cases}$

$\begin{cases} y \le x+3 \\ y \le -\dfrac{1}{2}x+2 \\ y \ge 0 \end{cases}$

Test Point: $(1, 1)$
$y \le x+3$
$1 \le 1+3$
$1 \le 4;$

$x+2y \le 4$
$1+2(1) \le 4$
$1+2 \le 4$
$3 \le 4;$

$y \ge 0$
$1 \ge 0$

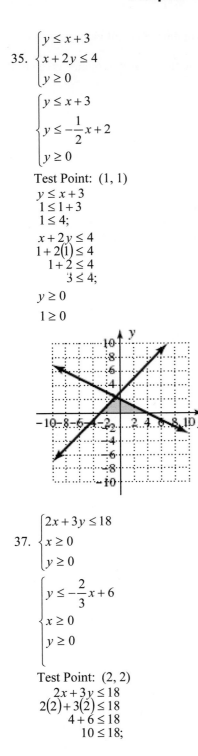

37. $\begin{cases} 2x+3y \le 18 \\ x \ge 0 \\ y \ge 0 \end{cases}$

$\begin{cases} y \le -\dfrac{2}{3}x+6 \\ x \ge 0 \\ y \ge 0 \end{cases}$

Test Point: $(2, 2)$
$2x+3y \le 18$
$2(2)+3(2) \le 18$
$4+6 \le 18$
$10 \le 18;$

$\begin{array}{l} x \ge 0 \\ 2 \ge 0; \\ y \ge 0 \\ 2 \ge 0 \end{array}$

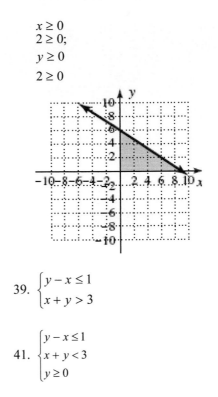

39. $\begin{cases} y-x \le 1 \\ x+y > 3 \end{cases}$

41. $\begin{cases} y-x \le 1 \\ x+y < 3 \\ y \ge 0 \end{cases}$

43.

Point	Objective Function $f(x,y)=12x+10y$	Result
(0, 0)	$f(0,0)=12(0)+10(0)$	0
(0, 8.5)	$f(0,8.5)=12(0)+10(8.5)$	85
(7, 0)	$f(7,0)=12(7)+10(0)$	84
(5, 3)	$f(5,3)=12(5)+10(3)$	90

Maximum value occurs at $(5, 3)$.

45.

Point	Objective Function $f(x,y)=8x+15y$	Result
(0, 20)	$f(0,20)=8(0)+15(20)$	300
(35, 0)	$f(35,0)=8(35)+15(0)$	280
(5, 15)	$f(5,15)=8(5)+15(15)$	265
(12, 11)	$f(12,11)=8(12)+15(11)$	261

Minimum value occurs at $(12, 11)$.

5.4 Exercises

47. $\begin{cases} x + 2y \le 6 \\ 3x + y \le 8 \\ x \ge 0 \\ y \ge 0 \end{cases}$

$\begin{cases} y \le -\dfrac{1}{2}x + 3 \\ y \le -3x + 8 \\ x \ge 0 \\ y \ge 0 \end{cases}$

Corner Point	Objective Function $f(x, y) = 8x + 5y$	Result
$(0, 0)$	$f(0,0) = 8(0) + 5(0)$	0
$(0, 3)$	$f(0,3) = 8(0) + 5(3)$	15
$\left(\dfrac{8}{3}, 0\right)$	$f\left(\dfrac{8}{3}, 0\right) = 8\left(\dfrac{8}{3}\right) + 5(0)$	$\dfrac{64}{3}$
$(2, 2)$	$f(2,2) = 8(2) + 5(2)$	26

Maximum value: $(2, 2)$

49. $\begin{cases} 3x + 2y \ge 18 \\ 3x + 4y \ge 24 \\ x \ge 0 \\ y \ge 0 \end{cases}$

$\begin{cases} y \ge -\dfrac{3}{2}x + 9 \\ y \ge -\dfrac{3}{4}x + 6 \\ x \ge 0 \\ y \ge 0 \end{cases}$

Corner Point	Objective Function $f(x, y) = 36x + 40y$	Result
$(0, 9)$	$f(0,9) = 36(0) + 40(9)$	360
$(4, 3)$	$f(4,3) = 36(4) + 40(3)$	264
$(8, 0)$	$f(8,0) = 36(8) + 40(0)$	288

Minimum value: $(4, 3)$

51. $\begin{cases} 20H < 200 \\ \dfrac{1}{2}(20)H > 50 \\ H > 0 \end{cases}$

$20H < 200$

$H < 10;$

$10H > 50$

$H > 5;$

$5 < H < 10$

53. Let J represent the amount of money given to Julius.
Let A represent the amount of money given to Anthony.

$\begin{cases} J + A \le 50000 \\ J \ge 20000 \\ A \le 25000 \end{cases}$

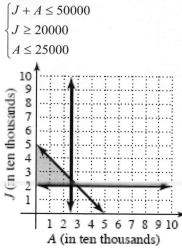

55. Let C represent the number of acres of corn.
Let S represent the number of acres of soybeans.

$\begin{cases} C + S \le 500 \\ 3C + 2S \le 1300 \end{cases}$

$P = 900C + 800S$

$\begin{cases} S \le -C + 500 \\ 2S \le -3C + 1300 \end{cases}$

$\begin{cases} S \le -C + 500 \\ S \le \dfrac{-3}{2}C + 650 \end{cases}$

Using a grapher, the corner points are:

$(0, 500), \left(433\dfrac{1}{3}, 0\right), (300, 200)$

$P = 900(0) + 800(500) = 400{,}000;$

$P = 900\left(433\dfrac{1}{3}\right) + 800(0) = 390{,}000;$

$P = 900(300) + 800(200) = 430{,}000;$

300 acres of corn, 200 acres of soybeans

57. Let x represent the number of sheet metal screws.
Let y represent the number of wood screws.

$$\begin{cases} 20x + 5y \le 3(60)(60) \\ 15x + 15y \le 3(60)(60) \\ 5x + 20y \le 3(60)(60) \end{cases}$$

$R = 0.10x + 0.12y$

$$\begin{cases} 5y \le -20x + 10800 \\ 15y \le -15x + 10800 \\ 20y \le -5x + 10800 \end{cases}$$

$$\begin{cases} y \le -4x + 2160 \\ y \le -1x + 720 \\ y \le \dfrac{-1}{4}x + 540 \end{cases}$$

Using a grapher, the corner points are:
$(0,540),(240,480),(480,240),(540,0)$

$R = 0.10(0) + 0.12(540) = 64.80;$

$R = 0.10(240) + 0.12(480) = 81.60;$

$R = 0.10(480) + 0.12(240) = 76.80;$

$R = 0.10(540) + 0.12(0) = 54;$

240 sheet metal screws; 480 wood screws

59. Let t represent the number of ounces of traditional sandwiches.
Let d represent the number of ounces of Double-T's.

$$\begin{cases} 2t + 4d \le 250 \\ 3t + 5d \le 345 \\ t \ge 0 \\ d \ge 0 \end{cases}$$

$R = 2t + 3.50d$

$$\begin{cases} 2t \le -4d + 250 \\ 3t \le -5d + 345 \end{cases}$$

$$\begin{cases} t \le -2d + 125 \\ t \le -\dfrac{5}{3}d + 115 \end{cases}$$

Using a grapher, the pt of intersection is $(30,65)$.

$R = 2(65) + 3.50(30) = 235$

65 traditionals, 30 Double-T's.

61. Let A represent the number of thousand gallons shipped from OK to CO.
Let B represent the number of thousand gallons shipped from OK to MS.
Let C represent the number of thousand gallons shipped from TX to CO.
Let D represent the number of thousand gallons shipped from TX to MS.

Cost $= 0.05A + 0.075C + 0.06B + 0.065D$

$$\begin{cases} A + C = 220 \Rightarrow C = 220 - A \\ B + D = 250 \Rightarrow D = 250 - B \end{cases}$$

Cost
$= 0.05A + 0.075(220 - A) + 0.06B + 0.065(250 - B)$
$= 0.05A + 16.5 - 0.75A + 0.06B + 16.25 - 0.065B$
$= -0.25A - 0.005B + 32.75 ;$

$A + B \le 320;$
$C + D \le 240$
but $C = 220 - A$ and $D = 250 - B$
Thus,
$220 - A + 250 - B \le 240$
$-A - B \le -230$
$A + B \ge 230;$
$A \ge 0, B \ge 0, C \ge 0, D \ge 0$
$220 - A \ge 0, 250 - B \ge 0$
$A \le 220, B \le 250;$

$$\begin{cases} A + B \le 320 \\ A + B \ge 230 \\ A \le 220 \\ B \le 250 \end{cases}$$

$$\begin{cases} B \le -A + 320 \\ B \ge -A + 230 \\ A \le 220 \\ B \le 250 \end{cases}$$

Using a grapher, the corner points are:
$(220,100),(220,10),(70,250),(0,250)$

Cost $= -0.7(220) - 0.005(100) + 32.75 = -121.75$

Cost $= -0.7(220) - 0.005(10) + 32.75 = -121.30;$

Cost $= -0.7(70) - 0.005(250) + 32.75 = -17.5;$

Cost $= -0.7(0) - 0.005(250) + 32.75 = 31.5$

$A = 220, B = 100,$
$C = 220 - A = 0, D = 250 - B = 150;$

220,000 gallons from OK to CO,
100,000 gallons from OK to MS,
0 thousand gallons from TX to CO,
150,000 gallons from TX to MS

5.4 Exercises

63. $\begin{cases} x \geq 0 \\ y \geq 0 \\ y \leq 3 \\ x \leq 3 \end{cases}$

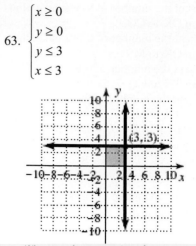

The graph is a rectangle.

Corner Point	Objective Function $f(x,y) = 4.5x + 7.2y$	Result
(0, 0)	$f(0,0) = 4.5(0) + 7.2(0)$	0
(0, 3)	$f(0,3) = 4.5(0) + 7.2(3)$	21.6
(3, 3)	$f(3,3) = 4.5(3) + 7.2(3)$	35.1
(3, 0)	$f(3,0) = 4.5(3) + 7.2(0)$	13.5

Maximum value: (3, 3)
Optimal solutions occur at vertices.

65. $x^3 - 5x^2 + 3x - 15 = 0$
$(x^3 - 5x^2) + (3x - 15) = 0$
$x^2(x - 5) + 3(x - 5) = 0$
$(x - 5)(x^2 + 3) = 0$
$x^2 + 3 = 0$
$x^2 = -3$
$x = \pm\sqrt{3}i$
$x = 5; \quad x = \pm\sqrt{3}i$

67. $r = \dfrac{kl}{d^2}$

$1500 = \dfrac{k(8)}{(0.004)^2}$

$0.024 = 8k$

$0.003 = k$

$r = \dfrac{0.003l}{d^2}$

$r = \dfrac{0.003(2.7)}{(0.005)^2}$

$r = \dfrac{0.0081}{0.000025}$

$r = 324 \ \Omega$

<u>Chapter 5 Summary and Review</u>

1. $\begin{cases} 3x - 2y = 4 \\ -x + 3y = 8 \end{cases}$

$\begin{cases} y = \dfrac{3}{2}x - 2 \\ y = \dfrac{1}{3}x + \dfrac{8}{3} \end{cases}$

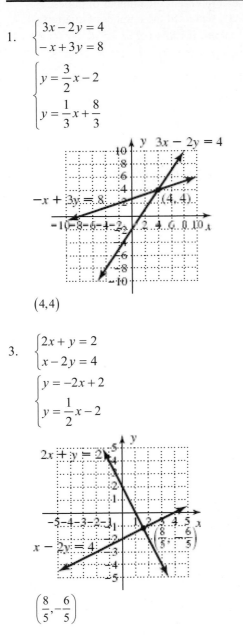

$(4, 4)$

3. $\begin{cases} 2x + y = 2 \\ x - 2y = 4 \end{cases}$

$\begin{cases} y = -2x + 2 \\ y = \dfrac{1}{2}x - 2 \end{cases}$

$\left(\dfrac{8}{5}, -\dfrac{6}{5} \right)$

5. $\begin{cases} x + y = 4 \\ 0.4x + 0.3y = 1.7 \end{cases}$

$x = 4 - y$

$0.4x + 0.3y = 1.7$
$0.4(4 - y) + 0.3y = 1.7$
$1.6 - 0.4y + 0.3y = 1.7$
$\qquad\qquad -0.1y = 0.1$
$\qquad\qquad\qquad y = -1;$

$\qquad x + y = 4$
$\qquad x + (-1) = 4$
$\qquad\quad x - 1 = 4$
$\qquad\qquad x = 5$
$(5, -1)$; consistent

7. $\begin{cases} 2x - 4y = 10 \\ 3x + 4y = 5 \end{cases}$

R1 + R2 = Sum
$2x - 4y = 10$
$\underline{3x + 4y = 5}$
$5x = 15$
$\quad x = 3;$
$\quad 2x - 4y = 10$
$\quad 2(3) - 4y = 10$
$\qquad 6 - 4y = 10$
$\qquad\quad -4y = 4$
$\qquad\qquad y = -1$
$(3, -1)$; consistent

9. $\begin{cases} 2x = 3y + 6 \\ 2.4x + 3.6y = 6 \end{cases}$

$\begin{cases} 2x - 3y = 6 \\ 24x + 36y = 60 \end{cases}$

12R1 + R2 = Sum
$24x - 36y = 72$
$\underline{24x + 36y = 60}$
$48x = 132$
$\quad x = \dfrac{11}{4};$

Chapter 5 Summary and Review

$2x = 3y + 6$

$2\left(\dfrac{11}{4}\right) = 3y + 6$

$\dfrac{11}{2} = 3y + 6$

$-\dfrac{1}{2} = 3y$

$-\dfrac{1}{6} = y$

$\left(\dfrac{11}{4}, -\dfrac{1}{6}\right)$; consistent

11. $\begin{cases} x + y - 2z = -1 \\ 4x - y + 3z = 3 \\ 3x + 2y - z = 4 \end{cases}$

R1 + R2 = Sum

$x + y - 2z = -1$

$4x - y + 3z = 3$

$\overline{5x + z = 2}$

-2R1 + R3 = Sum

$-2x - 2y + 4z = 2$

$3x + 2y - z = 4$

$\overline{x + 3z = 6}$

$\begin{cases} 5x + z = 2 \\ x + 3z = 6 \end{cases}$

-3R1 + R2 = Sum

$-15x - 3z = -6$

$x + 3z = 6$

$\overline{-14x = 0}$

$x = 0$;

$x + 3z = 6$

$0 + 3z = 6$

$3z = 6$

$z = 2$;

$x + y - 2z = -1$

$0 + y - 2(2) = -1$

$y - 4 = -1$

$y = 3$

$(0, 3, 2)$

13. $\begin{cases} 3x + y + 2z = 3 \\ x - 2y + 3z = 1 \\ 4x - 8y + 12z = 7 \end{cases}$

-4R2 + R3 = Sum

$-4x + 8y - 12z = -4$

$4x - 8y + 12z = 7$

$\overline{0 \ne 3}$

No solution; inconsistent

15. Let n represent the number of nickels.
Let d represent the number of dimes.
Let q represent the number of quarters.

$\begin{cases} 0.05n + 0.10d + 0.25q = 536 \\ n + d = 360 + q \\ q = 110 + 2n \end{cases}$

$\begin{cases} 0.05n + 0.10d + 0.25q = 536 \\ n + d - q = 360 \\ -2n + q = 110 \end{cases}$

R1 $-$ 0.10R2 = Sum

$-0.05n + 0.35q = 500$

$\begin{cases} -0.05n + 0.35q = 500 \\ -2n + q = 110 \end{cases}$

R1 $-$ 0.35R2 = Sum

$0.65n = 461.50$

$n = 710$;

$-2n + q = 110$

$-2(710) + q = 110$

$q = 1530$;

$n + d = 360 + q$

$710 + d = 360 + 1530$

$710 + d = 1890$

$d = 1180$;

710 nickels, 1180 dimes, 1530 quarters

17. $\begin{cases} x = y^2 - 1 & \text{Parabola} \\ x + 4y = -5 & \text{Line} \end{cases}$

$y^2 - 1 + 4y = -5$

$y^2 + 4y + 4 = 0$

$(y + 2)^2 = 0$

$y = -2$;

$x = (-2)^2 - 1 = 3$

Solution: $(3, -2)$

19. $\begin{cases} x^2 + y^2 = 10 & \text{Circle} \\ y - 3x^2 = 0 & \text{Parabola} \end{cases}$

$x^2 = \dfrac{y}{3}$

$\dfrac{y}{3} + y^2 = 10$

$3y^2 + y - 30 = 0$

$(3y + 10)(y - 3) = 0$

$3y + 10 = 0 \quad \text{or} \quad y - 3 = 0$

$y \neq -\dfrac{10}{3} \quad \text{or} \quad y = 3$

Solutions: $(1,3),(-1,3)$

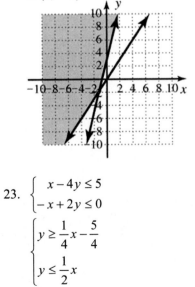

21. $\begin{cases} x^2 + y^2 > 9 \\ x^2 + y \le -3 \end{cases}$

Inequality 1 is a circle with center (0,0), radius 3.
Inequality 2 is a parabola with vertex
(0, − 3)
Note the open circle showing non-inclusion
at (0, − 3)

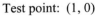

23. $\begin{cases} x - 4y \le 5 \\ -x + 2y \le 0 \end{cases}$

$\begin{cases} y \ge \dfrac{1}{4}x - \dfrac{5}{4} \\ y \le \dfrac{1}{2}x \end{cases}$

Test point: (1, 0)

$x - 4y \le 5$
$1 - 4(0) \le 5$
$1 \le 5$
$-x + 2y \le 0$
$-1 + 2(0) \le 0$
$-1 \le 0$

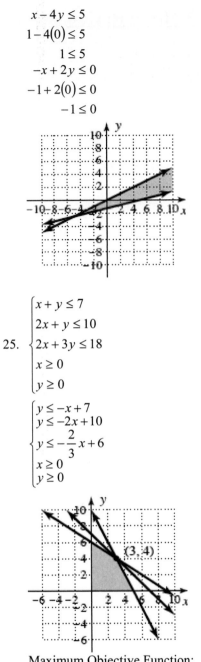

25. $\begin{cases} x + y \le 7 \\ 2x + y \le 10 \\ 2x + 3y \le 18 \\ x \ge 0 \\ y \ge 0 \end{cases}$

$\begin{cases} y \le -x + 7 \\ y \le -2x + 10 \\ y \le -\dfrac{2}{3}x + 6 \\ x \ge 0 \\ y \ge 0 \end{cases}$

Maximum Objective Function:
$f(x,y) = 30x + 45y$

Corner Point	Objective Function $f(x,y) = 30x + 45y$	Result
(0, 0)	$f(0,0) = 30(0) + 45(0)$	0
(0, 6)	$f(0,6) = 30(0) + 45(6)$	270
(3, 4)	$f(3,4) = 30(3) + 45(4)$	270
(5, 0)	$f(5,0) = 30(5) + 45(0)$	150

Maximum value: (3, 4) and (0, 6)

Chapter 5 Mixed Review

1. a. $\begin{cases} -3x+5y=10 \\ 6x+20=10y \end{cases}$

$\begin{cases} y=\dfrac{3}{5}x+2 \\ y=\dfrac{3}{5}x+2 \end{cases}$

Consistent/dependent

b. $\begin{cases} 4x-3y=9 \\ -2x+5y=-10 \end{cases}$

$\begin{cases} y=\dfrac{4}{3}x-3 \\ y=\dfrac{2}{5}x-2 \end{cases}$

Consistent/independent

c. $\begin{cases} x-3y=9 \\ -6y+2x=10 \end{cases}$

$\begin{cases} y=\dfrac{1}{3}x-3 \\ y=\dfrac{1}{3}x-\dfrac{5}{3} \end{cases}$

Inconsistent

3. $\begin{cases} 2x+3y=5 \\ -x+5y=17 \end{cases}$

$x=5y-17$

$2x+3y=5$

$2(5y-17)+3y=5$

$10y-34+3y=5$

$13y=39$

$y=3;$

$-x+5y=17$

$-x+5(3)=17$

$-x+15=17$

$-x=2$

$x=-2$

$(-2,3)$

5. Let v represent the number of veggie burritos sold.
Let b represent the number of beef burritos sold.

$\begin{cases} 2.45v+2.95b=148.80 \\ v+b=54 \end{cases}$

$v=54-b;$

$2.45(54-b)+2.95b=148.80$

$132.30-2.45b+2.95b=148.80$

$0.50b=16.50$

$b=33;$

$v=54-b=54-33=21;$

21 veggie, 33 beef

7. $\begin{cases} 0.1x-0.2y+z=1.7 \\ 0.3x+y-0.1z=3.6 \\ -0.2x-0.1y+0.2z=-1.7 \end{cases}$

$\begin{cases} x-2y+10z=17 \\ 3x+10y-z=36 \\ -2x-y+2z=-17 \end{cases}$

$-3R1+R2=$ Sum

$-3x+6y-30z=-51$

$\underline{3x+10y-z=36}$

$16y-31z=-15$

$2R1+R3=$ Sum

$2x-4y+20z=34$

$\underline{-2x-y+2z=-17}$

$-5y+22z=17$

$\begin{cases} 16y-31z=-15 \\ -5y+22z=17 \end{cases}$

$5R1+16R2=$ Sum

$80y-155z=-75$

$\underline{-80y+352z=272}$

$197z=197$

$z=1;$

$-5y+22z=17$

$-5y+22(1)=17$

$-5y+22=17$

$-5y=-5$

$y=1;$

$x-2y+10z=17$

$x-2(1)+10(1)=17$

$x-2+10=17$

$x+8=17$

$x=9$

$(9,1,1)$

9. $\begin{cases} x - 2y + 3z = 4 \\ 2x + y - z = 1 \\ 5x \quad + z = 6 \end{cases}$

R1 + 2R2 = Sum

$5x + z = 6$

$\begin{cases} 5x + z = 6 \\ 5x + z = 6 \end{cases}$

Dependent

$z = -5x + 6;$

Equation 1

$x - 2y + 3(-5x + 6) = 4$

$x - 2y - 15x + 18 = 4$

$-2y - 14x + 18 = 4$

$-2y = 14x - 14$

$y = -7x + 7;$

$\{(x, y, z) | x \in \mathbb{R}, y = -7x + 7, z = -5x + 6\}$

11. $\begin{cases} 2x + y \le 4 \\ x - 3y > 6 \end{cases}$

$\begin{cases} y \le -2x + 4 \\ -3y > -x + 6 \end{cases}$

$\begin{cases} y \le -2x + 4 \\ y < \dfrac{1}{3}x - 2 \end{cases}$

Inequality 1, line with y-intercept $(0, 4)$, slope -2.

Inequality 2, line with y-intercept $(0, -2)$,

slope $\dfrac{1}{3}$.

13. $\begin{cases} x - 2y \ge 5 \\ x \le 2y \end{cases}$

$\begin{cases} -2y \ge -x + 5 \\ 2y \ge x \end{cases}$

$\begin{cases} y \le \dfrac{1}{2}x - \dfrac{5}{2} \\ y \ge \dfrac{1}{2}x \end{cases}$

Inequality 1, line with y-intercept $\left(0, \dfrac{5}{2}\right)$,

slope $\dfrac{1}{2}$.

Inequality 2, line with y-intercept $(0,0)$,

slope $\dfrac{1}{2}$.

15. $P(x,y) = 2.5x + 3.75y$

$$\begin{cases} x + y \le 8 \\ x + 2y \le 14 \\ 4x + 3y \le 30 \\ x, y \ge 0 \end{cases}$$

$$\begin{cases} y \le -x + 8 \\ y \le -\dfrac{1}{2}x + 7 \\ y \le -\dfrac{4}{3}x + 10 \\ x, y \ge 0 \end{cases}$$

Corner Point	Objective Function $P(x,y) = 2.5x + 3.75y$	Result
(0, 0)	$P(0,0) = 2.5(0) + 3.75(0)$	0
(0, 7)	$P(0,7) = 2.5(0) + 3.75(7)$	26.25
(7.5, 0)	$P(7.5, 0)$ $= 2.5(7.5) + 3.75(0)$	18.75
(2, 6)	$P(2,6) = 2.5(2) + 3.75(6)$	27.5
(6, 2)	$P(6,2) = 2.5(6) + 3.75(2)$	22.5

Maximum value occurs at (2, 6): 27.5

17. $\begin{cases} 4x^2 - y^2 = -9 \\ x^2 + 3y^2 = 79 \end{cases}$

$\begin{cases} 12x^2 - 3y^2 = -27 \\ x^2 + 3y^2 = 79 \end{cases}$

$13x^2 = 52$

$x^2 = 4$

$x = \pm 2$;

$4(\pm 2)^2 + 9 = y^2$

$25 = y^2$

$y = \pm 5$;

$(2,5), (2,-5), (-2,5), (-2,-5)$

19. $\begin{cases} x + y > 1 \\ x^2 + y^2 \ge 16 \end{cases}$

Solve for y in Inequality 1.

$\begin{cases} y > -x + 1 \\ x^2 + y^2 \ge 16 \end{cases}$

Inequality 1 is a line with y-intercept $(0,1)$, slope -1.

Inequality 2 is a circle with center $(0,0)$, radius 4.

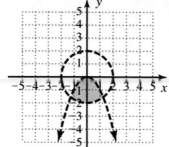

Chapter 6 Practice Test

1. $\begin{cases} 3x + 2y = 12 \\ -x + 4y = 10 \end{cases}$

$\begin{cases} y = -\dfrac{3}{2}x + 6 \\ y = \dfrac{1}{4}x + \dfrac{5}{2} \end{cases}$

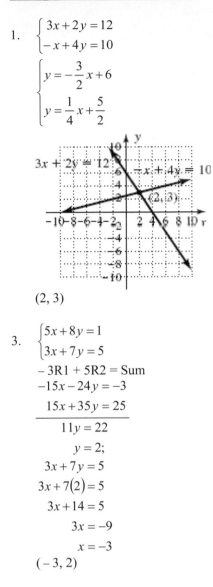

(2, 3)

3. $\begin{cases} 5x + 8y = 1 \\ 3x + 7y = 5 \end{cases}$

$-3R1 + 5R2 = \text{Sum}$

$-15x - 24y = -3$

$\underline{15x + 35y = 25}$

$11y = 22$

$y = 2;$

$3x + 7y = 5$

$3x + 7(2) = 5$

$3x + 14 = 5$

$3x = -9$

$x = -3$

$(-3, 2)$

5. $\begin{cases} 2x - y + z = 4 \\ -x \quad\quad + 2z = 1 \\ x - 2y + 8z = 11 \end{cases}$

$-2R1 + R3 = \text{Sum}$

$-3x + 6z = 3$

$\dfrac{R1}{-3}$

$x - 2z = -1$

$\begin{cases} x - 2z = -1 \\ -x + 2z = 1 \end{cases}$

$R1 + R2 = \text{Sum}$

$0 = 0;$

Dependent

$x + 2z = -1$

$-x = -2z + 1$

$x = 2z - 1;$

$2x - y + z = 4$

$2(2z - 1) - y + z = 4$

$4z - 2 - y + z = 4$

$-y = -5z + 6$

$y = 5z - 6;$

$\left\{ (x, y, z) \,\middle|\, x = 2z - 1, y = 5z - 6, z \in \mathbb{R} \right\}$

7. Let l represent the length of the paper.
Let w represent the width of the paper.

$\begin{cases} 2l + 2w = 114.3 \\ l = 2w - 7.62 \end{cases}$

$2l + 2w = 114.3$

$2(2w - 7.62) + 2w = 114.3$

$4w - 15.24 + 2w = 114.3$

$6w = 129.54$

$w = 21.59;$

$l = 2w - 7.62$

$l = 2(21.59) - 7.62$

$l = 43.18 - 7.62$

$l = 35.56$

21.59 cm by 35.56 cm

8. Let h represent the land area of Tahiti.
   Let n represent the land area of Tonga.
   $$\begin{cases} h+n=692 \\ h=n+112 \end{cases}$$
   $$h+n=692$$
   $$n+112+n=692$$
   $$2n+112=692$$
   $$2n=580$$
   $$n=290;$$
   $$h=290+112$$
   $$h=402$$
   Tahiti is 402 mi^2
   Tonga is 290 mi^2

11. $$\begin{cases} x-y\le 2 \\ x+2y\ge 8 \end{cases}$$
    $$\begin{cases} y\ge x-2 \\ y\ge -\dfrac{1}{2}x+4 \end{cases}$$

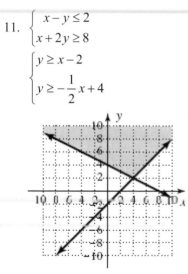

13. $P(x,y)=4.25x+5y$
    $$\begin{cases} x+y\le 50 \\ 2x+3y\le 120 \end{cases}$$

Corner Point	Objective Function $P(x,y)=4.25x+5y$	Result
(0, 0)	$P(0,0)=4.25(0)+5(0)$	0
(0, 40)	$P(0,40)=4.25(0)+5(40)$	200
(30, 20)	$P(30,20)=4.25(30)+5(20)$	227.5
(50, 0)	$P(50,0)=4.25(50)+5(0)$	212.50

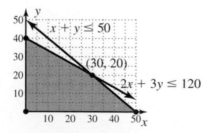

30 plain, 20 deluxe

15. $$\begin{cases} 4y-x^2=1 \\ y^2+x^2=4 \end{cases}$$
    R1 + R2
    $$y^2+4y=5$$
    $$y^2+4y-5=0$$
    $$(y+5)(y-1)=0$$
    $$y+5=0 \text{ or } y-1=0$$
    $$y=-5 \text{ or } y=1;$$
    If $y=-5, 4y-x^2=1$
    $$4(-5)-x^2=1$$
    $$-20-x^2=1$$
    $$x^2=-21 \text{ not real};$$
    If $y=1, 4y-x^2=1$
    $$4(1)-x^2=1$$
    $$4-x^2=1$$
    $$x^2=3$$
    $$x=\pm\sqrt{3};$$
    $$\left(\sqrt{3},1\right),\left(-\sqrt{3},1\right)$$

17. $$\begin{cases} x^2-y\le 2 \\ x-y^2\ge -2 \end{cases}$$
    $$\begin{cases} -y\le -x^2+2 \\ x\ge y^2-2 \end{cases}$$
    $$\begin{cases} y\ge x^2-2 \\ x\ge y^2-2 \end{cases}$$
    Inequality 1 is a parabola, vertex $(0,-2)$, opens up;
    Inequality 2 is a parabola, vertex $(-2,0)$, opens right.

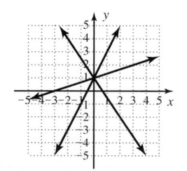

19. $\begin{cases} 2x - y \le -1 \\ 3x + 2y \ge 2 \\ x - 3y \ge -3 \end{cases}$

$\begin{cases} -y \le -2x - 1 \\ 2y \ge -3x + 2 \\ -3y \ge -x - 3 \end{cases}$

$\begin{cases} y \ge 2x + 1 \\ y \ge -\dfrac{3}{2}x + 1 \\ y \le \dfrac{1}{3}x + 1 \end{cases}$

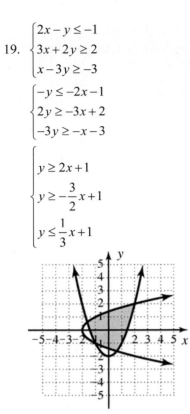

Calculator Exploration and Discovery

Exercise 1: $\begin{cases} 2x + 2y = 15 \\ x + y = 6 \\ x + 4y = 9 \\ x, y \ge 0 \end{cases}$

$(5, 1)$

Strengthening Core Skills

Exercise 1:

(b) $\begin{cases} 2x + y = 2 \\ 4x + 3y = 8 \end{cases}$

$-2R1$

$\begin{cases} -4x - 2y = -4 \\ 4x + 3y = 8 \end{cases}$

R1 + R2

$y = 4;$

$2x + y = 2$

$2x + 4 = 2$

$2x = -2$

$x = -1;$

$(-1, 4)$

(c) $\begin{cases} 2x + y = 2 \\ 4x + 3y = 8 \end{cases}$

$\begin{cases} y = -2x + 2 \\ 4x + 3y = 8 \end{cases}$

$4x + 3(-2x + 2) = 8$

$4x - 6x + 6 = 8$

$-2x = 2$

$x = -1;$

$y = -2x + 2$

$y = -2(-1) + 2$

$y = 2 + 2$

$y = 4;$

$(-1, 4)$

Elimination

Chapter 1-5 Cumulative Review

1. $y = \frac{2}{3}x + 2$

   x-intercept:

   $0 = \frac{2}{3}x + 2$

   $-2 = \frac{2}{3}x$

   $-3 = x$

   $(-3, 0)$

   y-intercept:

   $y = \frac{2}{3}(0) + 2$

   $y = 2$

   $(0, 2)$

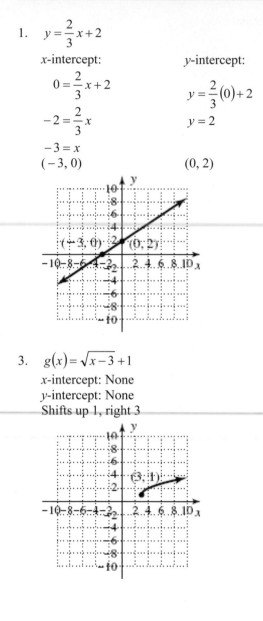

3. $g(x) = \sqrt{x-3} + 1$

   x-intercept: None
   y-intercept: None
   Shifts up 1, right 3

5. $g(x) = (x-3)(x+1)(x+4)$

   x-intercepts: $(3, 0)$, $(-1, 0)$, $(-4, 0)$
   y-intercept: $(0, -12)$

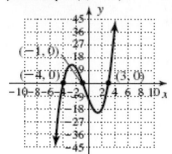

7. a) Domain: $x \in (-\infty, \infty)$
   b) Range: $y \in (-\infty, 4]$
   c) $f(x) \uparrow : (-\infty, -1)$
      $f(x) \downarrow : (-1, \infty)$
   d) N/A
   e) Maximum: $(-1, 4)$
   f) $f(x) > 0 : (-4, 2)$
      $f(x) < 0 : (-\infty, -4) \cup (2, \infty)$
   g) $\frac{7}{4}$

9. $g(v) = v^3 - 9v^2 + 2v - 18$

   $g(v) = v^2(v-9) + 2(v-9)$

   $g(v) = (v-9)(v^2 + 2);$

   $v - 9 = 0$ $v^2 + 2 = 0$
   $v = 9$ $v^2 = -2$
                        $v = \pm\sqrt{2}i$

   Zeroes: $9, \pm\sqrt{2}i$

11. $f(x) = 2x - 5; \quad g(x) = 3x^2 + 2x$

    $(g - f)(x) = g(x) - f(x)$

    $\qquad = 3x^2 + 2x - 2x + 5$

    $\qquad = 3x^2 + 5$

13. $f(x) = 2x - 5; \quad g(x) = 3x^2 + 2x$

$(g \circ f)(2) = g(f(2));$

$f(2) = 2(2) - 5 = 4 - 5 = -1;$

$g(-1) = 3(-1)^2 + 2(-1) = 3 - 2 = 1$

15. $x^4 - 6x^3 - 13x^2 + 24x + 36$

Possible roots:

$$\frac{\{\pm 1, \pm 36, \pm 2, \pm 18, \pm 3, \pm 12, \pm 4, \pm 9, \pm 6\}}{\{\pm 1\}}$$

$$\begin{array}{r|rrrrr}
2 & 1 & -6 & -13 & 24 & 36 \\
  &   & 2 & -8 & -42 & -36 \\
\hline
  & 1 & -4 & -21 & -18 & \boxed{0}
\end{array}$$

$$\begin{array}{r|rrrr}
-2 & 1 & -4 & -21 & -18 \\
   &   & -2 & 12 & 18 \\
\hline
   & 1 & -6 & -9 & \boxed{0}
\end{array}$$

$x^2 - 6x - 9$

$a = 1, b = -6, c = -9$

$$x = \frac{6 \pm \sqrt{(-6)^2 - 4(1)(-9)}}{2(1)}$$

$$x = \frac{6 \pm \sqrt{72}}{2}$$

$$x = \frac{6 \pm 6\sqrt{2}}{2}$$

$x = 3 \pm 3\sqrt{2}$

$(x-2)(x+2)(x-3-3\sqrt{2})(x-3+3\sqrt{2})$

17. $\dfrac{x-2}{x+3} \le 3$

Vertical asymptote: $x = -3$

$$\frac{x-2}{x+3} - 3 \le 0$$

$$\frac{x-2-3(x+3)}{x+3} \le 0$$

$$\frac{-2x-11}{x+3} \le 0$$

$x = -\dfrac{11}{2}; \quad x = -3$

$$\xrightarrow[\quad -\frac{11}{2} \qquad -3 \quad]{\text{neg} \qquad \text{pos} \qquad \text{neg}}$$

$x \in \left(-\infty, -\dfrac{11}{2}\right] \cup (-3, \infty)$

19. $A = Pe^{rt}$

$12000 = 5000e^{0.09t}$

$2.4 = e^{0.09t}$

$\ln 2.4 = \ln e^{0.09t}$

$\ln 2.4 = 0.09t$

$\dfrac{\ln 2.4}{0.09} = t$

$t \approx 9.7$ years

21. $h(x) = \dfrac{9 - x^2}{x^2 - 4} = \dfrac{(3-x)(3+x)}{(x-2)(x+2)}$

Vertical asymptotes: $x = -2, x = 2$

x-intercepts: $(3, 0), (-3, 0)$

y-intercept: $(0, -2.25)$

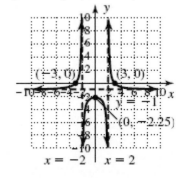

23. Let a represent the number of arrows.
Let b represent the number of bowling balls.
Let c represent the number of cricket bats.

$$\begin{cases} a+b+c=120 \\ a=2(b+c) \\ b=c+10 \end{cases}$$

$$\begin{cases} a+b+c=120 \\ a-2b-2c=0 \\ b-c=10 \end{cases}$$

$-R1+R2$

$-3b-3c=-120$

$\dfrac{R1}{-3}$

$b+c=40$

Sub-system

$$\begin{cases} b+c=40 \\ b-c=10 \end{cases}$$

$2b=50$

$b=25$;

$b+c=40$

$25+c=40$

$c=15$;

$a+b+c=120$

$a+25+15=120$

$a=80$;

80 arrows, 25 balls, 15 bats

25. $\begin{cases} y=\log x+4 \\ y=5-\log(x-3) \end{cases}$

$\log x+4=5-\log(x-3)$

$\log x+\log(x-3)=1$

$\log x(x-3)=1$

$10^1=x(x-3)$

$10=x^2-3x$

$0=x^2-3x-10$

$0=(x-5)(x+2)$

$x-5=0$ or $x+2=0$

$x=5$ or $x=-2$;

If $x=5, y=\log x+4$

$y=\log 5+4$

$y\approx 4.7$;

If $x=-2, y=\log x+4$

$y=\log(-2)+4$

not defined;

$x=5, y\approx 4.7$

6.1 Technology Highlight

Exercise 1: (10, 12)

6.1 Exercises

1. Square

3. 2 by 3, 1

5. Multiply R_1 by -2 and add that result to R_2. This sum will be the new R_2.

7. $\begin{bmatrix} 1 & 0 \\ 2.1 & 1 \\ -3 & 5.8 \end{bmatrix}$

   3×2, 5.8

9. $\begin{bmatrix} 1 & 0 & 4 \\ 1 & 3 & -7 \\ 5 & -1 & 2 \\ 2 & -3 & 9 \end{bmatrix}$

   4×3, -1

11. $\begin{cases} x + 2y - z = 1 \\ x + z = 3 \\ 2x - y + z = 3 \end{cases}$

   $\begin{bmatrix} 1 & 2 & -1 & \vdots & 1 \\ 1 & 0 & 1 & \vdots & 3 \\ 2 & -1 & 1 & \vdots & 3 \end{bmatrix}$

   Diagonal entries 1, 0, 1

13. $\begin{bmatrix} 1 & 4 & \vline & 5 \\ 0 & 1 & \vline & \frac{1}{2} \end{bmatrix}$

   $\begin{cases} x + 4y = 5 \\ y = \frac{1}{2} \end{cases}$

   $x + 4y = 5$

   $x + 4\left(\frac{1}{2}\right) = 5$

   $x + 2 = 5$

   $x = 3$

   $\left(3, \frac{1}{2}\right)$

15. $\begin{bmatrix} 1 & 2 & -1 & \vline & 0 \\ 0 & 1 & 2 & \vline & 2 \\ 0 & 0 & 1 & \vline & 3 \end{bmatrix}$

   $\begin{cases} x + 2y - z = 0 \\ y + 2z = 2 \\ z = 3 \end{cases}$

   $y + 2z = 2$

   $y + 2(3) = 2$

   $y + 6 = 2$

   $y = -4;$

   $x + 2y - z = 0$

   $x + 2(-4) - (3) = 0$

   $x - 8 - 3 = 0$

   $x - 11 = 0$

   $x = 11$

   $(11, -4, 3)$

17. $\begin{bmatrix} 1 & 3 & -4 & \vline & 29 \\ 0 & 1 & -\frac{3}{2} & \vline & \frac{21}{2} \\ 0 & 0 & 1 & \vline & 3 \end{bmatrix}$

   $\begin{cases} x + 3y - 4z = 29 \\ y - \frac{3}{2}z = \frac{21}{2} \\ z = 3 \end{cases}$

   $y - \frac{3}{2}z = \frac{21}{2}$

   $y - \frac{3}{2}(3) = \frac{21}{2}$

   $y - \frac{9}{2} = \frac{21}{2}$

   $y = \frac{30}{2}$

   $y = 15;$

   $x + 3y - 4z = 29$

   $x + 3(15) - 4(3) = 29$

   $x + 45 - 12 = 29$

   $x + 33 = 29$

   $x = -4$

   $(-4, 15, 3)$

19. $\begin{bmatrix} \frac{1}{2} & -3 & -1 \\ -5 & 2 & 4 \end{bmatrix}$ $2R1 \to R1$

$\begin{bmatrix} 1 & -6 & -2 \\ -5 & 2 & 4 \end{bmatrix}$ $5R1 + R2 \to R2$

$\begin{bmatrix} 1 & -6 & -2 \\ 0 & -28 & -6 \end{bmatrix}$

21. $\begin{bmatrix} -2 & 1 & 0 & 4 \\ 5 & 8 & 3 & -5 \\ 1 & -3 & 3 & 2 \end{bmatrix}$ $R1 \leftrightarrow R3$

$\begin{bmatrix} 1 & -3 & 3 & 2 \\ 5 & 8 & 3 & -5 \\ -2 & 1 & 0 & 4 \end{bmatrix}$ $-5R1 + R2 \to R2$

$\begin{bmatrix} 1 & -3 & 3 & 2 \\ 0 & 23 & -12 & -15 \\ -2 & 1 & 0 & 4 \end{bmatrix}$

23. $\begin{bmatrix} 3 & 1 & 1 & 8 \\ 6 & -1 & -1 & 10 \\ 4 & -2 & -3 & 22 \end{bmatrix}$ $-2R1 + R2 \to R2$

$\begin{bmatrix} 3 & 1 & 1 & 8 \\ 0 & -3 & -3 & -6 \\ 4 & -2 & -3 & 34 \end{bmatrix}$ $-4R1 + 3R3 \to R3$

$\begin{bmatrix} 3 & 1 & 1 & 8 \\ 0 & -3 & -3 & -6 \\ 0 & -10 & -13 & 34 \end{bmatrix}$

25. $\begin{bmatrix} 1 & 3 & 0 & 2 \\ -2 & 4 & 1 & 1 \\ 3 & -1 & -2 & 9 \end{bmatrix}$

$2R1 + R2 \to R2$

$-3R1 + R3 \to R3$

27. $\begin{bmatrix} 1 & 2 & 0 & 10 \\ 5 & 1 & 2 & 6 \\ -4 & 3 & -3 & 2 \end{bmatrix}$

$-5R1 + R2 \to R2$

$4R1 + R3 \to R3$

29. $\begin{cases} 0.15g - 0.35h = -0.5 \\ -0.12g + 0.25h = 0.1 \end{cases}$

$\begin{cases} 15g - 35h = -50 \\ -12g + 25h = 10 \end{cases}$

$\begin{bmatrix} 15 & -35 & -50 \\ -12 & 25 & 10 \end{bmatrix}$ $\frac{1}{15}R1 \to R1$

$\begin{bmatrix} 1 & -\frac{7}{3} & -\frac{10}{3} \\ -12 & 25 & 10 \end{bmatrix}$ $12R1 + R2 \to R2$

$\begin{bmatrix} 1 & -\frac{7}{3} & -\frac{10}{3} \\ 0 & -3 & -30 \end{bmatrix}$ $-\frac{1}{3}R2 \to R2$

$\begin{bmatrix} 1 & -\frac{7}{3} & -\frac{10}{3} \\ 0 & 1 & 10 \end{bmatrix}$

$h = 10$;

$0.15g - 0.35h = -0.5$

$0.15g - 0.35(10) = -0.5$

$0.15g - 3.5 = -0.5$

$0.15g = 3$

$g = 20$

$(20, 10)$

31. $\begin{cases} x - 2y + 2z = 7 \\ 2x + 2y - z = 5 \\ 3x - y + z = 6 \end{cases}$

$\begin{bmatrix} 1 & -2 & 2 & 7 \\ 2 & 2 & -1 & 5 \\ 3 & -1 & 1 & 6 \end{bmatrix}$ $-2R1 + R2 \rightarrow R2$

$\begin{bmatrix} 1 & -2 & 2 & 7 \\ 0 & 6 & -5 & -9 \\ 3 & -1 & 1 & 6 \end{bmatrix}$ $-3R1 + R3 \rightarrow R3$

$\begin{bmatrix} 1 & -2 & 2 & 7 \\ 0 & 6 & -5 & -9 \\ 0 & 5 & -5 & -15 \end{bmatrix}$ $-\dfrac{5}{6}R2 + R3 \rightarrow R3$

$\begin{bmatrix} 1 & -2 & 2 & 7 \\ 0 & 6 & -5 & -9 \\ 0 & 0 & -\dfrac{5}{6} & -\dfrac{15}{2} \end{bmatrix}$ $-\dfrac{6}{5}R3 \rightarrow R3$

$\begin{bmatrix} 1 & -2 & 2 & 7 \\ 0 & 6 & -5 & -9 \\ 0 & 0 & 1 & 9 \end{bmatrix}$

$z = 9$;
$6y - 5z = -9$
$6y - 5(9) = -9$
$6y - 45 = -9$
$6y = 36$
$y = 6;$
$x - 2y + 2z = 7$
$x - 2(6) + 2(9) = 7$
$x - 12 + 18 = 7$
$x + 6 = 7$
$x = 1$
$(1, 6, 9)$

33. $\begin{cases} x + 2y - z = 1 \\ x + z = 3 \\ 2x - y + z = 3 \end{cases}$

$\begin{bmatrix} 1 & 2 & -1 & 1 \\ 1 & 0 & 1 & 3 \\ 2 & -1 & 1 & 3 \end{bmatrix}$ $-R1 + R2 \rightarrow R2$

$\begin{bmatrix} 1 & 2 & -1 & 1 \\ 0 & -2 & 2 & 2 \\ 2 & -1 & 1 & 3 \end{bmatrix}$ $-2R1 + R3 \rightarrow R3$

$\begin{bmatrix} 1 & 2 & -1 & 1 \\ 0 & -2 & 2 & 2 \\ 0 & -5 & 3 & 1 \end{bmatrix}$ $-\dfrac{5}{2}R2 + R3 \rightarrow R3$

$\begin{bmatrix} 1 & 2 & -1 & 1 \\ 0 & -2 & 2 & 2 \\ 0 & 0 & -2 & -4 \end{bmatrix}$ $-\dfrac{1}{2}R3 \rightarrow R3$

$\begin{bmatrix} 1 & 2 & -1 & 1 \\ 0 & -2 & 2 & 2 \\ 0 & 0 & 1 & 2 \end{bmatrix}$

$z = 2$;
$x + z = 3$
$x + 2 = 3$
$x = 1;$
$x + 2y - z = 1$
$1 + 2y - 2 = 1$
$2y - 1 = 1$
$2y = 2$
$y = 1$
$(1, 1, 2)$

35. $\begin{cases} -x+y+2z=2 \\ x+y-z=1 \\ 2x+y+z=4 \end{cases}$

$\begin{bmatrix} -1 & 1 & 2 & | & 2 \\ 1 & 1 & -1 & | & 1 \\ 2 & 1 & 1 & | & 4 \end{bmatrix}$ $R1 \leftrightarrow R2$

$\begin{bmatrix} 1 & 1 & -1 & | & 1 \\ -1 & 1 & 2 & | & 2 \\ 2 & 1 & 1 & | & 4 \end{bmatrix}$ $R1+R2 \to R2$

$\begin{bmatrix} 1 & 1 & -1 & | & 1 \\ 0 & 2 & 1 & | & 3 \\ 2 & 1 & 1 & | & 4 \end{bmatrix}$ $-2R1+R3 \to R3$

$\begin{bmatrix} 1 & 1 & -1 & | & 1 \\ 0 & 2 & 1 & | & 3 \\ 0 & -1 & 3 & | & 2 \end{bmatrix}$ $\dfrac{1}{2}R2+R3 \to R3$

$\begin{bmatrix} 1 & 1 & -1 & | & 1 \\ 0 & 2 & 1 & | & 3 \\ 0 & 0 & \frac{7}{2} & | & \frac{7}{2} \end{bmatrix}$ $\dfrac{2}{7}R3 \to R3$

$\begin{bmatrix} 1 & 1 & -1 & | & 1 \\ 0 & 2 & 1 & | & 3 \\ 0 & 0 & 1 & | & 1 \end{bmatrix}$

$z=1$;
$2y+z=3$
$2y+1=3$
$2y=2$
$y=1$;
$x+y-z=1$
$x+1-1=1$
$x=1$
$(1,1,1)$

37. $\begin{cases} 4x-8y+8z=24 \\ 2x-6y+3z=13 \\ 3x+4y-z=-11 \end{cases}$

$\begin{bmatrix} 4 & -8 & 8 & | & 24 \\ 2 & -6 & 3 & | & 13 \\ 3 & 4 & -1 & | & -11 \end{bmatrix}$ $\dfrac{1}{4}R1 \to R1$

$\begin{bmatrix} 1 & -2 & 2 & | & 6 \\ 2 & -6 & 3 & | & 13 \\ 3 & 4 & -1 & | & -11 \end{bmatrix}$ $-2R1+R2 \to R2$

$\begin{bmatrix} 1 & -2 & 2 & | & 6 \\ 0 & -2 & -1 & | & 1 \\ 3 & 4 & -1 & | & -11 \end{bmatrix}$ $-3R1+R3 \to R3$

$\begin{bmatrix} 1 & -2 & 2 & | & 6 \\ 0 & -2 & -1 & | & 1 \\ 0 & 10 & -7 & | & -29 \end{bmatrix}$ $5R2+R3 \to R3$

$\begin{bmatrix} 1 & -2 & 2 & | & 6 \\ 0 & -2 & -1 & | & 1 \\ 0 & 0 & -12 & | & -24 \end{bmatrix}$ $-\dfrac{1}{12}R3 \to R3$

$\begin{bmatrix} 1 & -2 & 2 & | & 6 \\ 0 & -2 & -1 & | & 1 \\ 0 & 0 & 1 & | & 2 \end{bmatrix}$

$z=2$;
$-2y-z=1$
$-2y-2=1$
$-2y=3$
$y=-\dfrac{3}{2}$;
$3x+4y-z=-11$
$3x+4\left(-\dfrac{3}{2}\right)-2=-11$
$3x-6-2=-11$
$3x-8=-11$
$3x=-3$
$x=-1$
$\left(-1, \dfrac{-3}{2}, 2\right)$

39. $\begin{cases} x+3y+5z=20 \\ 2x+3y+4z=16 \\ x+2y+3z=12 \end{cases}$

$\begin{bmatrix} 1 & 3 & 5 & | & 20 \\ 2 & 3 & 4 & | & 16 \\ 1 & 2 & 3 & | & 12 \end{bmatrix}$ $R2-R1 \to R2$

$\begin{bmatrix} 1 & 3 & 5 & | & 20 \\ 1 & 0 & -1 & | & -4 \\ 1 & 2 & 3 & | & 12 \end{bmatrix}$ $R3-R1 \to R3$

$\begin{bmatrix} 1 & 3 & 5 & | & 20 \\ 1 & 0 & -1 & | & -4 \\ 0 & -1 & -2 & | & -8 \end{bmatrix}$ $R2 \leftrightarrow R3$

$\begin{bmatrix} 1 & 3 & 5 & | & 20 \\ 0 & -1 & -2 & | & -8 \\ 1 & 0 & -1 & | & -4 \end{bmatrix}$ $3R2+R1 \to R1$

$\begin{bmatrix} 1 & 0 & -1 & | & -4 \\ 0 & -1 & -2 & | & -8 \\ 1 & 0 & -1 & | & -4 \end{bmatrix}$ $-R1+R3 \to R3$

$\begin{bmatrix} 1 & 0 & -1 & | & -4 \\ 0 & -1 & -2 & | & -8 \\ 0 & 0 & 0 & | & 0 \end{bmatrix}$

$x-z=-4$

$x=z-4;$

$-y-2z=-8$

$-y=2z-8$

$y=-2z+8;$

Linear dependence; $\left(p-4, -2p+8, p \right)$

41. $\begin{cases} 3x-4y+2z=-2 \\ \dfrac{3}{2}x-2y+z=-1 \\ -6x+8y-4z=4 \end{cases}$

$\begin{bmatrix} 3 & -4 & 2 & | & -2 \\ \dfrac{3}{2} & -2 & 1 & | & -1 \\ -6 & 8 & -4 & | & 4 \end{bmatrix}$ $\dfrac{1}{3}R1 \to R1$

$\begin{bmatrix} 1 & -\dfrac{4}{3} & \dfrac{2}{3} & | & -\dfrac{2}{3} \\ \dfrac{3}{2} & -2 & 1 & | & -1 \\ -6 & 8 & -4 & | & 4 \end{bmatrix}$ $-\dfrac{3}{2}R1+R2 \to R2$

$\begin{bmatrix} 1 & -\dfrac{4}{3} & \dfrac{2}{3} & | & -\dfrac{2}{3} \\ 0 & 0 & 0 & | & 0 \\ -6 & 8 & -4 & | & 4 \end{bmatrix}$ $6R1+R3 \to R3$

$\begin{bmatrix} 1 & -\dfrac{4}{3} & \dfrac{2}{3} & | & -\dfrac{2}{3} \\ 0 & 0 & 0 & | & 0 \\ 0 & 0 & 0 & | & 0 \end{bmatrix}$ $3R1 \to R1$

$\begin{bmatrix} 3 & -4 & 2 & | & -2 \\ 0 & 0 & 0 & | & 0 \\ 0 & 0 & 0 & | & 0 \end{bmatrix}$

Coincident dependence;

$\left\{ (x,y,z) \mid 3x-4y+2z=-2 \right\}$

43. $\begin{cases} 2x-y+3z=1 \\ 2y+6z=2 \\ x-\dfrac{1}{2}y+\dfrac{3}{2}z=5 \end{cases}$

In terms of z:

$\begin{bmatrix} 2 & -1 & 3 & | & 1 \\ 0 & 2 & 6 & | & 2 \\ 1 & -\dfrac{1}{2} & \dfrac{3}{2} & | & 5 \end{bmatrix}$ $R1 \leftrightarrow R3$

$\begin{bmatrix} 1 & -\dfrac{1}{2} & \dfrac{3}{2} & | & 5 \\ 0 & 2 & 6 & | & 2 \\ 2 & -1 & 3 & | & 1 \end{bmatrix}$ $-2R1+R3 \to R3$

$\begin{bmatrix} 1 & -\dfrac{1}{2} & \dfrac{3}{2} & | & 5 \\ 0 & 2 & 6 & | & 2 \\ 0 & 0 & 0 & | & -9 \end{bmatrix}$

$0 \neq -9$

No solution

45. $\begin{cases} -2x+4y-3z = 4 \\ 5x-6y+7z = -12 \\ x+2y+z = -4 \end{cases}$

In terms of z:

$$\begin{bmatrix} -2 & 4 & -3 & | & 4 \\ 5 & -6 & 7 & | & -12 \\ 1 & 2 & 1 & | & -4 \end{bmatrix} R1 \leftrightarrow R3$$

$$\begin{bmatrix} 1 & 2 & 1 & | & -4 \\ 5 & -6 & 7 & | & -12 \\ -2 & 4 & -3 & | & 4 \end{bmatrix} -5R1 + R2 \leftrightarrow R2$$

$$\begin{bmatrix} 1 & 2 & 1 & | & -4 \\ 0 & -16 & 2 & | & 8 \\ -2 & 4 & -3 & | & 4 \end{bmatrix} 2R1 + R3 \to R3$$

$$\begin{bmatrix} 1 & 2 & 1 & | & -4 \\ 0 & -16 & 2 & | & 8 \\ 0 & 8 & -1 & | & -4 \end{bmatrix} -\frac{1}{16}R2 \to R2$$

$$\begin{bmatrix} 1 & 2 & 1 & | & -4 \\ 0 & 1 & -\frac{1}{8} & | & -\frac{1}{2} \\ 0 & 8 & -1 & | & -4 \end{bmatrix} -8R2 + R3 \to R3$$

$$\begin{bmatrix} 1 & 2 & 1 & | & -4 \\ 0 & 1 & -\frac{1}{8} & | & -\frac{1}{2} \\ 0 & 0 & 0 & | & 0 \end{bmatrix} -2R2 + R1 \to R1$$

$$\begin{bmatrix} 1 & 0 & \frac{5}{4} & | & -3 \\ 0 & 1 & -\frac{1}{8} & | & -\frac{1}{2} \\ 0 & 0 & 0 & | & 0 \end{bmatrix}$$

$x + \frac{5}{4}z = -3$

$x = -\frac{5}{4}z - 3;$

$y - \frac{1}{8}z = -\frac{1}{2}$

$y = \frac{1}{8}z - \frac{1}{2}$

$\left(-\frac{5}{4}p - 3, \frac{1}{8}p - \frac{1}{2}, p \right)$

47.

$A = \pm\frac{1}{2}\left(x_1 y_2 - x_2 y_1 + x_2 y_3 - x_3 y_2 + x_3 y_1 - x_1 y_3 \right)$

$(6, -2), (-5, 4), (-1, 7)$

$A = \pm\frac{1}{2}\left(6(4) - (-5)(-2) + (-5)(7) - (-1)(4) + (-1)(-2) - 6(7) \right)$

$= \pm\frac{1}{2}(24 - 10 - 35 + 4 + 2 - 42)$

$= \pm\frac{1}{2}(-57)$

$= 28.5$

28.5 units^2

49. Let x represent the Heat's score.
Let y represent the Maverick's score.

$\begin{cases} x - y = 3 \\ x + y = 187 \end{cases}$

$$\begin{bmatrix} 1 & -1 & | & 3 \\ 1 & 1 & | & 187 \end{bmatrix} -R1 + R2 \to R2$$

$$\begin{bmatrix} 1 & -1 & | & 1 \\ 0 & 2 & | & 184 \end{bmatrix} \frac{1}{2}R2 \to R2$$

$$\begin{bmatrix} 1 & -1 & | & 1 \\ 0 & 1 & | & 92 \end{bmatrix}$$

$y = 92$;

$x - y = 3$

$x - 92 = 3$

$x = 95$

Heat: 95, Mavericks: 92

Chapter 6: Matrices and Matrix Applications

51. Let x represent Poe's book.
Let y represent Baum's book.
Let z represent Wouk's book.

$$\begin{cases} x + y + z = 100000 \\ x + 2z = y \\ z = 2x \end{cases}$$

$$\begin{cases} x + y + z = 100000 \\ x - y + 2z = 0 \\ -2x + z = 0 \end{cases}$$

$$\begin{bmatrix} 1 & 1 & 1 & | & 100000 \\ 1 & -1 & 2 & | & 0 \\ -2 & 0 & 1 & | & 0 \end{bmatrix} -R1 + R2 \to R2$$

$$\begin{bmatrix} 1 & 1 & 1 & | & 100000 \\ 0 & -2 & 1 & | & -100000 \\ -2 & 0 & 1 & | & 0 \end{bmatrix} 2R1 + R3 \to R3$$

$$\begin{bmatrix} 1 & 1 & 1 & | & 100000 \\ 0 & -2 & 1 & | & -100000 \\ 0 & 2 & 3 & | & 200000 \end{bmatrix} -\frac{1}{2}R2 \to R2$$

$$\begin{bmatrix} 1 & 1 & 1 & | & 100000 \\ 0 & 1 & -\frac{1}{2} & | & 50000 \\ 0 & 2 & 3 & | & 200000 \end{bmatrix} -2R2 + R3 \to R3$$

$$\begin{bmatrix} 1 & 1 & 1 & | & 100000 \\ 0 & 1 & -\frac{1}{2} & | & 50000 \\ 0 & 0 & 4 & | & 100000 \end{bmatrix} \frac{1}{4}R3 \to R3$$

$$\begin{bmatrix} 1 & 1 & 1 & | & 100000 \\ 0 & 1 & -\frac{1}{2} & | & 50000 \\ 0 & 0 & 1 & | & 25000 \end{bmatrix}$$

$z = 25000$;

$$y - \frac{1}{2}z = 50000$$

$$y - \frac{1}{2}(25000) = 50000$$

$$y - 12500 = 50000$$

$$y = 62500;$$

$$x + y + z = 100000$$

$$x + 62500 + 25000 = 100000$$

$$x + 87500 = 100000$$

$$x = 12500$$

Poe: \$12,500
Baum: \$62,500
Wouk: \$25,000

53. Let A represent the measure of angle A.
Let B represent the measure of angle B.
Let C represent the measure of angle C.

$$\begin{cases} A + B + C = 180 \\ A + C = 3B \\ C = 2B + 10 \end{cases}$$

$$\begin{cases} A + B + C = 180 \\ -A + 3B - C = 0 \\ -2B + C = 10 \end{cases}$$

$$\begin{bmatrix} 1 & 1 & 1 & | & 180 \\ -1 & 3 & -1 & | & 0 \\ 0 & -2 & 1 & | & 10 \end{bmatrix} R1 + R2 \to R2$$

$$\begin{bmatrix} 1 & 1 & 1 & | & 180 \\ 0 & 4 & 0 & | & 180 \\ 0 & -2 & 1 & | & 10 \end{bmatrix} \frac{1}{2}R2 + R3 \to R3$$

$$\begin{bmatrix} 1 & 1 & 1 & | & 180 \\ 0 & 4 & 0 & | & 180 \\ 0 & 0 & 1 & | & 100 \end{bmatrix}$$

$C = 100$;
$4B = 180$
$B = 45$;

$$A + B + C = 180$$

$$A + 45 + 100 = 180$$

$$A + 145 = 180$$

$$A = 35$$

$A = 35°, B = 45°, C = 100°$

55. Let x represent the amount of money invested in the 4% savings fund. Let y represent the amount of money invested in the 7% money market. Let z represent the amount of money invested in the 8% government bonds.

$$\begin{cases} x + y + z = 2.5 \\ 0.04x + 0.07y + 0.08z = 0.178 \\ z = 2y + 0.3 \end{cases}$$

$$\begin{cases} x + y + z = 2.5 \\ 40x + 70y + 80z = 178 \\ -20y + 10z = 3 \end{cases}$$

$$\begin{bmatrix} 1 & 1 & 1 & | & 2.5 \\ 40 & 70 & 80 & | & 178 \\ 0 & -20 & 10 & | & 3 \end{bmatrix} \quad -40R1 + R2 \rightarrow R2$$

$$\begin{bmatrix} 1 & 1 & 1 & | & 2.5 \\ 0 & 30 & 40 & | & 78 \\ 0 & -20 & 10 & | & 3 \end{bmatrix} \quad \frac{1}{30}R2 \rightarrow R2$$

$$\begin{bmatrix} 1 & 1 & 1 & | & 2.5 \\ 0 & 1 & \frac{4}{3} & | & 2.6 \\ 0 & -20 & 10 & | & 3 \end{bmatrix} \quad 20R2 + R3 \rightarrow R3$$

$$\begin{bmatrix} 1 & 1 & 1 & | & 2.5 \\ 0 & 1 & \frac{4}{3} & | & 2.6 \\ 0 & 0 & \frac{110}{3} & | & 55 \end{bmatrix} \quad \frac{3}{110}R3 \rightarrow R3$$

$$\begin{bmatrix} 1 & 1 & 1 & | & 2.5 \\ 0 & 1 & \frac{4}{3} & | & 2.6 \\ 0 & 0 & 1 & | & 1.5 \end{bmatrix}$$

$z = 1.5$;

$$y + \frac{4}{3}z = 2.6$$

$$y + \frac{4}{3}(1.5) = 2.6$$

$$y + 2 = 2.6$$

$$y = 0.6;$$

$$x + y + z = 2.5$$

$$x + 0.6 + 1.5 = 2.5$$

$$x + 2.1 = 2.5$$

$$x = 0.4$$

$0.4 million at 4%
$0.6 million at 7%
$1.5 million at 8%

57. $$\begin{cases} x + y = 180 - 71 \\ x - 59 = y \end{cases}$$

$$\begin{cases} x + y = 109 \\ x - y = 59 \end{cases}$$

$$\begin{bmatrix} 1 & 1 & | & 109 \\ 1 & -1 & | & 59 \end{bmatrix} \quad -R1 + R2 \rightarrow R2$$

$$\begin{bmatrix} 1 & 1 & | & 109 \\ 0 & -2 & | & -50 \end{bmatrix} \quad -\frac{1}{2}R2 \rightarrow R2$$

$$\begin{bmatrix} 1 & 1 & | & 109 \\ 0 & 1 & | & 25 \end{bmatrix}$$

$y = 25$;

$$x + y = 109$$

$$x + 25 = 109$$

$$x = 84$$

$x = 84°; y = 25°$

59. $f(x) = x^3 - 8$; $g(x) = x - 2$

$f + g = x^3 - 8 + x - 2 = x^3 + x - 10$;

$f - g = x^3 - 8 - (x - 2)$

$\quad = x^3 - 8 - x + 2$

$\quad = x^3 - x - 6$,

$fg = (x^3 - 8)(x - 2)$

$\quad = x^4 - 2x^3 - 8x + 16$;

$$\frac{f}{g} = \frac{x^3 - 8}{x - 2} = \frac{(x - 2)(x^2 + 2x + 4)}{x - 2} = x^2 + 2x + 4$$

61. $C(t) = 15\ln(t + 1)$

$30 = 15\ln(t + 1)$

$2 = \ln(t + 1)$

$e^2 = e^{\ln(t+1)}$

$e^2 = t + 1$

$e^2 - 1 = t$

$t \approx 6.39$

$C > 30,000$ in the year 2011.

6.2 Exercises

1. $a_{ij} = b_{ij}$

3. Scalar

5. Answers will vary.

7. $\begin{bmatrix} 1 & -3 \\ 5 & -7 \end{bmatrix}$

   2×2, $a_{12} = -3$, $a_{21} = 5$

9. $\begin{bmatrix} 2 & -3 & 0.5 \\ 0 & 5 & 6 \end{bmatrix}$

   2×3, $a_{12} = -3$, $a_{23} = 6$, $a_{22} = 5$

11. $\begin{bmatrix} -2 & 1 & -7 \\ 0 & 8 & 1 \\ 5 & -1 & 4 \end{bmatrix}$

    3×3, $a_{12} = 1$, $a_{23} = 1$, $a_{31} = 5$

13. $\begin{bmatrix} \sqrt{1} & \sqrt{4} & \sqrt{8} \\ \sqrt{16} & \sqrt{32} & \sqrt{64} \end{bmatrix} = \begin{bmatrix} 1 & 2 & 2\sqrt{2} \\ 4 & 4\sqrt{2} & 8 \end{bmatrix}$

    True.

15. $\begin{bmatrix} -2 & 3 & a \\ 2b & -5 & 4 \\ 0 & -9 & 3c \end{bmatrix} = \begin{bmatrix} c & 3 & -4 \\ 6 & -5 & -a \\ 0 & -3b & -6 \end{bmatrix}$

    Conditional, $c = -2$, $a = -4$, $b = 3$

17. $A + H$

    $= \begin{bmatrix} 2 & 3 \\ 5 & 8 \end{bmatrix} + \begin{bmatrix} 8 & -3 \\ -5 & 2 \end{bmatrix}$

    $= \begin{bmatrix} 2+8 & 3+(-3) \\ 5+(-5) & 8+2 \end{bmatrix}$

    $= \begin{bmatrix} 10 & 0 \\ 0 & 10 \end{bmatrix}$

19. $F + H$

    $= \begin{bmatrix} 6 & -3 & 9 \\ 12 & 0 & -6 \end{bmatrix} + \begin{bmatrix} 8 & -3 \\ -5 & 2 \end{bmatrix}$

    Not possible, different order.

21. $3H - 2A$

    $= 3\begin{bmatrix} 8 & -3 \\ -5 & 2 \end{bmatrix} - 2\begin{bmatrix} 2 & 3 \\ 5 & 8 \end{bmatrix}$

    $= \begin{bmatrix} 24 & -9 \\ -15 & 6 \end{bmatrix} - \begin{bmatrix} 4 & 6 \\ 10 & 16 \end{bmatrix}$

    $= \begin{bmatrix} 24-4 & -9-6 \\ -15-10 & 6-16 \end{bmatrix}$

    $= \begin{bmatrix} 20 & -15 \\ -25 & -10 \end{bmatrix}$

23. $\dfrac{1}{2}E - 3D$

    $= \dfrac{1}{2}\begin{bmatrix} 1 & -2 & 0 \\ 0 & -1 & 2 \\ 4 & 3 & -6 \end{bmatrix} - 3\begin{bmatrix} 1 & 0 & 0 \\ 0 & 1 & 0 \\ 0 & 0 & 1 \end{bmatrix}$

    $= \begin{bmatrix} \dfrac{1}{2} & -1 & 0 \\ 0 & -\dfrac{1}{2} & 1 \\ 2 & \dfrac{3}{2} & -3 \end{bmatrix} - \begin{bmatrix} 3 & 0 & 0 \\ 0 & 3 & 0 \\ 0 & 0 & 3 \end{bmatrix}$

    $= \begin{bmatrix} \dfrac{1}{2}-3 & -1-0 & 0-0 \\ 0-0 & -\dfrac{1}{2}-3 & 1-0 \\ 2-0 & \dfrac{3}{2}-0 & -3-3 \end{bmatrix}$

    $= \begin{bmatrix} \dfrac{-5}{2} & -1 & 0 \\ 0 & \dfrac{-7}{2} & 1 \\ 2 & \dfrac{3}{2} & -6 \end{bmatrix}$

25. ED

    $= \begin{bmatrix} 1 & -2 & 0 \\ 0 & -1 & 2 \\ 4 & 3 & -6 \end{bmatrix}\begin{bmatrix} 1 & 0 & 0 \\ 0 & 1 & 0 \\ 0 & 0 & 1 \end{bmatrix}$

    $= \begin{bmatrix} 1+0+0 & 0+(-2)+0 & 0+0+0 \\ 0+0+0 & 0+(-1)+0 & 0+0+2 \\ 4+0+0 & 0+3+0 & 0+0+(-6) \end{bmatrix}$

    $= \begin{bmatrix} 1 & -2 & 0 \\ 0 & -1 & 2 \\ 4 & 3 & -6 \end{bmatrix}$

6.2 Exercises

27. *AH*

$$= \begin{bmatrix} 2 & 3 \\ 5 & 8 \end{bmatrix} \begin{bmatrix} 8 & -3 \\ -5 & 2 \end{bmatrix}$$

$$= \begin{bmatrix} 2(8)+3(-5) & 2(-3)+3(2) \\ 5(8)+8(-5) & 5(-3)+8(2) \end{bmatrix}$$

$$= \begin{bmatrix} 16-15 & -6+6 \\ 40-40 & -15+16 \end{bmatrix}$$

$$= \begin{bmatrix} 1 & 0 \\ 0 & 1 \end{bmatrix}$$

29. *FD*

$$= \begin{bmatrix} 6 & -3 & 9 \\ 12 & 0 & -6 \end{bmatrix} \begin{bmatrix} 1 & 0 & 0 \\ 0 & 1 & 0 \\ 0 & 0 & 1 \end{bmatrix}$$

$$= \begin{bmatrix} 6+0+0 & 0+(-3)+0 & 0+0+9 \\ 12+0+0 & 0+0+0 & 0+0+(-6) \end{bmatrix}$$

$$= \begin{bmatrix} 6 & -3 & 9 \\ 12 & 0 & -6 \end{bmatrix}$$

31. *HF*

$$= \begin{bmatrix} 8 & -3 \\ -5 & 2 \end{bmatrix} \begin{bmatrix} 6 & -3 & 9 \\ 12 & 0 & -6 \end{bmatrix}$$

$$= \begin{bmatrix} 8(6)+(-3)(12) & 8(-3)+(-3)(0) & 8(9)+(-3)(-6) \\ -5(6)+2(12) & -5(-3)+2(0) & -5(9)+2(-6) \end{bmatrix}$$

$$= \begin{bmatrix} 48-36 & -24+0 & 72+18 \\ -30+24 & 15+0 & -45-12 \end{bmatrix}$$

$$= \begin{bmatrix} 12 & -24 & 90 \\ -6 & 15 & -57 \end{bmatrix}$$

33. H^2

$$= \begin{bmatrix} 8 & -3 \\ -5 & 2 \end{bmatrix} \begin{bmatrix} 8 & -3 \\ -5 & 2 \end{bmatrix}$$

$$= \begin{bmatrix} 8(8)+(-3)(-5) & 8(-3)+(-3)(2) \\ -5(8)+2(-5) & -5(-3)+2(2) \end{bmatrix}$$

$$= \begin{bmatrix} 64+15 & -24-6 \\ -40-10 & 15+4 \end{bmatrix}$$

$$= \begin{bmatrix} 79 & -30 \\ -50 & 19 \end{bmatrix}$$

35. *FE*

$$= \begin{bmatrix} 6 & -3 & 9 \\ 12 & 0 & -6 \end{bmatrix} \begin{bmatrix} 1 & -2 & 0 \\ 0 & -1 & 2 \\ 4 & 3 & -6 \end{bmatrix}$$

$$= \begin{bmatrix} 6+0+36 & -12+3+27 & 0+-6-54 \\ 12+0-24 & -24+0-18 & 0+0+36 \end{bmatrix}$$

$$= \begin{bmatrix} 42 & 18 & -60 \\ -12 & -42 & 36 \end{bmatrix}$$

37. $C+H$

$$= \begin{bmatrix} \dfrac{\sqrt{3}}{2} & \sqrt{3} \\ \sqrt{3} & 2\sqrt{3} \end{bmatrix} + \begin{bmatrix} -\dfrac{3}{19} & \dfrac{4}{57} \\ \dfrac{1}{19} & \dfrac{5}{57} \end{bmatrix}$$

$$= \begin{bmatrix} \dfrac{\sqrt{3}}{2}+\left(-\dfrac{3}{19}\right) & \dfrac{\sqrt{3}}{3}+\dfrac{4}{57} \\ \sqrt{3}+\dfrac{1}{19} & 2\sqrt{3}+\dfrac{5}{57} \end{bmatrix}$$

$$\approx \begin{bmatrix} 0.71 & 0.65 \\ 1.78 & 3.55 \end{bmatrix}$$

39. $E+G$

$$= \begin{bmatrix} 1 & -2 & 0 \\ 0 & -1 & 2 \\ 4 & 3 & -6 \end{bmatrix} + \begin{bmatrix} 0 & \dfrac{3}{4} & \dfrac{1}{4} \\ -\dfrac{1}{2} & \dfrac{3}{8} & \dfrac{1}{8} \\ -\dfrac{1}{4} & \dfrac{11}{16} & \dfrac{1}{16} \end{bmatrix}$$

$$= \begin{bmatrix} 1+0 & -2+\dfrac{3}{4} & 0+\dfrac{1}{4} \\ 0+\left(-\dfrac{1}{2}\right) & -1+\dfrac{3}{8} & 2+\dfrac{1}{8} \\ 4+\left(-\dfrac{1}{4}\right) & 3+\dfrac{11}{16} & -6+\dfrac{1}{16} \end{bmatrix}$$

$$\approx \begin{bmatrix} 1 & -1.25 & 0.25 \\ -0.5 & -0.63 & 2.13 \\ 3.75 & 3.69 & -5.94 \end{bmatrix}$$

41. *AH*

$$= \begin{bmatrix} -5 & 4 \\ 3 & 9 \end{bmatrix} \begin{bmatrix} -\dfrac{3}{19} & \dfrac{4}{57} \\ \dfrac{1}{19} & \dfrac{5}{57} \end{bmatrix}$$

$$= \begin{bmatrix} -5\left(-\dfrac{3}{19}\right)+4\left(\dfrac{1}{19}\right) & -5\left(\dfrac{4}{57}\right)+4\left(\dfrac{5}{57}\right) \\ 3\left(-\dfrac{3}{19}\right)+9\left(\dfrac{1}{19}\right) & 3\left(\dfrac{4}{57}\right)+9\left(\dfrac{5}{57}\right) \end{bmatrix}$$

$$= \begin{bmatrix} 1 & 0 \\ 0 & 1 \end{bmatrix}$$

43. *EG*

$$= \begin{bmatrix} 1 & -2 & 0 \\ 0 & -1 & 2 \\ 4 & 3 & -6 \end{bmatrix} \begin{bmatrix} 0 & \dfrac{3}{4} & \dfrac{1}{4} \\ -\dfrac{1}{2} & \dfrac{3}{8} & \dfrac{1}{8} \\ -\dfrac{1}{4} & \dfrac{11}{16} & \dfrac{1}{16} \end{bmatrix}$$

$$= \begin{bmatrix} 0+1+0 & \dfrac{3}{4}-\dfrac{3}{4}+0 & \dfrac{1}{4}-\dfrac{1}{4}+0 \\ 0+\dfrac{1}{2}-\dfrac{1}{2} & 0-\dfrac{3}{8}+\dfrac{11}{8} & 0-\dfrac{1}{8}+\dfrac{1}{8} \\ 0-\dfrac{3}{2}+\dfrac{3}{2} & 3+\dfrac{9}{8}-\dfrac{33}{8} & 1+\dfrac{3}{8}-\dfrac{3}{8} \end{bmatrix}$$

$$= \begin{bmatrix} 1 & 0 & 0 \\ 0 & 1 & 0 \\ 0 & 0 & 1 \end{bmatrix}$$

45. *HB*

$$= \begin{bmatrix} -\dfrac{3}{19} & \dfrac{4}{57} \\ \dfrac{1}{19} & \dfrac{5}{57} \end{bmatrix} \begin{bmatrix} 1 & 0 \\ 0 & 1 \end{bmatrix}$$

$$= \begin{bmatrix} -\dfrac{3}{19} & \dfrac{4}{57} \\ \dfrac{1}{19} & \dfrac{5}{57} \end{bmatrix}$$

47. *DG*

$$= \begin{bmatrix} 1 & 0 & 0 \\ 0 & 1 & 0 \\ 0 & 0 & 1 \end{bmatrix} \begin{bmatrix} 0 & \dfrac{3}{4} & \dfrac{1}{4} \\ -\dfrac{1}{2} & \dfrac{3}{8} & \dfrac{1}{8} \\ -\dfrac{1}{4} & \dfrac{11}{16} & \dfrac{1}{16} \end{bmatrix}$$

$$= \begin{bmatrix} 0 & \dfrac{3}{4} & \dfrac{1}{4} \\ \dfrac{-1}{2} & \dfrac{3}{8} & \dfrac{1}{8} \\ \dfrac{-1}{4} & \dfrac{11}{16} & \dfrac{1}{16} \end{bmatrix}$$

49. C^2

$$= \begin{bmatrix} \dfrac{\sqrt{3}}{2} & \dfrac{\sqrt{3}}{3} \\ \sqrt{3} & 2\sqrt{3} \end{bmatrix} \begin{bmatrix} \dfrac{\sqrt{3}}{2} & \dfrac{\sqrt{3}}{3} \\ \sqrt{3} & 2\sqrt{3} \end{bmatrix}$$

$$= \begin{bmatrix} \dfrac{3}{4}+1 & \dfrac{1}{2}+2 \\ \dfrac{3}{2}+6 & 1+12 \end{bmatrix}$$

$$= \begin{bmatrix} 1.75 & 2.5 \\ 7.5 & 13 \end{bmatrix}$$

51. *FG*

$$= \begin{bmatrix} -0.52 & 0.002 & 1.032 \\ 1.021 & -1.27 & 0.019 \end{bmatrix} \begin{bmatrix} 0 & \dfrac{3}{4} & \dfrac{1}{4} \\ -\dfrac{1}{2} & \dfrac{3}{8} & \dfrac{1}{8} \\ -\dfrac{1}{4} & \dfrac{11}{16} & \dfrac{1}{16} \end{bmatrix}$$

$$\approx \begin{bmatrix} -0.26 & 0.32 & -0.07 \\ 0.63 & 0.30 & 0.10 \end{bmatrix}$$

6.2 Exercises

53. (a) $AB \neq BA$

$$AB = \begin{bmatrix} -1 & 3 & 5 \\ 2 & 7 & -1 \\ 4 & 0 & 6 \end{bmatrix}\begin{bmatrix} 0.3 & -0.4 & 1.2 \\ -2.5 & 2 & 0.9 \\ 1 & -0.5 & 0.2 \end{bmatrix}$$

$$= \begin{bmatrix} -2.8 & 3.9 & 2.5 \\ -17.9 & 13.7 & 8.5 \\ 7.2 & -4.6 & 6 \end{bmatrix}$$

$$BA = \begin{bmatrix} 0.3 & -0.4 & 1.2 \\ -2.5 & 2 & 0.9 \\ 1 & -0.5 & 0.2 \end{bmatrix}\begin{bmatrix} -1 & 3 & 5 \\ 2 & 7 & -1 \\ 4 & 0 & 6 \end{bmatrix}$$

$$= \begin{bmatrix} 3.7 & -1.9 & 9.1 \\ 10.1 & 6.5 & -9.1 \\ -1.2 & -0.5 & 6.7 \end{bmatrix}$$

$AB \neq BA$; Verified

(b) $AC \neq CA$

$$AC = \begin{bmatrix} -1 & 3 & 5 \\ 2 & 7 & -1 \\ 4 & 0 & 6 \end{bmatrix}\begin{bmatrix} 45 & -1 & 3 \\ -6 & 10 & -15 \\ 21 & -28 & 36 \end{bmatrix}$$

$$= \begin{bmatrix} 42 & -109 & 132 \\ 27 & 96 & -135 \\ 306 & -172 & 228 \end{bmatrix}$$

$$CA = \begin{bmatrix} 45 & -1 & 3 \\ -6 & 10 & -15 \\ 21 & -28 & 36 \end{bmatrix}\begin{bmatrix} -1 & 3 & 5 \\ 2 & 7 & -1 \\ 4 & 0 & 6 \end{bmatrix}$$

$$= \begin{bmatrix} -35 & 128 & 244 \\ -34 & 52 & -130 \\ 67 & -133 & 349 \end{bmatrix}$$

$AC \neq CA$; Verified

(c) $BC \neq CB$

BC

$$= \begin{bmatrix} 0.3 & -0.4 & 1.2 \\ -2.5 & 2 & 0.9 \\ 1 & -0.5 & 0.2 \end{bmatrix}\begin{bmatrix} 45 & -1 & 3 \\ -6 & 10 & -15 \\ 21 & -28 & 36 \end{bmatrix}$$

$$= \begin{bmatrix} 41.1 & -37.9 & 50.1 \\ -105.6 & -2.7 & -5.1 \\ 52.2 & -11.6 & 17.7 \end{bmatrix}$$

CB

$$= \begin{bmatrix} 45 & -1 & 3 \\ -6 & 10 & -15 \\ 21 & -28 & 36 \end{bmatrix}\begin{bmatrix} 0.3 & -0.4 & 1.2 \\ -2.5 & 2 & 0.9 \\ 1 & -0.5 & 0.2 \end{bmatrix}$$

$$= \begin{bmatrix} 19 & -21.5 & 53.7 \\ -41.8 & 29.9 & -1.2 \\ 112.3 & -82.4 & 7.2 \end{bmatrix}$$

$BC \neq CB$; Verified

55. $(B+C)A = BA + CA$

$$(B+C)A = \begin{bmatrix} 45.3 & -1.4 & 4.2 \\ -8.5 & 12 & -14.1 \\ 22 & -28.5 & 36.2 \end{bmatrix}A$$

$$= \begin{bmatrix} -31.3 & 126.1 & 253.1 \\ -23.9 & 58.5 & -139.1 \\ 65.8 & -133.5 & 355.7 \end{bmatrix}$$

$BA + CA$

$$= \begin{bmatrix} 3.7 & -1.9 & 9.1 \\ 10.1 & 6.5 & -9.1 \\ -1.2 & -0.5 & 6.7 \end{bmatrix} + \begin{bmatrix} -35 & 128 & 244 \\ -34 & 52 & -130 \\ 67 & -133 & 349 \end{bmatrix}$$

$$= \begin{bmatrix} -31.3 & 26.1 & 253.1 \\ -23.9 & 58.5 & -139.1 \\ 65.8 & -133.5 & 355.7 \end{bmatrix}$$

$(B+C)A = BA + CA$; Verified

57. $\begin{bmatrix} 2 & 2 \\ 4.35 & 0 \end{bmatrix} \cdot \begin{bmatrix} 6.374 \\ 4.35 \end{bmatrix}$

$= \begin{bmatrix} 2(6.374)+2(4.35) \\ 4.35(6.374)+0(4.35) \end{bmatrix}$

$= \begin{bmatrix} 21.448 \\ 27.7269 \end{bmatrix}$

$P = 2l + 2w$

$P = 2(6.374) + 2(4.35)$

$P = 12.748 + 8.7$

$P = 21.448;$

$A = lw$

$A = 6.374(4.35)$

$A = 27.7269$

$P = 21.448 \, \text{cm}, \quad A = 27.7269 \, \text{cm}^2$

59. a.
$$V \rightarrow \begin{array}{c} \\ S \\ D \\ P \end{array} \begin{array}{cc} T & S \\ \begin{bmatrix} 3820 & 1960 \\ 2460 & 1240 \\ 1540 & 920 \end{bmatrix} \end{array}$$

$$M \rightarrow \begin{array}{c} \\ S \\ D \\ P \end{array} \begin{array}{cc} T & S \\ \begin{bmatrix} 4220 & 2960 \\ 2960 & 3240 \\ 1640 & 820 \end{bmatrix} \end{array}$$

b. $\begin{bmatrix} 4220 & 2960 \\ 2960 & 3240 \\ 1640 & 820 \end{bmatrix} - \begin{bmatrix} 3820 & 1960 \\ 2460 & 1240 \\ 1540 & 920 \end{bmatrix}$

$= \begin{bmatrix} 4220-3820 & 2960-1960 \\ 2960-2460 & 3240-1240 \\ 1640-1540 & 820-920 \end{bmatrix}$

$= \begin{bmatrix} 400 & 1000 \\ 500 & 2000 \\ 100 & -100 \end{bmatrix}$

$400 + 1000 + 500 + 2000 + 100 - 100$

$= 3900$

3,900 more by Minsk

c. $V \rightarrow 1.04 \begin{bmatrix} 3820 & 1960 \\ 2460 & 1240 \\ 1540 & 920 \end{bmatrix}$

$= \begin{bmatrix} 3972.8 & 2038.4 \\ 2558.4 & 1289.6 \\ 1601.6 & 956.8 \end{bmatrix};$

$M \rightarrow 1.04 \begin{bmatrix} 4220 & 2960 \\ 2960 & 3240 \\ 1640 & 820 \end{bmatrix}$

$= \begin{bmatrix} 4388.8 & 3078.4 \\ 3078.4 & 3369.6 \\ 1705.6 & 852.8 \end{bmatrix}$

d. $\begin{bmatrix} 3972.8 & 2038.4 \\ 2558.4 & 1289.6 \\ 1601.6 & 956.8 \end{bmatrix} + \begin{bmatrix} 4388.8 & 3078.4 \\ 3078.4 & 3369.6 \\ 1705.6 & 852.8 \end{bmatrix}$

$= \begin{bmatrix} 8361.6 & 5116.8 \\ 5636.8 & 4659.2 \\ 3307.2 & 1809.6 \end{bmatrix}$

61. $\begin{bmatrix} 1500 & 500 & 2500 \end{bmatrix} \begin{bmatrix} 9 & 6 & 5 & 4 \\ 7 & 5 & 7 & 6 \\ 2 & 3 & 5 & 2 \end{bmatrix}$

$= \begin{bmatrix} 22000 & 19000 & 23500 & 14000 \end{bmatrix}$

Total profit for north: $22,000.
Total profit for south: $19,000.
Total profit for east: $23,500.
Total profit for west: $14,000.

63. a. $10(8) + 8(1.5) + 18(0.9) = \$108.20$

b. $8(7.5) + 12(1.75) + 20(1) = \$101.00$

c. $\begin{bmatrix} 8 & 12 & 20 \\ 10 & 8 & 18 \end{bmatrix} \begin{bmatrix} 8 & 7.5 & 10 \\ 1.5 & 1.75 & 2 \\ 0.9 & 1 & 0.75 \end{bmatrix}$

$= \begin{array}{c} \text{Science} \\ \text{Math} \end{array} \begin{bmatrix} 100 & 101 & 119 \\ 108.2 & 107 & 129.5 \end{bmatrix}$

1st row, total cost for science from each restaurant.
2nd row, Total cost for math from each restaurant.

65. $\begin{bmatrix} 25 & 18 & 21 \\ 22 & 19 & 18 \end{bmatrix} \begin{bmatrix} 0.6 & 0.1 & 0.3 \\ 0.5 & 0.2 & 0.3 \\ 0.4 & 0.2 & 0.4 \end{bmatrix}$

$= \begin{bmatrix} 32.4 & 10.3 & 21.3 \\ 29.9 & 9.6 & 19.5 \end{bmatrix}$

a. Approximately 10 females
b. Approximately 20 males
c. The approximate number of females expected to join the writing club

67. 1st, 3rd rows entries double, 2nd row entries double and increase by 1 in a_{21}, a_{23} positions, and a_{22} stays a 1.

$\begin{bmatrix} 2^{n-1} & 0 & 2^{n-1} \\ 2^{n}-1 & 1 & 2^{n}-1 \\ 2^{n-1} & 0 & 2^{n-1} \end{bmatrix}$

69. $\begin{bmatrix} 2 & 1 \\ -3 & -2 \end{bmatrix} \cdot \begin{bmatrix} a & b \\ c & d \end{bmatrix} = \begin{bmatrix} 1 & 0 \\ 0 & 1 \end{bmatrix}$

$\begin{cases} 2a+c=1 \\ -3a-2c=0 \end{cases}$ $\qquad$ $\begin{cases} 2b+d=0 \\ -3b-2d=1 \end{cases}$

$2R1 + R2 = Sum$ $\qquad$ $2R1 + R2 = Sum$

$4a+2c=2$ $\qquad\qquad$ $4b+2d=0$

$-3a-2c=0$ $\qquad\qquad$ $-3b-2d=1$

$a=2$ $\qquad\qquad\qquad$ $b=1$

$2a+c=1$ $\qquad\qquad\quad$ $2b+d=0$

$2(2)+c=1$ $\qquad\qquad$ $2(1)+d=0$

$4+c=1$ $\qquad\qquad\quad$ $2+d=0$

$c=-3$ $\qquad\qquad\quad$ $d=-2$

$a=2, b=1, c=-3, d=-2$

71. $f(x) \geq 0$

$x \in [-3,-1] \cup [1,3]$

73. $\dfrac{x^3 - 9x + 10}{x-2}$

$\begin{array}{r|rrrr} 2 & 1 & 0 & -9 & 10 \\ & & 2 & 4 & -10 \\ \hline & 1 & 2 & -5 & 0 \end{array}$

$= x^2 + 2x - 5$

Mid-Chapter Check

1. $\begin{bmatrix} 0.4 & 1.1 & 0.2 \\ -0.2 & 0.1 & -0.9 \\ 0.7 & 0.4 & 0.8 \end{bmatrix}$

   $3 \times 3, \ -0.9$

3. $\begin{cases} 2x + 3y = -5 \\ -5x - 4y = 2 \end{cases}$

   $\begin{bmatrix} 2 & 3 & | & -5 \\ -5 & -4 & | & 2 \end{bmatrix} R1 + R2 \to R2$

   $\begin{bmatrix} 2 & 3 & | & -5 \\ -3 & -1 & | & -3 \end{bmatrix} 3R2 + R1 \to R1$

   $\begin{bmatrix} -7 & 0 & | & -14 \\ -3 & -1 & | & -3 \end{bmatrix} -\dfrac{1}{7}R1 \to R1$

   $\begin{bmatrix} 1 & 0 & | & 2 \\ -3 & -1 & | & -3 \end{bmatrix} 3R1 + R2 \to R2$

   $\begin{bmatrix} 1 & 0 & | & 2 \\ 0 & -1 & | & 3 \end{bmatrix} -1R2 \to R2$

   $\begin{bmatrix} 1 & 0 & | & 2 \\ 0 & 1 & | & -3 \end{bmatrix}$

   $(2, -3)$

5. $\begin{cases} x + y - 3z = -11 \\ 4x - y - 2z = -4 \\ 3x - 2y + z = 7 \end{cases}$

   $\begin{bmatrix} 1 & 1 & -3 & | & -11 \\ 4 & -1 & -2 & | & -4 \\ 3 & -2 & 1 & | & 7 \end{bmatrix} -4R1 + R2 \to R2$

   $\begin{bmatrix} 1 & 1 & -3 & | & -11 \\ 0 & -5 & 10 & | & 40 \\ 3 & -2 & 1 & | & 7 \end{bmatrix} -\dfrac{1}{5}R2 \leftrightarrow R2$

   $\begin{bmatrix} 1 & 1 & -3 & | & -11 \\ 0 & 1 & -2 & | & -8 \\ 3 & -2 & 1 & | & 7 \end{bmatrix} -3R1 + R3 \to R3$

   $\begin{bmatrix} 1 & 1 & -3 & | & -11 \\ 0 & 1 & -2 & | & -8 \\ 0 & -5 & 10 & | & 40 \end{bmatrix} 5R2 + R3 \to R3$

$\begin{bmatrix} 1 & 1 & -3 & | & -11 \\ 0 & 1 & -2 & | & -8 \\ 0 & 0 & 0 & | & 0 \end{bmatrix} R1 - R2 \to R1$

$\begin{bmatrix} 1 & 0 & -1 & | & -3 \\ 0 & 1 & -2 & | & -8 \\ 0 & 0 & 0 & | & 0 \end{bmatrix}$

$x - z = -3$

$\quad x = z - 3;$

$y - 2z = -8$

$\quad y = 2z - 8$

$(p - 3, 2p - 8, p)$

7.

$C = \begin{bmatrix} -0.2 & 0 & 0.2 \\ 0.4 & 0.8 & 0 \\ 0.1 & -0.2 & -0.1 \end{bmatrix},$

$D = \begin{bmatrix} 5 & 2.5 & 10 \\ -2.5 & 0 & -5 \\ 10 & 2.5 & 10 \end{bmatrix}$

a. $C + \dfrac{1}{5}D =$

$\begin{bmatrix} -0.2 + \dfrac{5}{5} & 0 + \dfrac{2.5}{5} & 0.2 + \dfrac{10}{5} \\ 0.4 + \dfrac{-2.5}{5} & 0.8 + \dfrac{0}{5} & 0 + \dfrac{-5}{5} \\ 0.1 + \dfrac{10}{5} & -0.2 + \dfrac{2.5}{5} & -0.1 + \dfrac{10}{5} \end{bmatrix}$

$= \begin{bmatrix} 0.8 & 0.5 & 2.2 \\ -0.1 & 0.8 & -1 \\ 2.1 & 0.3 & 1.9 \end{bmatrix}$

b. $-0.6D = \begin{bmatrix} -3 & -1.5 & -6 \\ 1.5 & 0 & 3 \\ -6 & -1.5 & -6 \end{bmatrix}$

c. $CD =$

$\begin{bmatrix} -0.2 & 0 & 0.2 \\ 0.4 & 0.8 & 0 \\ 0.1 & -0.2 & -0.1 \end{bmatrix} \begin{bmatrix} 5 & 2.5 & 10 \\ -2.5 & 0 & -5 \\ 10 & 2.5 & 10 \end{bmatrix}$

$= \begin{bmatrix} 1 & 0 & 0 \\ 0 & 1 & 0 \\ 0 & 0 & 1 \end{bmatrix}$

9. $\begin{bmatrix} 14 & 10 & | & 2370 \\ 2 & 4 & | & 660 \end{bmatrix}$ $-7R2+R1 \rightarrow R1$

$\begin{bmatrix} 0 & -18 & | & -2250 \\ 2 & 4 & | & 660 \end{bmatrix}$ $\frac{1}{2}R2 \rightarrow R2$

$\begin{bmatrix} 0 & -18 & | & -2250 \\ 1 & 2 & | & 330 \end{bmatrix}$ $\frac{1}{-18}R1 \rightarrow R1$

$\begin{bmatrix} 0 & 1 & | & 125 \\ 1 & 2 & | & 330 \end{bmatrix}$ $-2R1+R2 \rightarrow R2$

$\begin{bmatrix} 0 & 1 & | & 125 \\ 1 & 0 & | & 80 \end{bmatrix}$ $R1 \leftrightarrow R2$

$\begin{bmatrix} 1 & 0 & | & 80 \\ 0 & 1 & | & 125 \end{bmatrix}$

Used: $80, New: $125

Reinforcing Basic Concepts

1. P_{32}

2. 1st row of A with erd column of B
   2nd row of A with 2nd column of B

3. $[A] \rightarrow 3\text{x}1; [B] \rightarrow 1\text{x}3;$
   $[A] \rightarrow 3\text{x}2; [B] \rightarrow 2\text{x}3;$
   $[A] \rightarrow 3\text{x}3; [B] \rightarrow 3\text{x}3;$
   $[A] \rightarrow 3\text{x}n; [B] \rightarrow n\text{x}3; n \in \mathbb{N}$

4. only $[F][E] = [P]$

6.3 Exercises

1. Main diagonal; zeroes

3. Identity

5. Answers will vary.

7. $A = \begin{bmatrix} 2 & 5 \\ -3 & -7 \end{bmatrix} \cdot \begin{bmatrix} a & b \\ c & d \end{bmatrix} = \begin{bmatrix} 2 & 5 \\ -3 & -7 \end{bmatrix}$

$\begin{bmatrix} 2a+5c & 2b+5d \\ -3a-7c & -3b-7d \end{bmatrix} = \begin{bmatrix} 2 & 5 \\ -3 & -7 \end{bmatrix}$

$\begin{cases} 2a+5c = 2 \\ -3a-7c = -3 \end{cases}$

3R1 + 2R2

$\begin{cases} 6a+15c = 6 \\ -6a-14c = -6 \end{cases}$

$c = 0$;

$2a+5c = 2$

$2a+5(0) = 2$

$2a = 2$

$a = 1$

$\begin{cases} 2b+5d = 5 \\ -3b-7d = -7 \end{cases}$

3R1 + 2R2

$\begin{cases} 6b+15d = 15 \\ -6b-14d = -14 \end{cases}$

$d = 1$;

$2b+5d = 5$

$2b+5(1) = 5$

$2b+5 = 5$

$2b = 0$

$b = 0$

$\begin{bmatrix} a & b \\ c & d \end{bmatrix} = \begin{bmatrix} 1 & 0 \\ 0 & 1 \end{bmatrix}$

9. $A = \begin{bmatrix} 0.4 & 0.6 \\ 0.3 & 0.2 \end{bmatrix} \cdot \begin{bmatrix} a & b \\ c & d \end{bmatrix} = \begin{bmatrix} 0.4 & 0.6 \\ 0.3 & 0.2 \end{bmatrix}$

$\begin{bmatrix} 0.4a+0.6c & 0.4b+0.6d \\ 0.3a+0.2c & 0.3b+0.2d \end{bmatrix} = \begin{bmatrix} 0.4 & 0.6 \\ 0.3 & 0.2 \end{bmatrix}$

$\begin{cases} 0.4a+0.6c = 0.4 \\ 0.3a+0.2c = 0.3 \end{cases}$

30R1 + (−40)R2

$\begin{cases} 12a+18c = 12 \\ -12a-8c = -12 \end{cases}$

$10c = 0$

$c = 0$;

$0.4a+0.6c = 0.4$

$0.4a+0.6(0) = 0.4$

$0.4a = 0.4$

$a = 1$;

$\begin{cases} 0.4b+0.6d = 0.6 \\ 0.3b+0.2d = 0.2 \end{cases}$

30R1 + (−40)R2

$\begin{cases} 12b+18d = 18 \\ -12b-8d = -8 \end{cases}$

$10d = 10$

$d = 1$;

$0.4b+0.6d = 0.6$

$0.4b+0.6(1) = 0.6$

$0.4b+0.6 = 0.6$

$0.4b = 0$

$b = 0$;

$\begin{bmatrix} a & b \\ c & d \end{bmatrix} = \begin{bmatrix} 1 & 0 \\ 0 & 1 \end{bmatrix}$

11. $\begin{bmatrix} -3 & 8 \\ -4 & 10 \end{bmatrix} \cdot \begin{bmatrix} 1 & 0 \\ 0 & 1 \end{bmatrix}$

$= \begin{bmatrix} -3(1)+8(0) & -3(0)+8(1) \\ -4(1)+10(0) & -4(0)+10(1) \end{bmatrix}$

$= \begin{bmatrix} -3 & 8 \\ -4 & 10 \end{bmatrix}$;

$\begin{bmatrix} 1 & 0 \\ 0 & 1 \end{bmatrix} \cdot \begin{bmatrix} -3 & 8 \\ -4 & 10 \end{bmatrix}$

$= \begin{bmatrix} 1(-3)+0(-4) & 1(8)+0(10) \\ 0(-3)+1(-4) & 0(8)+1(10) \end{bmatrix}$

$= \begin{bmatrix} -3 & 8 \\ -4 & 10 \end{bmatrix}$

$AI = IA = A$

13. $\begin{bmatrix} -4 & 1 & 6 \\ 9 & 5 & 3 \\ 0 & -2 & 1 \end{bmatrix} \cdot \begin{bmatrix} 1 & 0 & 0 \\ 0 & 1 & 0 \\ 0 & 0 & 1 \end{bmatrix}$

$= \begin{bmatrix} -4 & 1 & 6 \\ 9 & 5 & 3 \\ 0 & -2 & 1 \end{bmatrix}$;

$\begin{bmatrix} 1 & 0 & 0 \\ 0 & 1 & 0 \\ 0 & 0 & 1 \end{bmatrix} \cdot \begin{bmatrix} -4 & 1 & 6 \\ 9 & 5 & 3 \\ 0 & -2 & 1 \end{bmatrix}$

$= \begin{bmatrix} -4 & 1 & 6 \\ 9 & 5 & 3 \\ 0 & -2 & 1 \end{bmatrix}$

$AI = IA = A$

15. $\begin{bmatrix} 5 & -4 \\ 2 & 2 \end{bmatrix}$

$A^{-1} = \dfrac{1}{ad-bc} \begin{bmatrix} d & -b \\ -c & a \end{bmatrix}$

$A^{-1} = \dfrac{1}{5(2)-(-4)(2)} \begin{bmatrix} 2 & 4 \\ -2 & 5 \end{bmatrix}$

$A^{-1} = \dfrac{1}{18} \begin{bmatrix} 2 & 4 \\ -2 & 5 \end{bmatrix}$

$A^{-1} = \begin{bmatrix} \dfrac{1}{9} & \dfrac{2}{9} \\ \dfrac{-1}{9} & \dfrac{5}{18} \end{bmatrix}$

17. $\begin{bmatrix} 1 & -3 \\ 4 & -10 \end{bmatrix}$

$A^{-1} = \dfrac{1}{ad-bc} \begin{bmatrix} d & -b \\ -c & a \end{bmatrix}$

$A^{-1} = \dfrac{1}{1(-10)-(-3)(4)} \begin{bmatrix} -10 & 3 \\ -4 & 1 \end{bmatrix}$

$A^{-1} = \dfrac{1}{2} \begin{bmatrix} -10 & 3 \\ -4 & 1 \end{bmatrix}$

$A^{-1} = \begin{bmatrix} -5 & \dfrac{3}{2} \\ -2 & \dfrac{1}{2} \end{bmatrix}$

19. $A = \begin{bmatrix} 1 & 5 \\ -2 & -9 \end{bmatrix}$ $B = \begin{bmatrix} -9 & -5 \\ 2 & 1 \end{bmatrix}$

$AB = \begin{bmatrix} 1 & 5 \\ -2 & -9 \end{bmatrix} \begin{bmatrix} -9 & -5 \\ 2 & 1 \end{bmatrix}$

$= \begin{bmatrix} 1(-9)+5(2) & 1(-5)+5(1) \\ -2(-9)-9(2) & -2(-5)-9(1) \end{bmatrix}$

$= \begin{bmatrix} 1 & 0 \\ 0 & 1 \end{bmatrix}$;

$BA = \begin{bmatrix} -9 & -5 \\ 2 & 1 \end{bmatrix} \begin{bmatrix} 1 & 5 \\ -2 & -9 \end{bmatrix}$

$= \begin{bmatrix} -9(1)-5(-2) & -9(5)-5(-9) \\ 2(1)+1(-2) & 2(5)+1(-9) \end{bmatrix}$

$= \begin{bmatrix} 1 & 0 \\ 0 & 1 \end{bmatrix}$

$AB = BA = I$

21. $A = \begin{bmatrix} 4 & -5 \\ 0 & 2 \end{bmatrix}$ $B = \begin{bmatrix} \dfrac{1}{4} & \dfrac{5}{8} \\ 0 & \dfrac{1}{2} \end{bmatrix}$

$AB = \begin{bmatrix} 4 & -5 \\ 0 & 2 \end{bmatrix} \begin{bmatrix} \dfrac{1}{4} & \dfrac{5}{8} \\ 0 & \dfrac{1}{2} \end{bmatrix}$

$= \begin{bmatrix} 4\left(\dfrac{1}{4}\right)-5(0) & 4\left(\dfrac{5}{8}\right)-5\left(\dfrac{1}{2}\right) \\ 0\left(\dfrac{1}{4}\right)+2(0) & 0\left(\dfrac{5}{8}\right)+2\left(\dfrac{1}{2}\right) \end{bmatrix}$

$= \begin{bmatrix} 1 & 0 \\ 0 & 1 \end{bmatrix}$;

$BA = \begin{bmatrix} \dfrac{1}{4} & \dfrac{5}{8} \\ 0 & \dfrac{1}{2} \end{bmatrix} \begin{bmatrix} 4 & -5 \\ 0 & 2 \end{bmatrix}$

$= \begin{bmatrix} \dfrac{1}{4}(4)+\dfrac{5}{8}(0) & \dfrac{1}{4}(-5)+\dfrac{5}{8}(2) \\ 0(4)+\dfrac{1}{2}(0) & 0(-5)+\dfrac{1}{2}(2) \end{bmatrix}$

$= \begin{bmatrix} 1 & 0 \\ 0 & 1 \end{bmatrix}$

$AB = BA = I$

23. $A = \begin{bmatrix} -2 & 3 & 1 \\ 5 & 2 & 4 \\ 2 & 0 & -1 \end{bmatrix}$

$A^{-1} = B = \begin{bmatrix} -\dfrac{2}{39} & \dfrac{1}{13} & \dfrac{10}{39} \\ \dfrac{1}{3} & 0 & \dfrac{1}{3} \\ -\dfrac{4}{39} & \dfrac{2}{13} & -\dfrac{19}{39} \end{bmatrix}$;

$AB = \begin{bmatrix} -2 & 3 & 1 \\ 5 & 2 & 4 \\ 2 & 0 & -1 \end{bmatrix} \begin{bmatrix} -\dfrac{2}{39} & \dfrac{1}{13} & \dfrac{10}{39} \\ \dfrac{1}{3} & 0 & \dfrac{1}{3} \\ -\dfrac{4}{39} & \dfrac{2}{13} & -\dfrac{19}{39} \end{bmatrix}$

$= \begin{bmatrix} 1 & 0 & 0 \\ 0 & 1 & 0 \\ 0 & 0 & 1 \end{bmatrix}$

$BA = \begin{bmatrix} -\dfrac{2}{39} & \dfrac{1}{13} & \dfrac{10}{39} \\ \dfrac{1}{3} & 0 & \dfrac{1}{3} \\ -\dfrac{4}{39} & \dfrac{2}{13} & -\dfrac{19}{39} \end{bmatrix} \begin{bmatrix} -2 & 3 & 1 \\ 5 & 2 & 4 \\ 2 & 0 & -1 \end{bmatrix}$

$= \begin{bmatrix} 1 & 0 & 0 \\ 0 & 1 & 0 \\ 0 & 0 & 1 \end{bmatrix}$

$AB = BA = I$

25. $A = \begin{bmatrix} -7 & 5 & -3 \\ 1 & 9 & 0 \\ 2 & -2 & -5 \end{bmatrix}$

$A^{-1} = B = \begin{bmatrix} -\dfrac{9}{80} & \dfrac{31}{400} & \dfrac{27}{400} \\ \dfrac{1}{80} & \dfrac{41}{400} & -\dfrac{3}{400} \\ -\dfrac{1}{20} & -\dfrac{1}{100} & -\dfrac{17}{100} \end{bmatrix}$;

$AB = \begin{bmatrix} -7 & 5 & -3 \\ 1 & 9 & 0 \\ 2 & -2 & -5 \end{bmatrix} \begin{bmatrix} -\dfrac{9}{80} & \dfrac{31}{400} & \dfrac{27}{400} \\ \dfrac{1}{80} & \dfrac{41}{400} & -\dfrac{3}{400} \\ -\dfrac{1}{20} & -\dfrac{1}{100} & -\dfrac{17}{100} \end{bmatrix}$

$= \begin{bmatrix} 1 & 0 & 0 \\ 0 & 1 & 0 \\ 0 & 0 & 1 \end{bmatrix}$;

$BA = \begin{bmatrix} -\dfrac{9}{80} & \dfrac{31}{400} & \dfrac{27}{400} \\ \dfrac{1}{80} & \dfrac{41}{400} & -\dfrac{3}{400} \\ -\dfrac{1}{20} & -\dfrac{1}{100} & -\dfrac{17}{100} \end{bmatrix} \begin{bmatrix} -7 & 5 & -3 \\ 1 & 9 & 0 \\ 2 & -2 & -5 \end{bmatrix}$

$= \begin{bmatrix} 1 & 0 & 0 \\ 0 & 1 & 0 \\ 0 & 0 & 1 \end{bmatrix}$

$AB = BA = I$

27. $\begin{cases} 2x - 3y = 9 \\ -5x + 7y = 8 \end{cases}$

$\begin{bmatrix} 2 & -3 \\ -5 & 7 \end{bmatrix} \begin{bmatrix} x \\ y \end{bmatrix} = \begin{bmatrix} 9 \\ 8 \end{bmatrix}$

29. $\begin{cases} x + 2y - z = 1 \\ x + z = 3 \\ 2x - y + z = 3 \end{cases}$

$\begin{bmatrix} 1 & 2 & -1 \\ 1 & 0 & 1 \\ 2 & -1 & 1 \end{bmatrix} \begin{bmatrix} x \\ y \\ z \end{bmatrix} = \begin{bmatrix} 1 \\ 3 \\ 3 \end{bmatrix}$

31. $\begin{cases} -2w + x - 4y + 5 = -3 \\ 2w - 5x + y - 3z = 4 \\ -3w + x + 6y + z = 1 \\ w + 4x - 5y + z = -9 \end{cases}$

$\begin{bmatrix} -2 & 1 & -4 & 5 \\ 2 & -5 & 1 & -3 \\ -3 & 1 & 6 & 1 \\ 1 & 4 & -5 & 1 \end{bmatrix} \begin{bmatrix} w \\ x \\ y \\ z \end{bmatrix} = \begin{bmatrix} -3 \\ 4 \\ 1 \\ -9 \end{bmatrix}$

6.3 Exercises

33. $\begin{cases} 0.05x - 3.2y = -15.8 \\ 0.02x + 2.4y = 12.08 \end{cases}$

$$\begin{bmatrix} 0.05 & -3.2 \\ 0.02 & 2.4 \end{bmatrix}\begin{bmatrix} x \\ y \end{bmatrix} = \begin{bmatrix} -15.8 \\ 12.08 \end{bmatrix}$$

Using the grapher,

$$A^{-1} = \begin{bmatrix} \dfrac{300}{23} & \dfrac{400}{23} \\ -\dfrac{5}{46} & \dfrac{25}{92} \end{bmatrix};$$

$$A^{-1}\left(\begin{bmatrix} 0.05 & -3.2 \\ 0.02 & 2.4 \end{bmatrix}\begin{bmatrix} x \\ y \end{bmatrix}\right) = A^{-1}\begin{bmatrix} -15.8 \\ 12.08 \end{bmatrix}$$

$$\left(A^{-1}\begin{bmatrix} 0.05 & -3.2 \\ 0.02 & 2.4 \end{bmatrix}\right)\begin{bmatrix} x \\ y \end{bmatrix} = A^{-1}\begin{bmatrix} -15.8 \\ 12.08 \end{bmatrix}$$

$$\begin{bmatrix} 1 & 0 \\ 0 & 1 \end{bmatrix}\begin{bmatrix} x \\ y \end{bmatrix} = A^{-1}\begin{bmatrix} -15.8 \\ 12.08 \end{bmatrix}$$

$$\begin{bmatrix} x \\ y \end{bmatrix} = \begin{bmatrix} 4 \\ 5 \end{bmatrix}$$

$(4,5)$

35. $\begin{cases} -\dfrac{1}{6}u + \dfrac{1}{4}v = 1 \\ \dfrac{1}{2}u - \dfrac{2}{3}v = -2 \end{cases}$

$$\begin{bmatrix} -\dfrac{1}{6} & \dfrac{1}{4} \\ \dfrac{1}{2} & -\dfrac{2}{3} \end{bmatrix}\begin{bmatrix} u \\ v \end{bmatrix} = \begin{bmatrix} 1 \\ -2 \end{bmatrix}$$

$$A^{-1} = \begin{bmatrix} 48 & 18 \\ 36 & 12 \end{bmatrix};$$

$$\begin{bmatrix} 48 & 18 \\ 36 & 12 \end{bmatrix}\left(\begin{bmatrix} -\dfrac{1}{6} & \dfrac{1}{4} \\ \dfrac{1}{2} & -\dfrac{2}{3} \end{bmatrix}\begin{bmatrix} u \\ v \end{bmatrix}\right) = \begin{bmatrix} 48 & 18 \\ 36 & 12 \end{bmatrix}\begin{bmatrix} 1 \\ -2 \end{bmatrix}$$

$$\left(\begin{bmatrix} 48 & 18 \\ 36 & 12 \end{bmatrix}\begin{bmatrix} -\dfrac{1}{6} & \dfrac{1}{4} \\ \dfrac{1}{2} & -\dfrac{2}{3} \end{bmatrix}\right)\begin{bmatrix} u \\ v \end{bmatrix} = \begin{bmatrix} 48 & 18 \\ 36 & 12 \end{bmatrix}\begin{bmatrix} 1 \\ -2 \end{bmatrix}$$

$$\begin{bmatrix} 1 & 0 \\ 0 & 1 \end{bmatrix}\begin{bmatrix} u \\ v \end{bmatrix} = \begin{bmatrix} 48 & 18 \\ 36 & 12 \end{bmatrix}\begin{bmatrix} 1 \\ -2 \end{bmatrix}$$

$$\begin{bmatrix} u \\ v \end{bmatrix} = \begin{bmatrix} 12 \\ 12 \end{bmatrix}$$

$(12,12)$

37. $\begin{cases} \dfrac{-1}{8}a + \dfrac{3}{5}b = \dfrac{5}{6} \\ \dfrac{5}{16}a - \dfrac{3}{2}b = \dfrac{-4}{5} \end{cases}$

$$\begin{bmatrix} -\dfrac{1}{8} & \dfrac{3}{5} \\ \dfrac{5}{16} & -\dfrac{3}{2} \end{bmatrix}\begin{bmatrix} a \\ b \end{bmatrix} = \begin{bmatrix} \dfrac{5}{6} \\ -\dfrac{4}{5} \end{bmatrix}$$

No Solution; matrix is singular

39. $\begin{cases} 0.2x - 1.6y + 2z = -1.9 \\ -0.4x - y + 0.6z = -1 \\ 0.8x + 3.2y - 0.4z = 0.2 \end{cases}$

$$\begin{bmatrix} 0.2 & -1.6 & 2 \\ -0.4 & -1 & 0.6 \\ 0.8 & 3.2 & -0.4 \end{bmatrix}\begin{bmatrix} x \\ y \\ z \end{bmatrix} = \begin{bmatrix} -1.9 \\ -1 \\ 0.2 \end{bmatrix}$$

Using the grapher,

$$A^{-1} = \begin{bmatrix} \dfrac{95}{111} & -\dfrac{120}{37} & -\dfrac{65}{111} \\ -\dfrac{20}{111} & \dfrac{35}{37} & \dfrac{115}{222} \\ \dfrac{10}{37} & \dfrac{40}{37} & \dfrac{35}{74} \end{bmatrix};$$

$$A^{-1}\left(\begin{bmatrix} 0.2 & -1.6 & 2 \\ -0.4 & -1 & 0.6 \\ 0.8 & 3.2 & -0.4 \end{bmatrix}\begin{bmatrix} x \\ y \\ z \end{bmatrix}\right) = A^{-1}\begin{bmatrix} -1.9 \\ -1 \\ 0.2 \end{bmatrix}$$

$$\left(A^{-1}\begin{bmatrix} 0.2 & -1.6 & 2 \\ -0.4 & -1 & 0.6 \\ 0.8 & 3.2 & -0.4 \end{bmatrix}\right)\begin{bmatrix} x \\ y \\ z \end{bmatrix} = A^{-1}\begin{bmatrix} -1.9 \\ -1 \\ 0.2 \end{bmatrix}$$

$$\begin{bmatrix} 1 & 0 & 0 \\ 0 & 1 & 0 \\ 0 & 0 & 1 \end{bmatrix}\begin{bmatrix} x \\ y \\ z \end{bmatrix} = A^{-1}\begin{bmatrix} -1.9 \\ -1 \\ 0.2 \end{bmatrix}$$

$$\begin{bmatrix} x \\ y \\ z \end{bmatrix} = \begin{bmatrix} 1.5 \\ -0.5 \\ -1.5 \end{bmatrix}$$

$(1.5, -0.5, -1.5)$

41. $\begin{cases} x - 2y + 2z = 6 \\ 2x - 1.5y + 1.8z = 2.8 \\ -\dfrac{2}{3}x + \dfrac{1}{2}y - \dfrac{3}{5}z = -\dfrac{11}{30} \end{cases}$

$$\begin{bmatrix} 1 & -2 & 2 \\ 2 & -1.5 & 1.8 \\ -\dfrac{2}{3} & \dfrac{1}{2} & -\dfrac{3}{5} \end{bmatrix} \begin{bmatrix} x \\ y \\ z \end{bmatrix} = \begin{bmatrix} 6 \\ 2.8 \\ -\dfrac{11}{30} \end{bmatrix}$$

Singular; no solution

43. $\begin{cases} -2w + 3x - 4y + 5z = -3 \\ 0.2w - 2.6x + y - 0.4z = 2.4 \\ -3w + 3.2x + 2.8y + z = 6.1 \\ 1.6w + 4x - 5y + 2.6z = -9.8 \end{cases}$

$$\begin{bmatrix} -2 & 3 & -4 & 5 \\ 0.2 & -2.6 & 1 & -0.4 \\ -3 & 3.2 & 2.8 & 1 \\ 1.6 & 4 & -5 & 2.6 \end{bmatrix} \begin{bmatrix} w \\ x \\ y \\ z \end{bmatrix} = \begin{bmatrix} -3 \\ 2.4 \\ 6.1 \\ -9.8 \end{bmatrix}$$

Using the grapher,

$$A^{-1} = \begin{bmatrix} -0.35859 & 1.15741 & 0.36978 & 0.72543 \\ -0.10811 & -0.13514 & 0.13514 & 0.13514 \\ -0.23646 & 0.933507 & 0.44725 & 0.42633 \\ -0.06774 & 1.290852 & 0.42463 & 0.55015 \end{bmatrix}$$

$$A^{-1}\left(\begin{bmatrix} -2 & 3 & -4 & 5 \\ 0.2 & -2.6 & 1 & -0.4 \\ -3 & 3.2 & 2.8 & 1 \\ 1.6 & 4 & -5 & 2.6 \end{bmatrix} \begin{bmatrix} w \\ x \\ y \\ z \end{bmatrix} \right) = A^{-1} \begin{bmatrix} -3 \\ 2.4 \\ 6.1 \\ -9.8 \end{bmatrix}$$

$$\left(A^{-1} \begin{bmatrix} -2 & 3 & -4 & 5 \\ 0.2 & -2.6 & 1 & -0.4 \\ -3 & 3.2 & 2.8 & 1 \\ 1.6 & 4 & -5 & 2.6 \end{bmatrix} \right) \begin{bmatrix} w \\ x \\ y \\ z \end{bmatrix} = A^{-1} \begin{bmatrix} -3 \\ 2.4 \\ 6.1 \\ -9.8 \end{bmatrix}$$

$$\begin{bmatrix} 1 & 0 & 0 & 0 \\ 0 & 1 & 0 & 0 \\ 0 & 0 & 1 & 0 \\ 0 & 0 & 0 & 1 \end{bmatrix} \begin{bmatrix} w \\ x \\ y \\ z \end{bmatrix} = A^{-1} \begin{bmatrix} -3 \\ 2.4 \\ 6.1 \\ -9.8 \end{bmatrix}$$

$$\begin{bmatrix} w \\ x \\ y \\ z \end{bmatrix} = \begin{bmatrix} -1 \\ -0.5 \\ 1.5 \\ 0.5 \end{bmatrix}$$

$(-1, -0.5, 1.5, 0.5)$

45. $\begin{bmatrix} 4 & -7 \\ 3 & -5 \end{bmatrix}$

$\det A = 4(-5) - 3(-7) = -20 + 21 = 1$

yes

47. $\begin{bmatrix} 1.2 & -0.8 \\ 0.3 & -0.2 \end{bmatrix}$

$\det A = 1.2(-0.2) - (0.3)(-0.8)$

$= -0.24 + 0.24 = 0$

no

49. $A = \begin{bmatrix} 1 & 0 & -2 \\ 0 & -1 & -1 \\ 2 & 1 & -4 \end{bmatrix}$

$\det A$

$= 1\begin{vmatrix} -1 & -1 \\ 1 & -4 \end{vmatrix} - 0\begin{vmatrix} 0 & -1 \\ 2 & -4 \end{vmatrix} + (-2)\begin{vmatrix} 0 & -1 \\ 2 & 1 \end{vmatrix}$

$= 1(4+1) - 0 - 2(0+2)$

$= 5 - 0 - 4$

$= 1$

51. $C = \begin{bmatrix} -2 & 3 & 4 \\ 0 & 6 & 2 \\ 1 & -1.5 & -2 \end{bmatrix}$

$\det C$

$= 0\begin{vmatrix} 3 & 4 \\ -1.5 & -2 \end{vmatrix} + 6\begin{vmatrix} -2 & 4 \\ 1 & -2 \end{vmatrix} - 2\begin{vmatrix} -2 & 3 \\ 1 & -1.5 \end{vmatrix}$

$= 0 + 6(4 - 4) - 2(3 - 3)$

$= 0$

Singular Matrix

53. $A = \begin{bmatrix} 1 & 0 & 3 & -4 \\ 2 & 5 & 0 & 1 \\ 8 & 15 & 6 & -5 \\ 0 & 8 & -4 & 1 \end{bmatrix}$

$\det A = 0$

Singular

55. $\begin{bmatrix} 2 & -3 & 1 \\ 4 & -1 & 5 \\ 1 & 0 & -2 \end{bmatrix}$

$\begin{bmatrix} 2 & -3 & 1 \\ 4 & -1 & 5 \\ 1 & 0 & -2 \end{bmatrix}\begin{matrix} 2 & -3 \\ 4 & -1 \\ 1 & 0 \end{matrix}$

$2(-1)(-2) = 4,$
$(-3)(5)(1) = -15,$
$1(4)(0) = 0,$
$4 + (-15) + 0 = -11;$
$(1)(-1)(1) = -1,$
$(0)(5)(2) = 0,$
$(-2)(4)(-3) = 24,$
$-1 + 0 + 24 = -23;$
$-11 - 23 = -34$

57. $\begin{bmatrix} 1 & -1 & 2 \\ 3 & -2 & 4 \\ 4 & 3 & 1 \end{bmatrix}$

$\begin{bmatrix} 1 & -1 & 2 \\ 3 & -2 & 4 \\ 4 & 3 & 1 \end{bmatrix}\begin{matrix} 1 & -1 \\ 3 & -2 \\ 4 & 3 \end{matrix}$

$1(-2)(1) = -2,$
$(-1)(4)(4) = -16,$
$2(3)(3) = 18,$
$-2 + (-16) + 18 = 0;$
$(4)(-2)(2) = -16,$
$(3)(4)(1) = 12,$
$(1)(3)(-1) = -3,$
$-16 + 12 + (-3) = -7;$
$0 - (-7) = 7$

59. $\begin{cases} x - 2y + 2z = 7 \\ 2x + 2y - z = 5 \\ 3x - y + z = 6 \end{cases}$

(1) $\begin{bmatrix} 1 & -2 & 2 \\ 2 & 2 & -1 \\ 3 & -1 & 1 \end{bmatrix}\begin{bmatrix} x \\ y \\ z \end{bmatrix} = \begin{bmatrix} 7 \\ 5 \\ 6 \end{bmatrix}$

(2) $\det A$

$= 1\begin{vmatrix} 2 & -1 \\ -1 & 1 \end{vmatrix} - (-2)\begin{vmatrix} 2 & -1 \\ 3 & 1 \end{vmatrix} + 2\begin{vmatrix} 2 & 2 \\ 3 & -1 \end{vmatrix}$

$= 1(2-1) + 2(2+3) + 2(-2-6)$
$= 1 + 10 - 16$
$= -5$

(3) $A^{-1} = \begin{bmatrix} -0.2 & 0 & 0.4 \\ 1 & 1 & -1 \\ 1.6 & 1 & -1.2 \end{bmatrix}$

$X = A^{-1}B$

$X = \begin{bmatrix} -0.2 & 0 & 0.4 \\ 1 & 1 & -1 \\ 1.6 & 1 & -1.2 \end{bmatrix}\begin{bmatrix} 7 \\ 5 \\ 6 \end{bmatrix}$

$X = \begin{bmatrix} 1 \\ 6 \\ 9 \end{bmatrix}$

$(1, 6, 9)$

61. $\begin{cases} x - 3y + 4z = -1 \\ 4x - y + 5z = 7 \\ 3x + 2y + z = -3 \end{cases}$

(1) $\begin{bmatrix} 1 & -3 & 4 \\ 4 & -1 & 5 \\ 3 & 2 & 1 \end{bmatrix}\begin{bmatrix} x \\ y \\ z \end{bmatrix} = \begin{bmatrix} -1 \\ 7 \\ -3 \end{bmatrix}$

(2) $\det A$

$= 1\begin{vmatrix} -1 & 5 \\ 2 & 1 \end{vmatrix} - (-3)\begin{vmatrix} 4 & 5 \\ 3 & 1 \end{vmatrix} + 4\begin{vmatrix} 4 & -1 \\ 3 & 2 \end{vmatrix}$

$= 1(-1-10) + 3(4-15) + 4(8+3)$
$= -11 - 33 + 44$
$= 0$
Singular

63. $A = \begin{bmatrix} 3 & -5 \\ 2 & 1 \end{bmatrix}$

$A^{-1} = \dfrac{1}{ad-bc}\begin{bmatrix} d & -b \\ -c & a \end{bmatrix}$

$A^{-1} = \dfrac{1}{3(1)-(-5)(2)}\begin{bmatrix} 1 & 5 \\ -2 & 3 \end{bmatrix}$

$A^{-1} = \dfrac{1}{13}\begin{bmatrix} 1 & 5 \\ -2 & 3 \end{bmatrix}$

$A^{-1} = \begin{bmatrix} \dfrac{1}{13} & \dfrac{5}{13} \\ \dfrac{-2}{13} & \dfrac{3}{13} \end{bmatrix};$

$AA^{-1} = \begin{bmatrix} 3 & -5 \\ 2 & 1 \end{bmatrix}\begin{bmatrix} \dfrac{1}{13} & \dfrac{5}{13} \\ \dfrac{-2}{13} & \dfrac{3}{13} \end{bmatrix} = \begin{bmatrix} 1 & 0 \\ 0 & 1 \end{bmatrix}$

$A^{-1}A = \begin{bmatrix} \dfrac{1}{13} & \dfrac{5}{13} \\ \dfrac{-2}{13} & \dfrac{3}{13} \end{bmatrix}\begin{bmatrix} 3 & -5 \\ 2 & 1 \end{bmatrix} = \begin{bmatrix} 1 & 0 \\ 0 & 1 \end{bmatrix}$

$AA^{-1} = A^{-1}A = I$

65. $C = \begin{bmatrix} 0.3 & -0.4 \\ -0.6 & 0.8 \end{bmatrix}$

$C^{-1} = \dfrac{1}{ad-bc}\begin{bmatrix} d & -b \\ -c & a \end{bmatrix}$

$C^{-1} = \dfrac{1}{0.3(0.8)-(-0.4)(-0.6)}\begin{bmatrix} 0.8 & 0.4 \\ 0.6 & 0.3 \end{bmatrix}$

$C^{-1} = \dfrac{1}{0}\begin{bmatrix} 0.8 & 0.4 \\ 0.6 & 0.3 \end{bmatrix}$

Singular

67. Let B represent the number of behemoth slushies sold.
Let G represent the number of gargantuan slushies sold.
Let M represent the number of mammoth slushies sold.
Let J represent the number of jumbo slushies sold.

$\begin{cases} 2.59B + 2.29G + 1.99M + 1.59J = 402.29 \\ 60B + 48G + 36M + 24J = 7884 \\ B + G + M + J = 191 \\ B = J + 1 \end{cases}$

$\begin{cases} 2.59B + 2.29G + 1.99M + 1.59J = 402.29 \\ 60B + 48G + 36M + 24J = 7884 \\ B + G + M + J = 191 \\ B - J = 1 \end{cases}$

$\begin{bmatrix} 2.59 & 2.29 & 1.99 & 1.59 \\ 60 & 48 & 36 & 24 \\ 1 & 1 & 1 & 1 \\ 1 & 0 & 0 & -1 \end{bmatrix}\begin{bmatrix} B \\ G \\ M \\ J \end{bmatrix} = \begin{bmatrix} 402.29 \\ 7884 \\ 191 \\ 1 \end{bmatrix}$

$A^{-1} = \begin{bmatrix} -10 & \dfrac{1}{4} & \dfrac{109}{10} & 1 \\ 10 & -\dfrac{1}{6} & -\dfrac{139}{10} & -2 \\ 10 & -\dfrac{1}{3} & -\dfrac{69}{10} & 1 \\ -10 & \dfrac{1}{4} & \dfrac{109}{10} & 0 \end{bmatrix};$

$X = A^{-1}B$

$X = \begin{bmatrix} -10 & \dfrac{1}{4} & \dfrac{109}{10} & 1 \\ 10 & -\dfrac{1}{6} & -\dfrac{139}{10} & -2 \\ 10 & -\dfrac{1}{3} & -\dfrac{69}{10} & 1 \\ -10 & \dfrac{1}{4} & \dfrac{109}{10} & 0 \end{bmatrix}\begin{bmatrix} 402.29 \\ 7884 \\ 191 \\ 1 \end{bmatrix}$

$X = \begin{bmatrix} 31 \\ 52 \\ 78 \\ 30 \end{bmatrix}$

31 behemoth
52 gargantuan
78 mammoth
30 jumbo

69. Let J represent the playing time of *Jumpin' Jack Flash*.
Let T represent the playing time of *Tumbling Dice*.
Let W represent the playing time of *Wild Horses*.
Let Y represent the playing time of You Can't Always Get What You Want.

$$\begin{cases} J+T+W+Y=20.75 \\ J+T=Y \\ W=J+2 \\ Y=2T \end{cases}$$

$$\begin{cases} J+T+W+Y=20.75 \\ J+T-Y=0 \\ -J+W=2 \\ -2T+Y=0 \end{cases}$$

$$\begin{bmatrix} 1 & 1 & 1 & 1 \\ 1 & 1 & 0 & -1 \\ -1 & 0 & 1 & 0 \\ 0 & -2 & 0 & 1 \end{bmatrix} \begin{bmatrix} J \\ T \\ W \\ Y \end{bmatrix} = \begin{bmatrix} 20.75 \\ 0 \\ 2 \\ 0 \end{bmatrix}$$

$$A^{-1} = \begin{bmatrix} 0.2 & 0.6 & -0.2 & 0.4 \\ 0.2 & -0.4 & -0.2 & -0.6 \\ 0.2 & 0.6 & 0.8 & 0.4 \\ 0.4 & -0.8 & -0.4 & -0.2 \end{bmatrix};$$

$X = A^{-1}B$

$$X = \begin{bmatrix} 0.2 & 0.6 & -0.2 & 0.4 \\ 0.2 & -0.4 & -0.2 & -0.6 \\ 0.2 & 0.6 & 0.8 & 0.4 \\ 0.4 & -0.8 & -0.4 & -0.2 \end{bmatrix} \begin{bmatrix} 20.75 \\ 0 \\ 2 \\ 0 \end{bmatrix}$$

$$X = \begin{bmatrix} 3.75 \\ 3.75 \\ 5.75 \\ 7.5 \end{bmatrix}$$

Jumpin' Jack Flash: 3.75 min
Tumbling Dice: 3.75 min
Wild Horses: 5.75 min
You Can't Always Get What You Want: 7.5 min

71. Let A represent the number of Clock A manufactured.
Let B represent the number of Clock B manufactured.
Let C represent the number of Clock C manufactured.
Let D represent the number of Clock D manufactured.

$$\begin{cases} 2.2A+2.5B+2.75C+3D=262 \\ 1.2A+1.4B+1.8C+2D=160 \\ 0.2A+0.25B+0.3C+0.5D=29 \\ 0.5A+0.55B+0.75C+D=68 \end{cases}$$

$$\begin{bmatrix} 2.2 & 2.5 & 2.75 & 3 \\ 1.2 & 1.4 & 1.8 & 2 \\ 0.2 & 0.25 & 0.3 & 0.5 \\ 0.5 & 0.55 & 0.75 & 1 \end{bmatrix} \begin{bmatrix} A \\ B \\ C \\ D \end{bmatrix} = \begin{bmatrix} 262 \\ 160 \\ 29 \\ 68 \end{bmatrix}$$

$$A^{-1} = \begin{bmatrix} \dfrac{30}{11} & -\dfrac{205}{22} & -\dfrac{210}{11} & 20 \\ 0 & 5 & 20 & -20 \\ -\dfrac{20}{11} & \dfrac{50}{11} & -\dfrac{80}{11} & 0 \\ 0 & -\dfrac{3}{2} & 4 & 2 \end{bmatrix};$$

$X = A^{-1}B$

$$X = \begin{bmatrix} \dfrac{30}{11} & -\dfrac{205}{22} & -\dfrac{210}{11} & 20 \\ 0 & 5 & 20 & -20 \\ -\dfrac{20}{11} & \dfrac{50}{11} & -\dfrac{80}{11} & 0 \\ 0 & -\dfrac{3}{2} & 4 & 2 \end{bmatrix} \begin{bmatrix} 262 \\ 160 \\ 29 \\ 68 \end{bmatrix}$$

$$X = \begin{bmatrix} 30 \\ 20 \\ 40 \\ 12 \end{bmatrix}$$

30 of clock A
20 of clock B
40 of clock C
12 of clock D

73.

$$\begin{cases} p_1 = \dfrac{70+64+p_2+p_3}{4} \\[2mm] p_2 = \dfrac{80+64+p_4+p_1}{4} \\[2mm] p_3 = \dfrac{70+96+p_1+p_4}{4} \\[2mm] p_4 = \dfrac{96+80+p_2+p_3}{4} \end{cases}$$

$$\begin{cases} 4p_1 - p_2 - p_3 = 134 \\ -p_1 + 4p_2 - p_4 = 144 \\ -p_1 + 4p_3 - p_4 = 166 \\ -p_2 - p_3 + 4p_4 = 176 \end{cases}$$

$$\begin{bmatrix} 4 & -1 & -1 & 0 \\ -1 & 4 & 0 & -1 \\ -1 & 0 & 4 & -1 \\ 0 & -1 & -1 & 4 \end{bmatrix} \begin{bmatrix} p_1 \\ p_2 \\ p_3 \\ p_4 \end{bmatrix} = \begin{bmatrix} 134 \\ 144 \\ 166 \\ 176 \end{bmatrix}$$

$$A^{-1} = \begin{bmatrix} \dfrac{7}{24} & \dfrac{1}{12} & \dfrac{1}{12} & \dfrac{1}{24} \\[2mm] \dfrac{1}{12} & \dfrac{7}{24} & \dfrac{1}{24} & \dfrac{1}{12} \\[2mm] \dfrac{1}{12} & \dfrac{1}{24} & \dfrac{7}{24} & \dfrac{1}{12} \\[2mm] \dfrac{1}{24} & \dfrac{1}{12} & \dfrac{1}{12} & \dfrac{7}{24} \end{bmatrix}$$

$$A^{-1} \begin{bmatrix} 4 & -1 & -1 & 0 \\ -1 & 4 & 0 & -1 \\ -1 & 0 & 4 & -1 \\ 0 & -1 & -1 & 4 \end{bmatrix} \begin{bmatrix} p_1 \\ p_2 \\ p_3 \\ p_4 \end{bmatrix} = A^{-1} \begin{bmatrix} 134 \\ 144 \\ 166 \\ 176 \end{bmatrix}$$

$$\begin{bmatrix} 1 & 0 & 0 & 0 \\ 0 & 1 & 0 & 0 \\ 0 & 0 & 1 & 0 \\ 0 & 0 & 0 & 1 \end{bmatrix} \begin{bmatrix} p_1 \\ p_2 \\ p_3 \\ p_4 \end{bmatrix} = \begin{bmatrix} 72.25 \\ 74.75 \\ 80.25 \\ 82.75 \end{bmatrix}$$

$p_1 = 72.25^\circ, p_2 = 74.75^\circ, p_3 = 80.25^\circ, p_4 = 82.75^\circ$

75. $ax^3 + bx^2 + cx + d = 0$

$(-4, -6), (-1, 0), (1, -16)$ and $(3, 8)$

$a(-4)^3 + b(-4)^2 + c(-4) + d = -6$
$-64a + 16b - 4c + d = -6;$

$a(-1)^3 + b(-1)^2 + c(-1) + d = 0$
$-a + b - c + d = 0;$

$a(1)^3 + b(1)^2 + c(1) + d = -16$
$a + b + c + d = -16;$

$a(3)^3 + b(3)^2 + c(3) + d = 8$
$27a + 9b + 3c + d = 8;$

$$\begin{cases} -64a + 16b - 4c + d = -6 \\ -a + b - c + d = 0 \\ a + b + c + d = -16 \\ 27a + 9b + 3c + d = 8 \end{cases}$$

$$\begin{bmatrix} -64 & 16 & -4 & 1 \\ -1 & 1 & -1 & 1 \\ 1 & 1 & 1 & 1 \\ 27 & 9 & 3 & 1 \end{bmatrix} \begin{bmatrix} a \\ b \\ c \\ d \end{bmatrix} = \begin{bmatrix} -6 \\ 0 \\ -16 \\ 8 \end{bmatrix}$$

$$A^{-1} = \begin{bmatrix} -\dfrac{1}{105} & \dfrac{1}{24} & -\dfrac{1}{20} & \dfrac{1}{56} \\[2mm] \dfrac{1}{35} & 0 & -\dfrac{1}{10} & \dfrac{1}{14} \\[2mm] \dfrac{1}{105} & -\dfrac{13}{24} & \dfrac{11}{20} & -\dfrac{1}{56} \\[2mm] -\dfrac{1}{35} & \dfrac{1}{2} & \dfrac{3}{5} & -\dfrac{1}{14} \end{bmatrix};$$

$X = A^{-1}B$

$$X = \begin{bmatrix} -\dfrac{1}{105} & \dfrac{1}{24} & -\dfrac{1}{20} & \dfrac{1}{56} \\[2mm] \dfrac{1}{35} & 0 & -\dfrac{1}{10} & \dfrac{1}{14} \\[2mm] \dfrac{1}{105} & -\dfrac{13}{24} & \dfrac{11}{20} & -\dfrac{1}{56} \\[2mm] -\dfrac{1}{35} & \dfrac{1}{2} & \dfrac{3}{5} & -\dfrac{1}{14} \end{bmatrix} \begin{bmatrix} -6 \\ 0 \\ -16 \\ 8 \end{bmatrix}$$

$$X = \begin{bmatrix} 1 \\ 2 \\ -9 \\ -10 \end{bmatrix}$$

$y = x^3 + 2x^2 - 9x - 10$

77. Let x represent the number of ounces of food for Food I.
Let y represent the number of ounces of food for Food II.
Let z represent the number of ounces of food for Food III.

$$\begin{cases} 2x+4y+3z=20 \\ 4x+2y+5z=30 \\ 5x+6y+7z=44 \end{cases}$$

$$\begin{bmatrix} 2 & 4 & 3 \\ 4 & 2 & 5 \\ 5 & 6 & 7 \end{bmatrix}\begin{bmatrix} x \\ y \\ z \end{bmatrix}=\begin{bmatrix} 20 \\ 30 \\ 44 \end{bmatrix}$$

$$A^{-1}=\begin{bmatrix} 8 & 5 & -7 \\ 1.5 & 0.5 & 1 \\ -7 & -4 & 6 \end{bmatrix};$$

$$X=A^{-1}B$$

$$X=\begin{bmatrix} 8 & 5 & -7 \\ 1.5 & 0.5 & -1 \\ -7 & -4 & 6 \end{bmatrix}\begin{bmatrix} 20 \\ 30 \\ 44 \end{bmatrix}$$

$$X=\begin{bmatrix} 2 \\ 1 \\ 4 \end{bmatrix}$$

2 oz food I
1 oz food II
4 oz food III

79. Answers will vary.

81. a.
$$\begin{bmatrix} 1 & -2 & 3 \\ -4 & 5 & -6 \\ 2 & 5 & 3 \end{bmatrix}\;\; 4R1+R2\rightarrow R2$$

$$\begin{bmatrix} 1 & -2 & 3 \\ 0 & -3 & 6 \\ 2 & 5 & 3 \end{bmatrix}\;\; -2R1+R3\rightarrow R3$$

$$\begin{bmatrix} 1 & -2 & 3 \\ 0 & -3 & 6 \\ 0 & 9 & -3 \end{bmatrix}\;\; 3R2+R3\rightarrow R3$$

$$\begin{bmatrix} 1 & -2 & 3 \\ 0 & -3 & 6 \\ 0 & 0 & 15 \end{bmatrix}\;\; -\frac{2}{3}R2+R1\rightarrow R1$$

$$\begin{bmatrix} 1 & 0 & -1 \\ 0 & -3 & 6 \\ 0 & 0 & 15 \end{bmatrix}\;\; \frac{1}{15}R3+R1\rightarrow R1$$

$$\begin{bmatrix} 1 & 0 & 0 \\ 0 & -3 & 6 \\ 0 & 0 & 15 \end{bmatrix}\;\; -\frac{2}{5}R3+R2\rightarrow R2$$

$$\begin{bmatrix} 1 & 0 & 0 \\ 0 & -3 & 0 \\ 0 & 0 & 15 \end{bmatrix}$$

$(1)(-3)(15)=-45$

b.
$$\begin{bmatrix} 2 & 5 & -1 \\ -2 & -3 & 4 \\ 4 & 6 & 5 \end{bmatrix}\;\; R1+R2\rightarrow R2$$

$$\begin{bmatrix} 2 & 5 & -1 \\ 0 & 2 & 3 \\ 4 & 6 & 5 \end{bmatrix}\;\; -2R1+R3\rightarrow R3$$

$$\begin{bmatrix} 2 & 5 & -1 \\ 0 & 2 & 3 \\ 0 & -4 & 7 \end{bmatrix}\;\; 2R2+R3\rightarrow R3$$

$$\begin{bmatrix} 2 & 5 & -1 \\ 0 & 2 & 3 \\ 0 & 0 & 13 \end{bmatrix}\;\; -\frac{5}{2}R2+R1\rightarrow R1$$

$$\begin{bmatrix} 2 & 0 & -\frac{17}{2} \\ 0 & 2 & 3 \\ 0 & 0 & 13 \end{bmatrix}\;\; \frac{17}{26}R3+R1\rightarrow R1$$

$$\begin{bmatrix} 2 & 0 & 0 \\ 0 & 2 & 3 \\ 0 & 0 & 13 \end{bmatrix}\;\; -\frac{3}{13}R3+R2\rightarrow R2$$

$$\begin{bmatrix} 2 & 0 & 0 \\ 0 & 2 & 0 \\ 0 & 0 & 13 \end{bmatrix}$$

$(2)(2)(13)=52$

c. $\begin{bmatrix} -2 & 4 & 1 \\ 5 & 7 & -2 \\ 3 & -8 & -1 \end{bmatrix} R1 + R3 \rightarrow R1$

$\begin{bmatrix} 1 & -4 & 0 \\ 5 & 7 & -2 \\ 3 & -8 & -1 \end{bmatrix} -5R1 + R2 \rightarrow R2$

$\begin{bmatrix} 1 & -4 & 0 \\ 0 & 27 & -2 \\ 3 & -8 & -1 \end{bmatrix} -3R1 + R3 \rightarrow R3$

$\begin{bmatrix} 1 & -4 & 0 \\ 0 & 27 & -2 \\ 0 & 4 & -1 \end{bmatrix} -2R3 + R2 \rightarrow R2$

$\begin{bmatrix} 1 & -4 & 0 \\ 0 & 19 & 0 \\ 0 & 4 & -1 \end{bmatrix} -\dfrac{4}{19}R2 + R3 \rightarrow R3$

$\begin{bmatrix} 1 & -4 & 0 \\ 0 & 19 & 0 \\ 0 & 0 & -1 \end{bmatrix} \dfrac{4}{19}R2 + R1 \rightarrow R1$

$\begin{bmatrix} 1 & 0 & 0 \\ 0 & 19 & 0 \\ 0 & 0 & -1 \end{bmatrix}$

$(1)(19)(-1) = -19$

d. $\begin{bmatrix} 3 & -1 & 4 \\ 0 & -2 & 6 \\ -2 & 1 & -3 \end{bmatrix} R1 + R3 \rightarrow R1$

$\begin{bmatrix} 1 & 0 & 1 \\ 0 & -2 & 6 \\ -2 & 1 & -3 \end{bmatrix} 2R1 + R3 \rightarrow R3$

$\begin{bmatrix} 1 & 0 & 1 \\ 0 & -2 & 6 \\ 0 & 1 & -1 \end{bmatrix} \dfrac{1}{2}R2 + R3 \rightarrow R3$

$\begin{bmatrix} 1 & 0 & 1 \\ 0 & -2 & 6 \\ 0 & 0 & 2 \end{bmatrix} -3R3 + R2 \rightarrow R2$

$\begin{bmatrix} 1 & 0 & 1 \\ 0 & -2 & 0 \\ 0 & 0 & 2 \end{bmatrix} -\dfrac{1}{2}R3 + R1 \rightarrow R1$

$\begin{bmatrix} 1 & 0 & 0 \\ 0 & -2 & 0 \\ 0 & 0 & 2 \end{bmatrix}$

$(1)(-2)(2) = -4$

83. $x^3 - 7x^2 = -36$

$x^3 - 7x^2 + 36 = 0$

Possible roots:

$$\dfrac{\{\pm 1, \pm 36, \pm 2, \pm 18, \pm 3, \pm 12, \pm 4, \pm 9, \pm 6\}}{\{\pm 1\}}$$

$\underline{6|}\ \ \begin{array}{rrrr} 1 & -7 & 0 & 36 \\ & 6 & -6 & -36 \end{array}$
$\overline{\ \ \begin{array}{rrrr} 1 & -1 & -6 & |\underline{0} \end{array}}$

$(x - 6)(x^2 - x - 6) = 0$

$(x - 6)(x - 3)(x + 2) = 0$

$x = 6,\ x = 3,\ x = -2$

85. $y = \log_2(x - 2)$

Graph a; shifts right 2

$y = \log_2(x) - 2$

Graph b; shifts down 2

6.4 Exercises

6.4 Exercises

1. $a_{11}a_{22} - a_{21}a_{12}$

3. Constant

5. Answers will vary.

7. $\begin{cases} 2x + 5y = 7 \\ -3x + 4y = 1 \end{cases}$

$D = \begin{vmatrix} 2 & 5 \\ -3 & 4 \end{vmatrix}$; $D_x = \begin{vmatrix} 7 & 5 \\ 1 & 4 \end{vmatrix}$; $D_y = \begin{vmatrix} 2 & 7 \\ -3 & 1 \end{vmatrix}$

9. $\begin{cases} 4x + y = -11 \\ 3x - 5y = -60 \end{cases}$

$D = \begin{vmatrix} 4 & 1 \\ 3 & -5 \end{vmatrix} = -20 - 3 = -23$;

$D_x = \begin{vmatrix} -11 & 1 \\ -60 & -5 \end{vmatrix} = 55 + 60 = 115$;

$D_y = \begin{vmatrix} 4 & -11 \\ 3 & -60 \end{vmatrix} = -240 + 33 = -207$;

$x = \dfrac{D_x}{D} = \dfrac{115}{-23} = -5$;

$y = \dfrac{D_y}{D} = \dfrac{-207}{-23} = 9$

$(-5, 9)$

11. $\begin{cases} \dfrac{x}{8} + \dfrac{y}{4} = 1 \\ \dfrac{y}{5} = \dfrac{x}{2} + 6 \end{cases}$

$\begin{cases} \dfrac{x}{8} + \dfrac{y}{4} = 1 \\ -\dfrac{x}{2} + \dfrac{y}{5} = 6 \end{cases}$

$D = \begin{vmatrix} \dfrac{1}{8} & \dfrac{1}{4} \\ -\dfrac{1}{2} & \dfrac{1}{5} \end{vmatrix} = \dfrac{1}{40} + \dfrac{1}{8} = \dfrac{3}{20}$;

$D_x = \begin{vmatrix} 1 & \dfrac{1}{4} \\ 6 & \dfrac{1}{5} \end{vmatrix} = \dfrac{1}{5} - \dfrac{3}{2} = -\dfrac{13}{10}$;

$D_y = \begin{vmatrix} \dfrac{1}{8} & 1 \\ -\dfrac{1}{2} & 6 \end{vmatrix} = \dfrac{3}{4} + \dfrac{1}{2} = \dfrac{5}{4}$;

$x = \dfrac{D_x}{D} = \dfrac{-\dfrac{13}{10}}{\dfrac{3}{20}} = -\dfrac{260}{30} = -\dfrac{26}{3}$;

$y = \dfrac{D_y}{D} = \dfrac{\dfrac{5}{4}}{\dfrac{3}{20}} = \dfrac{100}{12} = \dfrac{25}{3}$

$\left(\dfrac{-26}{3}, \dfrac{25}{3} \right)$

13. $\begin{cases} 0.6x - 0.3y = 8 \\ 0.8x - 0.4y = -3 \end{cases}$

$D = \begin{vmatrix} 0.6 & -0.3 \\ 0.8 & -0.4 \end{vmatrix} = -0.24 + 0.24 = 0$

No Solution; determinant cannot be zero.

15. a. $\begin{cases} 4x - y + 2z = -5 \\ -3x + 2y - z = 8 \\ x - 5y + 3z = -3 \end{cases}$

$D = \begin{vmatrix} 4 & -1 & 2 \\ -3 & 2 & -1 \\ 1 & -5 & 3 \end{vmatrix}$

$D = 4\begin{vmatrix} 2 & -1 \\ -5 & 3 \end{vmatrix} + 1\begin{vmatrix} -3 & -1 \\ 1 & 3 \end{vmatrix} + 2\begin{vmatrix} -3 & 2 \\ 1 & -5 \end{vmatrix}$

$D = 4(1) + 1(-8) + 2(13) = 22$; solutions possible

$D_x = \begin{vmatrix} -5 & -1 & 2 \\ 8 & 2 & -1 \\ -3 & -5 & 3 \end{vmatrix}$;

$D_y = \begin{vmatrix} 4 & -5 & 2 \\ -3 & 8 & -1 \\ 1 & -3 & 3 \end{vmatrix}$;

$D_z = \begin{vmatrix} 4 & -1 & -5 \\ -3 & 2 & 8 \\ 1 & -5 & -3 \end{vmatrix}$

b. $\begin{cases} 4x - y + 2z = -5 \\ -3x + 2y - z = 8 \\ x + y + z = -3 \end{cases}$

$D = \begin{vmatrix} 4 & -1 & 2 \\ -3 & 2 & -1 \\ 1 & 1 & 1 \end{vmatrix}$

$D = 4\begin{vmatrix} 2 & -1 \\ 1 & 1 \end{vmatrix} + 1\begin{vmatrix} -3 & -1 \\ 1 & 1 \end{vmatrix} + 2\begin{vmatrix} -3 & 2 \\ 1 & 1 \end{vmatrix}$

$D = 4(3) + 1(-2) + 2(-5) = 0$

$D = 0$; Cramer's Rule cannot be used.

17. $\begin{cases} x + 2y + 5z = 10 \\ 3x + 4y - z = 10 \\ x - y - z = -2 \end{cases}$

$D = \begin{vmatrix} 1 & 2 & 5 \\ 3 & 4 & -1 \\ 1 & -1 & -1 \end{vmatrix} = -36$;

$D_x = \begin{vmatrix} 10 & 2 & 5 \\ 10 & 4 & -1 \\ -2 & -1 & -1 \end{vmatrix} = -36$;

$D_y = \begin{vmatrix} 1 & 10 & 5 \\ 3 & 10 & -1 \\ 1 & -2 & -1 \end{vmatrix} = -72$;

$D_z = \begin{vmatrix} 1 & 2 & 10 \\ 3 & 4 & 10 \\ 1 & -1 & -2 \end{vmatrix} = -36$;

$x = \dfrac{D_x}{D} = \dfrac{-36}{-36} = 1$;

$y = \dfrac{D_y}{D} = \dfrac{-72}{-36} = 2$;

$z = \dfrac{D_z}{D} = \dfrac{-36}{-36} = 1$

$(1, 2, 1)$

19. $\begin{cases} y + 2z = 1 \\ 4x - 5y + 8z = -8 \\ 8x - 9z = 9 \end{cases}$

$D = \begin{vmatrix} 0 & 1 & 2 \\ 4 & -5 & 8 \\ 8 & 0 & -9 \end{vmatrix} = 180$;

$D_x = \begin{vmatrix} 1 & 1 & 2 \\ -8 & -5 & 8 \\ 9 & 0 & -9 \end{vmatrix} = 135$;

$D_y = \begin{vmatrix} 0 & 1 & 2 \\ 4 & -8 & 8 \\ 8 & 9 & -9 \end{vmatrix} = 300$;

$D_z = \begin{vmatrix} 0 & 1 & 1 \\ 4 & -5 & -8 \\ 8 & 0 & 9 \end{vmatrix} = -60$;

$x = \dfrac{D_x}{D} = \dfrac{135}{180} = \dfrac{3}{4}$;

$y = \dfrac{D_y}{D} = \dfrac{300}{180} = \dfrac{5}{3}$;

$z = \dfrac{D_z}{D} = \dfrac{-60}{180} = -\dfrac{1}{3}$

$\left(\dfrac{3}{4}, \dfrac{5}{3}, -\dfrac{1}{3} \right)$

21.
$$\begin{cases} w+2x-3y=-8 \\ x-3y+5z=-22 \\ 4w-5x=5 \\ -y+3z=-11 \end{cases}$$

$$D = \begin{vmatrix} 1 & 2 & -3 & 0 \\ 0 & 1 & -3 & 5 \\ 4 & -5 & 0 & 0 \\ 0 & 0 & -1 & 3 \end{vmatrix} = -16\,;$$

$$D_w = \begin{vmatrix} -8 & 2 & -3 & 0 \\ -22 & 1 & -3 & 5 \\ 5 & -5 & 0 & 0 \\ -11 & 0 & -1 & 3 \end{vmatrix} = 0\,;$$

$$D_x = \begin{vmatrix} 1 & -8 & -3 & 0 \\ 0 & -22 & -3 & 5 \\ 4 & 5 & 0 & 0 \\ 0 & -11 & -1 & 3 \end{vmatrix} = 16\,;$$

$$D_y = \begin{vmatrix} 1 & 2 & -8 & 0 \\ 0 & 1 & -22 & 5 \\ 4 & -5 & 5 & 0 \\ 0 & 0 & -11 & 3 \end{vmatrix} = -32\,;$$

$$D_z = \begin{vmatrix} 1 & 2 & -3 & -8 \\ 0 & 1 & -3 & -22 \\ 4 & -5 & 0 & 5 \\ 0 & 0 & -1 & -11 \end{vmatrix} = 48\,;$$

$$w = \frac{D_w}{D} = \frac{0}{-16} = 0\,;$$

$$x = \frac{D_x}{D} = \frac{16}{-16} = -1\,;$$

$$y = \frac{D_y}{D} = \frac{-32}{-16} = 2\,;$$

$$z = \frac{D_z}{D} = \frac{48}{-16} = -3$$

$$(0,\ -1,\ 2,\ -3)$$

23. $\dfrac{3x+2}{(x+3)(x-2)}$

$$= \frac{A}{x+3} + \frac{B}{x-2}$$

25. $\dfrac{3x^2-2x+5}{(x-1)(x+2)(x-3)}$

$$= \frac{A}{x-1} + \frac{B}{x+2} + \frac{C}{x-3}$$

27. $\dfrac{x^2+5}{x(x-3)(x+1)}$

$$= \frac{A}{x} + \frac{B}{x-3} + \frac{C}{x+1}$$

29. $\dfrac{x^2+x-1}{x^2(x+2)}$

$$= \frac{A}{x} + \frac{B}{x^2} + \frac{C}{x+2}$$

31. $\dfrac{x^3+3x-2}{(x+1)(x^2+2)^2}$

$$= \frac{A}{x+1} + \frac{Bx+C}{(x^2+2)} + \frac{Dx+E}{(x^2+2)^2}$$

33. $\dfrac{4-x}{x^2+x} = \dfrac{4-x}{x(x+1)} = \dfrac{A}{x} + \dfrac{B}{x+1}$

$$4-x = A(x+1)+Bx$$

$$4-x = Ax+A+Bx$$

$$4-x = (A+B)x+A$$

$$\begin{cases} A+B=-1 \\ A=4 \end{cases}$$

$$\begin{bmatrix} 1 & 1 \\ 1 & 0 \end{bmatrix}\begin{bmatrix} A \\ B \end{bmatrix} = \begin{bmatrix} -1 \\ 4 \end{bmatrix}$$

$$A^{-1} = \begin{bmatrix} 0 & 1 \\ 1 & -1 \end{bmatrix}$$

$$X = A^{-1}B$$

$$X = \begin{bmatrix} 4 \\ -5 \end{bmatrix}$$

$$\frac{4}{x} - \frac{5}{x+1}$$

35. $\dfrac{2x-27}{2x^2+x-15} = \dfrac{2x-27}{(2x-5)(x+3)}$

$= \dfrac{A}{2x-5} + \dfrac{B}{x+3}$

$2x-27 = A(x+3) + B(2x-5)$

$2x-27 = Ax + 3A + B2x - 5B$

$2x-27 = (A+2B)x + 3A - 5B$

$\begin{cases} A+2B = 2 \\ 3A-5B = -27 \end{cases}$

$\begin{bmatrix} 1 & 2 \\ 3 & -5 \end{bmatrix}\begin{bmatrix} A \\ B \end{bmatrix} = \begin{bmatrix} 2 \\ -27 \end{bmatrix}$

$A^{-1} = \begin{bmatrix} \dfrac{5}{11} & \dfrac{2}{11} \\ \dfrac{3}{11} & \dfrac{-1}{11} \end{bmatrix}$

$X = A^{-1}B$

$X = \begin{bmatrix} -4 \\ 3 \end{bmatrix}$

$\dfrac{-4}{2x-5} + \dfrac{3}{x+3}$

37. $\dfrac{8x^2-3x-7}{x^3-x} = \dfrac{8x^2-3x-7}{x(x^2-1)}$

$= \dfrac{8x^2-3x-7}{x(x-1)(x+1)} = \dfrac{A}{x} + \dfrac{B}{x+1} + \dfrac{C}{x-1}$

$8x^2-3x-7 = A(x+1)(x-1) + Bx(x-1) + Cx(x+1)$

$8x^2-3x-7 = Ax^2 - A + Bx^2 - Bx + Cx^2 + Cx$

$8x^2-3x-7 = x^2(A+B+C) + x(-B+C) - A$

$\begin{cases} A+B+C = 8 \\ -B+C = -3 \\ -A = -7 \end{cases}$

$\begin{bmatrix} 1 & 1 & 1 \\ 0 & -1 & 1 \\ -1 & 0 & 0 \end{bmatrix}\begin{bmatrix} A \\ B \\ C \end{bmatrix} = \begin{bmatrix} 8 \\ -3 \\ -7 \end{bmatrix}$

$A^{-1} = \begin{bmatrix} 0 & 0 & -1 \\ \dfrac{1}{2} & \dfrac{-1}{2} & \dfrac{1}{2} \\ \dfrac{1}{2} & \dfrac{1}{2} & \dfrac{1}{2} \end{bmatrix}$

$X = A^{-1}B$

$X = \begin{bmatrix} 7 \\ 2 \\ -1 \end{bmatrix}$

$\dfrac{7}{x} + \dfrac{2}{x+1} - \dfrac{1}{x-1}$

39. $\dfrac{3x^2+7x-1}{x^3+2x^2+x} = \dfrac{3x^2+7x-1}{x(x^2+2x+1)}$

$= \dfrac{3x^2+7x-1}{x(x+1)^2} = \dfrac{A}{x} + \dfrac{B}{x+1} + \dfrac{C}{(x+1)^2}$

$3x^2+7x-1 = A(x+1)^2 + Bx(x+1) + Cx$

$3x^2+7x-1 = Ax^2 + 2Ax + A + Bx^2 + Bx + Cx$

$3x^2+7x-1 = x^2(A+B) + x(2A+B+C) + A$

$\begin{cases} A+B = 3 \\ 2A+B+C = 7 \\ A = -1 \end{cases}$

$\begin{bmatrix} 1 & 1 & 0 \\ 2 & 1 & 1 \\ 1 & 0 & 0 \end{bmatrix}\begin{bmatrix} A \\ B \\ C \end{bmatrix} = \begin{bmatrix} 3 \\ 7 \\ -1 \end{bmatrix}$

$A^{-1} = \begin{bmatrix} 0 & 0 & 1 \\ 1 & 0 & -1 \\ -1 & 1 & -1 \end{bmatrix}$

$X = A^{-1}B$

$X = \begin{bmatrix} -1 \\ 4 \\ 5 \end{bmatrix}$

$\dfrac{-1}{x} + \dfrac{4}{x+1} + \dfrac{5}{(x+1)^2}$

41. $\dfrac{3x^2+10x+4}{8-x^3}=\dfrac{3x^2+10x+4}{(2-x)(4+2x+x^2)}$

$=\dfrac{A}{2-x}+\dfrac{Bx+C}{4+2x+x^2}$

$3x^2+10x+4=A(4+2x+x^2)+(Bx+C)(2-x)$

$3x^2+10x+4$

$=4A+2Ax+Ax^2+2Bx-Bx^2+2C-Cx$

$3x^2+10x+4$

$=x^2(A-B)+x(2A+2B-C)+4A+2C$

$\begin{cases} A-B=3 \\ 2A+2B-C=10 \\ 4A+2C=4 \end{cases}$

$\begin{bmatrix} 1 & -1 & 0 \\ 2 & 2 & -1 \\ 4 & 0 & 2 \end{bmatrix}\begin{bmatrix} A \\ B \\ C \end{bmatrix}=\begin{bmatrix} 3 \\ 10 \\ 4 \end{bmatrix}$

$A^{-1}=\begin{bmatrix} \dfrac{1}{3} & \dfrac{1}{6} & \dfrac{1}{12} \\ \dfrac{-2}{3} & \dfrac{1}{6} & \dfrac{1}{12} \\ \dfrac{-2}{3} & \dfrac{-1}{3} & \dfrac{1}{3} \end{bmatrix}$

$X=A^{-1}B$

$X=\begin{bmatrix} 3 \\ 0 \\ -4 \end{bmatrix}$

$\dfrac{3}{2-x}-\dfrac{4}{4+2x+x^2}$

43. $\dfrac{6x^2+x+13}{x^3+2x^2+3x+6}=\dfrac{6x^2+x+13}{(x+2)(x^2+3)}$

$=\dfrac{A}{x+2}+\dfrac{Bx+C}{x^2+3}$

$6x^2+x+13=A(x^2+3)+(Bx+C)(x+2)$

$6x^2+x+13$

$=Ax^2+3A+Bx^2+2Bx+Cx+2C$

$6x^2+x+13$

$=x^2(A+B)+x(2B+C)+3A+2C$

$\begin{cases} A+B=6 \\ 2B+C=1 \\ 3A+2C=13 \end{cases}$

$\begin{bmatrix} 1 & 1 & 0 \\ 0 & 2 & 1 \\ 3 & 0 & 2 \end{bmatrix}\begin{bmatrix} A \\ B \\ C \end{bmatrix}=\begin{bmatrix} 6 \\ 1 \\ 13 \end{bmatrix}$

$A^{-1}=\begin{bmatrix} \dfrac{4}{7} & \dfrac{-2}{7} & \dfrac{1}{7} \\ \dfrac{3}{7} & \dfrac{2}{7} & \dfrac{-1}{7} \\ \dfrac{-6}{7} & \dfrac{3}{7} & \dfrac{2}{7} \end{bmatrix}$

$X=A^{-1}B$

$X=\begin{bmatrix} 5 \\ 1 \\ -1 \end{bmatrix}$

$\dfrac{5}{x+2}+\dfrac{x-1}{x^2+3}$

45. $\dfrac{x^4 - x^2 - 2x + 1}{x^5 + 2x^3 + x} = \dfrac{x^4 - x^2 - 2x + 1}{x\left(x^4 + 2x^2 + 1\right)}$

$= \dfrac{x^4 - x^2 - 2x + 1}{x\left(x^2 + 1\right)^2} = \dfrac{A}{x} + \dfrac{Bx + C}{x^2 + 1} + \dfrac{Dx + E}{\left(x^2 + 1\right)^2}$

$x^4 - x^2 - 2x + 1$

$= A\left(x^2 + 1\right)^2 + x(Bx + C)\left(x^2 + 1\right) + x(Dx + E)$

$\quad x^4 - x^2 - 2x + 1$

$\quad = Ax^4 + 2Ax^2 + A + Bx^4 + Bx^2$

$\qquad + Cx^3 + Cx + Dx^2 + Ex$

$\quad x^4 - x^2 - 2x + 1$

$\quad = x^4(A + B) + Cx^3 + x^2(2A + B + D)$

$\qquad + x(C + E) + A$

$\begin{cases} A + B = 1 \\ C = 0 \\ 2A + B + D = -1 \\ C + E = -2 \\ A = 1 \end{cases}$

$\begin{bmatrix} 1 & 1 & 0 & 0 & 0 \\ 0 & 0 & 1 & 0 & 0 \\ 2 & 1 & 0 & 1 & 0 \\ 0 & 0 & 1 & 0 & 1 \\ 1 & 0 & 0 & 0 & 0 \end{bmatrix} \begin{bmatrix} A \\ B \\ C \\ D \\ E \end{bmatrix} = \begin{bmatrix} 1 \\ 0 \\ -1 \\ -2 \\ 1 \end{bmatrix}$

$A^{-1} = \begin{bmatrix} 0 & 0 & 0 & 0 & 1 \\ 1 & 0 & 0 & 0 & -1 \\ 0 & 1 & 0 & 0 & 0 \\ -1 & 0 & 1 & 0 & -1 \\ 0 & -1 & 0 & 1 & 0 \end{bmatrix}$

$X = A^{-1}B$

$X = \begin{bmatrix} 1 \\ 0 \\ 0 \\ -3 \\ -2 \end{bmatrix}$

$\dfrac{1}{x} + \dfrac{-3x - 2}{\left(x^2 + 1\right)^2}$

47. $\dfrac{x^3 - 17x^2 + 76x - 98}{\left(x^2 - 6x + 9\right)\left(x^2 - 2x - 3\right)}$

$= \dfrac{x^3 - 17x^2 + 76x - 98}{(x - 3)(x - 3)(x - 3)(x + 1)}$

$= \dfrac{x^3 - 17x^2 + 76x - 98}{(x - 3)^3(x + 1)}$

$= \dfrac{A}{x + 1} + \dfrac{B}{(x - 3)} + \dfrac{C}{(x - 3)^2} + \dfrac{D}{(x - 3)^3}$

$x^3 - 17x^2 + 76x - 98$

$= A(x - 3)^3 + B(x + 1)(x - 3)^2$

$\quad + C(x + 1)(x - 3) + D(x + 1)$

$= Ax^3 - 9Ax^2 + 27Ax - 27A$

$\quad + Bx^3 - 5Bx^2 + 3Bx + 9B + Cx^2$

$\quad - 2Cx - 3C + Dx + D;$

$= x^3(A + B) + x^2(-9A - 5B + C)$

$\quad + x(27A + 3B - 2C + D)$

$\quad + (-27A + 9B - 3C + D);$

$\begin{cases} A + B = 1 \\ -9A - 5B + C = -17 \\ 27A + 3B - 2C + D = 76 \\ -27A + 9B - 3C + D = -98 \end{cases}$

$\begin{bmatrix} 1 & 1 & 0 & 0 \\ -9 & -5 & 1 & 0 \\ 27 & 3 & -2 & 1 \\ -27 & 9 & -3 & 1 \end{bmatrix} \begin{bmatrix} A \\ B \\ C \\ D \end{bmatrix} = \begin{bmatrix} 1 \\ -17 \\ 76 \\ -98 \end{bmatrix}$

$A^{-1} = \begin{bmatrix} \dfrac{1}{64} & \dfrac{-1}{64} & \dfrac{1}{64} & \dfrac{-1}{64} \\ \dfrac{63}{64} & \dfrac{1}{64} & \dfrac{-1}{64} & \dfrac{1}{64} \\ \dfrac{81}{16} & \dfrac{15}{16} & \dfrac{1}{16} & \dfrac{-1}{16} \\ \dfrac{27}{4} & \dfrac{9}{4} & \dfrac{3}{4} & \dfrac{1}{4} \end{bmatrix}$

$X = A^{-1}B$

$X = \begin{bmatrix} 3 \\ -2 \\ 0 \\ 1 \end{bmatrix}$

$= \dfrac{3}{x + 1} + \dfrac{-2}{(x - 3)} + \dfrac{0}{(x - 3)^2} + \dfrac{1}{(x - 3)^3}$

$= \dfrac{3}{x + 1} - \dfrac{2}{(x - 3)} + \dfrac{1}{(x - 3)^3}$

49. $A = \begin{vmatrix} L & r^2 \\ -\dfrac{\pi}{2} & W \end{vmatrix}$

$A = \begin{vmatrix} 20 & 8^2 \\ -\dfrac{\pi}{2} & 16 \end{vmatrix} = 320 - 64\left(-\dfrac{\pi}{2}\right)$

$= 320 + 32\pi \approx 420.5 \text{ in}^2$

51. $(2,1), (3,7), (5,3)$

$A = \dfrac{\begin{vmatrix} x_1 & y_1 & 1 \\ x_2 & y_2 & 1 \\ x_3 & y_3 & 1 \end{vmatrix}}{2}$

$A = \dfrac{\begin{vmatrix} 2 & 1 & 1 \\ 3 & 7 & 1 \\ 5 & 3 & 1 \end{vmatrix}}{2} = \left|\dfrac{-16}{2}\right| = 8 \text{ cm}^2$

53. $(-4,2), (-6,-1), (3,-1), (5,2)$

$A = \dfrac{\begin{vmatrix} x_1 & y_1 & 1 \\ x_2 & y_2 & 1 \\ x_3 & y_3 & 1 \end{vmatrix}}{2}$

For triangle use $(-4, 2)$, $(-6, -1)$ and $(5, 2)$.

$A = \dfrac{\begin{vmatrix} -4 & 2 & 1 \\ -6 & -1 & 1 \\ 5 & 2 & 1 \end{vmatrix}}{2} = \left|\dfrac{27}{2}\right| = \dfrac{27}{2}$

$2\left(\dfrac{27}{2}\right) = 27 \text{ ft}^2$

55. $h = 6\text{m}$; vertices $(3,5), (-4,2), (-1,6)$

$A = \dfrac{\begin{vmatrix} 3 & 5 & 1 \\ -4 & 2 & 1 \\ -1 & 6 & 1 \end{vmatrix}}{2} = \left|\dfrac{-19}{2}\right| = 9.5$

$V = \dfrac{1}{3}Bh$

$V = \dfrac{1}{3}(9.5)(6)$

$V = 19 \text{ m}^3$

57. $(1,5), (-2,-1), (4,11)$

$|A| = \begin{vmatrix} 1 & 5 & 1 \\ -2 & -1 & 1 \\ 4 & 11 & 1 \end{vmatrix} = 0$

Yes

59. $(-2.5, 5.2), (1.2, -5.6), (2.2, -8.5)$

$|A| = \begin{vmatrix} -2.5 & 5.2 & 1 \\ 1.2 & -5.6 & 1 \\ 2.2 & -8.5 & 1 \end{vmatrix} = 0.07$

No

61. $2x - 3y = 7$; $(2, -1), (-1.3, -3.2), (-3.1, -4.4)$

$(2, -1)$ Yes

$2(2) - 3(-1) = 7$

$4 + 3 = 7$

$7 = 7$;

$(-1.3, -3.2)$ Yes

$2(-1.3) - 3(-3.2) = 7$

$-2.6 + 9.6 = 7$

$7 = 7$;

$(-3.1, -4.4)$ Yes

$|A| = \begin{vmatrix} 2 & -1 & 1 \\ -1.3 & -3.2 & 1 \\ -3.1 & -4.4 & 1 \end{vmatrix} = 0$

63. Let x represent the rate of the $15000 investment.
Let y represent the rate of the $25000 investment.

$\begin{cases} 15000x + 25000y = 2900 \\ 25000x + 15000y = 2700 \end{cases}$

$D = \begin{vmatrix} 15000 & 25000 \\ 25000 & 15000 \end{vmatrix} = -400000000$;

$D_x = \begin{vmatrix} 2900 & 25000 \\ 2700 & 15000 \end{vmatrix} = -24000000$;

$D_y = \begin{vmatrix} 15000 & 2900 \\ 25000 & 2700 \end{vmatrix} = -32000000$;

$x = \dfrac{D_x}{D} = \dfrac{-24000000}{-400000000} = 0.06$;

$y = \dfrac{D_y}{D} = \dfrac{-32000000}{-400000000} = 0.08$

$15000 invested at 6%
$25000 invested at 8%

65. $\begin{cases} x+3y+5z=6 \\ 2x-4y+6z=14 \\ 9x-6y+3z=3 \end{cases}$

(1) $-2R1+R2$

$\begin{cases} -2x-6y-10z=-12 \\ \underline{2x-4y+6z=14} \end{cases}$

$-10y-4z=2$

$-9R1+R3$

$\begin{cases} -9x-27y-45z=-54 \\ \underline{9x-6y+3z=3} \end{cases}$

$-33y-42z=-51$

$\begin{cases} -10y-4z=2 \\ -33y-42z=-51 \end{cases}$

$-21R1+2R2$

$\begin{cases} 210y+84z=-42 \\ \underline{-66y-84z=-102} \end{cases}$

$144y=-144$

$y=-1;$

$210y+84z=-42$
$210(-1)+84z=-42$
$-210+84z=-42$
$84z=168$
$z=2;$

$x+3y+5z=6$
$x+3(-1)+5(2)=6$
$x-3+10=6$
$x=-1$

$(-1,\,-1,\,2)$

(2) $\begin{bmatrix} 1 & 3 & 5 & 6 \\ 2 & -4 & 6 & 14 \\ 9 & -6 & 3 & 3 \end{bmatrix}$ $-2R1+R2 \rightarrow R2$

$\begin{bmatrix} 1 & 3 & 5 & 6 \\ 0 & -10 & -4 & 2 \\ 9 & -6 & 3 & 3 \end{bmatrix}$ $-9R1+R3 \rightarrow R3$

$\begin{bmatrix} 1 & 3 & 5 & 6 \\ 0 & -10 & -4 & 2 \\ 0 & -33 & -42 & -51 \end{bmatrix}$ $-\dfrac{1}{10}R2 \rightarrow R2$

$\begin{bmatrix} 1 & 3 & 5 & 6 \\ 0 & 1 & \dfrac{2}{5} & -\dfrac{1}{5} \\ 0 & -33 & -42 & -51 \end{bmatrix}$ $33R2+R3 \rightarrow R3$

$\begin{bmatrix} 1 & 3 & 5 & 6 \\ 0 & 1 & \dfrac{2}{5} & -\dfrac{1}{5} \\ 0 & 0 & -\dfrac{144}{5} & -\dfrac{288}{5} \end{bmatrix}$ $-\dfrac{5}{144}R3 \rightarrow R3$

$\begin{bmatrix} 1 & 3 & 5 & 6 \\ 0 & 1 & \dfrac{2}{5} & -\dfrac{1}{5} \\ 0 & 0 & 1 & 2 \end{bmatrix}$

$z=2;$

$y+\dfrac{2}{5}z=-\dfrac{1}{5}$

$y+\dfrac{2}{5}(2)=-\dfrac{1}{5}$

$y+\dfrac{4}{5}=-\dfrac{1}{5}$

$y=-1;$

$x+3y+5z=6$
$x+3(-1)+5(2)=6$
$x-3+10=6$
$x+7=6$
$x=-1$

$(-1,\,-1,\,2)$

(3) $\begin{cases} x+3y+5z=6 \\ 2x-4y+6z=14 \\ 9x-6y+3z=3 \end{cases}$

$D=\begin{vmatrix} 1 & 3 & 5 \\ 2 & -4 & 6 \\ 9 & -6 & 3 \end{vmatrix}=288;$

$D_x=\begin{vmatrix} 6 & 3 & 5 \\ 14 & -4 & 6 \\ 3 & -6 & 3 \end{vmatrix}=-288;$

$D_y=\begin{vmatrix} 1 & 6 & 5 \\ 2 & 14 & 6 \\ 9 & 3 & 3 \end{vmatrix}=-288;$

$D_z=\begin{vmatrix} 1 & 3 & 6 \\ 2 & -4 & 14 \\ 9 & -6 & 3 \end{vmatrix}=576;$

$$x = \frac{D_x}{D} = \frac{-288}{288} = -1;$$

$$y = \frac{D_y}{D} = \frac{-288}{288} = -1;$$

$$z = \frac{D_z}{D} = \frac{576}{288} = 2$$

$$(-1, -1, 2)$$

$$(4) \quad \begin{bmatrix} 1 & 3 & 5 \\ 2 & -4 & 6 \\ 9 & -6 & 3 \end{bmatrix} \begin{bmatrix} x \\ y \\ z \end{bmatrix} = \begin{bmatrix} 6 \\ 14 \\ 3 \end{bmatrix}$$

$$A^{-1} = \begin{bmatrix} \dfrac{1}{12} & \dfrac{-13}{96} & \dfrac{19}{144} \\ \dfrac{1}{6} & \dfrac{-7}{48} & \dfrac{1}{72} \\ \dfrac{1}{12} & \dfrac{11}{96} & \dfrac{-5}{144} \end{bmatrix}$$

$$X = A^{-1}B$$

$$X = \begin{bmatrix} -1 \\ -1 \\ 2 \end{bmatrix}$$

$$(-1, -1, 2)$$

Answers will vary.

67. $x^2 + y^2 + Dx + Ey + F = 0$

$(-1, 7)$, $(2, 8)$ and $(5, -1)$

$x^2 + y^2 + Dx + Ey + F = 0$

$(-1)^2 + (7)^2 + D(-1) + E(7) + F = 0$

$1 + 49 - D + 7E + F = 0$

$-D + 7E + F = -50;$

$x^2 + y^2 + Dx + Ey + F = 0$

$(2)^2 + (8)^2 + D(2) + E(8) + F = 0$

$4 + 64 + 2D + 8E + F = 0$

$2D + 8E + F = -68;$

$x^2 + y^2 + Dx + Ey + F = 0$

$(5)^2 + (-1)^2 + D(5) + E(-1) + F = 0$

$25 + 1 + 5D - E + F = 0$

$5D - E + F = -26;$

$$\begin{cases} -D + 7E + F = -50 \\ 2D + 8E + F = -68 \\ 5D - E + F = -26 \end{cases}$$

$$\begin{bmatrix} -1 & 7 & 1 \\ 2 & 8 & 1 \\ 5 & -1 & 1 \end{bmatrix} \begin{bmatrix} D \\ E \\ F \end{bmatrix} = \begin{bmatrix} -50 \\ -68 \\ -26 \end{bmatrix}$$

$$A^{-1} = \begin{bmatrix} -\dfrac{3}{10} & \dfrac{4}{15} & \dfrac{1}{30} \\ \dfrac{-1}{10} & \dfrac{1}{5} & \dfrac{-1}{10} \\ \dfrac{7}{5} & \dfrac{-17}{15} & \dfrac{11}{15} \end{bmatrix}$$

$$X = A^{-1}B$$

$$X = \begin{bmatrix} -4 \\ -6 \\ -12 \end{bmatrix}$$

$$D = -4, \quad E = -6, \quad F = -12$$

$$x^2 + y^2 - 4x - 6y - 12 = 0$$

69. $3^{2x-1} = 9^{2-x}$

$$\ln 3^{2x-1} = \ln 9^{2-x}$$

$$(2x-1)\ln 3 = (2-x)\ln 9$$

$$2x \ln 3 - \ln 3 = 2 \ln 9 - x \ln 9$$

$$2x \ln 3 + x \ln 9 = 2 \ln 9 + \ln 3$$

$$x(2 \ln 3 + \ln 9) = 2 \ln 9 + \ln 3$$

$$x = \frac{2 \ln 9 + \ln 3}{2 \ln 3 + \ln 9}$$

$$x = 1.25;$$

$$3^{2x-1} = 9^{2-x}$$

$$3^{2x-1} = \left(3^2\right)^{2-x}$$

$$3^{2x-1} = 3^{4-2x}$$

$$2x - 1 = 4 - 2x$$

$$4x = 5$$

$$x = 1.25$$

71. Left graph; intersects the x-axis in three places. Right shows four roots.

Chapter 6 Summary and Review

1. Answers will vary.

3. $\begin{cases} x - 2y + 2z = 7 \\ 2x + 2y - z = 5 \\ 3x - y + z = 6 \end{cases}$

$\begin{bmatrix} 1 & -2 & 2 & | & 7 \\ 2 & 2 & -1 & | & 5 \\ 3 & -1 & 1 & | & 6 \end{bmatrix} \quad -2R1 + R2 \to R2$

$\begin{bmatrix} 1 & -2 & 2 & | & 7 \\ 0 & 6 & -5 & | & -9 \\ 3 & -1 & 1 & | & 6 \end{bmatrix} \quad -3R1 + R3 \to R3$

$\begin{bmatrix} 1 & -2 & 2 & | & 7 \\ 0 & 6 & -5 & | & -9 \\ 0 & 5 & -5 & | & -15 \end{bmatrix} \quad R2 - R3 \to R2$

$\begin{bmatrix} 1 & -2 & 2 & | & 7 \\ 0 & 1 & 0 & | & 6 \\ 0 & 5 & -5 & | & -15 \end{bmatrix} \quad \frac{1}{5}R3 \to R3$

$\begin{bmatrix} 1 & -2 & 2 & | & 7 \\ 0 & 1 & 0 & | & 6 \\ 0 & 1 & -1 & | & -3 \end{bmatrix} \quad R3 - R2 \to R3$

$\begin{bmatrix} 1 & -2 & 2 & | & 7 \\ 0 & 1 & 0 & | & 6 \\ 0 & 0 & -1 & | & -9 \end{bmatrix} \quad 2R2 + R1 \to R1$

$\begin{bmatrix} 1 & 0 & 2 & | & 19 \\ 0 & 1 & 0 & | & 6 \\ 0 & 0 & -1 & | & -9 \end{bmatrix} \quad 2R3 + R1 \to R1$

$\begin{bmatrix} 1 & 0 & 0 & | & 1 \\ 0 & 1 & 0 & | & 6 \\ 0 & 0 & -1 & | & -9 \end{bmatrix} \quad -1R3 \to R3$

$\begin{bmatrix} 1 & 0 & 0 & | & 1 \\ 0 & 1 & 0 & | & 6 \\ 0 & 0 & 1 & | & 9 \end{bmatrix}$

$(1, 6, 9)$

5. $\begin{cases} 2x - y + 2z = -1 \\ x + 2y + 2z = -3 \\ 3x - 4y + 2z = 1 \end{cases}$

$\begin{bmatrix} 2 & -1 & 2 & | & -1 \\ 1 & 2 & 2 & | & -3 \\ 3 & -4 & 2 & | & 1 \end{bmatrix} \quad R1 - R2 \to R1$

$\begin{bmatrix} 1 & -3 & 0 & | & 2 \\ 1 & 2 & 2 & | & -3 \\ 3 & -4 & 2 & | & 1 \end{bmatrix} \quad R2 - R1 \to R2$

$\begin{bmatrix} 1 & -3 & 0 & | & 2 \\ 0 & 5 & 2 & | & -5 \\ 3 & -4 & 2 & | & 1 \end{bmatrix} \quad -3R1 + R3 \to R3$

$\begin{bmatrix} 1 & -3 & 0 & | & 2 \\ 0 & 5 & 2 & | & -5 \\ 0 & 5 & 2 & | & -5 \end{bmatrix} \quad R3 - R2 \to R3$

$\begin{bmatrix} 1 & -3 & 0 & | & 2 \\ 0 & 5 & 2 & | & -5 \\ 0 & 0 & 0 & | & 0 \end{bmatrix}$

$x - 3y = 2$

$\quad x = 3y + 2;$

$5y + 2z = -5$

$\quad 2z = -5y - 5$

$\quad z = -\dfrac{5}{2}y - \dfrac{5}{2};$

$\left(x = 3y + 2, y \in \mathbb{R}, z = -\dfrac{5}{2}y - \dfrac{5}{2} \right)$

7. $B - A$

$$= \begin{bmatrix} -7 & 6 \\ 1 & -2 \end{bmatrix} - \begin{bmatrix} \dfrac{-1}{4} & \dfrac{-3}{4} \\ \dfrac{-1}{8} & \dfrac{-7}{8} \end{bmatrix}$$

$$= \begin{bmatrix} -7 - \left(\dfrac{-1}{4}\right) & 6 - \left(\dfrac{-3}{4}\right) \\ 1 - \left(\dfrac{-1}{8}\right) & -2 - \left(\dfrac{-7}{8}\right) \end{bmatrix}$$

$$= \begin{bmatrix} -6.75 & 6.75 \\ 1.125 & -1.125 \end{bmatrix}$$

9. $8A$

$$= 8 \begin{bmatrix} \dfrac{-1}{4} & \dfrac{-3}{4} \\ \dfrac{-1}{8} & \dfrac{-7}{8} \end{bmatrix}$$

$$= \begin{bmatrix} 8\left(\dfrac{-1}{4}\right) & 8\left(\dfrac{-3}{4}\right) \\ 8\left(\dfrac{-1}{8}\right) & 8\left(\dfrac{-7}{8}\right) \end{bmatrix}$$

$$= \begin{bmatrix} -2 & -6 \\ -1 & -7 \end{bmatrix}$$

11. $C + D$

$$= \begin{bmatrix} -1 & 3 & 4 \\ 5 & -2 & 0 \\ 6 & -3 & 2 \end{bmatrix} + \begin{bmatrix} 2 & -3 & 0 \\ 0.5 & 1 & -1 \\ 4 & 0.1 & 5 \end{bmatrix}$$

$$= \begin{bmatrix} -1+2 & 3+(-3) & 4+0 \\ 5+0.5 & -2+1 & 0+(-1) \\ 6+4 & -3+0.1 & 2+5 \end{bmatrix}$$

$$= \begin{bmatrix} 1 & 0 & 4 \\ 5.5 & -1 & -1 \\ 10 & -2.9 & 7 \end{bmatrix}$$

13. BC

$$= \begin{bmatrix} -7 & 6 \\ 1 & -2 \end{bmatrix} \begin{bmatrix} -1 & 3 & 4 \\ 5 & -2 & 0 \\ 6 & -3 & 2 \end{bmatrix}$$

Not possible

15. CD

$$= \begin{bmatrix} -1 & 3 & 4 \\ 5 & -2 & 0 \\ 6 & -3 & 2 \end{bmatrix} \begin{bmatrix} 2 & -3 & 0 \\ 0.5 & 1 & -1 \\ 4 & 0.1 & 5 \end{bmatrix}$$

$$= \begin{bmatrix} -1(2)+3(0.5)+4(4) & -1(-3)+3(1)+4(0.1) & -1(0)+3(-1)+4(5) \\ 5(2)+-2(0.5)+0(4) & 5(-3)+-2(1)+0(0.1) & 5(0)+-2(-1)+0(5) \\ 6(2)+-3(0.5)+2(4) & 6(-3)+-3(1)+2(0.1) & 6(0)+-3(-1)+2(5) \end{bmatrix}$$

$$= \begin{bmatrix} 15.5 & 6.4 & 17 \\ 9 & -17 & 2 \\ 18.5 & -20.8 & 13 \end{bmatrix}$$

17. $AB = \begin{bmatrix} 1 & 0 \\ 0 & 1 \end{bmatrix} \begin{bmatrix} 0.2 & 0.2 \\ -0.6 & 0.4 \end{bmatrix} = \begin{bmatrix} 0.2 & 0.2 \\ -0.6 & 0.4 \end{bmatrix}$

$BA = \begin{bmatrix} 0.2 & 0.2 \\ -0.6 & 0.4 \end{bmatrix} \begin{bmatrix} 1 & 0 \\ 0 & 1 \end{bmatrix} = \begin{bmatrix} 0.2 & 0.2 \\ -0.6 & 0.4 \end{bmatrix}$

It is an identity matrix

19. E

21. $EG = \begin{bmatrix} 1 & -2 & 3 \\ -2 & 1 & -5 \\ -1 & -1 & -2 \end{bmatrix} \begin{bmatrix} -1 & 0 & -1 \\ 0 & 1 & 0 \\ 2 & 1 & 1 \end{bmatrix}$

$= \begin{bmatrix} 5 & 1 & 2 \\ -8 & -4 & -3 \\ -3 & -3 & -1 \end{bmatrix}$

$GE = \begin{bmatrix} -1 & 0 & -1 \\ 0 & 1 & 0 \\ 2 & 1 & 1 \end{bmatrix} \begin{bmatrix} 1 & -2 & 3 \\ -2 & 1 & -5 \\ -1 & -1 & -2 \end{bmatrix}$

$= \begin{bmatrix} 0 & 3 & -1 \\ -2 & 1 & -5 \\ -1 & -4 & -1 \end{bmatrix}$

$EF = \begin{bmatrix} 1 & -2 & 3 \\ -2 & 1 & -5 \\ -1 & -1 & -2 \end{bmatrix} \begin{bmatrix} 1 & -1 & 1 \\ 0 & 1 & 0 \\ -2 & 1 & -1 \end{bmatrix}$

$= \begin{bmatrix} -5 & 0 & -2 \\ 8 & -2 & 3 \\ 3 & -2 & 1 \end{bmatrix}$

$FE = \begin{bmatrix} 1 & -1 & 1 \\ 0 & 1 & 0 \\ -2 & 1 & -1 \end{bmatrix} \begin{bmatrix} 1 & -2 & 3 \\ -2 & 1 & -5 \\ -1 & -1 & -2 \end{bmatrix}$

$= \begin{bmatrix} 2 & -4 & 6 \\ -2 & 1 & -5 \\ -3 & 6 & -9 \end{bmatrix}$

Matrix multiplication is not generally commutative.

23. $\begin{cases} 0.5x - 2.2y + 3z = -8 \\ -0.6x - y + 2z = -7.2 \\ x + 1.5y - 0.2z = 2.6 \end{cases}$

$\begin{bmatrix} 0.5 & -2.2 & 3 & | & -8 \\ -0.6 & -1 & 2 & | & -7.2 \\ 1 & 1.5 & -0.2 & | & 2.6 \end{bmatrix}$

$(2, 0, -3)$

25. $\begin{cases} 2x + y = -2 \\ -x + y + 5z = 12 \\ 3x - 2y + z = -8 \end{cases}$

$D = \begin{vmatrix} 2 & 1 & 0 \\ -1 & 1 & 5 \\ 3 & -2 & 1 \end{vmatrix} = 38 ;$

$D_x = \begin{vmatrix} -2 & 1 & 0 \\ 12 & 1 & 5 \\ -8 & -2 & 1 \end{vmatrix} = -74 ;$

$D_y = \begin{vmatrix} 2 & -2 & 0 \\ -1 & 12 & 5 \\ 3 & -8 & 1 \end{vmatrix} = 72 ;$

$D_z = \begin{vmatrix} 2 & 1 & -2 \\ -1 & 1 & 12 \\ 3 & -2 & -8 \end{vmatrix} = 62 ;$

$x = \dfrac{D_x}{D} = \dfrac{-74}{38} = -\dfrac{37}{19} ;$

$y = \dfrac{D_y}{D} = \dfrac{72}{38} = \dfrac{36}{19} ;$

$z = \dfrac{D_z}{D} = \dfrac{62}{38} = \dfrac{31}{19}$

$\left(\dfrac{-37}{19}, \dfrac{36}{19}, \dfrac{31}{19} \right)$

27. $(6, 1), (-1, -6)$ and $(-6, 2)$

$A = \dfrac{\begin{vmatrix} x_1 & y_1 & 1 \\ x_2 & y_2 & 1 \\ x_3 & y_3 & 1 \end{vmatrix}}{2}$

$A = \dfrac{\begin{vmatrix} 6 & 1 & 1 \\ -1 & -6 & 1 \\ -6 & 2 & 1 \end{vmatrix}}{2} = \left| \dfrac{-91}{2} \right| = \dfrac{91}{2} \text{ units}^2$

Chapter 6 Mixed Review

1.
$$\begin{cases} \dfrac{1}{2}x + \dfrac{2}{3}y = 3 \\ \dfrac{-2}{5}x - \dfrac{1}{4}y = 1 \end{cases}$$

$$\begin{bmatrix} \dfrac{1}{2} & \dfrac{2}{3} & 3 \\ \dfrac{-2}{5} & \dfrac{-1}{4} & 1 \end{bmatrix} \quad 2R1 \to R1$$

$$\begin{bmatrix} 1 & \dfrac{4}{3} & 6 \\ \dfrac{-2}{5} & \dfrac{-1}{4} & 1 \end{bmatrix} \quad \dfrac{2}{5}R1 + R2 \to R2$$

$$\begin{bmatrix} 1 & \dfrac{4}{3} & 6 \\ 0 & \dfrac{17}{60} & \dfrac{17}{5} \end{bmatrix} \quad \dfrac{60}{17}R2 \to R2$$

$$\begin{bmatrix} 1 & \dfrac{4}{3} & 6 \\ 0 & 1 & 12 \end{bmatrix}$$

$y = 12$;

$$x + \dfrac{4}{3}y = 6$$
$$x + \dfrac{4}{3}(12) = 6$$
$$x + 16 = 6$$
$$x = -10$$
$$(-10, 12)$$

3.
$$\begin{cases} 3x - 4y + 5z = 5 \\ -x + 2y - 3z = -3 \\ 3x - 2y + z = 1 \end{cases}$$

$$\begin{bmatrix} 3 & -4 & 5 & 5 \\ -1 & 2 & -3 & -3 \\ 3 & -2 & 1 & 1 \end{bmatrix} \quad R1 + R2 \to R1$$

$$\begin{bmatrix} 2 & -2 & 2 & 2 \\ -1 & 2 & -3 & -3 \\ 3 & -2 & 1 & 1 \end{bmatrix} \quad \dfrac{1}{2}R1 \to R1$$

$$\begin{bmatrix} 1 & -1 & 1 & 1 \\ -1 & 2 & -3 & -3 \\ 3 & -2 & 1 & 1 \end{bmatrix} \quad R1 + R2 \to R2$$

$$\begin{bmatrix} 1 & -1 & 1 & 1 \\ 0 & 1 & -2 & -2 \\ 3 & -2 & 1 & 1 \end{bmatrix} \quad 2R2 + R3 \to R3$$

$$\begin{bmatrix} 1 & -1 & 1 & 1 \\ 0 & 1 & -2 & -2 \\ 3 & 0 & -3 & -3 \end{bmatrix} \quad \dfrac{1}{3}R3 \to R3$$

$$\begin{bmatrix} 1 & -1 & 1 & 1 \\ 0 & 1 & -2 & -2 \\ 1 & 0 & -1 & -1 \end{bmatrix} \quad R1 + R2 \to R1$$

$$\begin{bmatrix} 1 & 0 & -1 & -1 \\ 0 & 1 & -2 & -2 \\ 1 & 0 & -1 & -1 \end{bmatrix} \quad R3 - R1 \to R3$$

$$\begin{bmatrix} 1 & 0 & -1 & -1 \\ 0 & 1 & -2 & -2 \\ 0 & 0 & 0 & 0 \end{bmatrix}$$

$$x - z = -1$$
$$x = z - 1;$$
$$y - 2z = -2$$
$$y = 2z - 2;$$
$$\{(x, y, z) \mid x = z - 1, y = 2z - 2, z \in \mathbb{R} \}$$

Chapter 6: Matrices and Matrix Applications

5. a. $-2AC$

$$= -2\begin{bmatrix} 2 & -1 \\ 0 & 3 \end{bmatrix}\begin{bmatrix} 1 & -4 & 2 \\ -2 & 0 & -1 \end{bmatrix}$$

$$= -2\begin{bmatrix} 4 & -8 & 5 \\ -6 & 0 & -3 \end{bmatrix}$$

$$= \begin{bmatrix} -8 & 16 & -10 \\ 12 & 0 & 6 \end{bmatrix}$$

b. CD

$$= \begin{bmatrix} 1 & -4 & 2 \\ -2 & 0 & -1 \end{bmatrix}\begin{bmatrix} 3 & 0 & 1 \\ -1 & 2 & 0 \\ 1 & 1 & -4 \end{bmatrix}$$

$$= \begin{bmatrix} 9 & -6 & -7 \\ -7 & -1 & 2 \end{bmatrix}$$

7. a. $a_{22} = 3$

b. $b_{21} = 3$

c. $c_{12} = -4$

d. $d_{32} = 1$

9. $\begin{cases} -x - 2z = 5 \\ 2y + z = -4 \\ -x + 2y = 3 \end{cases}$

$$\begin{bmatrix} -1 & 0 & -2 & | & 5 \\ 0 & 2 & 1 & | & -4 \\ -1 & 2 & 0 & | & 3 \end{bmatrix} -R1 \to R1$$

$$\begin{bmatrix} 1 & 0 & 2 & | & -5 \\ 0 & 2 & 1 & | & -4 \\ -1 & 2 & 0 & | & 3 \end{bmatrix} R1 + R3 \to R3$$

$$\begin{bmatrix} 1 & 0 & 2 & | & -5 \\ 0 & 2 & 1 & | & -4 \\ 0 & 2 & 2 & | & -2 \end{bmatrix} \frac{1}{2}R2 \to R2$$

$$\begin{bmatrix} 1 & 0 & 2 & | & -5 \\ 0 & 1 & \frac{1}{2} & | & -2 \\ 0 & 2 & 2 & | & -2 \end{bmatrix} -2R2 + R3 \to R3$$

$$\begin{bmatrix} 1 & 0 & 2 & | & -5 \\ 0 & 1 & \frac{1}{2} & | & -2 \\ 0 & 0 & 1 & | & 2 \end{bmatrix}$$

$z = 2$;

$$y + \frac{1}{2}z = -2$$
$$y + \frac{1}{2}(2) = -2$$
$$y + 1 = -2$$
$$y = -3;$$
$$x + 2z = -5$$
$$x + 2(2) = -5$$
$$x + 4 = -5$$
$$x = -9$$
$$(-9, -3, 2)$$

11. $A = \begin{bmatrix} 2 & 3 \\ 5 & 8 \end{bmatrix}$

$$\begin{bmatrix} 2 & 3 & | & 1 & 0 \\ 5 & 8 & | & 0 & 1 \end{bmatrix} -2R1 + R2 \to R2$$

$$\begin{bmatrix} 2 & 3 & | & 1 & 0 \\ 1 & 2 & | & -2 & 1 \end{bmatrix} R1 \leftrightarrow R2$$

$$\begin{bmatrix} 1 & 2 & | & -2 & 1 \\ 2 & 3 & | & 1 & 0 \end{bmatrix} -2R1 + R2 \to R2$$

$$\begin{bmatrix} 1 & 2 & | & -2 & 1 \\ 0 & -1 & | & 5 & -2 \end{bmatrix} 2R2 + R1 \to R1$$

$$\begin{bmatrix} 1 & 0 & | & 8 & -3 \\ 0 & -1 & | & 5 & -2 \end{bmatrix} -R2 \to R2$$

$$\begin{bmatrix} 1 & 0 & | & 8 & -3 \\ 0 & 1 & | & -5 & 2 \end{bmatrix}$$

$$A^{-1} = \begin{bmatrix} 8 & -3 \\ -5 & 2 \end{bmatrix}$$

13. $\begin{cases} x + y + 2z = 1 \\ 2x - z = 3 \\ -x + y + z = 3 \end{cases}$

Chapter 6 Mixed Review

15. $\begin{cases} -x + 5y - 2z = 1 \\ 2x + 3y - z = 3 \\ 3x - y + 3z = -2 \end{cases}$

$D = \begin{vmatrix} -1 & 5 & -2 \\ 2 & 3 & -1 \\ 3 & -1 & 3 \end{vmatrix} = -31;$

$D_x = \begin{vmatrix} 1 & 5 & -2 \\ 3 & 3 & -1 \\ -2 & -1 & 3 \end{vmatrix} = -33;$

$D_y = \begin{vmatrix} -1 & 1 & -2 \\ 2 & 3 & -1 \\ 3 & -2 & 3 \end{vmatrix} = 10;$

$D_z = \begin{vmatrix} -1 & 5 & 1 \\ 2 & 3 & 3 \\ 3 & -1 & -2 \end{vmatrix} = 57;$

$x = \dfrac{D_x}{D} = \dfrac{-33}{-31} = \dfrac{33}{31};$

$y = \dfrac{D_y}{D} = \dfrac{10}{-31} = \dfrac{-10}{31};$

$z = \dfrac{D_z}{D} = \dfrac{57}{-31} = \dfrac{-57}{31}$

$\left(\dfrac{33}{31}, \dfrac{-10}{31}, \dfrac{-57}{31} \right)$

17. $T = \begin{bmatrix} -1 & 2 & 1 \\ 3 & 1 & 1 \\ 0 & 4 & 1 \end{bmatrix}$, Area $= \left| \dfrac{\|T\|}{2} \right|$

$D = \begin{vmatrix} -1 & 2 & 1 \\ 3 & 1 & 1 \\ 0 & 4 & 1 \end{vmatrix} = 9;$

$Area = \left| \dfrac{9}{2} \right| = 4.5 \text{ units}^2$

19. $\begin{cases} 2l + 2w = 438 \\ l = w + 55 \end{cases}$

$\begin{cases} 2w + 2l = 438 \\ -w + l = 55 \end{cases}$

$\begin{bmatrix} 2 & 2 & | & 438 \\ -1 & 1 & | & 55 \end{bmatrix} \dfrac{1}{2}R1 \to R1$

$\begin{bmatrix} 1 & 1 & | & 219 \\ -1 & 1 & | & 55 \end{bmatrix} R1 + R2 \to R2$

$\begin{bmatrix} 1 & 1 & | & 219 \\ 0 & 2 & | & 274 \end{bmatrix} \dfrac{1}{2}R2 \to R2$

$\begin{bmatrix} 1 & 1 & | & 219 \\ 0 & 1 & | & 137 \end{bmatrix} R1 - R2 \to R1$

$\begin{bmatrix} 1 & 0 & | & 82 \\ 0 & 1 & | & 137 \end{bmatrix}$

137 m by 82 m

Chapter 6 Practice Test

1. $\begin{cases} 3x + 8y = -5 \\ x + 10y = 2 \end{cases}$

$\begin{bmatrix} 3 & 8 & | & -5 \\ 1 & 10 & | & 2 \end{bmatrix} -\dfrac{1}{3}R1 + R2 \to R2$

$\begin{bmatrix} 3 & 8 & | & -5 \\ 0 & \dfrac{22}{3} & | & \dfrac{11}{3} \end{bmatrix} \dfrac{3}{22}R2 \to R2$

$\begin{bmatrix} 3 & 8 & | & -5 \\ 0 & 1 & | & \dfrac{1}{2} \end{bmatrix} -8R2 + R1 \to R1$

$\begin{bmatrix} 3 & 0 & | & -9 \\ 0 & 1 & | & \dfrac{1}{2} \end{bmatrix} \dfrac{1}{3}R1 \to R1$

$\begin{bmatrix} 1 & 0 & | & -3 \\ 0 & 1 & | & \dfrac{1}{2} \end{bmatrix}$

$\left(-3, \dfrac{1}{2}\right)$

3. $\begin{cases} 4x - 5y - 6z = 5 \\ 2x - 3y + 3z = 0 \\ x + 2y - 3z = 5 \end{cases}$

$\begin{bmatrix} 4 & -5 & -6 & | & 5 \\ 2 & -3 & 3 & | & 0 \\ 1 & 2 & -3 & | & 5 \end{bmatrix} -2R3 + R2 \to R2$

$\begin{bmatrix} 4 & -5 & -6 & | & 5 \\ 0 & -7 & 9 & | & -10 \\ 1 & 2 & -3 & | & 5 \end{bmatrix} -4R3 + R1 \to R1$

$\begin{bmatrix} 0 & -13 & 6 & | & -15 \\ 0 & -7 & 9 & | & -10 \\ 1 & 2 & -3 & | & 5 \end{bmatrix} R3 + \dfrac{2}{7}R2 \to R3$

$\begin{bmatrix} 0 & -13 & 6 & | & -15 \\ 0 & -7 & 9 & | & -10 \\ 1 & 0 & -\dfrac{3}{7} & | & \dfrac{15}{7} \end{bmatrix} \dfrac{-13}{7}R2 + R1 \to R1$

$\begin{bmatrix} 0 & 0 & -\dfrac{75}{7} & | & \dfrac{25}{7} \\ 0 & -7 & 9 & | & -10 \\ 1 & 0 & -\dfrac{3}{7} & | & \dfrac{15}{7} \end{bmatrix} -\dfrac{7}{75}R1 \to R1$

$\begin{bmatrix} 0 & 0 & 1 & | & \dfrac{-1}{3} \\ 0 & -7 & 9 & | & -10 \\ 1 & 0 & -\dfrac{3}{7} & | & \dfrac{15}{7} \end{bmatrix}$

$z = -\dfrac{1}{3};$

$-7y + 9z = -10$

$-7y + 9\left(\dfrac{-1}{3}\right) = -10$

$-7y = -7$

$y = 1;$

$x - \dfrac{3}{7}z = \dfrac{15}{7}$

$x - \dfrac{3}{7}\left(\dfrac{-1}{3}\right) = \dfrac{15}{7}$

$x = 2$

$\left(2, 1, \dfrac{-1}{3}\right)$

5. a. $C - D$

$= \begin{bmatrix} 0.5 & 0 & 0.2 \\ 0.4 & -0.5 & 0 \\ 0.1 & -0.4 & -0.1 \end{bmatrix} - \begin{bmatrix} 0.5 & 0.1 & 0.2 \\ -0.1 & 0.1 & 0 \\ 0.3 & 0.4 & 0.8 \end{bmatrix}$

$= \begin{bmatrix} 0 & -0.1 & 0 \\ 0.5 & -0.6 & 0 \\ -0.2 & -0.8 & -0.9 \end{bmatrix}$

b. $-0.6D = -0.6 \begin{bmatrix} 0.5 & 0.1 & 0.2 \\ -0.1 & 0.1 & 0 \\ 0.3 & 0.4 & 0.8 \end{bmatrix}$

$= \begin{bmatrix} -0.3 & -0.06 & -0.12 \\ 0.06 & -0.06 & 0 \\ -0.18 & -0.24 & -0.48 \end{bmatrix}$

c. DC

$$= \begin{bmatrix} 0.5 & 0.1 & 0.2 \\ -0.1 & 0.1 & 0 \\ 0.3 & 0.4 & 0.8 \end{bmatrix} \begin{bmatrix} 0.5 & 0 & 0.2 \\ 0.4 & -0.5 & 0 \\ 0.1 & -0.4 & -0.1 \end{bmatrix}$$

$$= \begin{bmatrix} 0.31 & -0.13 & 0.08 \\ -0.01 & -0.05 & -0.02 \\ 0.39 & -0.52 & -0.02 \end{bmatrix}$$

d. $D^{-1} = \begin{bmatrix} \dfrac{40}{17} & 0 & \dfrac{-10}{17} \\ \dfrac{40}{17} & 10 & \dfrac{-10}{17} \\ \dfrac{-35}{17} & -5 & \dfrac{30}{17} \end{bmatrix}$

e. $|D| = \dfrac{17}{500}$

7. $\begin{cases} 2x - 3y = 2 \\ x - 6y = -2 \end{cases}$

$D = \begin{vmatrix} 2 & -3 \\ 1 & -6 \end{vmatrix} = -12 - (-3) = -9$;

$D_x = \begin{vmatrix} 2 & -3 \\ -2 & -6 \end{vmatrix} = -12 - 6 = -18$;

$D_y = \begin{vmatrix} 2 & 2 \\ 1 & -2 \end{vmatrix} = -4 - 2 = -6$;

$x = \dfrac{D_x}{D} = \dfrac{-18}{-9} = 2$;

$y = \dfrac{D_y}{D} = \dfrac{-6}{-9} = \dfrac{2}{3}$

$\left(2, \dfrac{2}{3}\right)$

9. $\begin{cases} 2x - 5y = 11 \\ 4x + 7y = 4 \end{cases}$

$\begin{bmatrix} 2 & -5 & | & 11 \\ 4 & 7 & | & 4 \end{bmatrix}$

$A = \begin{bmatrix} 2 & -5 \\ 4 & 7 \end{bmatrix}; \quad B = \begin{bmatrix} 11 \\ 4 \end{bmatrix}$

$A^{-1} = \begin{bmatrix} \dfrac{7}{34} & \dfrac{5}{34} \\ \dfrac{-2}{17} & \dfrac{1}{17} \end{bmatrix}$

$A^{-1}B = \begin{bmatrix} \dfrac{97}{34} \\ \dfrac{-18}{17} \end{bmatrix}$

$\left(\dfrac{97}{34}, -\dfrac{18}{17}\right)$

11. $\begin{bmatrix} 2x+y & 3 \\ x+z & 3x+2z \end{bmatrix} = \begin{bmatrix} z-1 & 3 \\ 2y+5 & y+8 \end{bmatrix}$

$\begin{cases} 2x + y = z - 1 \\ x + z = 2y + 5 \\ 3x + 2z = y + 8 \end{cases}$

$\begin{cases} 2x + y - z = -1 \\ x - 2y + z = 5 \\ 3x - y + 2z = 8 \end{cases}$

$\begin{bmatrix} 2 & 1 & -1 & | & -1 \\ 1 & -2 & 1 & | & 5 \\ 3 & -1 & 2 & | & 8 \end{bmatrix} \quad -3R2 + R3 \to R3$

$\begin{bmatrix} 2 & 1 & -1 & | & -1 \\ 1 & -2 & 1 & | & 5 \\ 0 & 5 & -1 & | & -7 \end{bmatrix} \quad -2R2 + R1 \to R1$

$\begin{bmatrix} 0 & 5 & -3 & | & -11 \\ 1 & -2 & 1 & | & 5 \\ 0 & 5 & -1 & | & -7 \end{bmatrix} \quad R1 - R3 \to R1$

$\begin{bmatrix} 0 & 0 & -2 & | & -4 \\ 1 & -2 & 1 & | & 5 \\ 0 & 5 & -1 & | & -7 \end{bmatrix} \quad \dfrac{-1}{2}R1 \to R1$

$\begin{bmatrix} 0 & 0 & 1 & | & 2 \\ 1 & -2 & 1 & | & 5 \\ 0 & 5 & -1 & | & -7 \end{bmatrix} \quad R1 + R3 \to R3$

$\begin{bmatrix} 0 & 0 & 1 & | & 2 \\ 1 & -2 & 1 & | & 5 \\ 0 & 5 & 0 & | & -5 \end{bmatrix} \quad \dfrac{1}{5}R3 \to R3$

$\begin{bmatrix} 0 & 0 & 1 & | & 2 \\ 1 & -2 & 1 & | & 5 \\ 0 & 1 & 0 & | & -1 \end{bmatrix} \quad R2 - R1 \to R2$

$$\begin{bmatrix} 0 & 0 & 1 & | & 2 \\ 1 & -2 & 0 & | & 3 \\ 0 & 1 & 0 & | & -1 \end{bmatrix} \quad R2 + 2R3 \to R2$$

$$\begin{bmatrix} 0 & 0 & 1 & | & 2 \\ 1 & 0 & 0 & | & 1 \\ 0 & 1 & 0 & | & -1 \end{bmatrix} \quad R1 \leftrightarrow R2$$

$$\begin{bmatrix} 1 & 0 & 0 & | & 1 \\ 0 & 0 & 1 & | & 2 \\ 0 & 1 & 0 & | & -1 \end{bmatrix} \quad R3 \leftrightarrow R2$$

$$\begin{bmatrix} 1 & 0 & 0 & | & 1 \\ 0 & 1 & 0 & | & -1 \\ 0 & 0 & 1 & | & 2 \end{bmatrix}$$

$(1, -1, 2)$

13. $\begin{vmatrix} -1 & 4 & 1 \\ 2 & 1 & 1 \\ 4 & -1 & 1 \end{vmatrix}$

$= -1\begin{vmatrix} 1 & 1 \\ -1 & 1 \end{vmatrix} - 4\begin{vmatrix} 2 & 1 \\ 4 & 1 \end{vmatrix} + 1\begin{vmatrix} 2 & 1 \\ 4 & -1 \end{vmatrix}$

$= -1(1+1) - 4(2-4) + 1(-2-4)$

$= -1(2) - 4(-2) + 1(-6) = 0$

$(-1, 4), (2, 1), (4, -1)$ are collinear.

15. $\begin{bmatrix} r & 2 \\ 3 & s \end{bmatrix}\begin{bmatrix} r & 2 \\ 3 & s \end{bmatrix} = \begin{bmatrix} 10 & -2 \\ -3 & 7 \end{bmatrix}$

$6 + s^2 = 7$

$s^2 = 1, s = \pm 1$

$r^2 + 6 = 10$

$r^2 = 4, r = \pm 2;$

$2r + 2s = -2$

$r + s = -1;$

If $r = -2, s = 1,$

If $r = -2, s = 1$

17. $\begin{cases} x + y = 23 \\ x = y + 8 \end{cases}$

$\begin{cases} x + y = 23 \\ x - y = 8 \end{cases}$

$\begin{bmatrix} 1 & 1 & | & 23 \\ 1 & -1 & | & 8 \end{bmatrix} \quad R2 - R1 \to R2$

$\begin{bmatrix} 1 & 1 & | & 23 \\ 0 & -2 & | & -15 \end{bmatrix} \quad \frac{1}{2}R2 + R1 \to R1$

$\begin{bmatrix} 1 & 0 & | & 15.5 \\ 0 & -2 & | & -15 \end{bmatrix} \quad -\frac{1}{2}R2 \to R2$

$\begin{bmatrix} 1 & 0 & | & 15.5 \\ 0 & 1 & | & 7.5 \end{bmatrix}$

7.5 hours, 15.5 hours

19. $\begin{cases} x & + y & + z = 1800000 \\ 0.02x + 0.05y + 0.085z = 94500 \\ 0.02x & + 0.085z = 29500 \end{cases}$

$$\begin{bmatrix} 1 & 1 & 1 & | & 1800000 \\ 0.02 & 0.05 & 0.085 & | & 94500 \\ 0.02 & 0 & 0.085 & | & 29500 \end{bmatrix} \quad R2 - R3 \to R2$$

$$\begin{bmatrix} 1 & 1 & 1 & | & 1800000 \\ 0 & 0.05 & 0 & | & 65000 \\ 0.02 & 0 & 0.085 & | & 29500 \end{bmatrix} \quad 20R2 \to R2$$

$$\begin{bmatrix} 1 & 1 & 1 & | & 1800000 \\ 0 & 1 & 0 & | & 1300000 \\ 0.02 & 0 & 0.085 & | & 29500 \end{bmatrix} \quad R1 - R2 \to R1$$

$$\begin{bmatrix} 1 & 0 & 1 & | & 500000 \\ 0 & 1 & 0 & | & 1300000 \\ 0.02 & 0 & 0.085 & | & 29500 \end{bmatrix} \quad -0.02R1 + R3 \to R3$$

$$\begin{bmatrix} 1 & 0 & 1 & | & 500000 \\ 0 & 1 & 0 & | & 1300000 \\ 0 & 0 & 0.065 & | & 19500 \end{bmatrix} \quad \frac{R3}{0.065} \to R3$$

$$\begin{bmatrix} 1 & 0 & 1 & | & 500000 \\ 0 & 1 & 0 & | & 1300000 \\ 0 & 0 & 1 & | & 300000 \end{bmatrix} \quad R1 - R3 \to R1$$

$$\begin{bmatrix} 1 & 0 & 0 & | & 200000 \\ 0 & 1 & 0 & | & 1300000 \\ 0 & 0 & 1 & | & 300000 \end{bmatrix}$$

Federal program: $200,000; municipal bonds: $1,300,000; bank loan: $300,000

Calculator Exploration and Discovery

Exercise 1: Answers will vary.

Strengthening Core Skills

Exercise 1:

$$\begin{cases} 2x + y = -2 \\ -x + 3y - 2z = -15 \\ 3x - y + 2z = 9 \end{cases}$$

$$A = \begin{bmatrix} 2 & 1 & 0 \\ -1 & 3 & -2 \\ 3 & -1 & 2 \end{bmatrix};$$

$$A^{-1} = \begin{bmatrix} 1 & -0.5 & -0.5 \\ -1 & 1 & 1 \\ -2 & 1.25 & 1.75 \end{bmatrix}; \quad B = \begin{bmatrix} -2 \\ -15 \\ 9 \end{bmatrix}$$

$$A^{-1}B = \begin{bmatrix} 1 \\ -4 \\ 1 \end{bmatrix}$$

$(1, -4, 1)$

Chapter 1-6 Cumulative Review

1. a. $(3 - 2i)(2 + i) = 6 - 4i + 3i - 2i^2 = 8 - i$

    b. $(5 - 3i)^2 = 25 - 30i + 9i^2 = 16 - 30i$

    c. $\dfrac{8-i}{2+i} = \dfrac{8-i}{2+i} \cdot \dfrac{2-i}{2-i} = \dfrac{16 - 10i + i^2}{4 - i^2}$

    $= \dfrac{15 - 10i}{5} = 3 - 2i$

    d. $i^{49} = i^{4(42)} i = i$

3. $2x - 4(3x + 1) = 5 - 4x$

    $2x - 12x - 4 = 5 - 4x$

    $-10x - 4 = 5 - 4x$

    $-6x = 9$

    $x = -\dfrac{3}{2}$

5. $9x^2 + 1 = 6x$

    $9x^2 - 6x + 1 = 0$

    $(3x - 1)(3x - 1) = 0$

    $x = \dfrac{1}{3}$

7. $(2, -2), (-3, 5)$

    $m = \dfrac{5 - (-2)}{-3 - 2} = \dfrac{7}{-5};$

    $y - (-2) = -\dfrac{7}{5}(x - 2)$

    $y + 2 = -\dfrac{7}{5}x + \dfrac{14}{5}$

    $y = -\dfrac{7}{5}x + \dfrac{4}{5}$

9. $f(x) = 3 - 4x - x^2, g(x) = \left| 4 - \sqrt{x + 2} \right|$

    a. $f(3) = 3 - 4(3) - (3)^2 = -18$

    b. $g(23) = \left| 4 - \sqrt{23 + 2} \right| = 1$

    c. $f(-2) = 3 - 4(-2) - (-2)^2 = 3 + 8 - 4 = 7$

    d. $g(-3) = \left| 4 - \sqrt{-3 + 2} \right| = \left| 4 - \sqrt{-1} \right|$

    -3 is not in the domain

11. $y = \dfrac{1}{2}|x| - 3$

    Shift $y = \dfrac{1}{2}|x| - 3$ down 2 units, compress vertically.

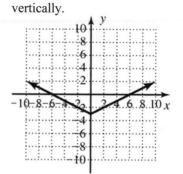

13. $y = \sqrt{-x} - 2$

    $y = \sqrt{x}$ reflected across the y-axis, shifted down 2 units.

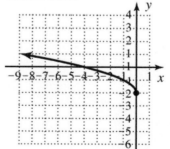

15. $x^2 + 5 > 6x$

$x^2 - 6x + 5 > 0$

$(x-5)(x-1) > 0$

Boundaries: $x = 5, x = 1$

$(-\infty, 1) \cup (5, \infty)$

17. $f(x) \uparrow : x \in (0,2) \cup (4, \infty)$

$f(x) \downarrow : x \in (-\infty, 0) \cup (2, 4)$

Constant: none

19. $\left(4y^2 - 3y + 10\right) \div (2y + 1)$

$$
\begin{array}{r}
2y - \dfrac{5}{2} \\[2pt]
2y+1 \overline{)\, 4y^2 - 3y + 10} \\
\underline{4y^2 + 2y} \\
-5y + 10 \\
\underline{-5y - \dfrac{5}{2}} \\
\dfrac{25}{2}
\end{array}
$$

$2y - \dfrac{5}{2} + \dfrac{25}{2(2y+1)}$

21. roots: $-2, 3-i, 3+i$

$(x+2)\big(x-(3-i)\big)\big(x-(3+i)\big)$

$(x+2)\big((x-3)+i\big)\big((x-3)-i\big)$

$(x+2)\big((x-3)^2 - i^2\big)$

$(x+2)\big(x^2 - 6x + 9 + 1\big)$

$(x+2)\big(x^2 - 6x + 10\big)$

$x^3 - 6x^2 + 10x + 2x^2 - 12x + 20$

$x^3 - 4x^2 - 2x + 20$

23. $\begin{cases} 5x + 6y = -30 \\ -3x + 2y = 4 \end{cases} \quad -3R2$

$\begin{cases} 5x + 6y = -30 \\ 9x - 6y = -12 \end{cases}$

$14x = -42$

$x = -3;$

$5(-3) + 6y = -30$

$-15 + 6y = -30$

$6y = -15$

$y = \dfrac{-5}{2};$

$\left(-3, \dfrac{-5}{2}\right)$

25. $\begin{cases} c + b + t = 600 \\ b = 2t + 20 \\ c = t + b + 200 \end{cases}$

$\begin{cases} c + b + t = 600 \\ b - 2t = 20 \\ c - b - t = 200 \end{cases}$

$\left[\begin{array}{ccc|c} 1 & 1 & 1 & 600 \\ 0 & 1 & -2 & 20 \\ 1 & -1 & -1 & 200 \end{array}\right] \quad R3 - R1 \to R3$

$\left[\begin{array}{ccc|c} 1 & 1 & 1 & 600 \\ 0 & 1 & -2 & 20 \\ 0 & -2 & -2 & -400 \end{array}\right] \quad 2R2 + R3 \to R3$

$\left[\begin{array}{ccc|c} 1 & 1 & 1 & 600 \\ 0 & 1 & -2 & 20 \\ 0 & 0 & -6 & -360 \end{array}\right] \quad -\dfrac{1}{6}R3 \to R3$

$\left[\begin{array}{ccc|c} 1 & 1 & 1 & 600 \\ 0 & 1 & -2 & 20 \\ 0 & 0 & 1 & 60 \end{array}\right] \quad R1 - R2 \to R1$

$\left[\begin{array}{ccc|c} 1 & 0 & 3 & 580 \\ 0 & 1 & -2 & 20 \\ 0 & 0 & 1 & 60 \end{array}\right] \quad R2 + 2R3 \to R2$

$\left[\begin{array}{ccc|c} 1 & 0 & 3 & 580 \\ 0 & 1 & 0 & 140 \\ 0 & 0 & 1 & 60 \end{array}\right] \quad R1 - 3R3 \to R1$

$\left[\begin{array}{ccc|c} 1 & 0 & 0 & 400 \\ 0 & 1 & 0 & 140 \\ 0 & 0 & 1 & 60 \end{array}\right]$

Condo: \$600,000; delivery truck: \$60000; business: \$140,000

MWT 3 Exercises

MWT III

1. $Y_1 = -2.25x + 1125$

$Y_2 = 0.75x - 75$;

Using grapher, $(400, 225)$

225 boards at $400 a piece.

3. $Y_1 = (-5.000\text{x}10^7)x + (2.435\text{x}10^8)$

$Y_2 = (5.000\text{x}10^7)x - (6.350\text{x}10^7)$;

Using grapher, $(3.07, 90000000)$

90,000,000 gallons at $3.07 per gallon.

5. $(:\ 410.07,:\ 226.58)$

or about 227 boards at approximately $410 a piece.

7. $(:\ 3.0442,:\ 8.9964)$

or about 90,000,000 gallons at approximately $3.04 per gallon.

9. Let x represent the number of 22" bass drums.
Let y represent the number of 12" toms.
Let z represent the number of 14" snare drums.

$$\begin{cases} 7x + 2y + 2.5z = \text{ft}^2 \text{ of skin} \\ 8.5x + 3y + 1.5z = \text{ft}^2 \text{ of wood} \\ 8x + 6y + 10z = \text{number of rods} \\ 11.5x + 6.5y + 7z = \text{number of hoops} \end{cases}$$

$$\begin{bmatrix} 7 & 2 & 2.5 \\ 8.5 & 3 & 1.5 \\ 8 & 6 & 10 \\ 11.5 & 6.5 & 7 \end{bmatrix} \begin{bmatrix} 15 \\ 21 \\ 27 \end{bmatrix} = \begin{bmatrix} s \\ w \\ r \\ h \end{bmatrix}$$

$$\begin{bmatrix} 214.5 \\ 231 \\ 516 \\ 498 \end{bmatrix}$$

214.5 ft^2 of skin, 231 ft^2 of wood veneer, 516 tension rods, and 498 feet of hoop.

11. Let x represent the number of 22" bass drums.
Let y represent the number of 12" toms.
Let z represent the number of 14" snare drums.

$$\begin{cases} 7x + 2y + 2.5z = \text{ft}^2 \text{ of skin} \\ 8.5x + 3y + 1.5z = \text{ft}^2 \text{ of wood} \\ 8x + 6y + 10z = \text{number of rods} \\ 11.5x + 6.5y + 7z = \text{number of hoops} \end{cases}$$

$23 + 21 + 17 + 14 = 75$;
$20 + 18 + 15 + 17 = 70$;
$29 + 35 + 27 + 25 = 116$;

$$\begin{bmatrix} 7 & 2 & 2.5 \\ 8.5 & 3 & 1.5 \\ 8 & 6 & 10 \\ 11.5 & 6.5 & 7 \end{bmatrix} \begin{bmatrix} 75 \\ 70 \\ 116 \end{bmatrix} = \begin{bmatrix} s \\ w \\ r \\ h \end{bmatrix}$$

$$\begin{bmatrix} 955 \\ 1021.5 \\ 2180 \\ 2129.5 \end{bmatrix}$$

955 ft^2 of skin, 1021.5 ft^2 of wood veneer, 2180 tension rods, and 2129.5 feet of hoop.

13. Let x represent the number of gallons of E10.
Let y represent the number of gallons of E85.
Let z represent the number of gallons of biodiesel.

$$\begin{cases} 0.9x + 0.15y + 0z = \text{number of gal gas} \\ 2x + 17y + 20z = \text{number of lbs corn} \\ 1x + 8.5y + 0z = \text{number of oz yeast} \\ 0.5x + 4.25y + 3z = \text{number of gal water} \end{cases}$$

$$\begin{bmatrix} 0.9 & 0.15 & 0 \\ 2 & 17 & 20 \\ 1 & 8.5 & 0 \\ 0.5 & 4.25 & 3 \end{bmatrix} \begin{bmatrix} 100000 \\ 15000 \\ 7000 \end{bmatrix} = \begin{bmatrix} g \\ c \\ y \\ w \end{bmatrix}$$

$$\begin{bmatrix} 92250 \\ 595000 \\ 227500 \\ 134750 \end{bmatrix}$$

92250 gallons of gasoline, 595000 lbs corn, 2275000 oz yeast, and 134750 gal water.

15. Let x represent the number of Silver Clams.
Let y represent the number of Gold Clams.
Let z represent the number of Platinum Clams.

$$\begin{cases} 1.2x + 0.5y + 0.2z = \text{number of oz silver} \\ 0.2x + 0.8y + 0.5z = \text{number of oz gold} \\ 0x + 0.1y + 0.7z = \text{number of oz platinum} \end{cases}$$

$$\begin{bmatrix} 1.2 & 0.5 & 0.2 \\ 0.2 & 0.8 & 0.5 \\ 0 & 0.1 & 0.7 \end{bmatrix} \begin{bmatrix} 10.9 \\ 9.2 \\ 2.3 \end{bmatrix} = \begin{bmatrix} s \\ p \\ g \end{bmatrix}$$

$$\begin{bmatrix} 5 \\ 9 \\ 2 \end{bmatrix}$$

5 Silver, 9 Gold, and 2 Platinum.

17. Let x represent the measures of grain in first class bundle.
Let y represent the measures of grain in second class bundle.
Let z represent the measures of grain in third class bundle.

$$\begin{cases} 3x + 2y + z = 39 \\ 2x + 3y + z = 34 \\ 1x + 2y + 3z = 26 \end{cases}$$

$$\begin{bmatrix} 3 & 2 & 1 \\ 2 & 3 & 1 \\ 1 & 2 & 3 \end{bmatrix} \begin{bmatrix} x \\ y \\ z \end{bmatrix} = \begin{bmatrix} 39 \\ 34 \\ 26 \end{bmatrix}$$

$$A^{-1} = \begin{bmatrix} \dfrac{7}{12} & \dfrac{-1}{3} & \dfrac{-1}{12} \\ \dfrac{-5}{12} & \dfrac{2}{3} & \dfrac{-1}{12} \\ \dfrac{1}{12} & \dfrac{-1}{3} & \dfrac{5}{12} \end{bmatrix}$$

$$\begin{bmatrix} \dfrac{7}{12} & \dfrac{-1}{3} & \dfrac{-1}{12} \\ \dfrac{-5}{12} & \dfrac{2}{3} & \dfrac{-1}{12} \\ \dfrac{1}{12} & \dfrac{-1}{3} & \dfrac{5}{12} \end{bmatrix} \begin{bmatrix} 39 \\ 34 \\ 26 \end{bmatrix} = \begin{bmatrix} 9.25 \\ 4.25 \\ 2.75 \end{bmatrix}$$

One bundle of first class = 9.25 measures of grain.
One bundle of second class = 4.25 measures of grain.
One bundle of third class = 2.75 measures of grain.

19. Answers will vary.

21. Answers will vary.

23. Answers will vary.

25. Answers will vary.

27. Answers will vary.

29. Answers will vary.

31. Answers will vary.

33. Answers will vary.

35. Answers will vary.

7.1 Exercises

1. Geometry, algebra

3. Perpendicular

5. Point, intersecting

7. Hypotenuse is from $P_2(1,2)$ to $P_3(-5,-6)$

   Midpoint $\left(\dfrac{1+(-5)}{2}, \dfrac{2+(-6)}{2}\right) = (-2,-2)$;

   Distance: $(-2,-2),(-5,2)$

   $\sqrt{(-2-(-5))^2 + (-2-2)^2}$

   $= \sqrt{9+16} = 5$;

   Distance: $(-5,-6),(-2,-2)$

   $\sqrt{(-5-(-2))^2 + (-6-(-2))^2}$

   $= \sqrt{9+16} = 5$;

   Distance: $(1,2),(-2,-2)$

   $\sqrt{(1-(-2))^2 + (2-(-2))^2}$

   $= \sqrt{9+16} = 5$;

   Verified.

9. Hypotenuse is from $P_1(-2,1)$ to $P_2(6,-5)$

   Midpoint $\left(\dfrac{-2+6}{2}, \dfrac{1+(-5)}{2}\right) = (2,-2)$;

   Distance: $(2,-7),(2,-2)$

   $\sqrt{(2-2)^2 + (-2-(-7))^2}$

   $= \sqrt{0+25} = 5$;

   Distance: $(-2,1),(2,-2)$

   $\sqrt{(2-(-2))^2 + (-2-1)^2}$

   $= \sqrt{16+9} = 5$;

   Distance: $(6,-5),(2,-2)$

   $\sqrt{(2-6)^2 + (-2-(-5))^2}$

   $= \sqrt{16+9} = 5$;

   Verified.

11. Hypotenuse is from $P_3(3,3)$ to $P_1(10,-21)$

    Midpoint $\left(\dfrac{3+10}{2}, \dfrac{3+(-21)}{2}\right) = \left(\dfrac{13}{2},-9\right)$;

    Distance: $(3,3),\left(\dfrac{13}{2},-9\right)$

    $\sqrt{\left(\dfrac{13}{2}-3\right)^2 + (-9-3)^2}$

    $= \sqrt{\left(\dfrac{7}{2}\right)^2 + 144} = \sqrt{\dfrac{625}{4}} = \dfrac{25}{2}$;

    Distance: $(-6,-9),\left(\dfrac{13}{2},-9\right)$

    $\sqrt{\left(\dfrac{13}{2}-(-6)\right)^2 + (-9-(-9))^2}$

    $= \sqrt{\left(\dfrac{25}{2}\right)^2 + 0} = \dfrac{25}{2}$;

    Distance: $(10,-21),\left(\dfrac{13}{2},-9\right)$

    $\sqrt{\left(\dfrac{13}{2}-10\right)^2 + (-9-(-21))^2}$

    $= \sqrt{\left(\dfrac{-7}{2}\right)^2 + 144} = \sqrt{\dfrac{625}{4}} = \dfrac{25}{2}$;

    Verified.

13. Center $(-2,-2)$, $d = 2(5) = 10$;

    $(x-(-2))^2 + (y-(-2))^2 = \left(\dfrac{10}{2}\right)^2$

    $(x+2)^2 + (y+2)^2 = 5^2$

15. Center $(2,-2)$, $d = 2(5) = 10$;

    $(x-2)^2 + (y-(-2))^2 = \left(\dfrac{10}{2}\right)^2$

    $(x-2)^2 + (y+2)^2 = 5^2$

17. Center $\left(\dfrac{13}{2},-9\right)$, $d = 2\left(\dfrac{25}{2}\right) = 25$;

    $\left(x-\dfrac{13}{2}\right)^2 + (y-(-9))^2 = \left(\dfrac{25}{2}\right)^2$

    $\left(x-\dfrac{13}{2}\right)^2 + (y+9)^2 = \left(\dfrac{25}{2}\right)^2$

19. (a) $A(2,3)$ to $B(7,15)$

$$\sqrt{(7-2)^2+(15-3)^2}$$

$$=\sqrt{5^2+12^2}=13;$$

$A(2,3)$ to $C(-10,8)$

$$\sqrt{(-10-2)^2+(8-3)^2}$$

$$=\sqrt{(-12)^2+5^2}=13;$$

$A(2,3)$ to $D(9,14)$

$$\sqrt{(9-2)^2+(14-3)^2}$$

$$=\sqrt{7^2+11^2}=\sqrt{170};$$

$A(2,3)$ to $E(-3,-9)$

$$\sqrt{(-3-2)^2+(-9-3)^2}$$

$$=\sqrt{(-5)^2+(-12)^2}=13;$$

$A(2,3)$ to $F\left(5,4+3\sqrt{10}\right)$

$$\sqrt{(5-2)^2+\left(4+3\sqrt{10}-3\right)^2}$$

$$=\sqrt{3^2+\left(1+3\sqrt{10}\right)^2}$$

$$=\sqrt{9+1+6\sqrt{10}+3(10)}$$

$$=\sqrt{40+6\sqrt{10}};$$

$A(2,3)$ to $G(2-2\sqrt{30},10)$

$$\sqrt{\left(2-2\sqrt{30}-2\right)^2+(10-3)^2}$$

$$=\sqrt{\left(-2\sqrt{30}\right)^2+(7)^2}=\sqrt{169}=13;$$

Points of equal distance from A are: B, C, E, and G. Distance is 13.

(b) To find other points, pick any x or y value and find the other.

Pick $x = 14$.

$A(2,3)$ to $(14,y)$

$$13=\sqrt{(14-2)^2+(y-3)^2}$$

$$13^2=12^2+(y-3)^2$$

$$169=144+y^2-6y+9$$

$$0=y^2-6y-16$$

$$0=(y-8)(y+2)$$

$$y-8=0 \text{ or } y+2=0$$

$$y=8 \quad \text{ or } y=-2;$$

$(14,8);(14,-2)$ are both the same distance away.

Pick $x = 13$.

$A(2,3)$ to $(13,y)$

$$13=\sqrt{(13-2)^2+(y-3)^2}$$

$$13^2=11^2+(y-3)^2$$

$$169=121+y^2-6y+9$$

$$0=y^2-6y-39$$

$$y=\frac{-(-6)\pm\sqrt{(-6)^2-4(1)(-39)}}{2(1)}$$

$$=\frac{6\pm\sqrt{36+156}}{2}=\frac{6\pm\sqrt{192}}{2}$$

$$=\frac{6\pm8\sqrt{3}}{2}=3\pm4\sqrt{3};$$

$\left(13,3+4\sqrt{3}\right);\left(13,3-4\sqrt{3}\right)$ are both the same distance away.

21. $d=\left|\dfrac{Ax_1+By_1+C}{\sqrt{A^2+B^2}}\right|$

$P(-6,2)$ to $y=-\dfrac{1}{2}x+3$

$$y=-\frac{1}{2}x+3$$

$$2y=-x+6$$

$$x+2y-6=0;$$

$$d=\left|\frac{(1)(-6)+2(2)-6}{\sqrt{1^2+2^2}}\right|=\left|\frac{8}{\sqrt{5}}\right|$$

$$d=\frac{8}{\sqrt{5}}\cdot\frac{\sqrt{5}}{\sqrt{5}}=\frac{8\sqrt{5}}{5}$$

$Q(6,4)$ to $y=-\dfrac{1}{2}x+3$

$$x+2y-6=0;$$

$$d=\left|\frac{(1)(6)+2(4)-6}{\sqrt{1^2+2^2}}\right|=\left|\frac{8}{\sqrt{5}}\right|$$

$$d=\frac{8}{\sqrt{5}}\cdot\frac{\sqrt{5}}{\sqrt{5}}=\frac{8\sqrt{5}}{5}$$

Verified.

23. a. $A(0, 1)$ and $y = -1$, $y + 1 = 0$

$A(0,1)$ to $B(-6,9)$

$$\sqrt{(-6-0)^2 + (9-1)^2}$$

$$= \sqrt{36+64} = 10;$$

$B(-6,9)$ to $y+1=0$

$$d = \left| \frac{0+9+1}{\sqrt{0^2+1^2}} \right| = 10;$$

$A(0,1)$ to $C(4,4)$

$$\sqrt{(4-0)^2 + (4-1)^2}$$

$$= \sqrt{16+9} = 5;$$

$C(4,4)$ to $y+1=0$

$$d = \left| \frac{0+4+1}{\sqrt{0^2+1^2}} \right| = 5;$$

$A(0,1)$ to $D(-2\sqrt{2},6)$

$$\sqrt{(-2\sqrt{2}-0)^2 + (6-1)^2}$$

$$= \sqrt{8+25} = \sqrt{33};$$

$D(-2\sqrt{2},6)$ to $y+1=0$

$$d = \left| \frac{0+6+1}{\sqrt{0^2+1^2}} \right| = 7;$$

$A(0,1)$ to $E(4\sqrt{2},8)$

$$\sqrt{(4\sqrt{2}-0)^2 + (8-1)^2}$$

$$= \sqrt{32+49} = 9;$$

$E(4\sqrt{2},8)$ to $y+1=0$

$$d = \left| \frac{0+8+1}{\sqrt{0^2+1^2}} \right| = 9;$$

Points B, C, and E

b. Answers will vary.

Pick $y = 5$

$A(0,1)$ to $(x,5)$

$$\sqrt{(x-0)^2 + (5-1)^2}$$

$$= \sqrt{x^2+16};$$

$(x,5)$ to $y+1=0$

$$d = \left| \frac{0+5+1}{\sqrt{0^2+1^2}} \right| = 6;$$

$$6 = \sqrt{x^2+16}$$

$$36 = x^2+16$$

$$20 = x^2$$

$$x = \pm\sqrt{20}$$

$$x = \pm 2\sqrt{5}$$

$$\left(2\sqrt{5},5\right), \left(-2\sqrt{5},5\right)$$

25. $(0, -4)$ and $y = 4$, $y - 4 = 0$

$A(4,-1)$ to $(0,-4)$

$$\sqrt{(0-4)^2 + (-4-(-1))^2}$$

$$= \sqrt{16+9} = 5;$$

$A(4,-1)$ to $y-4=0$

$$d = \left| \frac{0+(-1)-4}{\sqrt{0^2+1^2}} \right| = 5; \text{ verified}$$

$B\left(10,-\frac{25}{4}\right)$ to $(0,-4)$

$$\sqrt{(0-10)^2 + \left(-4-\frac{-25}{4}\right)^2}$$

$$= \sqrt{100+\frac{81}{16}} = \frac{41}{4};$$

$B\left(10,-\frac{25}{4}\right)$ to $y-4=0$

$$d = \left| \frac{0+\frac{-25}{4}-4}{\sqrt{0^2+1^2}} \right| = \frac{41}{4}; \text{ verified}$$

$C\left(4\sqrt{2},-2\right)$ to $(0,-4)$

$$d = \sqrt{(0-4\sqrt{2})^2 + (-4-(-2))^2}$$

$$= \sqrt{32+4} = 6;$$

$C\left(4\sqrt{2},-2\right)$ to $y-4=0$

$$d = \left| \frac{0-2-4}{\sqrt{0^2+1^2}} \right| = 6; \text{ verified}$$

$D\left(8\sqrt{5},-20\right)$ to $(0,-4)$

$$d = \sqrt{(0-8\sqrt{5})^2 + (-4-(-20))^2}$$

$$= \sqrt{320+256} = 24;$$

$D\left(8\sqrt{5},-20\right)$ to $y-4=0$

$$d = \left| \frac{0-20-4}{\sqrt{0^2+1^2}} \right| = 24; \text{ verified}$$

27. $(0,-4), y = 4$

$(x, 4)$ to (x, y) $(0, -4)$ to (x, y)

$$\sqrt{(x-x)^2 + (y-4)^2} = \sqrt{(x-0)^2 + (y-(-4))^2}$$

$$\sqrt{(y-4)^2} = \sqrt{x^2 + (y+4)^2}$$

$$y - 4 = \sqrt{x^2 + y^2 + 8y + 16}$$

$$y^2 - 8y + 16 = x^2 + y^2 + 8y + 16$$

$$-16y = x^2$$

$$y = -\frac{1}{16}x^2$$

29. Focus $(0, -2)$, directrix $y = -8$

$(0, -2)$ to (x, y) and $(x, -8)$ to (x, y)

$$\sqrt{(x-0)^2 + (y-(-2))^2} = \frac{1}{2}\sqrt{(x-x)^2 + (y-(-8))^2}$$

$$\sqrt{x^2 + (y+2)^2} = \frac{1}{2}\sqrt{(y+8)^2}$$

$$4\left(x^2 + (y+2)^2\right) = (y+8)^2$$

$$4x^2 + 4y^2 + 16y + 16 = y^2 + 16y + 64$$

$$4x^2 + 3y^2 = 48$$

31. $4x^2 + 3y^2 = 48, (-3, 2)$

$$4(-3)^2 + 3(2)^2 = 48$$

$$36 + 12 = 48$$

$$48 = 48 \text{ verified}$$

$$\left(\sqrt{12}, 0\right)$$

$$4\left(\sqrt{12}\right)^2 + 3(0)^2 = 48$$

$$48 + 0 = 48$$

$$48 = 48 \text{ verified}$$

$$d_1 = \sqrt{(0-(-3))^2 + (2-2)^2} = 3;$$

$$d_2 = \sqrt{(0-(-3))^2 + (-2-2)^2} = 5;$$

$$d_3 = \sqrt{\left(\sqrt{12}-0\right)^2 + (0-2)^2}$$

$$= \sqrt{12+4} = 4;$$

$$d_4 = \sqrt{\left(\sqrt{12}-0\right)^2 + (0-(-2))^2}$$

$$\sqrt{12+4} = 4;$$

$$d_1 + d_2 = d_3 + d_4$$

$$3 + 5 = 4 + 4$$

$$8 = 8 \text{ verified}$$

33. $x = \frac{1}{2}, (2, 0)$

$(2, 0)$ to (x, y) and $\left(\frac{1}{2}, y\right)$ to (x, y)

$$\sqrt{(x-2)^2 + (y-0)^2} = 2\sqrt{\left(x-\frac{1}{2}\right)^2 + (y-y)^2}$$

$$\sqrt{x^2 - 4x + 4 + y^2} = 2\sqrt{x^2 - x + \frac{1}{4} + 0}$$

$$x^2 - 4x + 4 + y^2 = 4\left(x^2 - x + \frac{1}{4}\right)$$

$$x^2 - 4x + 4 + y^2 = 4x^2 - 4x + 1$$

$$-3x^2 + y^2 = -3$$

$$3x^2 - y^2 = 3$$

35. $A(-8, 2), B(-2, -6), C(4, 0)$

a. orthocenter, point where altitudes meet

$$m_{\overline{AB}} = \frac{-6-2}{-2-(-8)} = \frac{-4}{3}$$

Perpendicular to $\overline{AB}: m = \frac{3}{4}$

$$m_{\overline{BC}} = \frac{0-(-6)}{4-(-2)} = 1$$

Perpendicular to $\overline{BC}: m = -1$

$$m_{\overline{CA}} = \frac{2-0}{-8-4} = -\frac{1}{6}$$

Perpendicular to $\overline{CA}: m = 6$

Equation of line from A to $\overline{BC}$

$$y - 2 = -1(x - (-8))$$

$$y - 2 = -x - 8$$

$$y = -x - 6 \text{ (Eq. 1)}$$

Equation of line from B to $\overline{CA}$

$$y - (-6) = 6(x - (-2))$$

$$y + 6 = 6x + 12$$

$$y = 6x + 6 \text{ (Eq. 2)}$$

Set Eq. 1 = Eq. 2

$$-x - 6 = 6x + 6$$

$$-7x = 12$$

$$x = -\frac{12}{7};$$

$$y = -\left(\frac{-12}{7}\right) - 6 = -\frac{30}{7}$$

$$\left(\frac{-12}{7}, \frac{-30}{7}\right)$$

b. Centroid, point where medians meet. Medians of triangle intersect in point that is two-thirds of the distance from each vertex to the midpoint of the opposite side.

$A(-8,2), B(-2,-6), C(4,0)$

Midpoint

$\overline{AB}$ $\left(\dfrac{-8+(-2)}{2}, \dfrac{2+(-6)}{2}\right) = (-5,-2)$

Midpoint $\overline{BC}$ $\left(\dfrac{-2+4}{2}, \dfrac{-6+0}{2}\right) = (1,-3)$

Midpoint $\overline{CA}$ $\left(\dfrac{4+-8}{2}, \dfrac{0+2}{2}\right) = (-2,1)$

B to midpoint of $\overline{CA}$ $(-2,-6),(-2,1)$

$\dfrac{2}{3}\sqrt{(-2-(-2))^2 + (1-(-6))^2}$

$= \dfrac{2}{3}\cdot 7 = \dfrac{14}{3}$

Using $B(-2,-6)$ to midpoint of $\overline{CA}$ $(-2,1)$,

Centroid at $(-2, y)$.

Because this is a vertical distance from B,

$y = -6 + \dfrac{14}{3} = -\dfrac{4}{3}$

$\left(-2, -\dfrac{4}{3}\right)$

39. a. $5 = \dfrac{10}{1+9e^{-0.5x}}$

$5\left(1+9e^{-0.5x}\right) = 10$

$1+9e^{-0.5x} = 2$

$9e^{-0.5x} = 1$

$e^{-0.5x} = \dfrac{1}{9}$

$\ln e^{-0.5x} = \ln\dfrac{1}{9}$

$-0.5x = \ln 9^{-1}$

$-0.5x = -\ln 9$

$x = \dfrac{-\ln 9}{-0.5}$

Exact: $x = 2\ln 9$

$x \approx 4.39$

b. $345 = 5e^{0.4x} + 75$

$270 = 5e^{0.4x}$

$54 = e^{0.4x}$

$\ln 54 = \ln e^{0.4x}$

$\ln 54 = 0.4x$

$\ln 54 = \dfrac{2}{5}x$

$\dfrac{5\ln 54}{2} = x$

$x \approx 9.97$

37. $A(-2,0), B(2,0), C(-2,3), D\left(2\sqrt{2}, \sqrt{6}\right)$

$AC = \sqrt{(-2-(-2))^2 + (3-0)^2}$

$\quad = \sqrt{0+9} = 3;$

$BC = \sqrt{(-2-2)^2 + (3-0)^2}$

$\quad = \sqrt{16+9} = 5;$

$AD = \sqrt{\left(2\sqrt{2}-(-2)\right)^2 + \left(\sqrt{6}-0\right)^2}$

$\quad = \sqrt{8\sqrt{2}+12+6} = \sqrt{8\sqrt{2}+18};$

$BD = \sqrt{\left(2\sqrt{2}-2\right)^2 + \left(\sqrt{6}-0\right)^2}$

$\quad = \sqrt{12-8\sqrt{2}+6} = \sqrt{18-8\sqrt{2}};$

$AC + BC = 3+5 = 8;$

$AD + BD$

$= \sqrt{8\sqrt{2}+18} + \sqrt{18-8\sqrt{2}} = 8$

Verified, both add to 8.

41. $h(x) = \dfrac{x^2-9}{x^2-4} = \dfrac{(x+3)(x-3)}{(x+2)(x-2)}$

HA: $y = 1$

x-intercepts $(-3, 0), (3, 0)$

y-intercept: $\left(0, \dfrac{9}{4}\right)$

VA: $x = -2, x = 2$

7.2 Exercises

1. $c^2 = \left| a^2 - b^2 \right|$

3. $2a, 2b$

5. Answers will vary.

7. $(x-0)^2 + (y-0)^2 = 7^2$
$x^2 + y^2 = 49$

9. $(x-5)^2 + (y-0)^2 = \left(\sqrt{3}\right)^2$
$(x-5)^2 + y^2 = 3$

11. Diameter endpoints: $(4, 9)$ and $(-2, 1)$
Center = Midpoint
$\left(\dfrac{4+(-2)}{2}, \dfrac{9+1}{2} \right) = (1,5)$
Radius: $\sqrt{(4-1)^2 + (9-5)^2} = \sqrt{9+16} = 5$
$(x-1)^2 + (y-5)^2 = (5)^2$
$(x-1)^2 + (y-5)^2 = 25$

13. $x^2 + y^2 - 12x - 10y + 52 = 0$
$x^2 - 12x + y^2 - 10y = -52$
$x^2 - 12x + 36 + y^2 - 10y + 25 = -52 + 36 + 25$
$(x-6)^2 + (y-5)^2 = 9$
Center: $(6, 5)$, radius: $\sqrt{9} = 3$

15. $x^2 + y^2 - 4x + 10y + 4 = 0$
$x^2 - 4x + y^2 + 10y = -4$
$x^2 - 4x + 4 + y^2 + 10y + 25 = -4 + 4 + 25$
$(x-2)^2 + (y+5)^2 = 25$
Center: $(2, -5)$, radius: $\sqrt{25} = 5$

17. $x^2 + y^2 + 6x - 5 = 0$
$x^2 + 6x + y^2 = 5$
$x^2 + 6x + 9 + y^2 = 5 + 9$
$(x+3)^2 + y^2 = 14$
Center: $(-3, 0)$, radius: $\sqrt{14} \approx 3.7$

19. $\dfrac{(x-1)^2}{9} + \dfrac{(y-2)^2}{16} = 1$
Center: $(1,2)$, $a = 3$, $b = 4$

7.2 Exercises

21. $\dfrac{(x-2)^2}{25} + \dfrac{(y+3)^2}{4} = 1$

    Center: $(2, -3)$, $a = 5$, $b = 2$

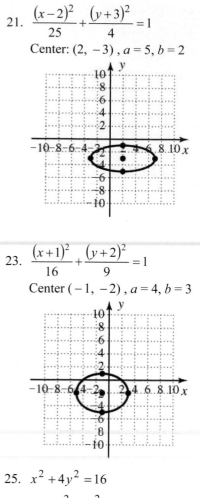

23. $\dfrac{(x+1)^2}{16} + \dfrac{(y+2)^2}{9} = 1$

    Center $(-1, -2)$, $a = 4$, $b = 3$

25. $x^2 + 4y^2 = 16$

    a. $\dfrac{x^2}{16} + \dfrac{y^2}{4} = 1$

       Center: $(0,0)$, $a = 4$, $b = 2$

    b. Vertices: $(-4,0), (4,0)$

       Endpts of minor axis: $(0,-2), (0,2)$

    c.

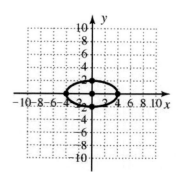

27. $16x^2 + 9y^2 = 144$

    a. $\dfrac{x^2}{9} + \dfrac{y^2}{16} = 1$

       Center: $(0,0)$, $a = 3$, $b = 4$

    b. Vertices: $(0,-4), (0,4)$

       Endpts of minor axis: $(-3,0), (3,0)$

    c.

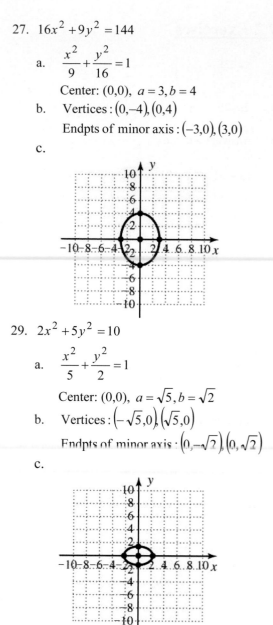

29. $2x^2 + 5y^2 = 10$

    a. $\dfrac{x^2}{5} + \dfrac{y^2}{2} = 1$

       Center: $(0,0)$, $a = \sqrt{5}$, $b = \sqrt{2}$

    b. Vertices: $\left(-\sqrt{5},0\right), \left(\sqrt{5},0\right)$

       Endpts of minor axis: $\left(0,-\sqrt{2}\right), \left(0,\sqrt{2}\right)$

    c.

31. $(x+1)^2 + 4(y-2)^2 = 16$

$$\frac{(x+1)^2}{16} + \frac{(y-2)^2}{4} = 1$$

$$\frac{(x+1)^2}{4^2} + \frac{(y-2)^2}{2^2} = 1$$

Ellipse

Center: $(-1,2)$, $a=4, b=2$

Vertices: $(-5,2),(3,2)$

Endpts of minor axis: $(-1,0),(-1,4)$

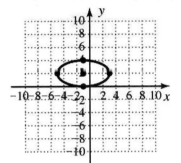

35. $4(x-1)^2 + 9(y-4)^2 = 36$

$$\frac{(x-1)^2}{9} + \frac{(y-4)^2}{4} = 1$$

$$\frac{(x-1)^2}{3^2} + \frac{(y-4)^2}{2^2} = 1$$

Ellipse

Center: $(1,4)$, $a=3, b=2$

Vertices: $(4,4),(-2,4)$

Endpts of minor axis: $(1,2),(1,6)$

(graph)

33. $2(x-2)^2 + 2(y+4)^2 = 18$

$$(x-2)^2 + (y+4)^2 = 9$$

Circle

Center: $(2,-4)$, Radius: 3

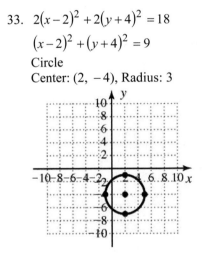

37. $4x^2 + y^2 + 6y + 5 = 0$

$$4x^2 + y^2 + 6y = -5$$

$$4x^2 + y^2 + 6y + 9 = -5 + 9$$

$$4(x)^2 + (y+3)^2 = 4$$

$$x^2 + \frac{(y+3)^2}{4} = 1$$

Center: $(0,-3)$

Vertices $(0,-3-2),(0,-3+2);$

$(0,-5),(0,-1)$

Endpts of minor axis: $(0-1,-3),(0+1,-3);$

$(-1,-3),(1,-3)$

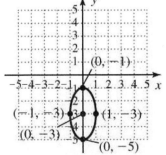

39. $x^2 + 4y^2 - 8y + 4x - 8 = 0$

$x^2 + 4x + 4y^2 - 8y = 8$

$x^2 + 4x + 4(y^2 - 2y) = 8$

$x^2 + 4x + 4 + 4(y^2 - 2y + 1) = 8 + 4 + 4$

$(x+2)^2 + 4(y-1)^2 = 16$

$\dfrac{(x+2)^2}{16} + \dfrac{(y-1)^2}{4} = 1$

$a = 4,\ b = 2$

Center: $(-2, 1)$

Vertices: $(-2-4, 1), (-2+4, 1)$;

$(-6, 1), (2, 1)$

Endpts of minor axis: $(-2, 1-2), (-2, 1+2)$;

$(-2, -1), (-2, 3)$

Center: $(3, -5)$

Vertices: $(3, -5-\sqrt{10}), (3, -5+\sqrt{10})$

Endpts of minor axis: $(3-2, -5), (3+2, -5)$

$(1, -5), (5, -5)$

41. $5x^2 + 2y^2 + 20y - 30x + 75 = 0$

$5x^2 - 30x + 2y^2 + 20y = -75$

$5(x^2 - 6x) + 2(y^2 + 10y) = -75$

$5(x^2 - 6x + 9) + 2(y^2 + 10y + 25) = -75 + 45 + 50$

$5(x-3)^2 + 2(y+5)^2 = 20$

$\dfrac{(x-3)^2}{4} + \dfrac{(y+5)^2}{10} = 1$

$a = 2,\ b = \sqrt{10}$

43. $2x^2 + 5y^2 - 12x + 20y - 12 = 0$

$2x^2 - 12x + 5y^2 + 20y = 12$

$2(x^2 - 6x) + 5(y^2 + 4y) = 12$

$2(x^2 - 6x + 9) + 5(y^2 + 4y + 4) = 12 + 18 + 20$

$2(x-3)^2 + 5(y+2)^2 = 50$

$\dfrac{(x-3)^2}{25} + \dfrac{(y+2)^2}{10} = 1$

$a - 5,\ b - \sqrt{10}$

Center: $(3, -2)$

Vertices $(3-5, -2), (3+5, -2)$;

$(-2, -2), (8, -2)$

Endpts of minor axis:

$(3, -2-\sqrt{10}), (3, -2+\sqrt{10})$

45. $c = 6, b = 8$;

$$a^2 - 8^2 = 6^2$$
$$a^2 = 100$$
$$a = 10;$$

$$2a = 20$$

47. $c = 8, b = 6$;

$$a^2 - 6^2 = 8^2$$
$$a^2 = 100$$
$$a = 10;$$

$$2a = 20$$

49. $4x^2 + 25y^2 - 16x - 50y - 59 = 0$

$$4(x^2 - 4x + 4) + 25(y^2 - 2y + 1) = 59 + 16 + 25$$
$$4(x - 2)^2 + 25(y - 1)^2 = 100$$
$$\frac{(x - 2)^2}{25} + \frac{(y - 1)^2}{4} = 1;$$

$$a = 5, \ b = 2$$
$$a^2 - b^2 = c^2$$
$$25 - 4 = c^2$$
$$21 = c^2$$
$$c = \sqrt{21};$$

a. Center: $(2, 1)$

b. Vertices: $(2 - 5, 1)$ and $(2 + 5, 1)$
   $(-3, 1)$ and $(7, 1)$

c. Foci: $\left(2 - \sqrt{21}, 1\right)$ and $\left(2 + \sqrt{21}, 1\right)$

d. Endpoint of minor axis:
   $(2, 1 + 2)$ and $(2, 1 - 2)$
   $(2, 3)$ and $(2, -1)$

e.

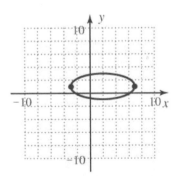

51. $25x^2 + 16y^2 - 200x + 96y + 144 = 0$

$$25(x^2 - 8x + 16) + 16(y^2 + 6y + 9) = -144 + 400 + 144$$
$$25(x - 4)^2 + 16(y + 3)^2 = 400$$
$$\frac{(x - 4)^2}{16} + \frac{(y + 3)^2}{25} = 1;$$

$$a = 4, \ b = 5$$

$$c^2 = 25 - 16 = 9$$
$$c = 3;$$

a. Center: $(4, -3)$

b. Vertices: $(4, -3 + 5)$ and $(4, -3 - 5)$
   $(4, 2)$ and $(4, -8)$

c. Foci: $(4, -3 + 3)$ and $(4, -3 - 3)$
   $(4, 0)$ and $(4, -6)$

d. Endpoint of minor axis:
   $(4 - 4, -3)$ and $(4 + 4, -3)$
   $(0, -3)$ and $(8, -3)$

e.

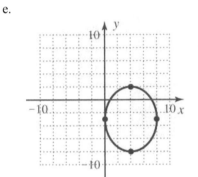

53. $6x^2 + 24x + 9y^2 + 36y + 6 = 0$

$6(x^2 + 4x + 4) + 9(y^2 + 4y + 4) = -6 + 24 + 36$

$6(x+2)^2 + 9(y+2)^2 = 54$

$\dfrac{(x+2)^2}{9} + \dfrac{(y+2)^2}{6} = 54;$

$a = 3, b = \sqrt{6}$

$c^2 = 9 - 6 = 3$

$c = \sqrt{3};$

a. Center: $(-2, -2)$

b. Vertices:

$(-2-3, -2)$ and $(-2+3, -2)$

$(-5, -2)$ and $(1, -2)$

c. Foci: $\left(-2-\sqrt{3}, -2\right)$ and $\left(-2+\sqrt{3}, -2\right)$

d. Endpoint of minor axis:

$\left(-2, -2+\sqrt{6}\right)$ and $\left(-2, -2-\sqrt{6}\right)$

e.

55. Vertices at $(-6,0)$ and $(6,0)$;

Foci at $(-4,0)$ and $(4,0), \leftrightarrow$;

$V: (-6,0), (6,0), a = 6$

$F: (-4,0), (4,0), c = 4$

$4^2 = 6^2 - b^2$

$b^2 = 36 - 16 = 20$

$b = \sqrt{20};$

Center: $(0,0)$

$\dfrac{x^2}{36} + \dfrac{y^2}{20} = 1$

56. Vertices at $(-8,0)$ and $(8,0)$;

Foci at $(-5,0)$ and $(5,0), \leftrightarrow$

$V: (-8,0), (8,0), a = 8$

$F: (-5,0)(5,0), c = 5$

$5^2 = 8^2 - b^2$

$b^2 = 64 - 25 = 39;$

Center: $(0, 0)$

$\dfrac{x^2}{64} + \dfrac{y^2}{39} = 1$

59. Center: $(0, 0)$

$a = 4, b = 3$

$c^2 = 4^2 - 3^2$

$c = \pm\sqrt{7};$

$\dfrac{x^2}{4^2} + \dfrac{y^2}{3^2} = 1$

$\dfrac{x^2}{16} + \dfrac{y^2}{9} = 1$

Foci: $\left(-\sqrt{7}, 0\right), \left(\sqrt{7}, 0\right)$

61. Center: $(-3, -1)$

$a = 2, b = 4$

$c^2 = 4^2 - 2^2$

$c = \pm\sqrt{12} = \pm 2\sqrt{3};$

$\dfrac{\left(x-(-3)\right)^2}{2^2} + \dfrac{\left(y-(-1)\right)^2}{4^2} = 1$

$\dfrac{(x+3)^2}{4} + \dfrac{(y+1)^2}{16} = 1$

Foci: $\left(-3, -1+2\sqrt{3}\right), \left(-3, -1-2\sqrt{3}\right)$

63. $A = \pi ab;$

$16x^2 + 9y^2 = 144$

$\dfrac{x^2}{9} + \dfrac{y^2}{16} = 1;$

$A = \pi(3)(4) = 12\pi$ units^2

65. $a = 4, b = 3$

$c^2 = a^2 - b^2 = 16 - 9 = 7$

$c = \sqrt{7}$;

Spines are $\sqrt{7}$ or ≈ 2.65 ft from center.

The height of the spine occurs at $\left(\sqrt{7}, y\right)$ on

the hyperbola $\dfrac{x^2}{16} + \dfrac{y^2}{9} = 1$

$\dfrac{\left(\sqrt{7}\right)^2}{16} + \dfrac{y^2}{9} = 1$

$\dfrac{7}{16} + \dfrac{y^2}{9} = 1$

$\dfrac{y^2}{9} = 1 - \dfrac{7}{16}$

$y^2 = 9\left(\dfrac{9}{16}\right)$

$y = \dfrac{9}{4} = 2.25$

Height of the spine: 2.25 ft.

67. $a = 12, b = 8$;

$c = \sqrt{12^2 - 8^2}$

$c = \sqrt{80} = 4\sqrt{5} \approx 8.9$ ft

8.9 ft from center

$2\left(4\sqrt{5}\right) \approx 17.9$ ft apart

69. $\dfrac{x^2}{15^2} + \dfrac{y^2}{8^2} = 1$

$\dfrac{9^2}{15^2} + \dfrac{y^2}{8^2} = 1$

$\dfrac{y^2}{8^2} = 0.64$

$y^2 = 40.96$

$y = 6.4$ ft

71. $a = \dfrac{72}{2} = 36; b = \dfrac{70.5}{2} = 35.25$

$\dfrac{x^2}{36^2} + \dfrac{y^2}{(35.25)^2} = 1$

73. $P = 2\pi\sqrt{\dfrac{a^2 + b^2}{2}}$

Aphelion (max)

$c - (-a) = 156$ million miles

Perihelion

$a - c = 128$

$\begin{cases} a + c = 156 \\ a - c = 128 \end{cases}$

$2a = 284$

Semi major $a = 142$ million miles;

$142 - c = 128$

$c = 142 - 128 = 14$;

$14^2 = 142^2 - b^2$

$b^2 = 142^2 - 14^2 = 19968$

$b \approx 141$;

Semi minor ≈ 141 million miles;

$P = 2\pi\sqrt{\dfrac{142^2 + 141^2}{2}}$

$= 2\pi\sqrt{20022.5}$ million miles;

$\dfrac{889.076 \text{ million miles}}{1.296 \text{ million miles/day}} \approx 686$ days

75. $4x^2 + 9y^2 = 900$

$\dfrac{x^2}{15^2} + \dfrac{y^2}{10^2} = 1$;

$9x^2 + 25y^2 = 900$

$\dfrac{x^2}{10^2} + \dfrac{y^2}{6^2} = 1$;

$A = \pi(15)(10) - \pi(10)(6) = 90\pi$

$= 9{,}000\pi$ yd^2

77. $\dfrac{x^2}{81} - \dfrac{y^2}{36} = 1$

$a = 9, b = 6;$

$L = \dfrac{2m^2}{n}$

$L = \dfrac{2(6)^2}{9} = 8 \text{ units};$

$81 - 36 = c^2$

$c = \sqrt{45} = 3\sqrt{5};$

$\dfrac{\left(\sqrt{45}\right)^2}{81} - \dfrac{y^2}{36} = 1$

$\dfrac{45}{81} - \dfrac{y^2}{36} = 1$

$-\dfrac{y^2}{36} = \dfrac{4}{9}$

$y^2 = 16$

$y = \pm 4$

Verified

$\left(3\sqrt{5}, 4\right), \left(3\sqrt{5}, -4\right),$

$\left(-3\sqrt{5}, 4\right), \left(-3\sqrt{5}, -4\right)$

79. Ellipse $\dfrac{x^2}{a^2} + \dfrac{y^2}{b^2} = 1;$

$c^2 = a^2 - b^2;$

$\dfrac{a^2 - b^2}{a^2} + \dfrac{y^2}{b^2} = 1$

$b^2 a^2 - b^2 b^2 + a^2 y^2 = a^2 b^2$

$a^2 y^2 - b^2 b^2 = -a^2 b^2 + a^2 b^2$

$a^2 y^2 = b^4$

$y^2 = \dfrac{b^4}{a^2}$

$y = \dfrac{b^2}{a};$

Focal Chord is $2y$ so $L = \dfrac{2b^2}{a}.$

Verified.

81. $z_1 \cdot z_2 = \left(2\sqrt{3} + 2\sqrt{3}i\right)\left(5\sqrt{3} - 5i\right)$

$= 10 \cdot 3 - 10\sqrt{3}\, i + 10 \cdot 3i - 10 \cdot \sqrt{3}\, i^2$

$= \left(30 + 10\sqrt{3}\right) + \left(30 - 10\sqrt{3}\right)i;$

$\dfrac{z_1}{z_2} = \dfrac{2\sqrt{3} + 2\sqrt{3}\, i}{5\sqrt{3} - 5i} \cdot \dfrac{5\sqrt{3} + 5i}{5\sqrt{3} + 5i}$

$= \dfrac{10 \cdot 3 + 10\sqrt{3}i + 10 \cdot 3i + 10\sqrt{3}\, i^2}{25 \cdot 3 - 25i^2}$

$= \dfrac{\left(30 - 10\sqrt{3}\right) + \left(30 + 10\sqrt{3}\right)i}{75 + 25}$

$= \dfrac{10\left[\left(3 - \sqrt{3}\right) + \left(3 + \sqrt{3}\right)\, i\right]}{100}$

$= \dfrac{\left(3 - \sqrt{3}\right) + \left(3 + \sqrt{3}\right)i}{10}$

83. a. $R = \dfrac{kL}{d^2}$

b. $240 = \dfrac{k(2)}{0.005^2}$

$k = 0.003$

c. $R = \dfrac{0.003(3)}{0.006^2} = 250 \ \Omega$

Chapter 7: Analytical Geometry and Conic Sections

Chapter 7 Mid-Chapter Check

1. $(x-4)^2 + (y+3)^2 = 9$

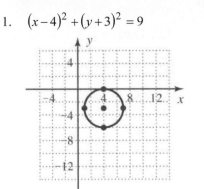

3. $\dfrac{(x-2)^2}{16} + \dfrac{(y+3)^2}{1} = 1$

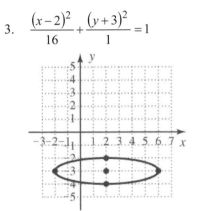

5. $\dfrac{(x+3)^2}{9} + \dfrac{(y-4)^2}{4} = 1$

Center: $(-3, 4)$

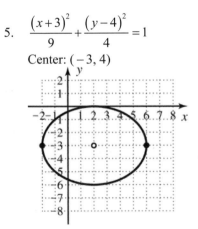

7. Focus: $(0,3)$, Directrix: $y = -3$

$\quad (0,3)$ to $(x,y) \qquad (x,-3)$ to (x,y)

$$\sqrt{(x-0)^2 + (y-3)^2} = \sqrt{(x-x)^2 + (y-(-3))^2}$$

$$\sqrt{x^2 + (y-3)^2} = \sqrt{(y+3)^2}$$

$$\sqrt{x^2 + y^2 - 6y + 9} = y + 3$$

$$x^2 + y^2 - 6y + 9 = y^2 + 6y + 9$$

$$x^2 = 12y$$

$$\frac{1}{12}x^2 = y$$

9. Foci: $(0, 13)$ and $(0, -13)$
   Minor axis length 10 units; center $(0,0)$,
   vertical hyperbola;
   $c = 13, 2a = 10, a = 5$;

$$13^2 = a^2 - 5^2$$

$$194 = b^2$$

$$b = \pm\sqrt{194}b$$

$$\frac{x^2}{5^2} + \frac{y^2}{\left(\sqrt{194}\right)^2} = 1$$

$$\frac{x^2}{25} + \frac{y^2}{194} = 1$$

Chapter 7-Reinforcing Basic Concepts

1. $100x^2 - 400x - 18y^2 - 108y + 230 = 0$

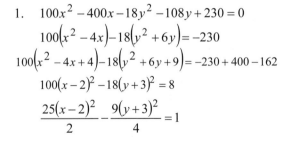

$$100\left(x^2 - 4x\right) - 18\left(y^2 + 6y\right) = -230$$

$$100\left(x^2 - 4x + 4\right) - 18\left(y^2 + 6y + 9\right) = -230 + 400 - 162$$

$$100(x-2)^2 - 18(y+3)^2 = 8$$

$$\frac{25(x-2)^2}{2} - \frac{9(y+3)^2}{4} = 1$$

Technology Highlight

1. $25y^2 - 4x^2 = 100$

   Vertices: $(0, 2)$, $(0, -2)$

   When $x = 4$, $y = \pm 2.5612497$

7.3 Exercises

1. transverse

3. midway

5. Answers will vary.

7. $\dfrac{x^2}{9} - \dfrac{y^2}{4} = 1$

   Center: $(0,0)$
   Vertices: $(-3,0)$, $(3,0)$

   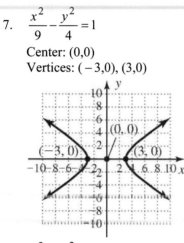

9. $\dfrac{x^2}{4} - \dfrac{y^2}{9} = 1$

   Center: $(0,0)$
   Vertices: $(-2,0)$, $(2,0)$

   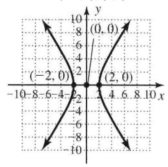

11. $\dfrac{x^2}{49} - \dfrac{y^2}{16} = 1$

    Center: $(0,0)$
    Vertices: $(-7,0)$, $(7,0)$

    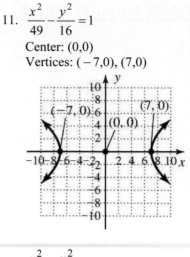

13. $\dfrac{x^2}{36} - \dfrac{y^2}{16} = 1$

    Center: $(0,0)$
    Vertices: $(-6,0)$, $(6,0)$

    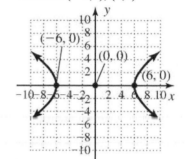

15. $\dfrac{y^2}{9} - \dfrac{x^2}{1} = 1$

    Center: $(0,0)$
    Vertices: $(0, -3)$, $(0,3)$

    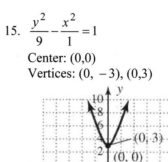

17. $\dfrac{y^2}{12} - \dfrac{x^2}{4} = 1$

Center: $(0,0)$

Vertices: $\left(0, -2\sqrt{3}\right), \left(0, 2\sqrt{3}\right)$

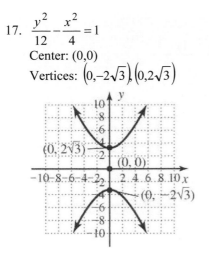

19. $\dfrac{y^2}{9} - \dfrac{x^2}{9} = 1$

Center: $(0,0)$

Vertices: $(0, -3), (0,3)$

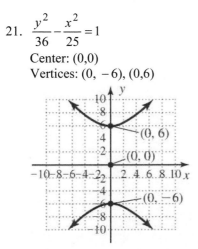

21. $\dfrac{y^2}{36} - \dfrac{x^2}{25} = 1$

Center: $(0,0)$

Vertices: $(0, -6), (0,6)$

23. Vertices: $(-4, -2), (2, -2)$

Transverse Axis: $y = -2$

$\dfrac{-4+2}{2} = -1$

Center: $(-1, -2)$

Conjugate Axis: $x = -1$

25. Vertices: $(4,1), (4, -3)$

Transverse Axis: $x = 4$

$\dfrac{1+-3}{2} = -1$

Center: $(4, -1)$

Conjugate Axis: $y = -1$

27. $\dfrac{(y+1)^2}{4} - \dfrac{x^2}{25} = 1$

Center: $(0, -1)$

$a = 5, \ b = 2$

Vertices: $\left(0, -1-2\right), \left(0, -1+2\right);$

$\left(0, -3\right), \left(0, 1\right)$

Transverse axis: $x = 0$

Conjugate axis: $y = -1$

Asymptotes: Slope $= \pm\dfrac{2}{5}$

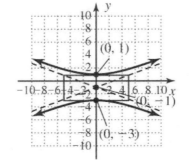

29. $\dfrac{(x-3)^2}{36} - \dfrac{(y+2)^2}{49} = 1$

Center: $(3, -2)$

$a = 6$, $b = 7$

Vertices: $(3-6,-2), (3+6,-2)$;

$(-3,-2), (9,-2)$

Transverse axis: $y = -2$

Conjugate axis: $x = 3$

Asymptotes: Slope $= \pm\dfrac{7}{6}$

31. $\dfrac{(y+1)^2}{7} - \dfrac{(x+5)^2}{9} = 1$

Center: $(-5, -1)$

$a = 3$, $b = \sqrt{7}$

Vertices: $\left(-5,-1+\sqrt{7}\right), \left(-5,-1-\sqrt{7}\right)$

Transverse axis: $x = -5$

Conjugate axis: $y = -1$

Asymptotes: Slope $= \pm\dfrac{\sqrt{7}}{3}$

33. $(x-2)^2 - 4(y+1)^2 = 16$

$\dfrac{(x-2)^2}{16} - \dfrac{(y+1)^2}{4} = 1$

Center: $(2, -1)$

$a = 4$, $b = 2$

Vertices: $(2-4,-1), (2+4,-1)$;

$(-2,-1), (6,-1)$

Transverse axis: $y = -1$

Conjugate axis: $x = 2$

Asymptotes: Slope $= \pm\dfrac{1}{2}$

35. $2(y+3)^2 - 5(x-1)^2 = 50$

$\dfrac{(y+3)^2}{25} - \dfrac{(x-1)^2}{10} = 1$

Center: $(1, \quad 3)$

$a = \sqrt{10}$, $b = 5$

Vertices: $(1,-3+5), (1,-3-5)$;

$(1,2), (1,-8)$

Transverse axis: $x = 1$

Conjugate axis: $y = -3$

Asymptotes: Slope $= \pm\dfrac{5}{\sqrt{10}} = \pm\dfrac{\sqrt{10}}{2}$

37. $12(x-4)^2 - 5(y-3)^2 = 60$

$$\frac{(x-4)^2}{5} - \frac{(y-3)^2}{12} = 1$$

Center: (4,3)

$a = \sqrt{5}, \ b = \sqrt{12} = 2\sqrt{3}$

Vertices: $\left(4+\sqrt{5}, 3\right), \left(4-\sqrt{5}, 3\right)$

Transverse axis: $y = 3$

Conjugate axis: $x = 4$

Asymptotes: Slope $= \pm\dfrac{2\sqrt{3}}{\sqrt{5}}$

$$= \pm\frac{2\sqrt{3}}{\sqrt{5}} \cdot \frac{\sqrt{5}}{\sqrt{5}} = \pm\frac{2\sqrt{15}}{5}$$

41. $9y^2 - 4x^2 = 36$

$$\frac{y^2}{4} - \frac{x^2}{9} = 1$$

Center: (0,0)

$a = 3, \ b = 2$

Vertices: $(0, 0-2), (0, 0+2)$;

$(0,-2), (0,2)$

Transverse axis: $x = 0$

Conjugate axis: $y = 0$

Asymptotes: Slope $= \pm\dfrac{2}{3}$

39. $16x^2 - 9y^2 = 144$

$$\frac{x^2}{9} - \frac{y^2}{16} = 1$$

Center: (0,0)

$a = 3, \ b = 4$

Vertices: $(0+3, 0), (0-3, 0)$;

$(3,0), (-3,0)$

Transverse axis: $y = 0$

Conjugate axis: $x = 0$

Asymptotes: Slope $= \pm\dfrac{4}{3}$

43. $12x^2 - 9y^2 = 72$

$$\frac{x^2}{6} - \frac{y^2}{8} = 1$$

Center: (0,0)

$a = \sqrt{6}, \ b = \sqrt{8} = 2\sqrt{2}$

Vertices: $\left(\sqrt{6}, 0\right), \left(-\sqrt{6}, 0\right)$

Transverse axis: $y = 0$

Conjugate axis: $x = 0$

Asymptotes: Slope $= \pm\dfrac{2\sqrt{2}}{\sqrt{6}}$

$$= \pm\frac{2\sqrt{2}}{\sqrt{6}} \cdot \frac{\sqrt{6}}{\sqrt{6}} = \pm\frac{2\sqrt{3}}{3}$$

45. $4x^2 - y^2 + 40x - 4y + 60 = 0$

$4x^2 + 40x - y^2 - 4y = -60$

$4\left(x^2 + 10x\right) - \left(y^2 + 4y\right) = -60$

$4\left(x^2 + 10x + 25\right) - \left(y^2 + 4y + 4\right) = -60 + 100 - 4$

$4(x+5)^2 - (y+2)^2 = 36$

$\dfrac{(x+5)^2}{9} - \dfrac{(y+2)^2}{36} = 1$

Center: $(-5, -2)$

$a = 3, \ b = 6$

Vertices: $\left(-5-3,-2\right),\left(-5+3,-2\right)$;

$\left(-8,-2\right),\left(-2,-2\right)$

Transverse axis: $y = -2$

Conjugate axis: $x = -5$

Asymptotes: Slope $= \pm\dfrac{6}{3} = \pm 2$

47. $x^2 - 4y^2 - 24y - 4x - 36 = 0$

$x^2 - 4x - 4y^2 - 24y = 36$

$\left(x^2 - 4x\right) - 4\left(y^2 + 6y\right) = 36$

$\left(x^2 - 4x + 4\right) - 4\left(y^2 + 6y + 9\right) = 36 + 4 - 36$

$(x-2)^2 - 4(y+3)^2 = 4$

$\dfrac{(x-2)^2}{4} - \dfrac{(y+3)^2}{1} = 1$

Center: $(2, -3)$

$a = 2, \ b = 1$

Vertices: $\left(2-2,-3\right),\left(2+2,-3\right)$;

$\left(0,-3\right),\left(4,-3\right)$

Transverse axis: $y = -3$

Conjugate axis: $x = 2$

Asymptotes: Slope $= \pm\dfrac{1}{2}$

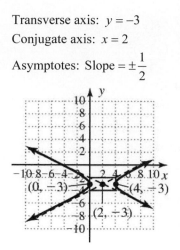

49. $-4x^2 - 4y^2 = -24$

$x^2 + y^2 = 6$

The equation contains a sum of second degree terms with equal coefficients. The equation represents a circle.

51. $x^2 + y^2 = 2x + 4y + 4$

The equation contains a sum of second degree terms with equal coefficients. The equation represents a circle.

53. $2x^2 - 4y^2 = 8$

The equation contains a difference of second degree terms. The equation represents a hyperbola.

55. $x^2 + 5 = 2y^2$

$x^2 - 2y^2 = -5$

$2y^2 - x^2 = 5$

The equation contains a difference of second degree terms. The equation represents a hyperbola.

57. $2x^2 = -2y^2 + x + 20$

$2x^2 - x + 2y^2 = 20$

The equation contains a sum of second degree terms with equal coefficients. The equation represents a circle.

59. $16x^2 + 5y^2 - 3x + 4y = 538$

The equation contains a sum of second degree terms with unequal coefficients. The equation represents an ellipse.

61. $\sqrt{(-5-5)^2+(0-2.25)^2}$

$-\sqrt{(5-5)^2+(0-2.25)^2}=2a$

$\sqrt{(-5-5)^2+(0-2.25)^2}-2.25=2a$

$\sqrt{100+5.0625}-2.25=2a$

$\qquad\qquad\qquad 8=2a$

$a=4, c=5$

$25=4^2+b^2$

$9=b^2$

$3=b$;

$2b=2(3)=6$;

Dimensions: 8 x 6

63.

$\sqrt{(-0-6)^2+(-10-7.5)^2}-\sqrt{(0-6)^2+(10-7.5)^2}=2b$

$\sqrt{36+(-17.5)^2}-\sqrt{36+(2.5)^2}=2b$

$\sqrt{342.25}-\sqrt{42.25}=2b$

$\qquad\qquad\qquad 12=2b$;

$b=6, c=10$

$100=a^2+36$

$a^2=64$

$a=8$;

$2a=2(8)=16$;

Dimensions: 16 x 12

65. $4x^2-9y^2-24x+72y-144=0, \leftrightarrow$

$4(x^2-6x+9)-9(y^2-8y+16)=144-144+36$

$4(x-3)^2-9(y-4)^2=36$

$\dfrac{(x-3)^2}{9}-\dfrac{(y-4)^2}{4}=1$;

$a=3, b=2$;

$c^2=9+4=13$

$c=\sqrt{13}$;

a. Center: $(3,4)$

b. Vertices: $(3-3,4)$ and $(3+3,4)$

$(0,4)$ and $(6,4)$

c. Foci: $(3-\sqrt{13},4)$ and $(3+\sqrt{13},4)$

d. $2a=6, 2b=4$;

e. Asymptotes: Slope $=\pm\dfrac{2}{3}$;

67. $16x^2-4y^2+24y-100=0, \leftrightarrow$

$16(x^2)-4(y^2-6y+9)=100-36$

$16(x^2)-4(y-3)^2=64$

$\dfrac{x^2}{4}-\dfrac{(y-3)^2}{16}=1$;

$a=2, b=4$;

$c^2=4+16$

$c^2=20$

$c=2\sqrt{5}$;

a. Center: $(0,3)$

b. Vertices: $(0-2,3)$ and $(0+2,3)$

$(-2,3)$ and $(2,3)$

c. Foci: $(-2\sqrt{5},3)$ and $(2\sqrt{5},3)$

d. $2a=4, 2b=8$;

e. Asymptotes: Slope $=\pm2$

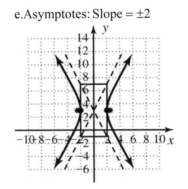

69. $9x^2 - 3y^2 - 54x - 12y + 33 = 0, \leftrightarrow$

$$9\left(x^2 - 6x + 9\right) - 3\left(y^2 + 4y + 4\right) = -33 + 81 - 12$$
$$9(x-3)^2 - 3(y+2)^2 = 36$$
$$\frac{(x-3)^2}{4} - \frac{(y+2)^2}{12} = 1;$$
$$a = 2, b = \sqrt{12} = 2\sqrt{3}$$
$$c^2 = 4 + 12 = 16$$
$$c = 4;$$

a. Center: $(3, -2)$

b. Vertices: $(3-2, -2)$ and $(3+2, -2)$
$$(1, -2) \text{ and } (5, -2)$$

c. Foci: $(-1, -2), (7, -2)$

d. $2a = 4, 2b = 4\sqrt{3};$

e. Asymptotes: Slope $= \pm\sqrt{3}$

71. Vertices: $(-6, 0)$ and $(6, 0)$

Foci: $(-8, 0)$ and $(8, 0)$

$$8^2 = 6^2 + b^2$$
$$64 = 36 + b^2$$
$$b^2 = 28;$$
$$\frac{x^2}{36} - \frac{y^2}{28} = 1$$

73. Foci: $\left(-2, -3\sqrt{2}\right)$ and $\left(-2, 3\sqrt{2}\right)$

Length of conjugate axis: 6 units

$$2a = 6$$
$$a = 3;$$

Center: $\left(\dfrac{-2 + (-2)}{2}, \dfrac{-3\sqrt{2} + 3\sqrt{2}}{2}\right) = (-2, 0)$

$$c = 3\sqrt{2};$$
$$\left(3\sqrt{2}\right)^2 = 3^2 + b^2$$
$$18 - 9 = b^2$$
$$b = 3;$$
$$\frac{(y-0)^2}{9} - \frac{(x-(-2))^2}{9} = 1$$
$$\frac{y^2}{9} - \frac{(x+2)^2}{9} = 1$$

75. Center $(0, 0)$

$$\frac{(x-0)^2}{2^2} - \frac{(y-0)^2}{3^2} = 1$$
$$\frac{x^2}{4} - \frac{y^2}{9} = 1;$$
$$c^2 = 4 + 9 = 13$$
$$c = \sqrt{13};$$

Foci: $\left(-\sqrt{13}, 0\right), \left(\sqrt{13}, 0\right)$

77. Center $(2, 1)$

$$c = 3;$$
$$3^2 = a^2 + 2^2$$
$$9 - 4 = a^2$$
$$a = \sqrt{5};$$
$$4 \times 2\sqrt{5}$$
$$\frac{(y-1)^2}{2^2} - \frac{(x-2)^2}{\left(\sqrt{5}\right)^2} = 1$$
$$\frac{(y-1)^2}{4} - \frac{(x-2)^2}{5} = 1;$$

79. $y = \sqrt{\dfrac{36 - 4x^2}{-9}}$

    a. $\quad y = \sqrt{\dfrac{-4\left(-9 + x^2\right)}{-9}}$

           $y = \sqrt{\dfrac{4\left(x^2 - 9\right)}{9}}$

           $y = \dfrac{2}{3}\sqrt{x^2 - 9}$

    b. $\quad x^2 - 9 \geq 0$

           $(x + 3)(x - 3) \geq 0$

           pos   neg   pos

           -3 3

           $x \in (-\infty, -3] \cup [3, \infty)$

    c. $\quad y = -\dfrac{2}{3}\sqrt{x^2 - 9}$

81. $25y^2 - 1600x^2 = 40000$

     $\dfrac{y^2}{1600} - \dfrac{x^2}{25} = 1$

     $\dfrac{y^2}{40^2} - \dfrac{x^2}{5^2} = 1$

     40 yards

83. $1600x^2 - 400(y - 50)^2 = 640000$

     $\dfrac{x^2}{400} - \dfrac{(y - 50)^2}{1600} = 1$

     $\dfrac{x^2}{20^2} - \dfrac{(y - 50)^2}{40^2} = 1$

     $20 + 20 = 40$ feet

85. 0.4 milliseconds to closer;

     0.5 milliseconds to farther;

     300 km/millisecond;

     $0.5(300) = 150$ km; $0.4(300) = 120$ km;

     $\sqrt{150^2} - \sqrt{120^2} = 2a$

           $150 - 120 = 2a$

               $30 = 2a$

                $a = 15$;

           $c = 50$;

           $50^2 = 15^2 + b^2$

         $2500 = 225 + b^2$

           $b^2 = 2275$;

     $\dfrac{x^2}{225} - \dfrac{y^2}{2275} = 1$;

     $\dfrac{x^2}{225} - \dfrac{(60)^2}{2275} = 1$

     $\dfrac{x^2}{225} = 1 + \dfrac{(60)^2}{2275}$

     $x^2 = \dfrac{52875}{91}$

     $x = \pm 24.1$;

     $(24.1, 60)$ or $(-24.1, 60)$

87. a. $\quad 4x^2 - 32x - y^2 + 4y + 60 = 0$

           $4\left(x^2 - 8x\right) - \left(y^2 - 4y\right) = -60$

           $4\left(x^2 - 8x + 16\right) - \left(y^2 - 4y + 4\right) = -60 + 64 - 4$

             $4(x - 4)^2 - (y - 2)^2 = 0$

             $\dfrac{(x - 4)^2}{\frac{1}{4}} - (y - 2)^2 = 0$

    b. $\quad x^2 - 4x + 5y^2 - 40y + 84 = 0$

           $\left(x^2 - 4x\right) + 5\left(y^2 - 8y\right) = -84$

     $\left(x^2 - 4x + 4\right) + 5\left(y^2 - 8y + 16\right) = -84 + 4 + 80$

           $(x - 2)^2 + 5(y - 4)^2 = 0$

           $(x - 2)^2 + \dfrac{(y - 4)^2}{\frac{1}{5}} = 0$

     Both equal 0.

89. a. $(x-5)^2 - (y+4)^2 = 57$

$$\frac{(x-5)^2}{57} - \frac{(y+4)^2}{57} = 1$$

Area of central rectangle:

$$A = \left(2\sqrt{57}\right)\left(2\sqrt{57}\right) = 228$$

b. $(x-5)^2 + (y+4)^2 = 57$

$$A = \pi\left(\sqrt{57}\right)^2 = 57\pi \approx 179.07$$

c. $9(x-5)^2 + 10(y+4)^2 = 570$

$$\frac{(x-5)^2}{190} + \frac{(y+4)^2}{57} = 1$$
$$\phantom{\frac{(x-5)^2}{}}{\scriptstyle 3}$$

$$A = \pi\left(\sqrt{\frac{190}{3}}\right)\left(\sqrt{57}\right) \approx 188.76$$

Choice a

91. $9(x-2)^2 - 25(y-3)^2 = 225$

$$\frac{(x-2)^2}{25} - \frac{(y-3)^2}{9} = 1;$$

$$c^2 = 25 + 9 = 34;$$
$$c^2 = a^2 - b^2$$
$$34 = a^2 - 9$$
$$a^2 = 43;$$

$$\frac{(x-2)^2}{43} + \frac{(y-3)^2}{9} = 1$$

93. $f(x) = \begin{cases} 4 - x^2 & -2 \le x < 3 \\ 5 & x \ge 3 \end{cases}$

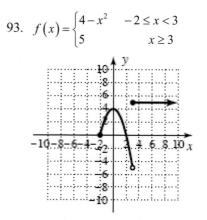

95. a. $x^4 + 4 = 0$

$$\left(1 + i\sqrt{2}\right)^4 = 0$$
$$\left(1 + i\sqrt{2}\right)\left(1 + i\sqrt{2}\right)\left(1 + i\sqrt{2}\right)\left(1 + i\sqrt{2}\right) = 0$$
$$\left(1 + 2i\sqrt{2} + 2i^2\right)\left(1 + 2i\sqrt{2} + 2i^2\right) = 0$$
$$\left(-1 + 2i\sqrt{2}\right)\left(-1 + 2i\sqrt{2}\right) = 0$$
$$1 - 4i\sqrt{2} + 4i^2(2) = 0$$
$$-7 - 4i\sqrt{2} \ne 0$$

b. $x^3 - 6x^2 + 11x - 12 = 0$

$$\left(1 + i\sqrt{2}\right)^3 - 6\left(1 + i\sqrt{2}\right)^2 + 11\left(1 + i\sqrt{2}\right) - 12 = 0$$
$$\left(1 + i\sqrt{2}\right)\left(1 + i\sqrt{2}\right)\left(1 + i\sqrt{2}\right) - 6\left(1 + i\sqrt{2}\right)\left(1 + i\sqrt{2}\right) + 11\left(1 + i\sqrt{2}\right) - 12 = 0$$
$$\left(1 + i\sqrt{2}\right)\left(1 + 2i\sqrt{2} + 2i^2\right) - 6\left(1 + 2i\sqrt{2} + 2i^2\right) + 11 + 11i\sqrt{2} - 12 = 0$$
$$\left(1 + i\sqrt{2}\right)\left(-1 + 2i\sqrt{2}\right) - 6 - 12i\sqrt{2} + 12 + 11 + 11i\sqrt{2} - 12 = 0$$
$$-1 + i\sqrt{2} + 2i^2(2) - 6 - 12i\sqrt{2} + 12 + 11 + 11i\sqrt{2} - 12 = 0$$
$$-1 + i\sqrt{2} - 4 - 6 - 12i\sqrt{2} + 12 + 11 + 11i\sqrt{2} - 12 = 0$$
$$0 = 0$$

c. $x^2 - 2x + 3 = 0$

$$\left(1 + i\sqrt{2}\right)^2 - 2\left(1 + i\sqrt{2}\right) + 3 = 0$$
$$1 + 2i\sqrt{2} + 2i^2 - 2 - 2i\sqrt{2} + 3 = 0$$
$$1 + 2i\sqrt{2} - 2 - 2 - 2i\sqrt{2} + 3 = 0$$
$$0 = 0$$

b and c

7.4 Exercises

1. Horizontal, right, $a < 0$

3. $(p,0)$, $x = -p$

5. Answers will vary.

7. $y = x^2 - 2x - 3$;
$0 = (x-3)(x+1)$
x-intercepts: $(-1,0), (3,0)$;
$y = (0)^2 - 2(0) - 3 = -3$
y-intercept: $(0,-3)$;
$x = \dfrac{-(-2)}{2(1)} = 1$
$y = (1)^2 - 2(1) - 3 = -4$
Vertex: $(1,-4)$;
Domain: $x \in (-\infty, \infty)$
Range: $y \in [-4, \infty)$
$y = (x^2 - 2x + 1) - 3 - 1$
$y = (x-1)^2 - 4$

9. $y = 2x^2 - 8x - 10$
$0 = 2(x-5)(x+1)$
x-intercepts: $(-1,0), (5,0)$;
$y = 2(0)^2 - 8(0) - 10 = -10$
y-intercept: $(0,-10)$;
$x = \dfrac{-(-8)}{2(2)} = 2$
$y = 2(2)^2 - 8(2) - 10 = -18$

Vertex: $(2,-18)$;
Domain: $x \in (-\infty, \infty)$
Range: $y \in [-18, \infty)$
$y = 2(x^2 - 4x + 4) - 10 - 8$
$y = 2(x-2)^2 - 18$

11. $y = 2x^2 + 5x - 7$;
$0 = (2x+7)(x-1)$
x-intercepts: $(-3.5,0), (1,0)$;
$y = 2(0)^2 + 5(0) - 7 = -7$
y-intercept: $(0,-7)$;
$x = \dfrac{-(5)}{2(2)} = -1.25$
$y = 2(-1.25)^2 + 5(-1.25) - 7 = -10.125$
Vertex: $(-1.25, -10.125)$;
Domain: $x \in (-\infty, \infty)$
Range: $y \in [-10.125, \infty)$
$y = 2\left(x^2 + \dfrac{5}{2}x + \dfrac{25}{16}\right) - 7 - \dfrac{25}{8}$
$y = 2\left(x + \dfrac{5}{4}\right)^2 - \dfrac{81}{8}$

13. $x = y^2 - 2y - 3$

$x = (0)^2 - 2(0) - 3 = -3$

x-intercept: $(-3, 0)$;

$0 = (y - 3)(y + 1)$

y-intercepts: $(0, 3), (0, -1)$;

$y = \dfrac{-(-2)}{2(1)} = 1$

$x = (1)^2 - 2(1) - 3 = -4$

Vertex: $(-4, 1)$;

Domain: $x \in [-4, \infty)$

Range: $y \in (-\infty, \infty)$

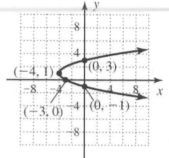

17. $x = -y^2 + 8y - 16$

$x = -(0)^2 + 8(0) - 16 = -16$

x-intercept: $(-16, 0)$;

$0 = (-y + 4)(y - 4)$

y-intercept: $(0, 4)$;

$y = \dfrac{-(8)}{2(-1)} = 4$

$x = -(4)^2 + 8(4) - 16 = 0$

Vertex: $(0, 4)$;

Domain: $x \in (-\infty, 0]$

Range: $y \in (-\infty, \infty)$

15. $x = -y^2 + 6y + 7$

$x = -(0)^2 + 6(0) + 7 = 7$

x-intercept: $(7, 0)$;

$0 = (-y + 7)(y + 1)$

y-intercepts: $(0, 7), (0, -1)$;

$y = \dfrac{-(6)}{2(-1)} = 3$

$x = -(3)^2 + 6(3) + 7 = 16$

Vertex: $(16, 3)$;

Domain: $x \in (-\infty, 16]$

Range: $y \in (-\infty, \infty)$

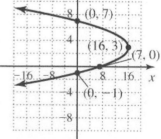

19. $x = y^2 - 6y$

$x = (y^2 - 6y + 9) - 9$

$x = (y - 3)^2 - 9$

Vertex: $(-9, 3)$;

$x = (0)^2 - 6(0) = 0$

x-intercept: $(0, 0)$;

$0 = y(y - 6)$

y-intercepts: $(0, 0), (0, 6)$;

Domain: $x \in [-9, \infty)$

Range: $y \in (-\infty, \infty)$

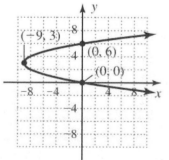

21. $x = y^2 - 4$

Vertex: $(-4, 0)$;

$x = (0)^2 - 4 = -4$

x-intercept: $(-4, 0)$;

$0 = (y + 2)(y - 2)$

y-intercepts: $(0, 2), (0, -2)$;

Domain: $x \in [-4, \infty)$

Range: $y \in (-\infty, \infty)$

23. $x = -y^2 + 2y - 1$

$x = -(y^2 - 2y) - 1$

$x = -(y^2 - 2y + 1) - 1 + 1$

$x = -(y - 1)^2 + 0$

Vertex: $(0, 1)$;

$x = -(0)^2 + 2(0) - 1 = -1$

x-intercept: $(-1, 0)$;

$0 = (-y + 1)(y - 1)$

y-intercept: $(0, 1)$;

Domain: $x \in (-\infty, 0]$

Range: $y \in (-\infty, \infty)$

25. $x = y^2 + y - 6$

$x = \left(y^2 + y + \dfrac{1}{4}\right) - 6 - \dfrac{1}{4}$

$x = \left(y + \dfrac{1}{2}\right)^2 - \dfrac{25}{4}$

Vertex: $(-6.25, -0.5)$;

$x = (0)^2 + (0) - 6 = -6$

x-intercept: $(-6, 0)$;

$0 = (y + 3)(y - 2)$

y-intercepts: $(0, 2), (0, -3)$;

Domain: $x \in [-6.25, \infty)$

Range: $y \in (-\infty, \infty)$

27. $x = y^2 - 10y + 4$

$x = (y^2 - 10y + 25) + 4 - 25$

$x = (y - 5)^2 - 21$

Vertex: $(-21, 5)$;

$x = (0)^2 - 10(0) + 4 = 4$

x-intercept: $(4, 0)$;

$a = 1, b = -10, c = 4$

$y = \dfrac{-(-10) \pm \sqrt{(-10)^2 - 4(1)(4)}}{2(1)}$

$y = \dfrac{10 \pm \sqrt{84}}{2} = \dfrac{10 \pm 2\sqrt{21}}{2} = 5 \pm \sqrt{21}$

y-intercepts: $(0, 5 - \sqrt{21}), (0, 5 + \sqrt{21})$;

Domain: $x \in [-21, \infty)$

Range: $y \in (-\infty, \infty)$

29. $x = 3 - 8y - 2y^2$

$x = -2y^2 - 8y + 3$

$x = -2(y^2 + 4y + 4) + 3 + 8$

$x = -2(y + 2)^2 + 11$

Vertex: $(11, -2)$;

$x = 3 - 8(0) - 2(0)^2 = 3$

x-intercept: $(3, 0)$;

$a = -2, b = -8, c = 3$

$y = \dfrac{-(-8) \pm \sqrt{(-8)^2 - 4(-2)(3)}}{2(-2)}$

$y = \dfrac{8 \pm \sqrt{88}}{-4} = \dfrac{8 \pm 2\sqrt{22}}{-4} = \dfrac{-4 \pm \sqrt{22}}{2}$

y-intercepts: $\left(0, \dfrac{-4 + \sqrt{22}}{2}\right), \left(0, \dfrac{-4 - \sqrt{22}}{2}\right)$;

Domain: $x \in (-\infty, 11]$

Range: $y \in (-\infty, \infty)$

31. $y = (x - 2)^2 + 3$

Vertex: $(2, 3)$;

$0 \neq (x - 2)^2 + 3$

x-intercept: None;

$y = (0 - 2)^2 + 3 = 7$

y-intercept: $(0, 7)$;

Domain: $x \in (-\infty, \infty)$

Range: $y \in [3, \infty)$

33. $x = (y-3)^2 + 2$

    Vertex: $(2,3)$;

    $x = (0-3)^2 + 2 = 11$

    x-intercept: $(11,0)$;

    $0 \neq (y-3)^2 + 2$

    y-intercept: None ;

    Domain: $x \in [2, \infty)$

    Range: $y \in (-\infty, \infty)$

35. $x = 2(y-3)^2 + 1$

    Vertex: $(1,3)$;

    $x = 2(0-3)^2 + 1 = 19$

    x-intercept: $(19,0)$;

    $0 \neq 2(y-3)^2 + 1$

    y-intercept: None;

    Domain: $x \in [1, \infty)$

    Range: $y \in (-\infty, \infty)$

37. $x^2 = 8y$

    Vertex: $(0,0)$

    $8 = 4p$

    $2 = p$

    Focus: $(0,2)$

    Directrix: $y = -2$

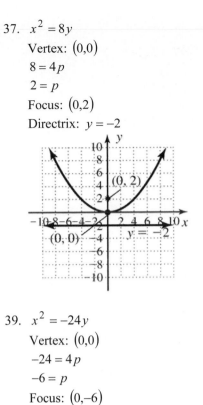

39. $x^2 = -24y$

    Vertex: $(0,0)$

    $-24 = 4p$

    $-6 = p$

    Focus: $(0,-6)$

    Directrix: $y = 6$

41. $x^2 = 6y$

Vertex: $(0,0)$

$6 = 4p$

$\dfrac{3}{2} = p$

Focus: $\left(0, \dfrac{3}{2}\right)$

Directrix: $y = \dfrac{-3}{2}$

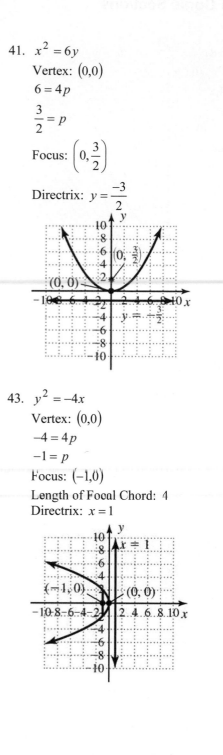

43. $y^2 = -4x$

Vertex: $(0,0)$

$-4 = 4p$

$-1 = p$

Focus: $(-1,0)$

Length of Focal Chord: 4

Directrix: $x = 1$

45. $y^2 = 18x$

Vertex: $(0,0)$

$18 = 4p$

$\dfrac{9}{2} = p$

Focus: $\left(\dfrac{9}{2}, 0\right)$

Length of Focal Chord: 18

Directrix: $x = -\dfrac{9}{2}$

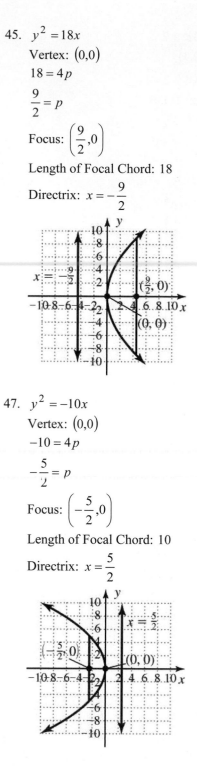

47. $y^2 = -10x$

Vertex: $(0,0)$

$-10 = 4p$

$-\dfrac{5}{2} = p$

Focus: $\left(-\dfrac{5}{2}, 0\right)$

Length of Focal Chord: 10

Directrix: $x = \dfrac{5}{2}$

49. $x^2 - 8x - 8y + 16 = 0$

$x^2 - 8x + 16 = 8y$

$(x-4)^2 = 8y$

Vertex: $(4,0)$

$8 = 4p$

$2 = p$

Focus: $(4,2)$

Length of Focal Chord: 8

Directrix: $y = -2$

51. $x^2 - 14x - 24y + 1 = 0$

$x^2 - 14x = 24y - 1$

$x^2 - 14x + 49 = 24y - 1 + 49$

$(x-7)^2 = 24y + 48$

$(x-7)^2 = 24(y+2)$

Vertex: $(7,-2)$

$24 = 4p$

$6 = p$

Focus: $(7,-2+6) = (7,4)$

Length of Focal Chord: 24

Directrix: $y = -2 - 6 = -8$

53. $3x^2 - 24x - 12y + 12 = 0$

$3x^2 - 24x = 12y - 12$

$3(x^2 - 8x + 16) = 12y - 12 + 48$

$3(x-4)^2 = 12y + 36$

$3(x-4)^2 = 12(y+3)$

$(x-4)^2 = 4(y+3)$

Vertex: $(4,-3)$

$4 = 4p$

$1 = p$

Focus: $(4,-3+1) = (4,-2)$

Length of Focal Chord: 4

Directrix: $y = -3 - 1 = -4$

55. $y^2 - 12y - 20x + 36 = 0$

$y^2 - 12y + 36 = 20x$

$(y-6)^2 = 20x$

Vertex: $(0,6)$

$4p = 20$

$p = 5$

Focus: $(5,6)$

Length of Focal Chord: 20

Directrix: $x = -5$

57. $y^2 - 6y + 4x + 1 = 0$

$y^2 - 6y = -4x - 1$

$y^2 - 6y + 9 = -4x - 1 + 9$

$(y-3)^2 = -4(x-2)$

Vertex: $(2,3)$

$4p = -4$

$p = -1$

Focus: $(2-1,3) = (1,3)$

Length of Focal Chord: 4

Directrix: $x = 2 - (-1) = 3$

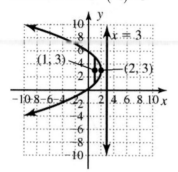

59. $2y^2 - 20y + 8x + 2 = 0$

$y^2 - 10y + 4x + 1 = 0$

$y^2 - 10y = -4x - 1$

$y^2 - 10y + 25 = -4x - 1 + 25$

$(y-5)^2 = -4(x-6)$

Vertex: $(6,5)$

$4p = -4$

$p = -1$

Focus: $(6-1,5) = (5,5)$

Length of Focal Chord: 4

Directrix: $x = 6 - (-1) = 7$

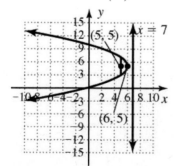

61. Focus $(0, 2)$

Directrix $y = -2$, pt $(0, -2)$

$2p = \sqrt{(0-0)^2 + (2-(-2))^2}$

$2p = 4$

$p = 2$, opens up

Vertex: $\left(\dfrac{0+0}{2}, \dfrac{2+(-2)}{2}\right) = (0,0)$;

$(x-0)^2 = 4(2)(y-0)$

$x^2 = 8y$

63. Focus $(4, 0)$

Directrix $x = -4$, pt $(-4, 0)$

$2p = \sqrt{(4-(-4))^2 + (0-0)^2}$

$2p = 8$

$p = 4$, opens right

Vertex: $\left(\dfrac{4+(-4)}{2}, \dfrac{0+0}{2}\right) = (0,0)$;

$(y-0)^2 = 4(4)(x-0)$

$y^2 = 16x$

65. Focus $(0, -5)$

Directrix $y = 5$, pt $(0, 5)$

$2p = \sqrt{(0-0)^2 + (-5-5)^2}$

$2p = 10$

$p = -5$, opens down

Vertex: $\left(\dfrac{0+0}{2}, \dfrac{-5+5}{2}\right) = (0,0)$;

$(x-0)^2 = 4(-5)(y-0)$

$x^2 = -20y$

67. Vertex: $(2, -2)$; Focus: $(-1, -2)$

$p = \sqrt{(2-(-1))^2 + (-2-(-2))^2}$

$p = -3$, opens left

$(y-(-2))^2 = 4(-3)(x-2)$

$(y+2)^2 = -12(x-2)$

69. Vertex: $(4, -7)$; Focus: $(4, -4)$

$$p = \sqrt{(4-4)^2 + (-7-(-4))^2}$$

$p = 3$, opens up

$$(x-4)^2 = 4(3)(y--7)$$

$$(x-4)^2 = 12(y+7)$$

71. Focus: $(3, 4)$; Directrix: $y = 0$, $(3,0)$

$$p = \sqrt{(3-3)^2 + (4-0)^2}$$

$p = 2$, opens up

Vertex: $(3, 2)$

$$(x-3)^2 = 4(2)(y-2)$$

$$(x-3)^2 = 8(y-2)$$

73. $4p = 8$, $p = 2$

Vertex: $(-1, 0)$; Focus: $(1, 0)$

$$(y-0)^2 = 4(2)(x-(-1))$$

$$y^2 = 8(x+1)$$

75. $p = -2$, Vertex: $(-2, 2)$; Focus: $(-4, 2)$

$$(y-2)^2 = 4(-2)(x-(-2))$$

$$(y-2)^2 = -8(x+2)$$

Directrix: $x = -2-(-2) = 0$

Endpoints of focal chord: $(-4, -2)$ and $(-4, 6)$

77. $\begin{cases} x^2 + y^2 = 25 \\ 2x^2 - 3y^2 = 5 \end{cases}$ $3R1$

$\begin{cases} 3x^2 + 3y^2 = 75 \\ 2x^2 - 3y^2 = 5 \end{cases}$

$$5x^2 = 80$$

$$x^2 = 16$$

$$x = \pm 4;$$

$$(4)^2 + y^2 = 25$$

$$y^2 = 25 - 16$$

$$y^2 = 9$$

$$y = \pm 3;$$

$$(4,3), (4,-3);$$

$$(-4)^2 + y^2 = 25$$

$$y^2 = 25 - 16$$

$$y^2 = 9$$

$$y = \pm 3;$$

$$(-4,3), (-4,-3)$$

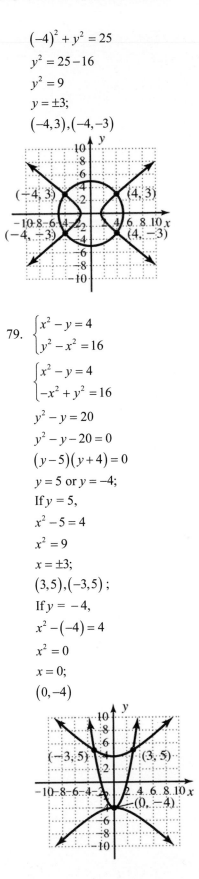

79. $\begin{cases} x^2 - y = 4 \\ y^2 - x^2 = 16 \end{cases}$

$\begin{cases} x^2 - y = 4 \\ -x^2 + y^2 = 16 \end{cases}$

$$y^2 - y = 20$$

$$y^2 - y - 20 = 0$$

$$(y-5)(y+4) = 0$$

$$y = 5 \text{ or } y = -4;$$

If $y = 5$,

$$x^2 - 5 = 4$$

$$x^2 = 9$$

$$x = \pm 3;$$

$$(3,5), (-3,5);$$

If $y = -4$,

$$x^2 - (-4) = 4$$

$$x^2 = 0$$

$$x = 0;$$

$$(0,-4)$$

81. $\begin{cases} 5x^2 - 2y^2 = 75 \\ 2x^2 + 3y^2 = 125 \end{cases}$

$\begin{cases} 15x^2 - 6y^2 = 225 \\ 4x^2 + 6y^2 = 250 \end{cases}$

$19x^2 = 475$

$x^2 = 25$

$x = \pm 5$;

If $x = 5$,

$2(5)^2 + 3y^2 = 125$

$50 + 3y^2 = 125$

$y^2 = 25$

$y = \pm 5$;

$(5,5), (5,-5)$;

If $x = -5$,

$2(-5)^2 + 3y^2 = 125$

$50 + 3y^2 = 125$

$y^2 = 25$

$y = \pm 5$;

$(-5,-5), (-5,5)$

83. $A = \dfrac{2}{3}ab$

$A = \dfrac{2}{3}(3)(8) = 16 \text{ units}^2$

85. $25x = 16y^2$

$\dfrac{25}{16}x = y^2$

Vertex: $(0,0)$

$4p = \dfrac{25}{16}$

$p = \dfrac{25}{64}$

Focus: $\left(\dfrac{25}{64}, 0\right)$

Length of Focal Chord: $\dfrac{25}{16}$

Directrix: $x = -\dfrac{25}{64}$

87. $y^2 = 54x$

$4p = 54$

$p = 13.5$

Focus: $(13.5, 0)$;

36 inch diameter, $y = 18$

$(18)^2 = 54x$

$x = 6$

Parabolic receiver: 6 inches

89. $x^2 = 167y$

$4p = 167$

$p = 41.75$

Focus: $(0, 41.75)$

100 feet diameter, $x = 50$

$(50)^2 = 167y$

$y \approx 14.97$

Parabolic receiver: 14.97 ft

91. Diameter = 10, depth = 5

$$x^2 = 4py$$
$$5^2 = 4p(5)$$
$$25 = 20p$$
$$\frac{5}{4} = p;$$
$$x^2 = 4\left(\frac{5}{4}\right)y$$
$$x^2 = 5y$$

Distance = 1.25 cm

93. $y = 2x^2 - 8x$

$$y = 2\left(x^2 - 4x\right)$$
$$y + 8 = 2\left(x^2 - 4x + 4\right)$$
$$y + 8 = 2(x - 2)^2$$
$$(x - 2)^2 = \frac{1}{2}(y + 8)$$

Vertex: $(2, -8)$

$$4p = \frac{1}{2}$$
$$p = \frac{1}{8}$$

95. $A = \frac{4}{3}T$;

Area of the triangle: $A = \left|\frac{\det(T)}{2}\right| = \left\|\frac{T}{2}\right\|$

where $T = \begin{bmatrix} -3 & 5 & 1 \\ -6 & -3 & 1 \\ 0 & 4 & 1 \end{bmatrix}$

$A = \left|\frac{\det(T)}{2}\right| = \left\|\frac{27}{2}\right\| = 13.5$;

Area of oblique parabolic segment:

$A = \frac{4}{3}(13.5) = 18$ units^2

97. $f(x) = x^5 + 2x^4 + 17x^3 + 34x^2 - 18x - 36$

Answers will vary.

99. Even: Symmetric with respect to the y - axis
Odd: Symmetric with respect to the origin

Chapter 7 Summary and Concept Review

1. $(-3,-4),(-5,4),(3,6),(5,-2)$

   $(-3,-4),(-5,4)$

   $m = \dfrac{4-(-4)}{-5-(-3)} = -4$;

   $d = \sqrt{(-3-(-5))^2 + (-4-4)^2}$

   $= \sqrt{4+64} = \sqrt{68} = 2\sqrt{17}$;

   $(-5,4),(3,6)$

   $m = \dfrac{6-4}{3-(-5)} = \dfrac{1}{4}$;

   $d = \sqrt{(3--5)^2 + (6-4)^2}$

   $= \sqrt{64+4} = \sqrt{68} = 2\sqrt{17}$;

   $(3,6),(5,-2)$

   $m = \dfrac{-2-6}{5-3} = -4$;

   $d = \sqrt{(5-3)^2 + (-2-6)^2}$

   $= \sqrt{4+64} = \sqrt{68} = 2\sqrt{17}$;

   $(5,-2),(-3,-4)$

   $m = \dfrac{-4-(-2)}{-3-5} = \dfrac{1}{4}$;

   $d = \sqrt{(-3-5)^2 + (-4-(-2))^2}$

   $= \sqrt{64+4} = \sqrt{68} = 2\sqrt{17}$;

   Figure is a square, all sides are equal in length and consecutive segments are perpendicular.
   Verified.

3. $(-3,6)$ to $(6,-9)$ and $(-5,-2)$ to $(5,4)$

   Midpoint $(-5,-2)$ to $(5,4)$

   $\left(\dfrac{-5+5}{2}, \dfrac{-2+4}{2}\right) = (0,1)$;

   Slope $(-5,-2)$ to $(5,4)$

   $m = \dfrac{4-(-2)}{5-(-5)} = \dfrac{3}{5}$;

   Slope $(-3,6)$ to $(6,-9)$

   $m = \dfrac{-9-6}{6-(-3)} = -\dfrac{5}{3}$;

   The two segments are perpendicular.
   The equation of the line through $(-3, 6)$ and $(6, -9)$:

   $y - 6 = -\dfrac{5}{3}(x-(-3))$

   $y - 6 = -\dfrac{5}{3}x - 5$

   $y = -\dfrac{5}{3}x + 1$;

   At $x = 0$, $y = 1$
   Line containing $(-3, 6)$ and $(6, -9)$
   contains $(0, 1)$.
   $(-3, 6)$ to $(6, -9)$ is perpendicular bisector
   to $(-5, -2)$ to $(-5, -4)$.

5. $x^2 + y^2 = 16$

   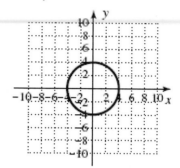

7. $9x^2 + y^2 - 18x - 27 = 0$

   $9x^2 - 18x + y^2 = 27$

   $9\left(x^2 - 2x\right) + y^2 = 27$

   $9\left(x^2 - 2x + 1\right) + y^2 = 27 + 9$

   $9(x-1)^2 + y^2 = 36$

   $\dfrac{(x-1)^2}{4} + \dfrac{y^2}{36} = 1$

   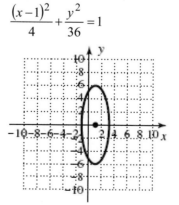

9. $\dfrac{(x+3)^2}{16}+\dfrac{(y-2)^2}{9}=1$

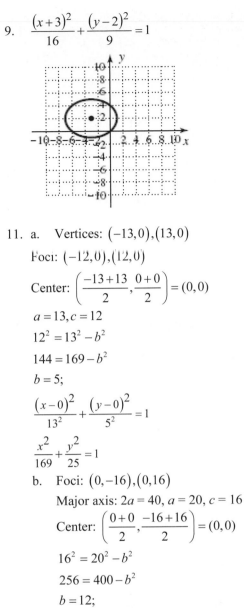

11. a. Vertices: $(-13,0),(13,0)$

Foci: $(-12,0),(12,0)$

Center: $\left(\dfrac{-13+13}{2},\dfrac{0+0}{2}\right)=(0,0)$

$a=13, c=12$

$12^2=13^2-b^2$

$144=169-b^2$

$b=5;$

$\dfrac{(x-0)^2}{13^2}+\dfrac{(y-0)^2}{5^2}=1$

$\dfrac{x^2}{169}+\dfrac{y^2}{25}=1$

b. Foci: $(0,-16),(0,16)$

Major axis: $2a=40,\ a=20,\ c=16$

Center: $\left(\dfrac{0+0}{2},\dfrac{-16+16}{2}\right)=(0,0)$

$16^2=20^2-b^2$

$256=400-b^2$

$b=12;$

$\dfrac{(x-0)^2}{20^2}+\dfrac{(y-0)^2}{12^2}=1$

$\dfrac{x^2}{400}+\dfrac{y^2}{144}=1$

13. $4y^2-25x^2=100$

$\dfrac{y^2}{25}-\dfrac{x^2}{4}=1$

$a=2,\ b=5$

Hyperbola

Center: $(0,0)$

Vertices: $(0,5),(0,-5)$

$a^2+b^2=c^2$

$25+4=c^2$

$\sqrt{29}=c$

Foci: $\left(0,\sqrt{29}\right),\left(0,-\sqrt{29}\right)$

Asymptotes: Slope $=\pm\dfrac{5}{2}$

15. $\dfrac{(x+2)^2}{9}-\dfrac{(y-1)^2}{4}=1$

Hyperbola

Center: $(-2,1)$

$a=3,\ b=2$

Vertices: $(-2-3,1),(-2+3,1)$

$(-5,1),(1,1);$

$a^2+b^2=c^2$

$9+4=c^2$

$\sqrt{13}=c$

Foci: $\left(-2-\sqrt{13},1\right),\left(-2+\sqrt{13},1\right)$

Asymptotes: Slope $=\pm\dfrac{2}{3}$

17. $x^2 - 4y^2 - 12x - 8y + 16 = 0$

$x^2 - 12x - 4y^2 - 8y = -16$

$\left(x^2 - 12x\right) - 4\left(y^2 + 2y\right) = -16$

$\left(x^2 - 12x + 36\right) - 4\left(y^2 + 2y + 1\right) = -16 + 36 - 4$

$(x-6)^2 - 4(y+1)^2 = 16$

$\dfrac{(x-6)^2}{16} - \dfrac{(y+1)^2}{4} = 1$

Hyperbola

Center: $(6, -1)$

$a = 4, \ b = 2$

Vertices: $(6+4, -1), (6-4, -1)$

$(10, -1), (2, -1)$;

$a^2 + b^2 = c^2$

$16 + 4 = c^2$

$2\sqrt{5} = c$

Foci: $\left(6 + 2\sqrt{5}, -1\right), \left(6 - 2\sqrt{5}, -1\right)$

Asymptotes: Slope $= \pm\dfrac{1}{2}$

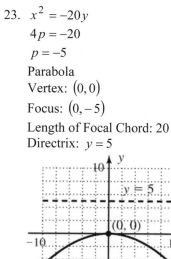

19. a.   Vertices: $(\pm 15, 0)$; Foci: $(\pm 17, 0)$

$a = 15, \ c = 17$

$17^2 = 15^2 + b^2$

$17^2 - 15^2 = b^2$

$64 = b^2$

$b = 8$;

Center $(0, 0)$;

$\dfrac{(x-0)^2}{15^2} - \dfrac{(y-0)^2}{8^2} = 1$

$\dfrac{x^2}{225} - \dfrac{y^2}{64} = 1$

b.   Foci: $(0, \pm 5)$

$2b = 8, \ b = 4$

$c = 5$

$5^2 = 4^2 + a^2$

$5^2 - 4^2 = a^2$

$9 = a^2$

$a = 3$;

Center $(0, 0)$;

$\dfrac{(y-0)^2}{4^2} - \dfrac{(x-0)^2}{3^2} = 1$

$\dfrac{y^2}{16} - \dfrac{x^2}{9} = 1$

21. $x = y^2 - 4$

Parabola

x intercept: $(-4, 0)$

y-intercepts: $(0, 2), (0, -2)$

Vertex: $(-4, 0)$

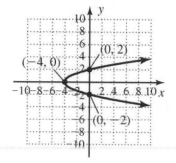

23. $x^2 = -20y$

$4p = -20$

$p = -5$

Parabola

Vertex: $(0, 0)$

Focus: $(0, -5)$

Length of Focal Chord: 20

Directrix: $y = 5$

Chapter 7 Mixed Review

1. $9x^2 + 9y^2 = 54$

    $x^2 + y^2 = 6$

    Circle

    Center: $(0,0)$, Radius : $r = \sqrt{6}$

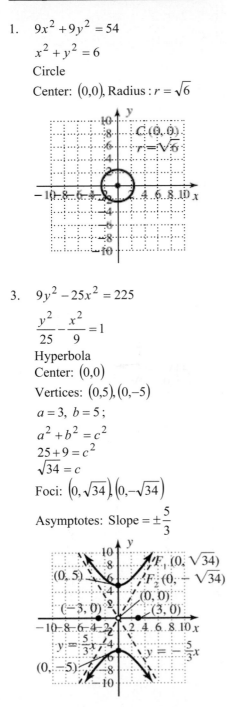

3. $9y^2 - 25x^2 = 225$

    $\dfrac{y^2}{25} - \dfrac{x^2}{9} = 1$

    Hyperbola

    Center: $(0,0)$

    Vertices: $(0,5), (0,-5)$

    $a = 3,\ b = 5$;

    $a^2 + b^2 = c^2$

    $25 + 9 = c^2$

    $\sqrt{34} = c$

    Foci: $\left(0, \sqrt{34}\right), \left(0, -\sqrt{34}\right)$

    Asymptotes: Slope $= \pm\dfrac{5}{3}$

5. $\dfrac{(y-2)^2}{25} - \dfrac{(x+3)^2}{16} = 1$

    Hyperbola

    Center: $(-3,2)$

    Vertices: $(-3,7), (-3,-3)$

    $a = 4,\ b = 5$;

    $a^2 + b^2 = c^2$

    $25 + 16 = c^2$

    $\sqrt{41} = c$

    Foci: $\left(-3, 2 + \sqrt{41}\right), \left(-3, 2 - \sqrt{41}\right)$

    Asymptotes: Slope $= \pm\dfrac{5}{4}$

7. $16(x+2)^2 + 4(y-1)^2 = 64$

    $\dfrac{(x+2)^2}{4} + \dfrac{(y-1)^2}{16} = 1$

    Ellipse

    Center: $(-2,1)$

    Vertices: $(-2, 1+4), (-2, 1-4)$

    $\qquad (-2,5), (-2,-3)$

    $a^2 = b^2 + c^2$

    $16 = 4 + c^2$

    $2\sqrt{3} = c$

    Foci: $\left(-2, 1 - 2\sqrt{3}\right), \left(-2, 1 + 2\sqrt{3}\right)$

9. $x^2 + y^2 - 8x + 12y + 16 = 0$

$x^2 - 8x + y^2 + 12y = -16$

$\left(x^2 - 8x + 16\right) + \left(y^2 + 12y + 36\right) = -16 + 16 + 36$

$(x-4)^2 + (y+6)^2 = 36$

Circle

Center: $(4,-6)$, Radius: $r = 6$

11. $y = 2x^2 - 8x - 10$

$y = 2\left(x^2 - 4x\right) - 10$

$y = 2\left(x^2 - 4x + 4\right) - 10 - 8$

$y = 2(x-2)^2 - 18$;

$y + 18 = 2(x-2)^2$

$\dfrac{1}{2}(y+18) = (x-2)^2$

$4p = \dfrac{1}{2}$

$p = \dfrac{1}{8}$

Parabola

Vertex: $(2,-18)$

Focus: $\left(2, -18 + \dfrac{1}{8}\right) = \left(2, \dfrac{-143}{8}\right)$

Directrix: $y = -\dfrac{145}{8}$

13. $x = -y^2 + 2y + 3$

$x = -\left(y^2 - 2y\right) + 3$

$x = -\left(y^2 - 2y + 1\right) + 3 + 1$

$x = -(y-1)^2 + 4$;

$x - 4 = -(y-1)^2$

$-(x-4) = (y-1)^2$

$4p = -1$

$p = -\dfrac{1}{4}$

Parabola

Vertex: $(4,1)$

Focus: $\left(4 - \dfrac{1}{4}, 1\right) = \left(\dfrac{15}{4}, 1\right)$

Directrix: $x = \dfrac{17}{4}$

15. $x = y^2 - 9$

$(x+9) = y^2$

$4p = 1$

$p = \dfrac{1}{4}$

Parabola

Vertex: $(-9,0)$

Focus: $\left(-9 + \dfrac{1}{4}, 0\right) = \left(-\dfrac{35}{4}, 0\right)$

Directrix: $x = -\dfrac{37}{4}$

17. $x^2 = -24y$

$4p = -24$

$p = -6$

Parabola

Vertex: $(0,0)$

Focus: $(0,-6)$

Directrix: $y = 6$

19. $4x^2 + 16y^2 - 12x - 48y - 19 = 0$

$4x^2 - 12x + 16y^2 - 48y = 19$

$4\left(x^2 - 3x\right) + 16\left(y^2 - 3y\right) = 19$

$4\left(x^2 - 3x + \dfrac{9}{4}\right) + 16\left(y^2 - 3y + \dfrac{9}{4}\right) = 19 + 9 + 36$

$4\left(x - \dfrac{3}{2}\right)^2 + 16\left(y - \dfrac{3}{2}\right)^2 = 64$

$\dfrac{\left(x - \dfrac{3}{2}\right)^2}{16} + \dfrac{\left(y - \dfrac{3}{2}\right)^2}{4} = 1$

Ellipse

Center: $\left(\dfrac{3}{2}, \dfrac{3}{2}\right)$

Vertices: $\left(\dfrac{3}{2} + 4, \dfrac{3}{2}\right), \left(\dfrac{3}{2} - 4, \dfrac{3}{2}\right)$

$\left(\dfrac{11}{2}, \dfrac{3}{2}\right), \left(-\dfrac{5}{2}, \dfrac{3}{2}\right)$

$a^2 = b^2 + c^2$

$16 = 4 + c^2$

$2\sqrt{3} = c$

Foci: $\left(\dfrac{3}{2} - 2\sqrt{3}, \dfrac{3}{2}\right), \left(\dfrac{3}{2} + 2\sqrt{3}, \dfrac{3}{2}\right)$

21. $x^2 + y^2 + 8x + 12y + 2 = 0$

$x^2 + 8x + y^2 + 12y = -2$

$\left(x^2 + 8x + 16\right) + \left(y^2 + 12y + 36\right) = -2 + 16 + 36$

$(x + 4)^2 + (y + 6)^2 = 50$

Circle

Center: $(-4,-6)$, Radius: $r = 5\sqrt{2}$

23. $d(P_1 P_3) = \sqrt{(-4 - (-6))^2 + (-9 - 3)^2} = \sqrt{148}$

$d(P_2 P_3) = \sqrt{(6 - (-6))^2 + (1 - 3)^2} = \sqrt{148}$;

$d(P_1 P_3) = d(P_2 P_3)$

25. apehelion(max) $c - (-a) = 70$ million ;

perihelion(min) $a - c = 46$ million ;

$\begin{cases} a + c = 70 \\ a - c = 46 \end{cases}$

a. $2a = 2(58) = 116$

$a = 58$

$58 - c = 46$

$c = 12$

$(12,0)$

b. $a^2 - b^2 = c^2$

$58^2 - b^2 = 12^2$

$\sqrt{3220} = b$

$b \approx 56.75$;

$2b \approx 113.50$

a. $\dfrac{x^2}{3364} + \dfrac{y^2}{3220} = 1$

Chapter 7 Practice Test

1. Circle (c)

3. Parabola (a)

5. $(x-4)^2+(y+3)^2=9$
   Center: $(4,-3)$, $r=3$

7. $\dfrac{(x+3)^2}{9}-\dfrac{(y-4)^2}{4}=1$
   Hyperbola
   Center: $(-3,4)$
   Vertices: $(-3+3,4),(-3-3,4)$
   $\qquad (0,4),(-6,4)$
   $a^2+b^2=c^2$
   $9+4=c^2$
   $\sqrt{13}=c$
   Foci: $\left(-3-\sqrt{13},4\right),\left(-3+\sqrt{13},4\right)$
   Asymptotes: Slope $=\pm\dfrac{2}{3}$

9. $9x^2+4y^2+18x-24y+9=0$
   $9x^2+18x+4y^2-24y=-9$
   $9\left(x^2+2x\right)+4\left(y^2-6y\right)=-9$
   $9\left(x^2+2x+1\right)+4\left(y^2-6y+9\right)=-9+9+36$

$9(x+1)^2+4(y-3)^2=36$
$\dfrac{(x+1)^2}{4}+\dfrac{(y-3)^2}{9}=1$
Ellipse
Center: $(-1,3)$
Vertices: $(-1,3+3),(-1,3-3)$
$\qquad (-1,6),(-1,0)$
$a^2=b^2+c^2$
$9=4+c^2$
$\sqrt{5}=c$
Foci: $\left(-1,3-\sqrt{5}\right),\left(-1,3+\sqrt{5}\right)$

11. $x=(y+3)^2-2$
    $(x+2)=(y+3)^2$
    $4p=1$
    $p=\dfrac{1}{4}$
    Parabola
    Vertex: $(-2,-3)$
    Focus: $\left(-2+\dfrac{1}{4},-3\right)=\left(-\dfrac{7}{4},-3\right)$
    Directrix: $x=-\dfrac{9}{4}$

13. a. $\begin{cases} 4x^2 - y^2 = 16 & \text{Hyperbola} \\ y - x = 2 & \text{Line} \end{cases}$

2nd equation: $y = x + 2$

$$4x^2 - (x+2)^2 = 16$$
$$4x^2 - x^2 - 4x - 4 = 16$$
$$3x^2 - 4x - 20 = 0$$
$$(3x - 10)(x + 2) = 0$$

$x = \dfrac{10}{3}, x = -2$;

$y = \dfrac{10}{3} + 2 = \dfrac{16}{3}$;

$y = -2 + 2 = 0$

Solutions: $\left(\dfrac{10}{3}, \dfrac{16}{3}\right), (-2, 0)$

b. $\begin{cases} 4y^2 - x^2 = 4 & \text{Hyperbola} \\ x^2 + y^2 = 4 & \text{Circle} \end{cases}$

$$5y^2 = 8$$
$$y^2 = \dfrac{8}{5}$$
$$y = \pm\sqrt{\dfrac{8}{5}} = \pm\dfrac{2\sqrt{10}}{5} ;$$

$$4\left(\pm\dfrac{2\sqrt{10}}{5}\right)^2 - x^2 = 4$$

$$-x^2 = 4 - 4\left(\dfrac{8}{5}\right)$$

$$x^2 = \dfrac{12}{5}$$

$$x = \pm\dfrac{2\sqrt{15}}{5} ;$$

Solutions: $\left(\dfrac{2\sqrt{15}}{5}, \dfrac{2\sqrt{10}}{5}\right), \left(\dfrac{2\sqrt{15}}{5}, -\dfrac{2\sqrt{10}}{5}\right),$

$\left(-\dfrac{2\sqrt{15}}{5}, \dfrac{2\sqrt{10}}{5}\right), \left(-\dfrac{2\sqrt{15}}{5}, -\dfrac{2\sqrt{10}}{5}\right)$

15. $(x+2)^2 + (y-5)^2 = r^2$;

$$(0+2)^2 + (3-5)^2 = r^2$$

$$8 = r^2 ;$$

$$(x+2)^2 + (y-5)^2 = 8$$

17. $\dfrac{x^2}{(141.65)^2} + \dfrac{y^2}{(141.03)^2} = 1$

$a^2 = b^2 + c^2$

$(141.65)^2 = (141.03)^2 + c^2$

$175.2616 = c^2$

$13.23864041 = c$;

aphelion:

$a + c = 141.65 + 13.24 = 154.89$ million miles

perihelion:

$a - c = 141.65 - 13.24 = 128.41$ million miles

19. $(x-1)^2 + (y-1)^2 = 25$

Center: (1,1), Radius = 5

$D: x \in [-4, 6]$

$R: y \in [-4, 6]$

Chapter 7 Calculator Exploration

1. $e > 1$

Chapter 7 Strengthening Core Skills

1. $100x^2 - 400x - 18y^2 - 108y + 230 = 0$

$100(x^2 - 4x) - 18(y^2 + 6y) = -230$

$100(x^2 - 4x + 4) - 18(y^2 + 6y + 9) = -230 + 400 - 162$

$100(x-2)^2 - 18(y+3)^2 = 8$

$\dfrac{(x-2)^2}{\dfrac{2}{25}} - \dfrac{(y+3)^2}{\dfrac{4}{9}} = 1$

$\dfrac{(x-2)^2}{\left(\dfrac{\sqrt{2}}{5}\right)^2} - \dfrac{(y+3)^2}{\left(\dfrac{2}{3}\right)^2} = 1$

$a = \dfrac{\sqrt{2}}{5}, b = \dfrac{2}{3}$

3. $\dfrac{4(x+3)^2}{49} + \dfrac{25(y-1)^2}{36} = 1$

$\dfrac{(x+3)^2}{\dfrac{49}{4}} + \dfrac{(y-1)^2}{\dfrac{36}{25}} = 1$

$\dfrac{(x+3)^2}{\left(\dfrac{7}{2}\right)^2} + \dfrac{(y-1)^2}{\left(\dfrac{6}{5}\right)^2} = 1$

Center: $(-3, 1)$

$a = \dfrac{7}{2}, b = \dfrac{6}{5}$

Vertices: $\left(-3 + \dfrac{7}{2}, 1\right), \left(-3 - \dfrac{7}{2}, 1\right)$

$\left(\dfrac{1}{2}, 1\right), \left(-\dfrac{13}{2}, 1\right)$

$(0.5, 1), (-6.5, 1)$

Cumulative Review: Chapters 1-7

1. $x^3 - 2x^2 + 4x - 8 = 0$

$x^2(x-2) + 4(x-2) = 0$

$(x-2)(x^2 + 4) = 0$

$x - 2 = 0$ or $x^2 + 4 = 0$

$x = 2$ or $x = \pm 2i$

3. $\sqrt{x+2} - 2 = \sqrt{3x+4}$

$\left(\sqrt{x+2} - 2\right)^2 = \left(\sqrt{3x+4}\right)^2$

$x + 2 - 4\sqrt{x+2} + 4 = 3x + 4$

$-4\sqrt{x+2} = 2x - 2$

$\left(-4\sqrt{x+2}\right)^2 = (2x-2)^2$

$16(x+2) = 4x^2 - 8x + 4$

$16x + 32 = 4x^2 - 8x + 4$

$0 = 4x^2 - 24x - 28$

$0 = 4(x^2 - 6x - 7)$

$(x-7)(x+1) = 0$

$x - 7 = 0$ or $x + 1 = 0$

$x = 7$ or $x = -1$

Both roots are extraneous. Thus, no solution.

5. $x^2 - 6x + 13 = 0$

$a = 1, b = -6, c = 13$

$x = \dfrac{-(-6) \pm \sqrt{(-6)^2 - 4(1)(13)}}{2(1)}$

$x = \dfrac{6 \pm \sqrt{-16}}{2} = \dfrac{6 \pm 4i}{2} = 3 \pm 2i$

7. $3^{x-2} = 7$

$\ln 3^{x-2} = \ln 7$

$(x-2)\ln 3 = \ln 7$

$x - 2 = \dfrac{\ln 7}{\ln 3}$

$x = 2 + \dfrac{\ln 7}{\ln 3}$

9. $\log_3 x + \log_3 (x-2) = 1$

    $\log_3 x(x-2) = 1$

    $3^1 = x(x-2)$

    $3 = x^2 - 2x$

    $0 = x^2 - 2x - 3$

    $(x-3)(x+1) = 0$

    $x - 3 = 0$ or $x + 1 = 0$

    $x = 3$ or $x = -1$

    $x = 3$, $x = -1$ is extraneous.

11. $y = |x-2| + 3$

    Vertex: (2,3)

    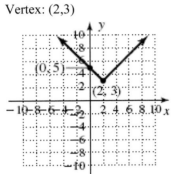

13. $y = \sqrt{x-3} + 1$

    Initial Point: (3,1)

    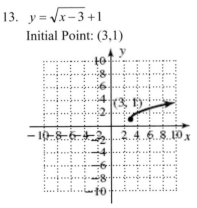

15. $h(x) = \dfrac{x-2}{x^2-9}$

    Horizontal asymptote: $y = 0$
    (deg num < deg den)

    $h(x) = \dfrac{x-2}{(x+3)(x-3)}$

    Vertical asymptotes: $x = 3$ or $x = -3$;

    $h(0) = \dfrac{0-2}{0^2-9} = \dfrac{2}{9} \approx 0.22$

    y-intercept: (0,0.22);

    $0 = \dfrac{x-2}{x^2-9}$

    x-intercept: (2,0);

    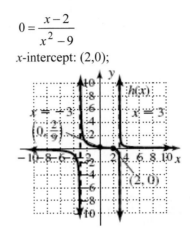

17. $f(x) = \log_2(x+1)$

    $f(0) = \log_2(0+1) = 0$

    y-intercept: (0,0);

    $x + 1 = 0$

    Vertical asymptote: $x = -1$

    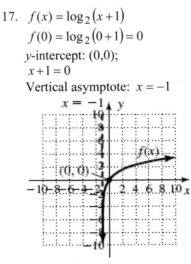

19. $x^2 + y^2 + 10x - 4y + 20 = 0$

$x^2 + 10x + y^2 - 4y = -20$

$\left(x^2 + 10x + 25\right) + \left(y^2 - 4y + 4\right) = -20 + 25 + 4$

$(x+5)^2 + (y-2)^2 = 9$

Center : $(-5,2)$, Radius $= 3$

21. (a) $x \in (-\infty, \infty)$

(b) $y \in (-\infty, 4]$

(c) $f(x) \uparrow (-\infty, -1), f(x) \downarrow (-1, \infty)$

(d) none

(e) max $(-1, 4)$

(f) $f(x) > 0 : x \in (-4, 2)$

(g) $f(x) < 0 : x \in (-\infty, -4) \cup (2, \infty)$

23. $\begin{cases} x^2 + y^2 = 25 \\ 64x^2 + 12y^2 = 768 \end{cases}$

$\begin{cases} -12x^2 - 12y^2 = -300 \\ 64x^2 + 12y^2 = 768 \end{cases}$

$52x^2 = 468$

$x^2 = 9$

$x = \pm 3 ;$

$(\pm 3)^2 + y^2 = 25$

$y^2 = 16$

$y = \pm 4 ;$

$(3,4), (3,-4), (-3,4), (-3,-4)$

25. Let x represent the number of liters of pure antifreeze.

Let y represent the number of liters of 40% antifreeze drained off.

$\begin{cases} y = x \\ 0.40(10) - 0.40y + x = 0.60(10) \end{cases}$

$4 - 0.4x + x = 6$

$0.6x = 2$

$x = 3\frac{1}{3}$ liters

410

8.1 Technology Highlight

Exercise 1: $a_n = \dfrac{1}{3^n}$

$$\frac{1}{3}, \frac{1}{9}, \frac{1}{27}, \frac{1}{81}, \frac{1}{243}, \frac{1}{729}, \frac{1}{2187},$$
$$\frac{1}{6561}, \frac{1}{19683}, \frac{1}{59049}, \frac{1}{177,147},$$
$$\frac{1}{531,441}, \frac{1}{1,594,323}, \frac{1}{4,782,969},$$
$$\frac{1}{14,348,907}, \frac{1}{43,046,721}, \frac{1}{129,140,163},$$
$$\frac{1}{387,420,489}, \frac{1}{1,162,261,467},$$
$$\frac{1}{3,486,784,401}$$

Sum of first 10 terms: sum $\to 0.5$
sum of first 20 terms: sum $\to 0.5$,
sum of seq $\to 0.5$

Exercise 3: $a_n = \dfrac{1}{(2n-1)(2n+1)}$

$$\frac{1}{3}, \frac{1}{15}, \frac{1}{35}, \frac{1}{63}, \frac{1}{99}, \frac{1}{143}, \frac{1}{195}, \frac{1}{255}, \frac{1}{323}, \frac{1}{399},$$
$$\frac{1}{483}, \frac{1}{575}, \frac{1}{675}, \frac{1}{783}, \frac{1}{899}, \frac{1}{1023}, \frac{1}{1155}, \frac{1}{1295},$$
$$\frac{1}{1443}, \frac{1}{1599}$$

Sum of first 10 terms: sum $\to 0.5$
Sum of first 20 terms: sum $\to 0.5$
sum of seq $\to 0.5$

8.1 Exercises

1. Pattern; order

3. Increasing

5. Formula defining the sequence uses the preceding term. Answers will vary.

7. $a_n = 2n - 1$
$a_1 = 2(1) - 1 = 2 - 1 = 1$;
$a_2 = 2(2) - 1 = 4 - 1 = 3$;
$a_3 = 2(3) - 1 = 6 - 1 = 5$;
$a_4 = 2(4) - 1 = 8 - 1 = 7$;
$a_8 = 2(8) - 1 = 16 - 1 = 15$;
$a_{12} = 2(12) - 1 = 24 - 1 = 23$;
1, 3, 5, 7; $a_8 = 15$; $a_{12} = 23$

9. $a_n = 3n^2 - 3$
$a_1 = 3(1)^2 - 3 = 3(1) - 3 = 3 - 3 = 0$;
$a_2 = 3(2)^2 - 3 = 3(4) - 3 = 12 - 3 = 9$;
$a_3 = 3(3)^2 - 3 = 3(9) - 3 = 27 - 3 = 24$;
$a_4 = 3(4)^2 - 3 = 3(16) - 3 = 48 - 3 = 45$;
$a_8 = 3(8)^2 - 3 = 3(64) - 3 = 192 - 3 = 189$;
$a_{12} = 3(12)^2 - 3 = 3(144) - 3 = 432 - 3 = 429$;
0, 9, 24, 45; $a_8 = 189$; $a_{12} = 429$

11. $a_n = (-1)^n n$
$a_1 = (-1)^1 (1) = -1(1) = -1$;
$a_2 = (-1)^2 (2) = 1(2) = 2$;
$a_3 = (-1)^3 (3) = -1(3) = -3$;
$a_4 = (-1)^4 (4) = 1(4) = 4$;
$a_8 = (-1)^8 (8) = 1(8) = 8$;
$a_{12} = (-1)^{12} (12) = 1(12) = 12$;
$-1, 2, -3, 4$; $a_8 = 8$; $a_{12} = 12$

13. $a_n = \dfrac{n}{n+1}$
$a_1 = \dfrac{1}{1+1} = \dfrac{1}{2}$;
$a_2 = \dfrac{2}{2+1} = \dfrac{2}{3}$;
$a_3 = \dfrac{3}{3+1} = \dfrac{3}{4}$;
$a_4 = \dfrac{4}{4+1} = \dfrac{4}{5}$;
$a_8 = \dfrac{8}{8+1} = \dfrac{8}{9}$;
$a_{12} = \dfrac{12}{12+1} = \dfrac{12}{13}$;
$\dfrac{1}{2}, \dfrac{2}{3}, \dfrac{3}{4}, \dfrac{4}{5}$; $a_8 = \dfrac{8}{9}$; $a_{12} = \dfrac{12}{13}$

15. $a_n = \left(\dfrac{1}{2}\right)^n$

    $a_1 = \left(\dfrac{1}{2}\right)^1 = \dfrac{1}{2}$;

    $a_2 = \left(\dfrac{1}{2}\right)^2 = \dfrac{1}{4}$;

    $a_3 = \left(\dfrac{1}{2}\right)^3 = \dfrac{1}{8}$;

    $a_4 = \left(\dfrac{1}{2}\right)^4 = \dfrac{1}{16}$;

    $a_8 = \left(\dfrac{1}{2}\right)^8 = \dfrac{1}{256}$;

    $a_{12} = \left(\dfrac{1}{2}\right)^{12} = \dfrac{1}{4096}$;

    $\dfrac{1}{2}, \dfrac{1}{4}, \dfrac{1}{8}, \dfrac{1}{16}; a_8 = \dfrac{1}{256}; a_{12} = \dfrac{1}{4096}$

17. $a_n = \dfrac{1}{n}$

    $a_1 = \dfrac{1}{1} = 1$;

    $a_2 = \dfrac{1}{2}$;

    $a_3 = \dfrac{1}{3}$;

    $a_4 = \dfrac{1}{4}$;

    $a_8 = \dfrac{1}{8}$;

    $a_{12} = \dfrac{1}{12}$;

    $1, \dfrac{1}{2}, \dfrac{1}{3}, \dfrac{1}{4}; a_8 = \dfrac{1}{8}; a_{12} = \dfrac{1}{12}$

19. $a_n = \dfrac{(-1)^n}{n(n+1)}$

    $a_1 = \dfrac{(-1)^1}{1(1+1)} = \dfrac{-1}{2}$;

    $a_2 = \dfrac{(-1)^2}{2(2+1)} = \dfrac{1}{2(3)} = \dfrac{1}{6}$;

    $a_3 = \dfrac{(-1)^3}{3(3+1)} = \dfrac{-1}{3(4)} = \dfrac{-1}{12}$;

    $a_4 = \dfrac{(-1)^4}{4(4+1)} = \dfrac{1}{4(5)} = \dfrac{1}{20}$;

    $a_8 = \dfrac{(-1)^8}{8(8+1)} = \dfrac{1}{8(9)} = \dfrac{1}{72}$;

    $a_{12} = \dfrac{(-1)^{12}}{12(12+1)} = \dfrac{1}{12(13)} = \dfrac{1}{156}$;

    $\dfrac{-1}{2}, \dfrac{1}{6}, \dfrac{-1}{12}, \dfrac{1}{20}; a_8 = \dfrac{1}{72}; a_{12} = \dfrac{1}{156}$

21. $a_n = (-1)^n 2^n$

    $a_1 = (-1)^1 2^1 = -1(2) = -2$;

    $a_2 = (-1)^2 2^2 = 1(4) = 4$;

    $a_3 = (-1)^3 2^3 = -1(8) = -8$;

    $a_4 = (-1)^4 2^4 = 1(16) = 16$;

    $a_8 = (-1)^8 2^8 = 1(256) = 256$;

    $a_{12} = (-1)^{12} 2^{12} = 1(4096) = 4096$;

    $-2, 4, -8, 16; a_8 = 256; \ a_{12} = 4096$

23. $a_n = n^2 - 2$

    $a_9 = (9)^2 - 2 = 81 - 2 = 79$

25. $a_n = \dfrac{(-1)^{n+1}}{n}$

    $a_5 = \dfrac{(-1)^{5+1}}{5} = \dfrac{(-1)^6}{5} = \dfrac{1}{5}$

27. $a_n = 2\left(\dfrac{1}{2}\right)^{n-1}$

$a_7 = 2\left(\dfrac{1}{2}\right)^{7-1} = 2\left(\dfrac{1}{2}\right)^6 = 2\left(\dfrac{1}{64}\right) = \dfrac{1}{32}$

29. $a_n = \left(1+\dfrac{1}{n}\right)^n$

$a_{10} = \left(1+\dfrac{1}{10}\right)^{10} = \left(\dfrac{11}{10}\right)^{10}$

31. $a_n = \dfrac{1}{(n)(2n+1)}$

$a_4 = \dfrac{1}{(4)[2(4)+1]} = \dfrac{1}{4(9)} = \dfrac{1}{36}$

33. $\begin{cases} a_1 = 2 \\ a_n = 5a_{n-1} - 3 \end{cases}$

$a_2 = 5a_{2-1} - 3 = 5a_1 - 3$
$= 5(2) - 3 = 10 - 3 = 7;$
$a_3 = 5a_{3-1} - 3 = 5a_2 - 3$
$= 5(7) - 3 = 35 - 3 = 32;$
$a_4 = 5a_{4-1} - 3 = 5a_3 - 3$
$= 5(32) - 3 = 160 - 3 = 157;$
$a_5 = 5a_{5-1} - 3 = 5a_4 - 3$
$= 5(157) - 3 = 785 - 3 = 782;$
$2, 7, 32, 157, 782$

35. $\begin{cases} a_1 = -1 \\ a_n = (a_{n-1})^2 + 3 \end{cases}$

$a_2 = (a_{2-1})^2 + 3 = (a_1)^2 + 3$
$= (-1)^2 + 3 = 1 + 3 = 4;$
$a_3 = (a_{3-1})^2 + 3 = (a_2)^2 + 3$
$= (4)^2 + 3 = 16 + 3 = 19;$
$a_4 = (a_{4-1})^2 + 3 = (a_3)^2 + 3$
$= (19)^2 + 3 = 361 + 3 = 364;$
$a_5 = (a_{5-1})^2 + 3 = (a_4)^2 + 3$
$= (364)^2 + 3 = 132,496 + 3 = 132,499;$
$-1, 4, 19, 364, 132,499$

37. $\begin{cases} c_1 = 64, c_2 = 32 \\ c_n = \dfrac{c_{n-2} - c_{n-1}}{2} \end{cases}$

$c_3 = \dfrac{c_{3-2} - c_{3-1}}{2} = \dfrac{c_1 - c_2}{2}$
$= \dfrac{64 - 32}{2} = \dfrac{32}{2} = 16;$

$c_4 = \dfrac{c_{4-2} - c_{4-1}}{2} = \dfrac{c_2 - c_3}{2}$
$= \dfrac{32 - 16}{2} = \dfrac{16}{2} = 8;$

$c_5 = \dfrac{c_{5-2} - c_{5-1}}{2} = \dfrac{c_3 - c_4}{2}$
$= \dfrac{16 - 8}{2} = \dfrac{8}{2} = 4;$
$64, 32, 16, 8, 4$

39. $\dfrac{8!}{5!} = \dfrac{8 \cdot 7 \cdot 6 \cdot 5!}{5!} = 8 \cdot 7 \cdot 6 = 336$

41. $\dfrac{9!}{7! \, 2!} = \dfrac{9 \cdot 8 \cdot 7!}{7! \, 2!} = \dfrac{9 \cdot 8}{2 \cdot 1} = \dfrac{72}{2} = 36$

43. $\dfrac{8!}{2! \, 6!} = \dfrac{8 \cdot 7 \cdot 6!}{2! \, 6!} = \dfrac{8 \cdot 7}{2 \cdot 1} = \dfrac{56}{2} = 28$

45. $a_n = \dfrac{n!}{(n+1)!}$

$a_1 = \dfrac{1!}{(1+1)!} = \dfrac{1!}{2!} = \dfrac{1!}{2 \cdot 1!} = \dfrac{1}{2}$

$a_2 = \dfrac{2!}{(2+1)!} = \dfrac{2!}{3!} = \dfrac{2!}{3 \cdot 2!} = \dfrac{1}{3}$

$a_3 = \dfrac{3!}{(3+1)!} = \dfrac{3!}{4!} = \dfrac{3!}{4 \cdot 3!} = \dfrac{1}{4}$

$a_4 = \dfrac{4!}{(4+1)!} = \dfrac{4!}{5!} = \dfrac{4!}{5 \cdot 4!} = \dfrac{1}{5}$

$\dfrac{1}{2}, \dfrac{1}{3}, \dfrac{1}{4}, \dfrac{1}{5}$

8.1 Exercises

47. $a_n = \dfrac{(n+1)!}{(3n)!}$

$a_1 = \dfrac{(1+1)!}{(3\cdot1)!} = \dfrac{2!}{3!} = \dfrac{2!}{3\cdot2!} = \dfrac{1}{3}$

$a_2 = \dfrac{(2+1)!}{(3\cdot2)!} = \dfrac{3!}{6!} = \dfrac{3!}{6\cdot5\cdot4\cdot3!} = \dfrac{1}{120}$

$a_3 = \dfrac{(3+1)!}{(3\cdot3)!} = \dfrac{4!}{9!} = \dfrac{4!}{9\cdot8\cdot7\cdot6\cdot5\cdot4!} = \dfrac{1}{15120}$

$a_4 = \dfrac{(4+1)!}{(3\cdot4)!} = \dfrac{5!}{12!}$

$= \dfrac{5!}{12\cdot11\cdot10\cdot9\cdot8\cdot7\cdot6\cdot5!} = \dfrac{1}{3,991,680}$

$\dfrac{1}{3}, \dfrac{1}{120}, \dfrac{1}{15,120}, \dfrac{1}{3,991,680}$

49. $a_n = \dfrac{n^n}{n!}$

$a_1 = \dfrac{1^1}{1!} = \dfrac{1}{1} = 1$

$a_2 = \dfrac{2^2}{2!} = \dfrac{4}{2!} = \dfrac{4}{2\cdot1} = \dfrac{4}{2} = 2$

$a_3 = \dfrac{3^3}{3!} = \dfrac{27}{3\cdot2\cdot1} = \dfrac{27}{6} = \dfrac{9}{2}$

$a_4 = \dfrac{4^4}{4!} = \dfrac{256}{4\cdot3\cdot2\cdot1} = \dfrac{256}{24} = \dfrac{32}{3}$

$1, 2, \dfrac{9}{2}, \dfrac{32}{3}$

51. $a_n = n$
$S_5 = a_1 + a_2 + a_3 + a_4 + a_5$
$S_5 = 1+2+3+4+5$
$S_5 = 15$

53. $a_n = 2n-1$
$S_8 = a_1 + a_2 + a_3 + a_4 + a_5 + a_6 + a_7 + a_8$
$S_8 = 1+3+5+7+9+11+13+15$
$S_8 = 64$

55. $a_n = \dfrac{1}{n}$
$S_5 = a_1 + a_2 + a_3 + a_4 + a_5$
$S_5 = 1+\dfrac{1}{2}+\dfrac{1}{3}+\dfrac{1}{4}+\dfrac{1}{5}$
$S_5 = \dfrac{137}{60}$

57. $\displaystyle\sum_{i=1}^{4}(3i-5)$
$= (3(1)-5)+(3(2)-5)+(3(3)-5)+(3(4)-5)$
$= -2+1+4+7 = 10$

59. $\displaystyle\sum_{k=1}^{5}(2k^2-3)$
$= (2(1)^2-3)+(2(2)^2-3)+(2(3)^2-3)+$
$(2(4)^2-3)+(2(5)^2-3)$
$= (2-3)+(8-3)+(18-3)+(32-3)+(50-3)$
$= -1+5+15+29+47 = 95$

61. $\displaystyle\sum_{k=1}^{7}(-1)^k k$
$= (-1)^1(1)+(-1)^2(2)+(-1)^3(3)+(-1)^4(4)+$
$(-1)^5(5)+(-1)^6(6)+(-1)^7(7)$
$= -1+2-3+4-5+6-7 = -4$

63. $\displaystyle\sum_{i=1}^{4}\dfrac{i^2}{2}$
$= \dfrac{1^2}{2}+\dfrac{2^2}{2}+\dfrac{3^2}{2}+\dfrac{4^2}{2}$
$= \dfrac{1}{2}+\dfrac{4}{2}+\dfrac{9}{2}+\dfrac{16}{2} = 15$

65. $\displaystyle\sum_{j=3}^{7}2j$
$= 2(3)+2(4)+2(5)+2(6)+2(7)$
$= 6+8+10+12+14 = 50$

67. $\displaystyle\sum_{k=3}^{8}\dfrac{(-1)^k}{k(k-2)}$
$= \dfrac{(-1)^3}{3(3-2)}+\dfrac{(-1)^4}{4(4-2)}+\dfrac{(-1)^5}{5(5-2)}+\dfrac{(-1)^6}{6(6-2)}+$
$\dfrac{(-1)^7}{7(7-2)}+\dfrac{(-1)^8}{8(8-2)}$
$= \dfrac{-1}{3}+\dfrac{1}{8}-\dfrac{1}{15}+\dfrac{1}{24}-\dfrac{1}{35}+\dfrac{1}{48}$
$= \dfrac{-27}{112}$

69. $4 + 8 + 12 + 16 + 20$

$$\sum_{n=1}^{5}(4n)$$

71. $-1 + 4 - 9 + 16 - 25 + 36$

$$\sum_{n=1}^{6}(-1)^{n}n^{2}$$

73. $a_n = n + 3; S_5$

$$\sum_{n=1}^{5}(n+3)$$

75. $a_n = \dfrac{n^2}{3}$; third partial sum

$$\sum_{n=1}^{3}\frac{n^2}{3}$$

77. $a_n = \dfrac{n}{2^n}$; sum for $n = 3$ to 7

$$\sum_{n=3}^{7}\frac{n}{2^n}$$

79. $\displaystyle\sum_{i=1}^{5}(4i-5) = \sum_{i=1}^{5}4i + \sum_{i=1}^{5}(-5)$

$$= 4\sum_{i=1}^{5}i + \sum_{i=1}^{5}(-5)$$

$$= 4(15) + (-5)(5) = 35$$

81. $\displaystyle\sum_{k=1}^{4}\left(3k^2 + k\right) = \sum_{k=1}^{4}3k^2 + \sum_{k=1}^{4}k$

$$= 3\sum_{k=1}^{4}k^2 + \sum_{k=1}^{4}k$$

$$= 3(30) + 10 = 100$$

83. $a_n = 3n - 2 : S_n = \dfrac{n(3n-1)}{2}$

$a_n = 3n - 2 = 1, 4, 7, 10 \ldots, (3n-2), \ldots$

$S_5 = \dfrac{5(3(5)-1)}{2} = \dfrac{5(14)}{2} = \dfrac{70}{2} = 35$;

$a_1 = 3(1) - 2 = 1$;

$a_2 = 3(2) - 2 = 4$;

$a_3 = 3(3) - 2 = 7$;

$a_4 = 3(4) - 2 = 10$;

$a_5 = 3(5) - 2 = 13$;

$1 + 4 + 7 + 10 + 13 = 35$

85. $a_n = (0.8)^{n-1}(6000)$

$a_1 = (0.8)^{1-1}(6000) = 6000$;

$a_2 = (0.8)^{2-1}(6000) = 0.8(6000) = 4800$;

$a_3 = (0.8)^{3-1}(6000) = (0.8)^2(6000) = 3840$;

$a_4 = (0.8)^{4-1}(6000) = (0.8)^3(6000) = 3072$;

$a_5 = (0.8)^{5-1}(6000) = (0.8)^4(6000) = 2457.60$;

$a_6 = (0.8)^{6-1}(6000) = (0.8)^5(6000) = 1966.08$

$6000; \$4800; \$3840; \$3072; \$2457.60;$
1966.08

87. $5.20, 5.70, 6.20, 6.70, 7.20$

$8(7.20)(240) = \$13{,}824$

89. $b_0 = 1500; \quad b_n = 1.05b_{n-1} + 100$

$b_1 = 1.05b_{1-1} + 100 = 1.05(1500) + 100 = 1675$;

$b_2 = 1.05b_{2-1} + 100 = 1.05b_1 + 100$
$\qquad = 1.05(1675) + 100 = 1858.75$;

$b_3 = 1.05b_{3-1} + 100 = 1.05b_2 + 100$
$\qquad = 1.05(1858.75) + 100 = 2051.69$;

$b_4 = 1.05b_{4-1} + 100 = 1.05b_3 + 100$
$\qquad = 1.05(2051.69) + 100 = 2254.27$;

$b_5 = 1.05b_{5-1} + 100 = 1.05b_4 + 100$
$\qquad = 1.05(2254.27) + 100 = 2466.98$;

$b_6 = 1.05b_{6-1} + 100 = 1.05b_6 + 100$
$\qquad = 1.05(2466.98) + 100 = 2690.33$

Approximately 2690

91. $\displaystyle\sum_{i=1}^{n}(a_i \pm b_i) = \sum_{i=1}^{n} a_i \pm \sum_{i=1}^{n} b_i$

$$\sum_{i=1}^{n}(a_i \pm b_i)$$
$$=[a_1 \pm b_1]+[a_2 \pm b_2]+...+[a_n \pm b_n]$$
$$=[a_1 + a_2 +...+a_n] \pm [b_1 + b_2 +...+b_n]$$
$$=\sum_{i=1}^{n} a_i \pm \sum_{i=1}^{n} b_i$$

Verified

93. $\displaystyle\sum_{h=1}^{n} \frac{1}{3^k}$

$$\sum_{k=1}^{4} \frac{1}{3^k}$$
$$S_4 = a_1 + a_2 + a_3 + a_4$$
$$=\frac{1}{3}+\frac{1}{9}+\frac{1}{27}+\frac{1}{81}=\frac{40}{81}$$

$$\sum_{k=1}^{8} \frac{1}{3^k}$$
$$S_8 = S_4 + a_5 + a_6 + a_7 + a_8$$
$$=\frac{40}{81}+\frac{1}{243}+\frac{1}{729}+\frac{1}{2187}+\frac{1}{6561} \approx 0.5 ;$$

$$\sum_{k=1}^{12} \frac{1}{3^k}$$
$$S_{12} = S_8 + a_9 + a_{10} + a_{11} + a_{12}$$
$$= 0.5 + \frac{1}{19,683}+\frac{1}{59,049}+\frac{1}{177,147}+\frac{1}{531,441}$$
$$\approx \frac{1}{2}$$

Approaches $\dfrac{1}{2}$

95. $\log_3 \dfrac{1}{81} = -x$

$$3^{-x} = \frac{1}{81}$$
$$3^{-x} = \left(\frac{1}{3}\right)^4$$
$$3^{-x} = 3^{-4}$$
$$-x = -4$$
$$x = 4$$

97. $\begin{cases} x^2 + y^2 = 9 \\ 9y^2 - 4x^2 = 16 \end{cases}$

$$\begin{cases} x^2 + y^2 = 9 \\ -4x^2 + 9y^2 = 16 \end{cases}$$

4R1 + R2

$$\begin{array}{r} 4x^2 + 4y^2 = 36 \\ -4x^2 + 9y^2 = 16 \\ \hline 13y^2 = 52 \end{array}$$

$$y^2 = 4$$
$$y = \pm 2;$$
$$x^2 + (2)^2 = 9$$
$$x^2 + 4 = 9$$
$$x^2 = 5$$
$$x = \pm\sqrt{5};$$
$$\left(\sqrt{5},-2\right),\left(\sqrt{5},2\right),\left(-\sqrt{5},-2\right),\left(-\sqrt{5},2\right)$$

8.2 Exercises

1. Common difference

3. $\dfrac{n(a_1 + a_n)}{2}$; n^{th}

5. Answers will vary.

7. $-5, -2, 1, 4, 7, 10, \dots$
   $-2 - (-5) = -2 + 5 = 3$;
   $1 - (-2) = 1 + 2 = 3$;
   $4 - 1 = 3$;
   $7 - 4 = 3$;
   $10 - 7 = 3$;
   Arithmetic; $d = 3$

9. $0.5, 3, 5.5, 8, 10.5, \dots$
   $3 - (0.5) = 2.5$;
   $5.5 - 3 = 2.5$;
   $8 - 5.5 = 2.5$;
   $10.5 - 8 = 2.5$;
   Arithmetic; $d = 2.5$

11. $2, 3, 5, 7, 11, 13, 17, \dots$
    $3 - 2 = 1$;
    $5 - 3 = 2$;
    Not arithmetic; all prime.

13. $\dfrac{1}{24}, \dfrac{1}{12}, \dfrac{1}{8}, \dfrac{1}{6}, \dfrac{5}{24}, \dots$
    $\dfrac{1}{12} - \dfrac{1}{24} = \dfrac{2}{24} - \dfrac{1}{24} = \dfrac{1}{24}$;
    $\dfrac{1}{8} - \dfrac{1}{12} = \dfrac{3}{24} - \dfrac{2}{24} = \dfrac{1}{24}$;
    $\dfrac{1}{6} - \dfrac{1}{8} = \dfrac{4}{24} - \dfrac{3}{24} = \dfrac{1}{24}$;
    $\dfrac{5}{24} - \dfrac{1}{6} = \dfrac{5}{24} - \dfrac{4}{24} = \dfrac{1}{24}$;
    Arithmetic; $d = \dfrac{1}{24}$

15. $1, 4, 9, 16, 25, 36, \dots$
    $4 - 1 = 3$;
    $9 - 4 = 5$;
    Not arithmetic; $a_n = n^2$

17. $\pi, \dfrac{5\pi}{6}, \dfrac{2\pi}{3}, \dfrac{\pi}{2}, \dfrac{\pi}{3}, \dfrac{\pi}{6}, \dots$
    $\dfrac{5\pi}{6} - \pi = \dfrac{5\pi}{6} - \dfrac{6\pi}{6} = \dfrac{-\pi}{6}$;
    $\dfrac{2\pi}{3} - \dfrac{5\pi}{6} = \dfrac{4\pi}{6} - \dfrac{5\pi}{6} = \dfrac{-\pi}{6}$;
    $\dfrac{\pi}{2} - \dfrac{2\pi}{3} = \dfrac{3\pi}{6} - \dfrac{4\pi}{6} = \dfrac{-\pi}{6}$;
    $\dfrac{\pi}{3} - \dfrac{\pi}{2} = \dfrac{2\pi}{6} - \dfrac{3\pi}{6} = \dfrac{-\pi}{6}$;
    $\dfrac{\pi}{6} - \dfrac{\pi}{3} = \dfrac{\pi}{6} - \dfrac{2\pi}{6} = \dfrac{-\pi}{6}$;
    Arithmetic; $d = \dfrac{-\pi}{6}$

19. $a_1 = 2, d = 3$
    $2 + 3 = 5$;
    $5 + 3 = 8$;
    $8 + 3 = 11$;
    $2, 5, 8, 11$

21. $a_1 = 7, d = -2$
    $7 - 2 = 5$;
    $5 - 2 = 3$;
    $3 - 2 = 1$;
    $7, 5, 3, 1$

23. $a_1 = 0.3, d = 0.03$
    $0.3 + 0.03 = 0.33$;
    $0.33 + 0.03 = 0.36$;
    $0.36 + 0.03 = 0.39$;
    $0.3, 0.33, 0.36, 0.39$

25. $a_1 = \dfrac{3}{2}, d = \dfrac{1}{2}$
    $\dfrac{3}{2} + \dfrac{1}{2} = \dfrac{4}{2} = 2$;
    $2 + \dfrac{1}{2} = \dfrac{4}{2} + \dfrac{1}{2} = \dfrac{5}{2}$;
    $\dfrac{5}{2} + \dfrac{1}{2} = \dfrac{6}{2} = 3$;
    $\dfrac{3}{2}, 2, \dfrac{5}{2}, 3$

27. $a_1 = \dfrac{3}{4}, d = -\dfrac{1}{8}$

$\dfrac{3}{4} - \dfrac{1}{8} = \dfrac{6}{8} - \dfrac{1}{8} = \dfrac{5}{8}$;

$\dfrac{5}{8} - \dfrac{1}{8} = \dfrac{4}{8} = \dfrac{1}{2}$;

$\dfrac{1}{2} - \dfrac{1}{8} = \dfrac{4}{8} - \dfrac{1}{8} = \dfrac{3}{8}$;

$\dfrac{3}{4}, \dfrac{5}{8}, \dfrac{1}{2}, \dfrac{3}{8}$

29. $a_1 = -2, d = -3$

$-2 - 3 = -5$;
$-5 - 3 = -8$;
$-8 - 3 = -11$;
$-2, -5, -8, -11$

31. $2, 7, 12, 17, \ldots$

$a_1 = 2, d = 5$
$a_n = a_1 + (n-1)d$
$a_n = 2 + (n-1)5$
$a_n = 2 + 5n - 5$
$a_n = 5n - 3$;
$a_6 = 5(6) - 3 = 30 - 3 = 27$;
$a_{10} = 5(10) - 3 = 50 - 3 = 47$;
$a_{12} = 5(12) - 3 = 60 - 3 = 57$

33. $5.10, 5.25, 5.40, \ldots$

$a_1 = 5.10, d = 0.15$
$a_n = a_1 + (n-1)d$
$a_n = 5.10 + (n-1)(0.15)$
$a_n = 5.10 + 0.15n - 0.15$
$a_n = 0.15n + 4.95$;
$a_6 = 0.15(6) + 4.95 = 0.90 + 4.95 = 5.85$;
$a_{10} = 0.15(10) + 4.95 = 1.50 + 4.95 = 6.45$
$a_{12} = 0.15(12) + 4.95 = 1.80 + 4.95 = 6.75$
$5.85, 6.45, 6.75$

35. $\dfrac{3}{2}, \dfrac{9}{4}, 3, \dfrac{15}{4}, \ldots$

$a_1 = \dfrac{3}{2}, d = \dfrac{3}{4}$
$a_n = a_1 + (n-1)d$
$a_n = \dfrac{3}{2} + (n-1)\left(\dfrac{3}{4}\right)$
$a_n = \dfrac{3}{2} + \dfrac{3}{4}n - \dfrac{3}{4}$
$a_n = \dfrac{3}{4}n + \dfrac{3}{4}$;
$a_6 = \dfrac{3}{4}(6) + \dfrac{3}{4} = \dfrac{18}{4} + \dfrac{3}{4} = \dfrac{21}{4}$;
$a_{10} = \dfrac{3}{4}(10) + \dfrac{3}{4} = \dfrac{30}{4} + \dfrac{3}{4} = \dfrac{33}{4}$;
$a_{12} = \dfrac{3}{4}(12) + \dfrac{3}{4} = \dfrac{36}{4} + \dfrac{3}{4} = \dfrac{39}{4}$

37. $a_1 = 5, d = 4$; Find a_{15}

$a_n = a_1 + (n-1)d$
$a_n = 5 + (n-1)4$
$a_n = 5 + 4n - 4$
$a_n = 4n + 1$;
$a_{15} = 4(15) + 1 = 60 + 1 = 61$

39. $a_1 = \dfrac{3}{2}, d = -\dfrac{1}{12}$; Find a_7

$a_n = a_1 + (n-1)d$
$a_n = \dfrac{3}{2} + (n-1)\left(-\dfrac{1}{12}\right)$
$a_n = \dfrac{3}{2} - \dfrac{1}{12}n + \dfrac{1}{12}$
$a_n = \dfrac{18}{12} - \dfrac{1}{12}n + \dfrac{1}{12}$
$a_n = -\dfrac{1}{12}n + \dfrac{19}{12}$;
$a_7 = -\dfrac{1}{12}(7) + \dfrac{19}{12} = \dfrac{-7}{12} + \dfrac{19}{12} = \dfrac{12}{12} = 1$

41. $a_1 = -0.025, d = 0.05$; Find a_{50}

$a_n = a_1 + (n-1)d$
$a_n = -0.025 + (n-1)(0.05)$
$a_n = -0.025 + 0.05n - 0.05$
$a_n = 0.05n - 0.075$;
$a_{50} = 0.05(50) - 0.075 = 2.5 - 0.075 = 2.425$

43. $a_1 = 2, a_n = -22, d = -3$

$a_n = a_1 + (n-1)d$

$-22 = 2 + (n-1)(-3)$

$-22 = 2 - 3n + 3$

$-22 = -3n + 5$

$-27 = -3n$

$9 = n$

45. $a_1 = 0.4, a_n = 10.9, d = 0.25$

$a_n = a_1 + (n-1)d$

$10.9 = 0.4 + (n-1)(0.25)$

$10.9 = 0.4 + 0.25n - 0.25$

$10.9 = 0.15 + 0.25n$

$10.75 = 0.25n$

$43 = n$

47. $-3, -0.5, 2, 4.5, 7, \ldots, 47$

$a_1 = -3, a_n = 47, d = 2.5$

$a_n = a_1 + (n-1)d$

$47 = -3 + (n-1)(2.5)$

$47 = -3 + 2.5n - 2.5$

$47 = -5.5 + 2.5n$

$52.5 = 2.5n$

$21 = n$

49. $\dfrac{1}{12}, \dfrac{1}{8}, \dfrac{1}{6}, \dfrac{5}{24}, \dfrac{1}{4}, \ldots, \dfrac{9}{8}$

$a_1 = \dfrac{1}{12}, a_n = \dfrac{9}{8}, d = \dfrac{1}{24}$

$a_n = a_1 + (n-1)d$

$\dfrac{9}{8} = \dfrac{1}{12} + (n-1)\left(\dfrac{1}{24}\right)$

$\dfrac{9}{8} = \dfrac{1}{12} + \dfrac{1}{24}n - \dfrac{1}{24}$

$\dfrac{9}{8} = \dfrac{1}{24} + \dfrac{1}{24}n$

$\dfrac{13}{12} = \dfrac{1}{24}n$

$26 = n$

51. $a_3 = 7, a_7 = 19$

$a_7 = a_3 + 4d$

$19 = 7 + 4d$

$12 = 4d$

$3 = d;$

$a_7 = a_1 + 6d$

$19 = a_1 + 6(3)$

$19 = a_1 + 18$

$1 = a_1;$

$d = 3, a_1 = 1$

53. $a_2 = 1.025, a_{26} = 10.025$

$a_{26} = a_2 + 24d$

$10.025 = 1.025 + 24d$

$9 = 24d$

$0.375 = d;$

$a_{26} = a_1 + 25d$

$10.025 = a_1 + 25(0.375)$

$10.025 = a_1 + 9.375$

$0.65 = a_1;$

$d = 0.375, \ a_1 = 0.65$

55. $a_{10} = \dfrac{13}{18}, a_{24} = \dfrac{27}{2}$

$a_{24} = a_{10} + 14d$

$\dfrac{27}{2} = \dfrac{13}{18} + 14d$

$\dfrac{115}{9} = 14d$

$\dfrac{115}{126} = d;$

$a_{24} = a_1 + 23d$

$\dfrac{27}{2} = a_1 + 23\left(\dfrac{115}{126}\right)$

$\dfrac{27}{2} = a_1 + \dfrac{2645}{126}$

$\dfrac{-472}{63} = a_1;$

$d = \dfrac{115}{126}, a_1 = \dfrac{-472}{63}$

8.2 Exercises

57. $\displaystyle\sum_{n=1}^{30} (3n-4)$

Initial terms: $-1, 2, 5, \ldots$
$a_1 = -1; d = 3; n = 30$
$a_{30} = a_1 + 29d$
$a_{30} = -1 + 29(3) = -1 + 87 = 86 \ ;$
$S_{30} = n\left(\dfrac{a_1 + a_n}{2}\right)$
$S_{30} = 30\left(\dfrac{-1 + 86}{2}\right)$
$S_{30} = 30\left(\dfrac{85}{2}\right)$
$S_{30} = 1275$

59. $\displaystyle\sum_{n=1}^{37} \left(\dfrac{3}{4}n + 2\right)$

Initial terms: $\dfrac{11}{4}, \dfrac{7}{2}, \dfrac{17}{4}, \ldots$

$a_1 = \dfrac{11}{4}; d = \dfrac{3}{4}; n = 37$

$a_{37} = a_1 + 36d$

$a_{37} = \dfrac{11}{4} + 36\left(\dfrac{3}{4}\right) = \dfrac{11}{4} + 27 = \dfrac{119}{4} \ ;$

$S_{37} = n\left(\dfrac{a_1 + a_{37}}{2}\right)$

$S_{37} = 37\left(\dfrac{\dfrac{11}{4} + \dfrac{119}{4}}{2}\right)$

$S_{37} = 37\left(\dfrac{\dfrac{65}{2}}{2}\right)$

$S_{37} = 601.25$

61. $\displaystyle\sum_{n=4}^{15} (3 - 5n)$

$= \displaystyle\sum_{n=4}^{15}(3) - 5\sum_{n=4}^{15}(n)$

$= 3(12) - 5(114) = -534$

63. $-12 + (-9.5) + (-7) + (-4.5) + \ldots$

Find S_{15}; $a_1 = -12$; $d = 2.5$; $n = 15$

$S_n = \dfrac{n}{2}[2a_1 + (n-1)d]$

$S_{15} = \dfrac{15}{2}[2(-12) + (15-1)(2.5)]$

$S_{15} = \dfrac{15}{2}[-24 + 14(2.5)]$

$S_{15} = \dfrac{15}{2}[-24 + 35]$

$S_{15} = \dfrac{15}{2}[11]$

$S_{15} = 82.5$

65. $0.003 + 0.173 + 0.343 + 0.513 + \ldots$

Find S_{30}; $a_1 = 0.003$; $d = 0.17$; $n = 30$

$S_n = \dfrac{n}{2}[2a_1 + (n-1)d]$

$S_{30} = \dfrac{30}{2}[2(0.003) + (30-1)(0.17)]$

$S_{30} = 15[0.006 + (29)(0.17)]$

$S_{30} = 15[0.006 + 4.93]$

$S_{30} = 15[4.936]$

$S_{30} - 74.04$

67. $\sqrt{2} + 2\sqrt{2} + 3\sqrt{2} + 4\sqrt{2} + \ldots$

Find S_{20}; $a_1 = \sqrt{2}$; $d = \sqrt{2}$; $n = 20$

$S_n = \dfrac{n}{2}[2a_1 + (n-1)d]$

$S_{20} = \dfrac{20}{2}[2\sqrt{2} + (20-1)\sqrt{2}]$

$S_{20} = 10(2\sqrt{2} + 19\sqrt{2})$

$S_{20} = 10(21\sqrt{2})$

$S_{20} = 210\sqrt{2}$

69. $S_n = \dfrac{n(n+1)}{2}$

$1 + 2 + 3 + 4 + 5 + 6 = 21 \ ;$

$S_6 = \dfrac{6(6+1)}{2} = \dfrac{6(7)}{2} = \dfrac{42}{2} = 21 \ ;$

$S_{75} = \dfrac{75(75+1)}{2} = \dfrac{75(76)}{2} = \dfrac{5700}{2} = 2850$

71. $a_0 = 36, a_1 = 33;\ d = -3;\ a_n = 0$

$a_n = a_1 + (n-1)d$

$0 = 33 + (n-1)(-3)$

$0 = 33 - 3n + 3$

$0 = 36 - 3n$

$-36 = -3n$

$12 = n$

12 half-hours after 5 P.M. is 11 P.M.

73. a. $a_1 = 10;\ d = \dfrac{-3}{4};\ n = 7$

$a_n = a_1 + (n-1)d$

$a_7 = 10 + (7-1)\left(\dfrac{-3}{4}\right)$

$a_7 = 10 + 6\left(\dfrac{-3}{4}\right)$

$a_7 = 10 - \dfrac{9}{2}$

$a_7 = 5.5$;

5.5 inches

b. $S_n = \dfrac{n(a_1 + a_n)}{2}$

$S_7 = \dfrac{7(10 + 5.5)}{2} = \dfrac{7(15.5)}{2} = 54.25$;

54.25 inches

75. $a_1 = 100;\ d = 20;\ a_n = 2500$

$a_n = a_1 + (n-1)d$

$a_7 = 100 + (7-1)(20) = 100 + 6(20) = 220$;

$220;

$a_{12} = 100 + (12-1)(20) = 100 + 11(20) = 320$;

$S_n = \dfrac{n(a_1 + a_n)}{2}$

$S_{12} = \dfrac{12(100 + 320)}{2} = \dfrac{12(420)}{2} = 2520$

$2520; yes

77. a. $19, 11.8, 4.6, -2.6, -9.8, -17, -24.2$

1st difference:

$11.8 - 19 = -7.2$

$4.6 - 11.8 = -7.2$

$-2.6 - 4.6 = -7.2$

$-9.8 - (-2.6) = -7.2$

$-17 - (-9.8) = -7.2$

$-24.2 - (-17) = -7.2$

Linear function.

b. $-10.31, -10.94, -11.99, -13.46, -15.35\ldots$

1st difference:

$-10.94 + 10.31 = -0.63$

$-11.99 + 10.94 = -1.05$

$-13.46 + 11.99 = -1.47$

$-15.35 + 13.46 = -1.89$;

$-0.63,\ -1.05,\ -1.47,\ -1.89$

2nd difference:

$-1.05 + 0.63 = -0.42$

$-1.47 + 1.05 = -0.42$

$-1.89 + 1.47 = -0.42$

Quadratic.

79. $2530 = 500e^{0.45t}$

$5.06 = e^{0.45t}$

$\ln 5.06 = \ln e^{0.45t}$

$\ln 5.06 = 0.45t$

$\dfrac{\ln 5.06}{0.45} = t$

$3.6 \approx t$

81. $(0, 972), (5, 1217)$

$m = \dfrac{1217 - 972}{5 - 0} = \dfrac{245}{5} = 49$

$y - y_1 = m(x - x_1)$

$y - 972 = 49(x - 0)$

$y - 972 = 49x$

$y = 49x + 972$;

$f(x) = 49x + 972$;

$f(8) = 49(8) + 972 = 392 + 972 = 1364$

8.3 Exercises

1. Multiplying.

3. $a_1 r^{n-1}$

5. Answers will vary.

7. $4, 8, 16, 32, \ldots$

$\dfrac{8}{4} = 2; \quad \dfrac{16}{8} = 2; \quad \dfrac{32}{16} = 2$

$r = 2$

9. $3, -6, 12, -24, 48, \ldots$

$\dfrac{-6}{3} = -2; \quad \dfrac{12}{-6} = -2; \quad \dfrac{-24}{12} = -2; \quad \dfrac{48}{-24} = -2$

$r = -2$

11. $2, 5, 10, 17, 26, \ldots$

$\dfrac{5}{2}; \quad \dfrac{10}{5} = 2; \quad \dfrac{17}{10}; \quad \dfrac{26}{17}$; not geometric

$a_n = n^2 + 1$

13. $3, 0.3, 0.03, 0.003, \ldots$

$\dfrac{0.3}{3} = 0.1; \quad \dfrac{0.03}{0.3} = 0.1; \quad \dfrac{0.003}{0.03} = 0.1$

$r = 0.1$

15. $-1, 3, -12, 60, -360, \ldots$

$\dfrac{3}{-1} = -3; \quad \dfrac{-12}{3} = -4; \quad \dfrac{60}{-12} = -5; \quad \dfrac{-360}{60} = -6$

Not geometric; ratio of terms decreases by 1.

17. $25, 10, 4, \dfrac{8}{5}, \ldots$

$\dfrac{10}{25} = \dfrac{2}{5}; \quad \dfrac{4}{10} = \dfrac{2}{5}; \quad \dfrac{\frac{8}{5}}{4} = \dfrac{8}{20} = \dfrac{2}{5}$

$r = \dfrac{2}{5}$

19. $\dfrac{1}{2}, \dfrac{1}{4}, \dfrac{1}{8}, \dfrac{1}{16}, \ldots$

$\dfrac{\frac{1}{4}}{\frac{1}{2}} = \dfrac{1}{2}; \quad \dfrac{\frac{1}{8}}{\frac{1}{4}} = \dfrac{1}{2}; \quad \dfrac{\frac{1}{16}}{\frac{1}{8}} = \dfrac{1}{2}$

$r = \dfrac{1}{2}$

21. $3, \dfrac{12}{x}, \dfrac{48}{x^2}, \dfrac{192}{x^3}, \ldots$

$\dfrac{\frac{12}{x}}{3} = \dfrac{12}{3x} = \dfrac{4}{x}; \quad \dfrac{\frac{48}{x^2}}{\frac{12}{x}} = \dfrac{48x}{12x^2} = \dfrac{4}{x};$

$\dfrac{\frac{192}{x^3}}{\frac{48}{x^2}} = \dfrac{192x^2}{48x^3} = \dfrac{4}{x}$

$r = \dfrac{4}{x}$

23. $240, 120, 40, 10, 2, \ldots$

$\dfrac{120}{240} = \dfrac{1}{2}; \quad \dfrac{40}{120} = \dfrac{1}{3}; \quad \dfrac{10}{40} = \dfrac{1}{4}; \quad \dfrac{2}{10} = \dfrac{1}{5}$

Not geometric $a_n = \dfrac{240}{n!}$

25. $a_1 = 5, r = 2$

$a_2 = 5 \cdot 2 = 10$;

$a_3 = 10 \cdot 2 = 20$;

$a_4 = 20 \cdot 2 = 40$;

$5, 10, 20, 40$

27. $a_1 = -6, r = -\dfrac{1}{2}$

$a_2 = -6\left(\dfrac{-1}{2}\right) = 3$;

$a_3 = 3\left(\dfrac{-1}{2}\right) = \dfrac{-3}{2}$;

$a_4 = \dfrac{-3}{2}\left(\dfrac{-1}{2}\right) = \dfrac{3}{4}$;

$-6, 3, \dfrac{-3}{2}, \dfrac{3}{4}$

29. $a_1 = 4, r = \sqrt{3}$

$a_2 = 4\sqrt{3}$;

$a_3 = 4\sqrt{3}\left(\sqrt{3}\right) = 12$;

$a_4 = 12\sqrt{3}$;

$4, 4\sqrt{3}, 12, 12\sqrt{3}$

31. $a_1 = 0.1, r = 0.1$

$a_2 = 0.1(0.1) = 0.01$;

$a_3 = 0.01(0.1) = 0.001$;

$a_4 = 0.001(0.1) = 0.0001$;

$0.1, 0.01, 0.001, 0.0001$

33. $a_1 = -24, \ r = \dfrac{1}{2}$; find a_7

$a_n = a_1 r^{n-1}$

$a_7 = -24\left(\dfrac{1}{2}\right)^{7-1} = -24\left(\dfrac{1}{2}\right)^6$

$= -24\left(\dfrac{1}{64}\right) = -\dfrac{3}{8}$

35. $a_1 = -\dfrac{1}{20}, \ r = -5$; find a_4

$a_n = a_1 r^{n-1}$

$a_4 = -\dfrac{1}{20}(-5)^{4-1} = -\dfrac{1}{20}(-5)^3$

$= -\dfrac{1}{20}(-125) = \dfrac{25}{4}$

37. $a_1 = 2, \ r = \sqrt{2}$; find a_7

$a_n = a_1 r^{n-1}$

$a_7 = 2\left(\sqrt{2}\right)^{7-1} = 2\left(\sqrt{2}\right)^6 = 2(8) = 16$

39. $\dfrac{1}{27}, -\dfrac{1}{9}, \dfrac{1}{3}, -1, 3, \ldots$

$r = \dfrac{-\dfrac{1}{9}}{\dfrac{1}{27}} = \dfrac{-27}{9} = -3$

$a_1 = \dfrac{1}{27}; \ r = -3$

$a_n = \dfrac{1}{27}(-3)^{n-1}$

$a_6 = \dfrac{1}{27}(-3)^{6-1} = \dfrac{1}{27}(-3)^5 = \dfrac{1}{27}(-243) = -9$;

$a_{10} = \dfrac{1}{27}(-3)^{10-1} = \dfrac{1}{27}(-3)^9$

$= \dfrac{1}{27}(-19683) = -729;$

$a_{12} = \dfrac{1}{27}(-3)^{12-1} = \dfrac{1}{27}(-3)^{11}$

$= \dfrac{1}{27}(-177147) = -6561$

41. $729, 243, 81, 27, 9, \ldots$

$r = \dfrac{243}{729} = \dfrac{1}{3}$

$a_1 = 729; \ r = \dfrac{1}{3}$

$a_n = 729\left(\dfrac{1}{3}\right)^{n-1}$

$a_6 = 729\left(\dfrac{1}{3}\right)^{6-1} = 729\left(\dfrac{1}{3}\right)^5 = 729\left(\dfrac{1}{243}\right) = 3$;

$a_{10} = 729\left(\dfrac{1}{3}\right)^{10-1} = 729\left(\dfrac{1}{3}\right)^9$

$= 729\left(\dfrac{1}{19683}\right) = \dfrac{1}{27};$

$a_{12} = 729\left(\dfrac{1}{3}\right)^{12-1} = 729\left(\dfrac{1}{3}\right)^{11}$

$= 729\left(\dfrac{1}{177147}\right) = \dfrac{1}{243}$

8.3 Exercises

43. $\dfrac{1}{2}, \dfrac{\sqrt{2}}{2}, 1, \sqrt{2}, 2, \ldots$

$r = \dfrac{\frac{\sqrt{2}}{2}}{\frac{1}{2}} = \dfrac{2\sqrt{2}}{2} = \sqrt{2}$

$a_1 = \dfrac{1}{2}; \ r = \sqrt{2}$

$a_n = \dfrac{1}{2}\left(\sqrt{2}\right)^{n-1}$

$a_6 = \dfrac{1}{2}\left(\sqrt{2}\right)^{6-1} = \dfrac{1}{2}\left(\sqrt{2}\right)^5 = \dfrac{1}{2}\left(4\sqrt{2}\right) = 2\sqrt{2}$;

$a_{10} = \dfrac{1}{2}\left(\sqrt{2}\right)^{10-1} = \dfrac{1}{2}\left(\sqrt{2}\right)^9 = \dfrac{1}{2}\left(16\sqrt{2}\right) = 8\sqrt{2}$;

$a_{12} = \dfrac{1}{2}\left(\sqrt{2}\right)^{12-1} = \dfrac{1}{2}\left(\sqrt{2}\right)^{11} = \dfrac{1}{2}\left(32\sqrt{2}\right) = 16\sqrt{2}$

45. $0.2, 0.08, 0.032, 0.0128, \ldots$

$r = \dfrac{0.08}{0.2} = 0.4$

$a_1 = 0.2; \ r = 0.4$

$a_n = 0.2(0.4)^{n-1}$

$a_6 = 0.2(0.4)^{6-1} = 0.2(0.4)^5$
$= 0.2(0.01024) = 0.002048$;

$a_{10} = 0.2(0.4)^{10-1} = 0.2(0.4)^9$
$= 0.2(0.000262144) = 0.0000524288$;

$a_{12} = 0.2(0.4)^{12-1} = 0.2(0.4)^{11}$
$= 0.2(0.00004194304)$
$= 0.000008388608$

47. $a_1 = 9, \ a_n = 729, r = 3$
$a_n = a_1 r^{n-1}$
$729 = 9(3)^{n-1}$
$81 = 3^{n-1}$
$3^4 = 3^{n-1}$
$4 = n-1$
$5 = n$

49. $a_1 = 16, a_n = \dfrac{1}{64}, r = \dfrac{1}{2}$
$a_n = a_1 r^{n-1}$
$\dfrac{1}{64} = 16\left(\dfrac{1}{2}\right)^{n-1}$
$\dfrac{1}{1024} = \left(\dfrac{1}{2}\right)^{n-1}$
$\left(\dfrac{1}{2}\right)^{10} = \left(\dfrac{1}{2}\right)^{n-1}$
$10 = n-1$
$11 = n$

51. $a_1 = -1, a_n = -1296, r = \sqrt{6}$
$a_n = a_1 r^{n-1}$
$-1296 = -1\left(\sqrt{6}\right)^{n-1}$
$1296 = \left(\sqrt{6}\right)^{n-1}$
$\left(\sqrt{6}\right)^8 = \left(\sqrt{6}\right)^{n-1}$
$8 = n-1$
$9 = n$

53. $2, -6, 18, -54, \ldots, -4374$
$r = \dfrac{-6}{2} = -3; \ a_1 = 2; \ a_n = -4374$
$a_n = a_1 r^{n-1}$
$-4374 = 2(-3)^{n-1}$
$-2187 = (-3)^{n-1}$
$(-3)^7 = (-3)^{n-1}$
$7 = n-1$
$8 = n$

55. $64, 32\sqrt{2}, 32, 16\sqrt{2}, \ldots, 1$

$r = \dfrac{32\sqrt{2}}{64} = \dfrac{\sqrt{2}}{2}; \ a_1 = 64; \ a_n = 1$

$a_n = a_1 r^{n-1}$

$1 = 64\left(\dfrac{\sqrt{2}}{2}\right)^{n-1}$

$\dfrac{1}{64} = \left(\dfrac{\sqrt{2}}{2}\right)^{n-1}$

$\left(\dfrac{\sqrt{2}}{2}\right)^{12} = \left(\dfrac{\sqrt{2}}{2}\right)^{n-1}$

$12 = n - 1$

$13 = n$

57. $\dfrac{3}{8}, -\dfrac{3}{4}, \dfrac{3}{2}, -3, \ldots, 96$

$r = \dfrac{-\frac{3}{4}}{\frac{3}{8}} = \dfrac{-24}{12} = -2; \ a_1 = \dfrac{3}{8}; \ a_n = 96$

$a_n = a_1 r^{n-1}$

$96 = \dfrac{3}{8}(-2)^{n-1}$

$256 = (-2)^{n-1}$

$(-2)^8 = (-2)^{n-1}$

$8 = n - 1$

$9 = n$

59. $a_3 = 324, \ a_7 = 64$

$a_7 = a_3 \cdot r^4$

$64 = 324 r^4$

$\dfrac{64}{324} = r^4$

$\sqrt[4]{\dfrac{16}{81}} = r$

$\dfrac{2}{3} = r;$

$a_7 = a_1 \cdot \left(\dfrac{2}{3}\right)^6$

$64 = a_1\left(\dfrac{2}{3}\right)^6$

$64 = a_1\left(\dfrac{64}{729}\right)$

$729 = a_1$

$r = \dfrac{2}{3}, \ a_1 = 729$

61. $a_4 = \dfrac{4}{9}, \ a_8 = \dfrac{9}{4}$

$a_8 = a_4 \cdot r^4$

$\dfrac{9}{4} = \dfrac{4}{9}(r)^4$

$\dfrac{81}{16} = r^4$

$\sqrt[4]{\dfrac{81}{16}} = r$

$\dfrac{3}{2} = r;$

$a_8 = a_1\left(\dfrac{3}{2}\right)^7$

$\dfrac{9}{4} = a_1\left(\dfrac{3}{2}\right)^7$

$\dfrac{9}{4} = a_1\left(\dfrac{2187}{128}\right)$

$\dfrac{32}{243} = a_1$

$r = \dfrac{3}{2}, \ a_1 = \dfrac{32}{243}$

63. $a_4 = \dfrac{32}{3}$, $a_8 = 54$

$$a_8 = a_4 \cdot r^4$$

$$54 = \left(\dfrac{32}{3}\right) r^4$$

$$\dfrac{81}{16} = r^4$$

$$\sqrt[4]{\dfrac{81}{16}} = r$$

$$\dfrac{3}{2} = r;$$

$$a_8 = a_1 \left(\dfrac{3}{2}\right)^7$$

$$54 = a_1 \left(\dfrac{3}{2}\right)^7$$

$$54 = a_1 \left(\dfrac{2187}{128}\right)$$

$$\dfrac{256}{81} = a_1$$

$$r = \dfrac{3}{2}, \; a_1 = \dfrac{256}{81}$$

65. $a_1 - 8$, $r - -2$, find S_{12}

$$S_n = \dfrac{a_1\left(1 - r^n\right)}{1 - r}$$

$$S_{12} = \dfrac{8\left(1 - (-2)^{12}\right)}{1 - (-2)} = \dfrac{8(1 - 4096)}{1 + 2}$$

$$= \dfrac{8(-4095)}{3} = -10{,}920$$

67. $a_1 = 96$, $r = \dfrac{1}{3}$, find S_5

$$S_n = \dfrac{a_1\left(1 - r^n\right)}{1 - r}$$

$$S_5 = \dfrac{96\left(1 - \left(\dfrac{1}{3}\right)^5\right)}{1 - \dfrac{1}{3}} = \dfrac{96\left(1 - \dfrac{1}{243}\right)}{\dfrac{2}{3}}$$

$$= \dfrac{96\left(\dfrac{242}{243}\right)}{\dfrac{2}{3}} = \dfrac{3872}{27} \approx 143.41$$

69. $a_1 = 8$, $r = \dfrac{3}{2}$, find S_7

$$S_n = \dfrac{a_1\left(1 - r^n\right)}{1 - r}$$

$$S_7 = \dfrac{8\left(1 - \left(\dfrac{3}{2}\right)^7\right)}{1 - \dfrac{3}{2}} = \dfrac{8\left(1 - \dfrac{2187}{128}\right)}{-\dfrac{1}{2}}$$

$$= \dfrac{8\left(\dfrac{-2059}{128}\right)}{-\dfrac{1}{2}} = \dfrac{2059}{8} = 257.375$$

71. $2 + 6 + 18 + \ldots$; find S_6

$$a_1 = 2; \; r = \dfrac{6}{2} = 3$$

$$S_n = \dfrac{a_1\left(1 - r^n\right)}{1 - r}$$

$$S_6 = \dfrac{2\left(1 - 3^6\right)}{1 - 3} = \dfrac{2(1 - 729)}{-2} = \dfrac{2(-728)}{-2} = 728$$

73. $16 - 8 + 4 - \ldots$; find S_8

$$a_1 = 16; \; r = \dfrac{-8}{16} = \dfrac{-1}{2}$$

$$S_n = \dfrac{a_1\left(1 - r^n\right)}{1 - r}$$

$$S_8 = \dfrac{16\left(1 - \left(\dfrac{-1}{2}\right)^8\right)}{1 - \left(-\dfrac{1}{2}\right)} = \dfrac{16\left(1 - \dfrac{1}{256}\right)}{\dfrac{3}{2}}$$

$$= \dfrac{16\left(\dfrac{255}{256}\right)}{\dfrac{3}{2}} = \dfrac{85}{8} = 10.625$$

75. $\dfrac{4}{3}+\dfrac{2}{9}+\dfrac{1}{27}+...;$ find S_9

$a_1=\dfrac{4}{3};\ r=\dfrac{1}{6}$

$S_n=\dfrac{a_1\left(1-r^n\right)}{1-r}$

$S_9=\dfrac{\dfrac{4}{3}\left(1-\left(\dfrac{1}{6}\right)^9\right)}{1-\dfrac{1}{6}}=\dfrac{\dfrac{4}{3}\left(1-\dfrac{1}{10077696}\right)}{\dfrac{5}{6}}$

$=\dfrac{\dfrac{4}{3}\left(\dfrac{10077695}{10077696}\right)}{\dfrac{5}{6}}\approx 1.60$

77. $\displaystyle\sum_{j=1}^{5}4^j$

Initial terms: 4, 16, 64, ...

$a_1=4;\ r=\dfrac{16}{4}=4;\ n=5$

$S_n=\dfrac{a_1\left(1-r^n\right)}{1-r}$

$S_5=\dfrac{4\left(1-4^5\right)}{1-4}=\dfrac{4(1-1024)}{-3}$

$=\dfrac{4(-1023)}{-3}=1,364$

79. $\displaystyle\sum_{k=1}^{8}5\left(\dfrac{2}{3}\right)^{k-1}$

$5\left(\dfrac{2}{3}\right)^0=5\ ;$

$5\left(\dfrac{2}{3}\right)^1=\dfrac{10}{3}\ ;$

$5\left(\dfrac{2}{3}\right)^2=\dfrac{20}{9}\ ;$

$5\left(\dfrac{2}{3}\right)^3=\dfrac{40}{27}$

Initial terms: $5,\dfrac{10}{3},\dfrac{20}{9},\dfrac{40}{27},....$

$a_1=5;\ r=\dfrac{\dfrac{20}{9}}{\dfrac{10}{3}}=\dfrac{2}{3};\ n=8$

$S_n=\dfrac{a_1\left(1-r^n\right)}{1-r}$

$S_8=\dfrac{5\left(1-\left(\dfrac{2}{3}\right)^8\right)}{1-\dfrac{2}{3}}=\dfrac{5\left(1-\dfrac{256}{6561}\right)}{\dfrac{1}{3}}$

$=\dfrac{5\left(\dfrac{6305}{6561}\right)}{\dfrac{1}{3}}=\dfrac{31525}{2187}\approx 14.41$

81. $\displaystyle\sum_{i=4}^{10}9\left(-\dfrac{1}{2}\right)^{i-1}$

$9\left(\dfrac{-1}{2}\right)^3=\dfrac{-9}{8}\ ;$

$9\left(\dfrac{-1}{2}\right)^4=\dfrac{9}{16}\ ;$

$9\left(\dfrac{-1}{2}\right)^5=\dfrac{-9}{32}\ ;$

$9\left(\dfrac{-1}{2}\right)^6=\dfrac{9}{64}$

Initial terms: $\dfrac{-9}{8},\dfrac{9}{16},\dfrac{-9}{32},\dfrac{9}{64},....$

$a_1=\dfrac{-9}{8};\ r=\dfrac{\dfrac{-9}{32}}{\dfrac{9}{16}}=\dfrac{-1}{2};\ n=7$

$S_n=\dfrac{a_1\left(1-r^n\right)}{1-r}$

$S_7=\dfrac{\dfrac{-9}{8}\left(1-\left(\dfrac{-1}{2}\right)^7\right)}{1-\left(\dfrac{-1}{2}\right)}=\dfrac{\dfrac{-9}{8}\left(1+\dfrac{1}{128}\right)}{\dfrac{3}{2}}$

$=\dfrac{\dfrac{-9}{8}\left(\dfrac{129}{128}\right)}{\dfrac{3}{2}}=\dfrac{-387}{512}\approx -0.76$

83. $a_2 = -5, a_5 = \dfrac{1}{25}$, find S_5

Find r:

$a_5 = a_2 r^3$

$\dfrac{1}{25} = -5r^3$

$\dfrac{-1}{125} = r^3$

$\sqrt[3]{\dfrac{-1}{125}} = r$

$\dfrac{-1}{5} = r;$

Find a_1:

$a_5 = a_1 r^4$

$\dfrac{1}{25} = a_1 \left(\dfrac{-1}{5} \right)^4$

$\dfrac{1}{25} = a_1 \left(\dfrac{1}{625} \right)$

$25 = a_1;$

$a_1 = 25; \quad r = \dfrac{-1}{5}$

$S_n = \dfrac{a_1 \left(1 - r^n \right)}{1 - r}$

$S_5 = \dfrac{25 \left(1 - \left(\dfrac{-1}{5} \right)^5 \right)}{1 - \left(\dfrac{-1}{5} \right)} = \dfrac{25 \left(1 + \dfrac{1}{3125} \right)}{\dfrac{6}{5}}$

$= \dfrac{25 \left(\dfrac{3126}{3125} \right)}{\dfrac{6}{5}} = \dfrac{521}{25}$

85. $a_3 = \dfrac{4}{9}, a_7 = \dfrac{9}{64}$, find S_6

Find r:

$a_7 = a_3 r^4$

$\dfrac{9}{64} = \dfrac{4}{9} r^4$

$\dfrac{81}{256} = r^4$

$\sqrt[4]{\dfrac{81}{256}} = r$

$\dfrac{3}{4} = r;$

Find a_1:

$a_7 = a_1 r^6$

$\dfrac{9}{64} = a_1 \left(\dfrac{3}{4} \right)^6$

$\dfrac{9}{64} = a_1 \left(\dfrac{729}{4096} \right)$

$\dfrac{64}{81} = a_1;$

$a_1 = \dfrac{64}{81}; \quad r = \dfrac{3}{4}$

$S_n = \dfrac{a_1 \left(1 - r^n \right)}{1 - r}$

$S_6 = \dfrac{\dfrac{64}{81} \left(1 - \left(\dfrac{3}{4} \right)^6 \right)}{1 - \dfrac{3}{4}} = \dfrac{\dfrac{64}{81} \left(1 - \dfrac{729}{4096} \right)}{\dfrac{1}{4}}$

$= \dfrac{\dfrac{64}{81} \left(\dfrac{3367}{4096} \right)}{\dfrac{1}{4}} = \dfrac{3367}{1296}$

87. $a_3 = 2\sqrt{2}, a_6 = 8$, find S_7

Find r:

$a_6 = a_3 r^3$

$8 = 2\sqrt{2}(r)^3$

$\dfrac{4}{\sqrt{2}} = r^3$

$2\sqrt{2} = r^3$

$2 \cdot 2^{\frac{1}{2}} = r^3$

$2^{\frac{3}{2}} = r^3$

$\left(2^{\frac{3}{2}}\right)^{\frac{1}{3}} = r$

$\sqrt{2} = r$;

Find a_1:

$a_6 = a_1 r^5$

$8 = a_1 \left(\sqrt{2}\right)^5$

$8 = a_1 \left(2^{\frac{5}{2}}\right)$

$\dfrac{2^3}{2^{\frac{5}{2}}} = a_1$

$2^{3-\frac{5}{2}} = a_1$

$\sqrt{2} = a_1$

$a_1 = \sqrt{2}$; $r = \sqrt{2}$

$S_n = \dfrac{a_1\left(1 - r^n\right)}{1 - r}$

$S_7 = \dfrac{\sqrt{2}\left(1 - \left(\sqrt{2}\right)^7\right)}{1 - \sqrt{2}} = \dfrac{\sqrt{2}\left(1 - \left(\sqrt{2}\right)^6\left(\sqrt{2}\right)\right)}{1 - \sqrt{2}}$

$= \dfrac{\sqrt{2}\left(1 - 8\left(\sqrt{2}\right)\right)}{1 - \sqrt{2}} = \dfrac{\sqrt{2} - 16}{1 - \sqrt{2}}$

$= \dfrac{\sqrt{2} - 16}{1 - \sqrt{2}} \cdot \dfrac{1 + \sqrt{2}}{1 + \sqrt{2}}$

$= \dfrac{\sqrt{2} + 2 - 16 - 16\sqrt{2}}{1 - 2}$

$= \dfrac{-14 - 15\sqrt{2}}{-1} = 14 + 15\sqrt{2}$

89. $3 + 6 + 12 + 24 + \dots$

$\dfrac{6}{3} = 2$; No

91. $9 + 3 + 1 + \dots$

$\dfrac{3}{9} = \dfrac{1}{3}$; $\left|\dfrac{1}{3}\right| < 1$

$a_1 = 9$; $r = \dfrac{1}{3}$

$S_\infty = \dfrac{a_1}{1 - r}$

$S_\infty = \dfrac{9}{1 - \dfrac{1}{3}} = \dfrac{9}{\dfrac{2}{3}} = \dfrac{27}{2}$

93. $25 + 10 + 4 + \dfrac{8}{5} + \dots$

$\dfrac{10}{25} = \dfrac{2}{5}$; $\left|\dfrac{2}{5}\right| < 1$

$a_1 = 25$; $r = \dfrac{2}{5}$

$S_\infty = \dfrac{a_1}{1 - r}$

$S_\infty = \dfrac{25}{1 - \dfrac{2}{5}} = \dfrac{25}{\dfrac{3}{5}} = \dfrac{125}{3}$

95. $6 + 3 + \dfrac{3}{2} + \dfrac{3}{4} + \dots$

$\dfrac{3}{6} = \dfrac{1}{2}$; $\left|\dfrac{1}{2}\right| < 1$

$a_1 = 6$; $r = \dfrac{1}{2}$

$S_\infty = \dfrac{a_1}{1 - r}$

$S_\infty = \dfrac{6}{1 - \dfrac{1}{2}} = \dfrac{6}{\dfrac{1}{2}} = 12$

97. $6 - 3 + \dfrac{3}{2} - \dfrac{3}{4} + \ldots$

$\dfrac{-3}{6} = \dfrac{-1}{2};\ \left|\dfrac{-1}{2}\right| < 1$

$a_1 = 6;\ r = \dfrac{-1}{2}$

$S_\infty = \dfrac{a_1}{1-r}$

$S_\infty = \dfrac{6}{1 - \dfrac{-1}{2}} = \dfrac{6}{\dfrac{3}{2}} = \dfrac{12}{3} = 4$

99. $0.3 + 0.03 + 0.003 + \ldots$

$\dfrac{0.03}{0.3} = \dfrac{1}{10};\ \left|\dfrac{1}{10}\right| < 1$

$a_1 = 0.3;\ r = \dfrac{1}{10}$

$S_\infty = \dfrac{a_1}{1-r}$

$S_\infty = \dfrac{0.3}{1 - \dfrac{1}{10}} = \dfrac{0.3}{\dfrac{9}{10}} = \dfrac{3}{9} = \dfrac{1}{3}$

101. $\displaystyle\sum_{k=1}^{\infty} \dfrac{3}{4}\left(\dfrac{2}{3}\right)^k$

$\dfrac{3}{4}\left(\dfrac{2}{3}\right)^1 = \dfrac{1}{2};$

$\dfrac{3}{4}\left(\dfrac{2}{3}\right)^2 = \dfrac{3}{4}\left(\dfrac{4}{9}\right) = \dfrac{1}{3};$

$\dfrac{3}{4}\left(\dfrac{2}{3}\right)^3 = \dfrac{3}{4}\left(\dfrac{8}{27}\right) = \dfrac{2}{9};$

Initial terms: $\dfrac{1}{2} + \dfrac{1}{3} + \dfrac{2}{9} + \ldots$

$\dfrac{\dfrac{1}{3}}{\dfrac{1}{2}} = \dfrac{2}{3};\ \left|\dfrac{2}{3}\right| < 1$

$a_1 = \dfrac{1}{2};\ r = \dfrac{2}{3}$

$S_\infty = \dfrac{a_1}{1-r}$

$S_\infty = \dfrac{\dfrac{1}{2}}{1 - \dfrac{2}{3}} = \dfrac{\dfrac{1}{2}}{\dfrac{1}{3}} = \dfrac{3}{2}$

103. $\displaystyle\sum_{j=1}^{\infty} 9\left(-\dfrac{2}{3}\right)^j$

$9\left(\dfrac{-2}{3}\right)^1 = -6;$

$9\left(\dfrac{-2}{3}\right)^2 = 9\left(\dfrac{4}{9}\right) = 4;$

$9\left(\dfrac{-2}{3}\right)^3 = 9\left(\dfrac{-8}{27}\right) = \dfrac{-8}{3}$

Initial terms: $-6 + 4 - \dfrac{8}{3} + \ldots$

$\dfrac{4}{-6} = \dfrac{2}{-3};\ \left|\dfrac{-2}{3}\right| < 1$

$a_1 = -6;\ r = \dfrac{-2}{3}$

$S_\infty = \dfrac{a_1}{1-r}$

$S_\infty = \dfrac{-6}{1 - \dfrac{-2}{3}} = \dfrac{-6}{\dfrac{5}{3}} = \dfrac{-18}{5}$

105. $S_n = \dfrac{n^2(n+1)^2}{4};\ 1^3 + 2^3 + 3^3 + \ldots 8^3$

$S_8 = \dfrac{8^2(8+1)^2}{4} = \dfrac{64(9)^2}{4} = \dfrac{64(81)}{4} = 1296$

$1^3 + 2^3 + 3^3 + 4^3 + 5^3 + 6^3 + 7^3 + 8^3$
$= 1 + 8 + 27 + 64 + 125 + 216 + 343 + 512$
$= 1296$

107. $a_1 = 24;\ r = 0.8;\ n = 7$

Initial terms: $24, 19.2, 15.36, \ldots$

$a_n = a_1(r)^{n-1}$

$a_7 = 24(0.8)^6 = 24(0.262144) \approx 6.3;$

$S_\infty = \dfrac{a_1}{1-r}$

$S_\infty = \dfrac{24}{1 - 0.8} = \dfrac{24}{0.2} = 120$

about 6.3 ft; 120 ft

Chapter 8: Additional Topics in Algebra

109. $a_1 = 46000;\ r = 0.20;\ n = 4$

$a_4 = 46000(1 - 0.2)^4$

$a_4 = 46000(0.8)^4$

$a_4 = 18841.60$;

$5000 = 46000(1 - 0.2)^n$

$0.1086956522 = 0.8^n$

$\ln 0.1086956522 = n \ln 0.8$

$\dfrac{\ln 0.1086956522}{\ln 0.8} = n$

$10 \approx n$;

about \$18,841.60; 10 years

111. $a_0 = 160;\quad a_1 = 160(0.97) = 155.2$;

$r = 0.03;\ n = 8$

$a_n = a_1(1 - r)^{n-1}$

$a_8 = 155.2(1 - 0.03)^{8-1}$

$a_8 = 155.2(0.97)^7$

$a_8 \approx 125.4$;

$118 = 155.2(0.97)^{n-1}$

$\dfrac{118}{155.2} = 0.97^{n-1}$

$\ln \dfrac{118}{155.2} = (n-1)\ln 0.97$

$\dfrac{\ln \dfrac{118}{155.2}}{\ln 0.97} = n - 1$

$9 \approx n - 1$

$10 \approx n$

about 125.4 gpm; about 10 months

113. $a_1 = 277;\ r = 0.023;\ n = 10$

$a_n = a_1(1 + r)^n$

$a_{10} = 277(1 + 0.023)^{10}$

$a_{10} = 277(1.023)^{10}$

$a_{10} \approx 347.7$

about 347.7 million

115. $a_1 = 50;\ r = 2;\ n = 10$

10 half-hours in 5 hours

$a_n = a_1(r)^n$

$a_{10} = 50(2)^{10} = 50(1024) = 51200$;

$204800 = 50(2)^n$

$4096 = 2^n$

$(2)^{12} = 2^n$

$12 = n$

51,200 bacteria;

12 half-hours later or 6 hours

117. $a_1 = \dfrac{4}{5}(20) = 16;\ r = \dfrac{4}{5};\ n = 7$

$a_n = a_1(r)^{n-1}$

$a_7 = 16\left(\dfrac{4}{5}\right)^6 = 16\left(\dfrac{4096}{15625}\right) \approx 4.2$;

$S_\infty = \dfrac{a_1}{1 - r}$

$S_\infty = \dfrac{16}{1 - \dfrac{4}{5}} = \dfrac{16}{\dfrac{1}{5}} = 80$ up

Down: $S = \dfrac{20}{1 - \dfrac{4}{5}} = \dfrac{20}{\dfrac{1}{5}} = 100$

approximately 4.2 m; 180 m

119. $a_1 = 462; \quad r = \dfrac{2}{5}; \quad n = 5$

$a_n = a_1(1-r)^n$

$a_5 = 462\left(1 - \dfrac{2}{5}\right)^5$

$a_5 = 462\left(\dfrac{3}{5}\right)^5$

$a_5 = 462\left(\dfrac{243}{3125}\right)$

$a_5 \approx 35.9$

$12.9 = 462\left(\dfrac{3}{5}\right)^n$

$0.0279 = \left(\dfrac{3}{5}\right)^n$

$\ln 0.0279 = n \ln 0.6$

$\dfrac{\ln 0.0279}{\ln 0.6} = n$

35.9 in³; about 7 strokes

121. $a_0 = 40000; \quad d = 1750; \quad r = 0.96$

$40000 + 1750n = 40000(1.04)^n$

$Y_1 = Y_2$

Using a grapher: about 6 years

123. $S_n = \displaystyle\sum_{k=1}^{n} \log(k)$

$S_n = \log n!$

125. $f(x) = x^2 + 5x + 9$

$x = \dfrac{-5 \pm \sqrt{(5)^2 - 4(1)(9)}}{2(1)}$

$x = \dfrac{-5 \pm \sqrt{25 - 36}}{2}$

$x = \dfrac{-5 \pm \sqrt{-11}}{2}$

$x = \dfrac{-5}{2} \pm \dfrac{\sqrt{11}}{2}i$

127. $h(x) = \dfrac{x^2}{x-1}$

Vertical asymptote: $x = 1$
Horizontal asymptote: none
(deg num > deg den)
Oblique asymptote: $y = x$

$$\require{enclose}\begin{array}{r} x \\ x-1 \enclose{longdiv}{x^2 } \\ -\underline{(x^2 - x)} \\ x \end{array}$$

y-intercept: (0,0)

$h(0) = \dfrac{0^2}{0-1} = 0$

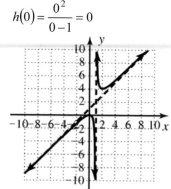

8.4 Exercises

1. Finite; universally

3. Induction hypothesis

5. Answers will vary.

7. $a_n = 10n - 6$
$a_4 = 10(4) - 6 = 40 - 6 = 34$;
$a_5 = 10(5) - 6 = 50 - 6 = 44$;
$a_k = 10k - 6$;
$a_{k+1} = 10(k+1) - 6 = 10k + 10 - 6 = 10k + 4$

9. $a_n = n$
$a_4 = 4$;
$a_5 = 5$;
$a_k = k$;
$a_{k+1} = k + 1$

11. $a_n = 2^{n-1}$
$a_4 = 2^{4-1} = 2^3 = 8$;
$a_5 = 2^{5-1} = 2^4 = 16$;
$a_k = 2^{k-1}$;
$a_{k+1} = 2^{k+1-1} = 2^k$

13. $S_n = n(5n - 1)$
$S_4 = 4(5(4) - 1) = 4(20 - 1) = 4(19) = 76$;
$S_5 = 5(5(5) - 1) = 5(25 - 1) = 5(24) = 120$;
$S_k = k(5k - 1)$;
$S_{k+1} = (k+1)(5(k+1) - 1) = (k+1)(5k + 5 - 1)$
$= (k+1)(5k + 4)$

15. $S_n = \dfrac{n(n+1)}{2}$
$S_4 = \dfrac{4(4+1)}{2} = \dfrac{4(5)}{2} = 10$;
$S_5 = \dfrac{5(5+1)}{2} = \dfrac{5(6)}{2} = 15$;
$S_k = \dfrac{k(k+1)}{2}$;
$S_{k+1} = \dfrac{(k+1)(k+1+1)}{2} = \dfrac{(k+1)(k+2)}{2}$

17. $S_n = 2^n - 1$
$S_4 = 2^4 - 1 = 16 - 1 = 15$;
$S_5 = 2^5 - 1 = 32 - 1 = 31$;
$S_k = 2^k - 1$;
$S_{k+1} = 2^{k+1} - 1$

19. $a_n = 10n - 6$; $S_n = n(5n - 1)$
$S_4 = 4(5(4) - 1) = 4(20 - 1) = 4(19) = 76$;
$a_5 = 10(5) - 6 = 50 - 6 = 44$;
$S_5 = 5(5(5) - 1) = 5(25 - 1) = 5(24) = 120$;
$S_4 + a_5 = S_5$
$76 + 44 = 120$
$120 = 120$
Verified

21. $a_n = n$; $S_n = \dfrac{n(n+1)}{2}$
$S_4 = \dfrac{4(4+1)}{2} = \dfrac{4(5)}{2} = 10$;
$a_5 = 5$;
$S_5 = \dfrac{5(5+1)}{2} = \dfrac{5(6)}{2} = 15$;
$S_4 + a_5 = S_5$
$10 + 5 = 15$
$15 = 15$
Verified

23. $a_n = 2^{n-1}$; $S_n = 2^n - 1$
$S_4 = 2^4 - 1 = 16 - 1 = 15$;
$a_5 = 2^{5-1} = 2^4 = 16$;
$S_5 = 2^5 - 1 = 32 - 1 = 31$;
$S_4 + a_5 = S_5$
$15 + 16 = 31$
$31 = 31$
Verified

8.4 Exercises

25. $a_n = n^3$; $S_n = (1+2+3+4+...+n)^2$

    a. for $n = 1$, $\quad 1 = (1)^2$;

    for $n = 5$,

    $1+8+27+64+125 = (1+2+3+4+5)^2$

    $225 == (15)^2$;

    for $n = 9$,

    $1+8+27+64+125+216+343+512+729$

    $\quad = (1+2+3+4+5+6+7+8+9)^2$

    $2025 = (45)^2$

    $2025 = 2025$

    b. The needed components are:

    $a_n = n^3$; $a_k = k^3$; $a_{k+1} = (k+1)^3$

    $S_k = (1+2+3+4+...+k)^2$;

    $S_{k+1} = (1+2+3+4+...+k+(k+1))^2$

    1. Show S_n is true for $n = 1$.

        $S_1 = (1)^2 = 1$

        Verified

    2. Assume S_k is true:

    $1+8+27+...+k^3 = (1+2+3+4+...+k)^2$

    and use it to show the truth of S_{k+1} follows. That is:

        $1+8+27+...+k^3+(k+1)^3$

        $= (1+2+3+...+k+(k+1))^2$

        $S_k + a_{k+1} = S_{k+1}$

        Working with the left hand side:

        $1+8+27+...+k^3+(k+1)^3$

        $= 1+8+27+...+k^3+(k+1)^2(k+1)$

    $= 1+8+27+...+k^3+k(k+1)^2+(k+1)^2$

    $= 1+8+27+...+k^3+k(k+1)(k+1)+(k+1)^2$

    $= 1+8+27+...+k^3$

    $+2\dfrac{k(k+1)}{2}(k+1)+(k+1)^2$

    $= (1+2+3+...+k)^2$

    $+2(1+2+3+...+k)(k+1)+(k+1)^2$

    Factoring as a trinomial:

    $= ((1+2+3+...+k)+(k+1))^2$

    Since the truth of S_{k+1} follows from S_k, the formula is true for all n.

27. $2+4+6+8+10+...+2n$;

    The needed components are:

    $a_n = 2n$; $a_k = 2k$; $a_{k+1} = 2(k+1)$

    $S_n = n(n+1)$; $S_k = k(k+1)$;

    $S_{k+1} = (k+1)(k+1+1) = (k+1)(k+2)$

    1. Show S_n is true for $n = 1$.

        $S_1 = 1(1+1) = 1(2) = 2$

        Verified

    2. Assume S_k is true:

        $2+4+6+8+10+...+2k = k(k+1)$

        and use it to show the truth of S_{k+1} follows. That is:

    $2+4+6+...+2k+2(k+1) = (k+1)(k+2)$

        $S_k + a_{k+1} = S_{k+1}$

        Working with the left hand side:

        $2+4+6+...+2k+2(k+1)$

        $= k(k+1)+2(k+1)$

        $= k^2+k+2k+2$

        $= k^2+3k+2$

        $= (k+1)(k+2)$

        $= S_{k+1}$

    Since the truth of S_{k+1} follows from S_k, the formula is true for all n.

29. $5+10+15+20+25+...+5n$

The needed components are:

$a_n = 5n$; $a_k = 5k$; $a_{k+1} = 5(k+1)$;

$S_n = \dfrac{5n(n+1)}{2}$; $S_k = \dfrac{5k(k+1)}{2}$;

$S_{k+1} = \dfrac{5(k+1)(k+1+1)}{2} = \dfrac{5(k+1)(k+2)}{2}$

1. Show S_n is true for $n = 1$.

$S_1 = \dfrac{5(1)(1+1)}{2} = \dfrac{5(2)}{2} = 5$

Verified

2. Assume S_k is true:

$5+10+15+...+5k = \dfrac{5k(k+1)}{2}$

and use it to show the truth of S_{k+1} follows. That is:

$5+10+15+...+5k+5(k+1)$

$= \dfrac{5(k+1)(k+1+1)}{2}$

$S_k + a_{k+1} = S_{k+1}$

Working with the left hand side:

$5+10+15+...+5k+5(k+1)$

$= \dfrac{5k(k+1)}{2} + 5(k+1)$

$= \dfrac{5k(k+1)+10(k+1)}{2}$

$= \dfrac{(k+1)(5k+10)}{2}$

$= \dfrac{5(k+1)(k+2)}{2}$

$= S_{k+1}$

Since the truth of S_{k+1} follows from S_k, the formula is true for all n.

31. $5+9+13+17+...+4n+1$

The needed components are:

$a_n = 4n+1$; $a_k = 4k+1$;

$a_{k+1} = 4(k+1)+1 = 4k+4+1 = 4k+5$;

$S_n = n(2n+3)$; $S_k = k(2k+3)$;

$S_{k+1} = (k+1)(2(k+1)+3) = (k+1)(2k+5)$

1. Show S_n is true for $n = 1$.

$S_1 = 1(2(1)+3) = 5$

Verified

2. Assume S_k is true:

$5+9+13+17+...+4k+1 = k(2k+3)$

and use it to show the truth of S_{k+1} follows. That is:

$5+9+13+17+...+4k+1+4(k+1)+1$

$= (k+1)(2(k+1)+3)$

$S_k + a_{k+1} = S_{k+1}$

Working with the left hand side:

$5+9+13+17+...+4k+1+4k+5$

$= k(2k+3)+4k+5$

$= 2k^2 + 3k + 4k + 5 = 2k^2 + 7k + 5$

$= (k+1)(2k+5) = S_{k+1}$

Since the truth of S_{k+1} follows from S_k, the formula is true for all n.

33. $3+9+27+81+243+...+3^n$

The needed components are:

$a_n = 3^n$; $a_k = 3^k$; $a_{k+1} = 3^{k+1}$;

$S_n = \dfrac{3(3^n - 1)}{2}$; $S_k = \dfrac{3(3^k - 1)}{2}$;

$S_{k+1} = \dfrac{3(3^{k+1} - 1)}{2}$

1. Show S_n is true for $n = 1$.

$S_1 = \dfrac{3(3^1 - 1)}{2} = \dfrac{3(3 - 1)}{2} = \dfrac{3(2)}{2} = 3$

Verified

2. Assume S_k is true:

$3+9+27+...+3^k = \dfrac{3(3^k - 1)}{2}$

and use it to show the truth of S_{k+1} follows. That is:

$3+9+27+...+3^k + 3^{k+1} = \dfrac{3(3^{k+1} - 1)}{2}$

$S_k + a_{k+1} = S_{k+1}$

Working with the left hand side:

$3+9+27+...+3^k + 3^{k+1}$

$= \dfrac{3(3^k - 1)}{2} + 3^{k+1}$

$= \dfrac{3(3^k - 1) + 2(3^{k+1})}{2}$

$= \dfrac{3^{k+1} - 3 + 2(3^{k+1})}{2}$

$= \dfrac{3(3^{k+1}) - 3}{2}$

$= \dfrac{3(3^{k+1} - 1)}{2}$

$= S_{k+1}$

Since the truth of S_{k+1} follows from S_k, the formula is true for all n.

35. $2+4+8+16+32+64+...+2^n$

The needed components are:

$a_n = 2^n$; $a_k = 2^k$; $a_{k+1} = 2^{k+1}$;

$S_n = 2^{n+1} - 2$; $S_k = 2^{k+1} - 2$;

$S_{k+1} = 2^{k+1+1} - 2 = 2^{k+2} - 2$

1. Show S_n is true for $n = 1$.

$S_n = 2^{n+1} - 2$

$S_1 = 2^{1+1} - 2 = 2^2 - 2 = 4 - 2 = 2$

Verified

2. Assume S_k is true:

$2+4+8+...+2^k = 2^{k+1} - 2$

and use it to show the truth of S_{k+1} follows. That is:

$2+4+8+...+2^k + 2^{k+1} = 2^{k+2} - 2$

$S_k + a_{k+1} = S_{k+1}$

Working with the left hand side:

$2+4+8+...+2^k + 2^{k+1}$

$= 2^{k+1} - 2 + 2^{k+1}$

$= 2(2^{k+1}) - 2$

$= 2^{k+2} - 2$

$= S_{k+1}$

Since the truth of S_{k+1} follows from S_k, the formula is true for all n.

37. $\dfrac{1}{1(3)} + \dfrac{1}{3(5)} + \dfrac{1}{5(7)} + ... + \dfrac{1}{(2n-1)(2n+1)}$

The needed components are:

$a_n = \dfrac{1}{(2n-1)(2n+1)}$; $a_k = \dfrac{1}{(2k-1)(2k+1)}$;

$a_{k+1} = \dfrac{1}{(2(k+1)-1)(2(k+1)+1)} = \dfrac{1}{(2k+1)(2k+3)}$;

$S_n = \dfrac{n}{2n+1}$; $S_k = \dfrac{k}{2k+1}$;

$S_{k+1} = \dfrac{k+1}{2(k+1)+1} = \dfrac{k+1}{2k+3}$

1. Show S_n is true for $n = 1$.

$S_n = \dfrac{n}{2n+1}$

$S_1 = \dfrac{1}{2(1)+1} = \dfrac{1}{2+1} = \dfrac{1}{3}$

Verified

2. Assume S_k is true:

$$\frac{1}{3}+\frac{1}{15}+\frac{1}{35}+...+\frac{1}{(2k-1)(2k+1)}=\frac{k}{2k+1}$$

and use it to show the truth of S_{k+1} follows. That is:

$$\frac{1}{3}+\frac{1}{15}+\frac{1}{35}+...+\frac{1}{(2k-1)(2k+1)}$$

$$+\frac{1}{(2(k+1)-1)(2(k+1)+1)}=\frac{k+1}{2(k+1)+1}$$

$$S_k+a_{k+1}=S_{k+1}$$

Working with the left hand side:

$$\frac{1}{3}+\frac{1}{15}+\frac{1}{35}+...+\frac{1}{(2k-1)(2k+1)}+\frac{1}{(2k+1)(2k+3)}$$

$$=\frac{k}{2k+1}+\frac{1}{(2k+1)(2k+3)}$$

$$=\frac{k(2k+3)+1}{(2k+1)(2k+3)}$$

$$=\frac{2k^2+3k+1}{(2k+1)(2k+3)}$$

$$=\frac{(2k+1)(k+1)}{(2k+1)(2k+3)}$$

$$=\frac{k+1}{2k+3}$$

$$=S_{k+1}$$

Since the truth of S_{k+1} follows from S_k, the formula is true for all n.

39. $S_n:3^n\geq 2n+1$

$S_k:3^k\geq 2k+1$

$S_{k+1}:3^{k+1}\geq 2(k+1)+1$

1. Show S_n is true for $n=1$.

$S_1:$

$3^1\geq 2(1)+1$

$3\geq 2+1$

$3\geq 3$

Verified

2. Assume $S_k:3^k\geq 2k+1$ is true and use it to show the truth of S_{k+1} follows. That is: $3^{k+1}\geq 2k+3$.
Working with the left hand side:

$3^{k+1}=3(3^k)$

$\geq 3(2k+1)$

$\geq 6k+3$

Since k is a positive integer,

$6k+3\geq 2k+3$

Showing $S_{k+1}:3^{k+1}\geq 2k+3$

Verified

41. $S_n:3\cdot 4^{n-1}\leq 4^n-1$

$S_k:3\cdot 4^{k-1}\leq 4^k-1$

$S_{k+1}:3\cdot 4^k\leq 4^{k+1}-1$

1. Show S_n is true for $n=1$.

$S_1:$

$3\cdot 4^{1-1}\leq 4^1-1$

$3\cdot 4^0\leq 4-1$

$3\cdot 1\leq 3$

$3\leq 3$

Verified

2. Assume $S_k:3\cdot 4^{k-1}\leq 4^k-1$ is true. and use it to show the truth of S_{k+1} follows. That is: $3\cdot 4^k\leq 4^{k+1}-1$.
Working with the left hand side:

$3\cdot 4^k=3\cdot 4\left(4^{k-1}\right)$

$=4\cdot 3\left(4^{k-1}\right)$

$\leq 4\left(4^k-1\right)$

$=4^{k+1}-4$

Since k is a positive integer,

$4^{k+1}-4\leq 4^{k+1}-1$

Showing that $3\cdot 4^k\leq 4^{k+1}-1$

8.4 Exercises

43. $n^2 - 7n$ is divisible by 2

1. Show S_n is true for $n = 1$.

$$S_n : n^2 - 7n = 2m$$

$$S_1 :$$

$$(1)^2 - 7(1) = 1 - 7 = -6 = 2(-3)$$

Verified

2. Assume $S_k : k^2 - 7k = 2m$ for $m \in Z$.
and use it to show the truth of S_{k+1}
follows. That is:

$$(k+1)^2 - 7(k+1) = 2p \text{ for } p \in Z.$$

Working with the left hand side:

$$= (k+1)^2 - 7(k+1)$$

$$= k^2 + 2k + 1 - 7k - 7$$

$$= k^2 - 7k + 2k - 6$$

$$= 2m + 2k - 6$$

$$= 2(m + k - 3)$$

is divisible by 2.

45. $n^3 + 3n^2 + 2n$ is divisible by 3

1. Show S_n is true for $n = 1$.

$$S_n : n^3 + 3n^2 + 2n = 3m$$

$$S_1 :$$

$$(1)^3 + 3(1)^2 + 2(1) = 3m$$

$$1 + 3 + 2 = 3m$$

$$6 = 3m$$

$$2 = m$$

Verified

2. Assume $S_k : k^3 + 3k^2 + 2k = 3m$ for
$m \in Z$ and use it to show the truth of S_{k+1}
follows. That is:

$$S_{k+1} : (k+1)^3 + 3(k+1)^2 + 2(k+1) = 3p \text{ for }$$
$p \in Z.$

Working with the left hand side:

$(k+1)^3 + 3(k+1)^2 + 2(k+1)$ is true.

$$= k^3 + 3k^2 + 3k + 1 + 3(k^2 + 2k + 1) + 2k + 2$$

$$= k^3 + 3k^2 + 2k + 3(k^2 + 2k + 1) + 3k + 3$$

$$= k^3 + 3k^2 + 2k + 3(k^2 + 2k + 1) + 3(k+1)$$

$$= 3m + 3(k^2 + 2k + 1) + 3(k+1)$$

is divisible by 3.

47. $6^n - 1$ is divisible by 5

1. Show S_n is true for $n = 1$.

$$S_n : 6^n - 1 = 5m$$

$$S_1 :$$

$$6^1 - 1 = 5m$$

$$6 - 1 = 5m$$

$$5 = 5m$$

$$1 = m$$

Verified

2. Assume $S_k : 6^k - 1 = 5m$ for $m \in Z$ and
use it to show the truth of S_{k+1} follows.

That is: $S_{k+1} : 6^{k+1} - 1 = 5p$ for $p \in Z.$

Working with the left hand side:

$$= 6^{k+1} - 1$$

$$= 6(6^k) - 1$$

$$= 6(5m + 1) - 1$$

$$= 30m + 6 - 1$$

$$= 30m + 5$$

$$= 5(6m + 1)$$

is divisible by 5.

Verified

49. $\dfrac{x^n - 1}{x - 1} = \left(1 + x + x^2 + x^3 + \ldots + x^{n-1}\right)$

The needed components are:

$$a_k = x^{k-1}, \ S_k = \frac{x^k - 1}{x - 1}, \ S_{k+1} = \frac{x^{k+1} - 1}{x - 1}$$

1. Show S_n is true for $n = 1$.

$$S_k = \frac{x^k - 1}{x - 1}$$

$$S_1 = \frac{x^1 - 1}{x - 1} = 1$$

Verified

2. Assume S_k is true:

$$1 + x + x^2 + x^3 + \ldots + x^{k-1}$$

$$= \frac{x^k - 1}{x - 1}$$

and use it to show the truth of S_{k+1}
follows. That is:

$$1 + x + x^2 + x^3 + \ldots + x^{k-1} + x^{k+1-1}$$

$$= \frac{x^{k+1} - 1}{x - 1}$$

$$S_k + a_{k+1} = S_{k+1}$$

Working with the left hand side:

$$1 + x + x^2 + x^3 + \ldots + x^{k-1} + x^{k+1-1}$$

$$= \frac{x^k - 1}{x - 1} + x^k$$

$$= \frac{x^k - 1}{x - 1} + \frac{x^k(x-1)}{x-1}$$

$$= \frac{x^k - 1 + x^k(x-1)}{x-1}$$

$$= \frac{x^k - 1 + x^{k+1} - x^k}{x-1}$$

$$= \frac{x^{k+1} - 1}{x - 1}$$

$$= S_{k+1}$$

Since the truth of S_{k+1} follows from S_k, the formula is true for all n.

51. $A = \begin{bmatrix} -1 & 2 \\ 3 & 1 \end{bmatrix}$; $B = \begin{bmatrix} 2 & -1 \\ 4 & 3 \end{bmatrix}$

$A + B$

$$\begin{bmatrix} -1 & 2 \\ 3 & 1 \end{bmatrix} + \begin{bmatrix} 2 & -1 \\ 4 & 3 \end{bmatrix}$$

$$= \begin{bmatrix} -1+2 & 2+(-1) \\ 3+4 & 1+3 \end{bmatrix}$$

$$= \begin{bmatrix} 1 & 1 \\ 7 & 4 \end{bmatrix}$$

$A - B$

$$\begin{bmatrix} -1 & 2 \\ 3 & 1 \end{bmatrix} - \begin{bmatrix} 2 & -1 \\ 4 & 3 \end{bmatrix}$$

$$= \begin{bmatrix} -1-2 & 2-(-1) \\ 3-4 & 1-3 \end{bmatrix}$$

$$= \begin{bmatrix} -3 & 3 \\ -1 & -2 \end{bmatrix}$$

$2A - 3B$

$$2\begin{bmatrix} -1 & 2 \\ 3 & 1 \end{bmatrix} - 3\begin{bmatrix} 2 & -1 \\ 4 & 3 \end{bmatrix}$$

$$= \begin{bmatrix} -2 & 4 \\ 6 & 2 \end{bmatrix} - \begin{bmatrix} 6 & -3 \\ 12 & 9 \end{bmatrix}$$

$$= \begin{bmatrix} -2-6 & 4-(-3) \\ 6-12 & 2-9 \end{bmatrix}$$

$$= \begin{bmatrix} -8 & 7 \\ -6 & -7 \end{bmatrix}$$

AB

$$\begin{bmatrix} -1 & 2 \\ 3 & 1 \end{bmatrix}\begin{bmatrix} 2 & -1 \\ 4 & 3 \end{bmatrix}$$

$$= \begin{bmatrix} -1(2)+2(4) & -1(-1)+2(3) \\ 3(2)+1(4) & 3(-1)+1(3) \end{bmatrix}$$

$$= \begin{bmatrix} -2+8 & 1+6 \\ 6+4 & -3+3 \end{bmatrix}$$

$$= \begin{bmatrix} 6 & 7 \\ 10 & 0 \end{bmatrix}$$

BA

$$\begin{bmatrix} 2 & -1 \\ 4 & 3 \end{bmatrix}\begin{bmatrix} -1 & 2 \\ 3 & 1 \end{bmatrix}$$

$$= \begin{bmatrix} 2(-1)+(-1)(3) & 2(2)+(-1)(1) \\ 4(-1)+3(3) & 4(2)+3(1) \end{bmatrix}$$

$$= \begin{bmatrix} -2-3 & 4-1 \\ -4+9 & 8+3 \end{bmatrix}$$

$$= \begin{bmatrix} -5 & 3 \\ 5 & 11 \end{bmatrix}$$

B^{-1}

$$\begin{bmatrix} 2 & -1 & \vdots & 1 & 0 \\ 4 & 3 & \vdots & 0 & 1 \end{bmatrix} \quad \frac{1}{2}R_1 \Rightarrow R_1$$

$$\begin{bmatrix} 1 & -\frac{1}{2} & \vdots & \frac{1}{2} & 0 \\ 4 & 3 & \vdots & 0 & 1 \end{bmatrix} \quad -4R_1 + R_2 \Rightarrow R_2$$

$$\begin{bmatrix} 1 & -\frac{1}{2} & \vdots & \frac{1}{2} & 0 \\ 0 & 5 & \vdots & -2 & 1 \end{bmatrix} \quad \frac{1}{5}R_2 \Rightarrow R_2$$

$$\begin{bmatrix} 1 & -\frac{1}{2} & \vdots & \frac{1}{2} & 0 \\ 0 & 1 & \vdots & -\frac{2}{5} & \frac{1}{5} \end{bmatrix} \quad \frac{1}{2}R_2 + R_1 \Rightarrow R_1$$

$$\begin{bmatrix} 1 & 0 & \vdots & \frac{3}{10} & \frac{1}{10} \\ 0 & 1 & \vdots & -\frac{2}{5} & \frac{1}{5} \end{bmatrix}$$

$$\begin{bmatrix} \frac{3}{10} & \frac{1}{10} \\ -\frac{2}{5} & \frac{1}{5} \end{bmatrix}$$

$$\begin{bmatrix} 0.3 & 0.1 \\ -0.4 & 0.2 \end{bmatrix}$$

53. Center: (4, 3); Point on circle: (1, 7);

$$d = \sqrt{(4-1)^2 + (3-7)^2} = 5, r = 5$$

$$(x-4)^2 + (y-3)^2 = 25$$

Mid-Chapter Check

1. $a_n = 7n - 4$
$a_1 = 7(1) - 4 = 7 - 4 = 3$;
$a_2 = 7(2) - 4 = 14 - 4 = 10$;
$a_3 = 7(3) - 4 = 21 - 4 = 17$;
$a_9 = 7(9) - 4 = 63 - 4 = 59$

3. $a_n = (-1)^n (2n - 1)$
$a_1 = (-1)^1 (2(1) - 1) = -1(2 - 1) = -1(1) = -1$;
$a_2 = (-1)^2 (2(2) - 1) = 1(4 - 1) = 1(3) = 3$;
$a_3 = (-1)^3 (2(3) - 1) = -1(6 - 1) = -1(5) = -5$;
$a_9 = (-1)^9 (2(9) - 1) = -1(18 - 1) = -17$

5. $1 + 4 + 7 + 10 + 13 + 16$
$\sum\limits_{n=1}^{6} (3k - 2)$

7. $a_n = a_1 r^{n-1}$; e
nth term formula for a geometric series

9. $a_n = a_1 + (n - 1)d$; b
nth term formula for an arithmetic series

11. (a) 2, 5, 8, 11, ...
$a_1 = 2$; $d = 5 - 2 = 3$
$a_n = 2 + (n - 1)3$
$a_n = 2 + 3n - 3$
$a_n = 3n - 1$

(b) $\dfrac{3}{2}, \dfrac{9}{4}, 3, \dfrac{15}{4}, ...$
$a_1 = \dfrac{3}{2}$; $d = \dfrac{9}{4} - \dfrac{3}{2} = \dfrac{3}{4}$
$a_n = \dfrac{3}{2} + (n - 1)\dfrac{3}{4}$
$a_n = \dfrac{3}{2} + \dfrac{3}{4}n - \dfrac{3}{4}$
$a_n = \dfrac{3}{4}n + \dfrac{3}{4}$

13. $\dfrac{1}{2} + \dfrac{3}{2} + \dfrac{5}{2} + \dfrac{7}{2} + ... + \dfrac{31}{2}$
$a_1 = \dfrac{1}{2}$; $d = 1$
$\dfrac{31}{2} = \dfrac{1}{2} + (n - 1)(1)$
$\dfrac{31}{2} = \dfrac{1}{2} + n - 1$
$\dfrac{31}{2} = n - \dfrac{1}{2}$
$16 = n$;
$S_{16} = \dfrac{16\left(\dfrac{1}{2} + \dfrac{31}{2}\right)}{2} = \dfrac{16(16)}{2} = 128$

15. $a_3 = -81$; $a_7 = -1$

$a_7 = a_3 \cdot r^4$

$-1 = -81r^4$

$\dfrac{1}{81} = r^4$

$\dfrac{1}{3} = r$;

$a_7 = a_1 r^6$

$-1 = a_1\left(\dfrac{1}{3}\right)^6$

$-1 = \dfrac{1}{729}a_1$

$-729 = a_1$;

$S_{10} = \dfrac{-729\left(1-\left(\dfrac{1}{3}\right)^{10}\right)}{1-\dfrac{1}{3}}$

$= \dfrac{-729\left(1-\dfrac{1}{59049}\right)}{\dfrac{2}{3}}$

$= \dfrac{-729\left(\dfrac{59048}{59049}\right)}{\dfrac{2}{3}}$

$= \dfrac{-29524}{27}$

17. $\dfrac{1}{54} + \dfrac{1}{18} + \dfrac{1}{6} + \dots + \dfrac{81}{2}$

$a_1 = \dfrac{1}{54}$; $r = \dfrac{\dfrac{1}{18}}{\dfrac{1}{54}} = 3$

$\dfrac{81}{2} = \dfrac{1}{54}(3)^{n-1}$

$2187 = 3^{(n-1)}$

$3^7 = 3^{(n-1)}$

$7 = n - 1$

$n = 8$;

$S_8 = \dfrac{\dfrac{1}{54}\left(1-3^8\right)}{1-3} = \dfrac{\dfrac{1}{54}\left(1-6561\right)}{-2}$

$= \dfrac{\dfrac{1}{54}(-6560)}{-2} = \dfrac{6560}{108} = \dfrac{1640}{27}$

19. $60, 59, 58, \dots, 10$

$a_1 = 60$; $d = -1$

$10 = 60 + (n-1)(-1)$

$10 = 60 - n + 1$

$10 = 61 - n$

$-51 = -n$

$51 = n$;

$S_{51} = \dfrac{51(60+10)}{2} = \dfrac{51(70)}{2} = 1785$

Reinforcing Basic Concepts

1. $a_n = 0.125x^3 - 2.5x^2 + 12x$

Sum(seq(Y1,X,1,12) = \$71,500

8.5 Technology Highlight

Exercise 1:
  (a) $_9C_2 = 36$
  (b) $_9C_3 = 84$
  (c) $_9C_4 = 126$
  (d) $_9C_5 = 126$

Exercise 3: $_6C_3 = 20$

8.5 Exercises

1. Experiment, well–defined.

3. Distinguishable

5. Answers will vary.

7. (a)

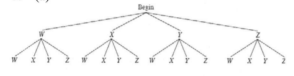

   (b) $WW, WX, WY, WZ,$
       $XW, XX, XY, XZ,$
       $YW, YX, YY, YZ,$
       ZW, ZX, ZY, ZZ

9. 32

11. $25^3 = 15,625$

13. $26 \cdot 26 \cdot 4 \cdot 10 \cdot 10 \cdot 10 = 2,704,000$

15. (a) $9^5 = 59,049$
    (b) $9 \cdot 8 \cdot 7 \cdot 6 \cdot 5 = 15,120$

17. $4 \cdot 6 \cdot 5 \cdot 3 = 360$
    360 if double vegetables are not allowed,
    $4 \cdot 6 \cdot 6 \cdot 3 = 432$
    432 if double vegetables are allowed.

19. (a) $5 \cdot 4 \cdot 3 \cdot 2 = 120$
    (b) $5^4 = 625$
    (c) $2 \cdot 3 \cdot 2 \cdot 1 = 12$

21. $4 \cdot 3 \cdot 2 \cdot 1 = 24$

23. $2 \cdot 2 \cdot 1 \cdot 1 = 4$

25. $5 \cdot 4 \cdot 3 \cdot 2 \cdot 1 = 120$

27. $1 \cdot 3 \cdot 2 \cdot 1 = 6$

29. $_{10}P_3 = 10 \cdot 9 \cdot 8 = 720$;

$$_nP_r = \frac{n!}{(n-r)!}$$

$$_{10}P_3 = \frac{10!}{(10-3)!} = \frac{10 \cdot 9 \cdot 8 \cdot 7!}{7!} = 720$$

31. $_9P_4 = 9 \cdot 8 \cdot 7 \cdot 6 = 3024$;

$$_nP_r = \frac{n!}{(n-r)!}$$

$$_9P_4 = \frac{9!}{(9-4)!} = \frac{9 \cdot 8 \cdot 7 \cdot 6 \cdot 5!}{5!} = 3024$$

33. $_8P_7 = 8 \cdot 7 \cdot 6 \cdot 5 \cdot 4 \cdot 3 \cdot 2 = 40320$

$$_nP_r = \frac{n!}{(n-r)!}$$

$$_8P_7 = \frac{8!}{(8-7)!}$$

$$= \frac{8 \cdot 7 \cdot 6 \cdot 5 \cdot 4 \cdot 3 \cdot 2 \cdot 1!}{1!} = 40320$$

35. T, R and A
    $_3P_3 = 3 \cdot 2 \cdot 1 = 6$
    TRA, TAR, RTA, RAT, ART, ATR
    3 actual words

37. $_{10}P_2 = 10 \cdot 9 = 90$

39. $_8P_3 = 8 \cdot 7 \cdot 6 = 336$

41. (a) $_6P_6 = 6 \cdot 5 \cdot 4 \cdot 3 \cdot 2 \cdot 1 = 720$
    (b) $_6P_3 = 6 \cdot 5 \cdot 4 = 120$
    (c) $_4P_4 = \cdot 4 \cdot 3 \cdot 2 \cdot 1 = 24$

43. $\dfrac{_nP_n}{p!} = \dfrac{_6P_6}{2!} = \dfrac{6 \cdot 5 \cdot 4 \cdot 3 \cdot 2 \cdot 1}{2 \cdot 1} = 360$

45. $\dfrac{_nP_n}{p!q!} = \dfrac{_6P_6}{2!3!} = \dfrac{6 \cdot 5 \cdot 4 \cdot 3 \cdot 2 \cdot 1}{2 \cdot 1 \cdot 3 \cdot 2 \cdot 1} = \dfrac{120}{2} = 60$

47. $\dfrac{_nP_n}{p!\,q!} = \dfrac{_6P_6}{2!3!} = \dfrac{6\cdot5\cdot4\cdot3\cdot2\cdot1}{2\cdot1\cdot3\cdot2\cdot1} = \dfrac{120}{2} = 60$

49. Logic

$\dfrac{_nP_n}{p!} = \dfrac{_5P_5}{1!} = \dfrac{5\cdot4\cdot3\cdot2\cdot1}{1} = 120$

51. Lotto

$\dfrac{_nP_n}{p!\,q!} = \dfrac{_5P_5}{2!2!} = \dfrac{5\cdot4\cdot3\cdot2\cdot1}{2\cdot1\cdot2\cdot1} = \dfrac{60}{2} = 30$

53. A, A, A, N, N, B

$\dfrac{_nP_n}{p!\,q!} = \dfrac{_6P_6}{3!2!} = \dfrac{6\cdot5\cdot4\cdot3\cdot2\cdot1}{3\cdot2\cdot1\cdot2\cdot1} = \dfrac{120}{2} = 60$

BANANA

55. $_9C_4$

(a) $_nC_r = \dfrac{_nP_r}{r!}$

$_9C_4 = \dfrac{_9P_4}{4!} = \dfrac{9\cdot8\cdot7\cdot6}{4\cdot3\cdot2\cdot1} = \dfrac{3024}{24} = 126$;

(b) $_nC_r = \dfrac{n!}{r!(n-r)!}$

$_9C_4 = \dfrac{9!}{4!(9-4)!} = \dfrac{9\cdot8\cdot7\cdot6\cdot5!}{4\cdot3\cdot2\cdot1\cdot5!}$

$= \dfrac{3024}{24} = 126$

57. $_8C_5$

(a) $_nC_r = \dfrac{_nP_r}{r!}$

$_8C_5 = \dfrac{_8P_5}{5!} = \dfrac{8\cdot7\cdot6\cdot5\cdot4}{5\cdot4\cdot3\cdot2\cdot1}$

$= \dfrac{6720}{120} = 56$;

(b) $_nC_r = \dfrac{n!}{r!(n-r)!}$

$_8C_5 = \dfrac{8!}{5!(8-5)!} = \dfrac{8\cdot7\cdot6\cdot5!}{5!\cdot3\cdot2\cdot1}$

$= \dfrac{336}{6} = 56$

59. $_6C_6$

(a) $_nC_r = \dfrac{_nP_r}{r!}$

$_6C_6 = \dfrac{_6P_6}{6!} = \dfrac{6\cdot5\cdot4\cdot3\cdot2\cdot1}{6\cdot5\cdot4\cdot3\cdot2\cdot1} = 1$;

(b) $_nC_r = \dfrac{n!}{r!(n-r)!}$

$_6C_6 = \dfrac{6!}{6!(6-6)!} = \dfrac{6!}{6!} = 1$

61. $_9C_4$, $_9C_5$

$_9C_4 = \dfrac{9!}{4!(9-4)!} = \dfrac{9!}{4!5!} = \dfrac{9\cdot8\cdot7\cdot6\cdot5!}{4\cdot3\cdot2\cdot1\cdot5!}$

$_9C_4 = \dfrac{3024}{24} = 126$;

$_9C_5 = \dfrac{9!}{5!(9-5)!} = \dfrac{9!}{5!4!} = \dfrac{9\cdot8\cdot7\cdot6\cdot5!}{5!\cdot4\cdot3\cdot2\cdot1}$

$_9C_5 = \dfrac{3024}{24} = 126$

Verified

63. $_8C_5$, $_8C_3$

$_8C_5 = \dfrac{8!}{5!(8-5)!} = \dfrac{8!}{5!3!}$

$= \dfrac{8\cdot7\cdot6\cdot5!}{5!\cdot3\cdot2\cdot1} = \dfrac{336}{6} = 56$;

$_8C_3 = \dfrac{8!}{3!(8-3)!} = \dfrac{8!}{3!5!}$

$= \dfrac{8\cdot7\cdot6\cdot5!}{3\cdot2\cdot1\cdot5!} = \dfrac{336}{6} = 56$

Verified

65. $_{12}C_4 = \dfrac{12!}{4!(12-4)!} = \dfrac{12!}{4!8!} = \dfrac{12\cdot11\cdot10\cdot9\cdot8!}{4\cdot3\cdot2\cdot1\cdot8!}$

$= \dfrac{11880}{24} = 495$

67. $_{14}C_3 = \dfrac{14!}{3!(14-3)!} = \dfrac{14!}{3!11!} = \dfrac{14\cdot13\cdot12\cdot11!}{3\cdot2\cdot1\cdot11!}$

$= \dfrac{2184}{6} = 364$

69. $_{10}C_5 = \dfrac{10!}{5!(10-5)!} = \dfrac{10!}{5!5!} = \dfrac{10\cdot9\cdot8\cdot7\cdot6\cdot5!}{5\cdot4\cdot3\cdot2\cdot1\cdot5!}$

$= \dfrac{30240}{120} = 252$

8.5 Exercises

71. $8! = 40,320$

73. $_nP_r = \dfrac{n!}{(n-r)!}$

$_8P_3 = \dfrac{8!}{(8-3)!} = \dfrac{8!}{5!} = \dfrac{8 \cdot 7 \cdot 6 \cdot 5!}{5!} = 336$

75. $_{20}C_5 = \dfrac{20!}{5!(20-5)!} = \dfrac{20!}{5!\,15!}$

$= \dfrac{20 \cdot 19 \cdot 18 \cdot 17 \cdot 16 \cdot 15!}{5 \cdot 4 \cdot 3 \cdot 2 \cdot 1 \cdot 15!} = \dfrac{1860480}{120} = 15,504$

77. $_8C_4 = \dfrac{8!}{4!(8-4)!} = \dfrac{8 \cdot 7 \cdot 6 \cdot 5 \cdot 4!}{4 \cdot 3 \cdot 2 \cdot 1 \cdot 4!}$

$= \dfrac{1680}{24} = 70$

79. (a) $7! = 5,040$;

$n! \approx \sqrt{2\pi} \cdot \left(n^{n+0.5}\right) \cdot e^{-n}$

$7! \approx \sqrt{2\pi} \cdot \left(7^{7+0.5}\right) \cdot e^{-7}$

$7! \approx \sqrt{2\pi} \cdot \left(7^{7.5}\right)\left(e^{-7}\right)$

$7! \approx 4980.395832$;

$\dfrac{5040 - 4980}{5040} = 0.0119 \approx 1.2\%$

(b) $10! = 3,628,800$;

$10! \approx \sqrt{2\pi} \cdot \left(10^{10+0.5}\right) \cdot e^{-10}$

$10! \approx \sqrt{2\pi} \cdot \left(10^{10.5}\right) \cdot e^{-10}$

$10! \approx 3598695.619$;

$\dfrac{3,628,800 - 3,598,696}{3,628,800} \approx 0.83\%$

81. $6^5 = 7776$

83. $9 \cdot 6 \cdot 6 = 324$

85. $8 \cdot 10 \cdot 10 = 800$

87. Exchanges: $8 \cdot 10 \cdot 10 = 800$
Area Codes: $8 \cdot 10 \cdot 10 = 800 - 16 = 784$
Final digits: $10 \cdot 10 \cdot 10 \cdot 10 = 10000$
$784 \cdot 800 \cdot 10000 = 6,272,000,000$

89. $9 \cdot 10 \cdot 10 \cdot 24 \cdot 24 = 518,400$

91. $9 \cdot 9 \cdot 8 \cdot 24 \cdot 23 = 357,696$

93. Five

$_8P_5 = \dfrac{8!}{(8-5)!} = \dfrac{8 \cdot 7 \cdot 6 \cdot 5 \cdot 4 \cdot 3!}{3!} = 6,720$

95. One

$_8P_1 = \dfrac{8!}{(8-1)!} = \dfrac{8 \cdot 7!}{7!} = 8$

97. $7 \cdot 2 \cdot 6! = 10,080$

7 ways they can sit side by side;
2 ways they can sit together, teacher 1 on the
left and teacher 2 on the right, or teacher 2
on the left and teacher 1 on the right;
the students can be seated randomly.

99. $1 \cdot 7! = 1 \cdot 7 \cdot 6 \cdot 5 \cdot 4 \cdot 3 \cdot 2 \cdot 1 = 5,040$

101. $2 \cdot 2 \cdot 6 \cdot 5 \cdot 4 \cdot 3 \cdot 2 \cdot 1 = 2880$

103. $_{15}C_6 = \dfrac{15!}{6!(15-6)!}$

$= \dfrac{15 \cdot 14 \cdot 13 \cdot 12 \cdot 11 \cdot 10 \cdot 9!}{6 \cdot 5 \cdot 4 \cdot 3 \cdot 2 \cdot 1 \cdot 9!}$

$_{15}C_6 = \dfrac{3603600}{720} = 5005$

105. $_{10}P_3 = \dfrac{10 \cdot 9 \cdot 8 \cdot 7!}{7!} = 720$

107. $26 \cdot 25 \cdot 9 \cdot 9 = 52,650$; no

109.(a) $_{10}C_3 \cdot _7 C_2 = _{10} C_2 \cdot _8 C_5$

$$\frac{10!}{7!3!} \cdot \frac{7!}{5!2!} = \frac{10!}{2!3!5!}$$

$$\frac{10!}{2!8!} \cdot \frac{8!}{5!3!} = \frac{10!}{2!3!5!}$$

(b) $_9 C_3 \cdot _6 C_2 = _9 C_2 \cdot _7 C_4$

$$\frac{9!}{3!6!} \cdot \frac{6!}{2!4!} = \frac{9!}{2!3!4!}$$

$$\frac{9!}{2!7!} \cdot \frac{7!}{4!3!} = \frac{9!}{2!3!4!}$$

(c) $_{11}C_4 \cdot _7 C_5 = _{11} C_5 \cdot _6 C_4$

$$\frac{11!}{4!7!} \cdot \frac{7!}{5!2!} = \frac{11!}{2!4!5!}$$

$$\frac{11!}{5!6!} \cdot \frac{6!}{4!2!} = \frac{11!}{2!4!5!}$$

(d) $_8 C_3 \cdot _5 C_2 = _8 C_2 \cdot _6 C_3$

$$\frac{8!}{3!5!} \cdot \frac{5!}{2!3!} = \frac{8!}{2!3!3!}$$

$$\frac{8!}{2!6!} \cdot \frac{6!}{3!3!} = \frac{8!}{2!3!3!}$$

111.
$$\begin{cases} 2x+y<6 \\ x+2y<6 \\ x\geq 0 \\ y\geq 0 \end{cases}$$

$$\begin{cases} y<-2x+6 \\ y<-\dfrac{1}{2}x+3 \\ x\geq 0 \\ y\geq 0 \end{cases}$$

113. $A = \begin{bmatrix} 1 & 0 & 3 \\ -2 & 5 & 1 \\ 2 & 1 & 4 \end{bmatrix}$; $B = \begin{bmatrix} 0.5 & 0.2 & -7 \\ -9 & 0.1 & 8 \\ 1.2 & 0 & 6 \end{bmatrix}$

Using a grapher:

$$A+B = \begin{bmatrix} 1.5 & 0.2 & -4 \\ -11 & 5.1 & 9 \\ 3.2 & 1 & 10 \end{bmatrix};$$

$$AB = \begin{bmatrix} 4.1 & 0.2 & 11 \\ -44.8 & 0.1 & 60 \\ -3.2 & 0.5 & 18 \end{bmatrix};$$

$$A^{-1} = \begin{bmatrix} \dfrac{-19}{17} & \dfrac{-3}{17} & \dfrac{15}{17} \\ \dfrac{-10}{17} & \dfrac{2}{17} & \dfrac{7}{17} \\ \dfrac{12}{17} & \dfrac{1}{17} & \dfrac{-5}{17} \end{bmatrix}$$

8.6 Technology Highlight

Exercise 1: The "middle values" are repeated for $n = 7$ and $n = 5$ but not $n = 6$ because of the symmetry of $_nC_r$.

8.6 Exercises

1. $P(E) = \dfrac{n(E)}{n(S)}$

3. $0 \le P(E) \le 1$
   $P(S) = 1$, and $P(\sim S) = 0$

5. Answers will vary.

7. $S = \{HH, HT, TH, TT\}$; $\dfrac{1}{4}$

9. $S = \{$coach of Patriots, Cougars, Angels, Sharks, Eagles, Stars$\}$; $\dfrac{1}{6}$

11. $S = \{$nine index cards 1-9$\}$
    $P(E) = \dfrac{4}{9}$

13. $S = \{52$ cards$\}$
    a. drawing a Jack: $\dfrac{4}{52} = \dfrac{1}{13}$
    b. drawing a spade: $\dfrac{13}{52} = \dfrac{1}{4}$
    c. drawing a black card: $\dfrac{26}{52} = \dfrac{1}{2}$
    d. drawing a red three: $\dfrac{2}{52} = \dfrac{1}{26}$

15. $S = \{$three males, five females$\}$
    $P(E_1) = \dfrac{1}{8}$;
    $P(E_2) = \dfrac{5}{8}$;
    $P(E_3) = \dfrac{6}{8} = \dfrac{3}{4}$

17. $S = \{$Spinner 1-4$\}$
    a. P(shaded): $\dfrac{3}{4}$
    b. P(less than 5): $\dfrac{4}{4} = 1$
    c. P(2): $\dfrac{1}{4}$
    d. P(prime number): $\dfrac{2}{4} = \dfrac{1}{2}$
       (1 is not considered a prime number)

19. $P(\sim C) = 1 - P(C) = 1 - \dfrac{13}{52} = \dfrac{39}{52} = \dfrac{3}{4}$

21. $10! = 3,628,800$
    $P(\sim 2) = 1 - P(2) = 1 - \dfrac{1}{7} = \dfrac{6}{7}$

23. $P(\sim F) = 1 - P(F) = 1 - 0.009 = 0.991$

25. a. $P(\text{Sum} < 4) = \dfrac{3}{36} = \dfrac{1}{12}$
    b. $P(\text{Sum} < 11) = 1 - P(\text{Sum} > 10)$
       $= 1 - \dfrac{3}{36} = \dfrac{11}{12}$
    c. $P(\sim \text{Sum } 9) = 1 - P(\text{Sum} 9)$
       $= 1 - \dfrac{4}{36} = \dfrac{8}{9}$
    d. $P(\sim D) = 1 - P(D) = 1 - \dfrac{6}{36} = \dfrac{30}{36} = \dfrac{5}{6}$

27. $n(E) = {}_6C_3 \cdot {}_4C_2 ; n(S) = {}_{10}C_5$
    $P(E) = \dfrac{{}_6C_3 \cdot {}_4C_2}{{}_{10}C_5} = \dfrac{120}{252} = \dfrac{10}{21}$

29. $n(E) = {}_9C_6 \cdot {}_5C_3 ; n(S) = {}_{14}C_9$
    $P(E) = \dfrac{{}_9C_6 \cdot {}_5C_3}{{}_{14}C_9} = \dfrac{840}{2002} = \dfrac{60}{143}$

Chapter 8: Additional Topics in Algebra

31. a. $P(\text{all red}) = \dfrac{_{26}C_5}{_{52}C_5}$

$= \dfrac{65780}{2598960} = \dfrac{253}{9996} \approx 0.025$

b. $P(\text{all numbered}) = \dfrac{_{36}C_5}{_{52}C_5}$

$= \dfrac{376992}{2598960} = \dfrac{66}{455} \approx 0.145$

b; $0.145 - 0.025 = 0.12$

about 12 %

33. a. exactly two vegetarians

$P(E) = \dfrac{_9C_2 \cdot _{15}C_4}{_{24}C_6} = \dfrac{49140}{134596} = 0.3651$

b. exactly four non-vegetarians

$P(E) = \dfrac{_{15}C_4 \cdot _9C_2}{_{24}C_6} = \dfrac{49140}{134596} = 0.3651$

c. at least three vegetarians

$1 - (P(0\text{veg}) + P(1\text{veg}) + P(2\text{veg}))$

$= 1 - \left(\dfrac{_9C_0 \cdot _{15}C_6 + _9C_1 \cdot _{15}C_5 + _9C_2 \cdot _{15}C_4}{_{24}C_6} \right)$

$= 1 - \left(\dfrac{81172}{134596} \right) = 0.3969$

35. $P(E_1) = 0.7; P(E_2) = 0.5; P(E_1 \cap E_2) = 0.3$

$P(E_1 \cup E_2) = P(E_1) + P(E_2) - P(E_1 \cap E_2)$

$P(E_1 \cup E_2) = 0.7 + 0.5 - 0.3 = 0.9$

37. $P(E_1) = \dfrac{3}{8}; P(E_2) = \dfrac{3}{4}; P(E_1 \cup E_2) = \dfrac{15}{18}$

$P(E_1 \cup E_2) = P(E_1) + P(E_2) - P(E_1 \cap E_2)$

$\dfrac{15}{18} = \dfrac{3}{8} + \dfrac{3}{4} - P(E_1 \cap E_2)$

$\dfrac{15}{18} = \dfrac{9}{8} - P(E_1 \cap E_2)$

$-\dfrac{7}{24} = -P(E_1 \cap E_2)$

$\dfrac{7}{24} = P(E_1 \cap E_2)$

39. $P(E_1 \cup E_2) = 0.72; P(E_2) = 0.56;$

$P(E_1 \cap E_2) = 0.43$

$P(E_1 \cup E_2) = P(E_1) + P(E_2) - P(E_1 \cap E_2)$

$0.72 = P(E_1) + 0.56 - 0.43$

$0.72 = P(E_1) + 0.13$

$0.59 = P(E_1)$

41. a. $P(\text{multiple of 3 and odd}) = \dfrac{6}{36} = \dfrac{1}{6}$

b. $P(\text{sum} > 5 \text{ and a } 3) = \dfrac{7}{36}$

c. $P(\text{even and} > 9) = \dfrac{4}{36} = \dfrac{1}{9}$

d. $P(\text{odd and} < 10) = \dfrac{16}{36} = \dfrac{4}{9}$

43. a. $P(\text{woman and sergeant}) = \dfrac{4}{50} = \dfrac{2}{25}$

b. $P(\text{man and private}) = \dfrac{9}{50}$

c. $P(\text{private and sergeant}) = \dfrac{0}{50} = 0$

d. $P(\text{woman and officer}) = \dfrac{4}{50} = \dfrac{2}{25}$

e. $P(\text{person in military}) = \dfrac{50}{50} = 1$

45. $\dfrac{9 \cdot 5 \cdot 10 + 9 \cdot 5 \cdot 10 - 9 \cdot 5 \cdot 5}{9 \cdot 10 \cdot 10} = \dfrac{3}{4}$

47. A number greater than 4000 can have digits 4, 5, 6, 7, 8, or 9 in the first position. (6 choices)
Multiples of 5 must end in 0 or 5. (2 choices)

$\dfrac{6 \cdot 10 \cdot 10 \cdot 10}{9 \cdot 10 \cdot 10 \cdot 10} + \dfrac{9 \cdot 10 \cdot 10 \cdot 2}{9 \cdot 10 \cdot 10 \cdot 10} - \dfrac{6 \cdot 10 \cdot 10 \cdot 2}{9 \cdot 10 \cdot 10 \cdot 10}$

$= \dfrac{6}{9} + \dfrac{1}{5} - \dfrac{2}{15} = \dfrac{11}{15}$

49. a. E_1 = boxcars; E_2 = snake eyes
$$P(E_1 \cup E_2) = P(E_1) + P(E_2)$$
$$P(E_1 \cup E_2) = \frac{1}{36} + \frac{1}{36} = \frac{2}{36} = \frac{1}{18}$$

b. E_1 = sum of 7; E_2 = sum of 11
$$P(E_1 \cup E_2) = P(E_1) + P(E_2)$$
$$P(E_1 \cup E_2) = \frac{6}{36} + \frac{2}{36} = \frac{8}{36} = \frac{2}{9}$$

c. E_1 = even numbered sum;
E_2 = prime sum
$$P(E_1 \cup E_2)$$
$$= P(E_1) + P(E_2) - P(E_1 \cap E_2)$$
$$P(E_1 \cup E_2) = \frac{18}{36} + \frac{15}{36} - \frac{1}{36} = \frac{32}{36} = \frac{8}{9}$$

d. E_1 = odd numbered sum;
E_2 = multiple of four
$$P(E_1 \cup E_2) = P(E_1) + P(E_2)$$
$$P(E_1 \cup E_2) = \frac{18}{36} + \frac{9}{36} = \frac{27}{36} = \frac{3}{4}$$

e. E_1 = a sum of 15; E_2 = multiple of 12
$$P(E_1 \cup E_2) = P(E_1) + P(E_2)$$
$$P(E_1 \cup E_2) = \frac{0}{36} + \frac{1}{36} = \frac{1}{36}$$

f. E = prime number sum
$$P(E) = \frac{15}{36} = \frac{5}{12}$$

51. $P(n) = \left(\frac{1}{4}\right)^n$

a. $P(\text{spins a 2}) = \left(\frac{1}{4}\right)^1 = \frac{1}{4}$

b. $P(\text{all 4 spin a 2}) = \left(\frac{1}{4}\right)^4 = \frac{1}{256}$

c. Answers will vary.

53. a. $P(x \geq 2) = 0.25 + 0.08 = 0.33$

b. $P(x < 2) = 0.07 + 0.28 + 0.32 = 0.67$

c. $P(x \leq 4)$
$= 0.08 + 0.25 + 0.32 + 0.28 + 0.07 = 1$

d. $P(x > 4) = 0$

e. $P(x < 2 \text{ or } x > 4) = 0.67 + 0 = 0.67$

f. $P(x \geq 3) = 0.08$

55. Total = 200

a. $P(\text{Isosceles}) = \dfrac{\frac{1}{2}(200)}{200} = \dfrac{1}{2}$

b. $P(\text{Right triangle}) = \dfrac{\frac{1}{2}(200)}{200} = \dfrac{1}{2}$

c. $5^2 + h^2 = 10^2$
$h^2 = 75$
$h = 5\sqrt{3};$
$$P(\text{Equilateral}) = \frac{\frac{1}{2}(10)5\sqrt{3}}{200} = \frac{\sqrt{3}}{8}$$
≈ 0.2165

57. a. $P(x \geq 4) = \dfrac{\pi(6)^2}{\pi(8)^2} = \dfrac{36}{64} = \dfrac{9}{16}$

b. $P(x \geq 6) = \dfrac{\pi(4)^2}{\pi(8)^2} = \dfrac{16}{64} = \dfrac{1}{4}$

c. $P(\text{exactly 8}) = \dfrac{\pi(2)^2}{\pi(8)^2} = \dfrac{4}{64} = \dfrac{1}{16}$

d. $P(x = 4) = \dfrac{\pi(6)^2 - \pi(4)^2}{\pi(8)^2} = \dfrac{20\pi}{64\pi} = \dfrac{5}{16}$

59. n = 13, 3R, 6B, 4W

a. $P(\text{red, blue}) = \dfrac{3}{13} \cdot \dfrac{6}{12} = \dfrac{3}{26}$

b. $P(\text{blue, red}) = \dfrac{6}{13} \cdot \dfrac{3}{12} = \dfrac{3}{26}$

c. $P(\text{white, white}) = \dfrac{4}{13} \cdot \dfrac{3}{12} = \dfrac{1}{13}$

d. $P(\text{blue, not red}) = \dfrac{6}{13} \cdot \dfrac{9}{12} = \dfrac{9}{26}$

e. $P(\text{white, not blue}) = \dfrac{4}{13} \cdot \dfrac{6}{12} = \dfrac{2}{13}$

f. $P(\text{not red, not blue})$

$= \dfrac{6}{13} \cdot \dfrac{3}{12} + \dfrac{6}{13} \cdot \dfrac{4}{12} + \dfrac{4}{13} \cdot \dfrac{3}{12} + \dfrac{4}{13} \cdot \dfrac{3}{12}$

$= \dfrac{3}{26} + \dfrac{4}{26} + \dfrac{2}{26} + \dfrac{2}{26} = \dfrac{11}{26}$

61. Let C represent correct.
Let W represent wrong.

a. $P(\text{Grade} \geq 80\%) = \left(\dfrac{1}{2}\right)^3 = \dfrac{1}{8}$

With 3 questions, the only grade greater than or equal to 80% would be a 100% since 2 out of three questions correct only give 67%.

Possible outcomes: CCC, CCW, CWW, CWC, WCC, WCW, WWC, WWW

b. $P(\text{Grade} \geq 80\%) = \left(\dfrac{1}{2}\right)^4 = \dfrac{1}{16}$

With 4 questions, the only grade greater than or equal to 80% would be a 100% since 3 out of four questions correct only give 75%. Possible outcomes: CCCC, CCCW, CCWC, CCWW, CWCC, CWCW, CWWC, CWWW, WCCW, WCCC, WCWC, WCWW, WWCC, WWCW, WWWC, WWWW.

c. $P(\text{Grade} \geq 80\%) = \dfrac{1}{32} + \dfrac{5}{32} = \dfrac{6}{32} = \dfrac{3}{16}$

With 5 questions, the only grade greater than 80% would be a 100% but 4 out of five questions would give 80%. Thus, we need 5 or 4 correct answers.

Possible outcomes: CCCCC, CCCCW, CCCWC, CCWW, CCWCC, CCWCW, CCWWC, CCWWW, CWCCC, CWCCW, CWCWC, CWCWW, CWWCC, CWWCW, CWWWC, CWWWW, WCCCC, WCCCW, WCCWC, WCWW, WCWCC, WCWCW, WCWWC, WCWWW, WWCCC, WWCCW, WWCWC, WWCWW, WWWCC, WWWCW, WWWWC, WWWWW.

63. a. $P(\text{career and opposed}) = \dfrac{47}{100}$

b. $P(\text{medical and supported}) = \dfrac{8}{100} = \dfrac{2}{25}$

c. $P(\text{military and opposed}) = \dfrac{3}{100}$

d. $P(\text{legal or business and opposed})$

$= \dfrac{18}{100} = \dfrac{9}{50}$

e. $P(\text{academic or medical and supported})$

$= \dfrac{11}{100}$

65. a. $\dfrac{_6C_4 \cdot _5C_4}{_{15}C_8} = \dfrac{5}{429}$

b. $\dfrac{_4C_3 \cdot _6C_5}{_{15}C_8} = \dfrac{8}{2145}$

67. $\dfrac{8!}{2!3!} = \dfrac{8 \cdot 7 \cdot 6 \cdot 5 \cdot 4 \cdot 3!}{2 \cdot 3!} = 3360$;

$P(\text{parallel}) = \dfrac{1}{3360}$

69. $\left(\dfrac{1}{2}\right)^x$ where $x =$ number of flips

$$P(\text{exactly 20 heads}) = \left(\dfrac{1}{2}\right)^{20} = \dfrac{1}{1048576}$$

P(winning the lottery) = will vary; but P(exactly 20 heads) > P(winning the lottery).

71. $\begin{cases} x-2y+3z=10 \\ 2x+y-z=18 \\ 3x-2y+z=26 \end{cases}$

$\begin{bmatrix} 1 & 2 & 3 & | & 10 \\ 2 & 1 & -1 & | & 18 \\ 3 & -2 & 1 & | & 26 \end{bmatrix}$ $-2R_1+R_2 \Rightarrow R_2$

$\begin{bmatrix} 1 & -2 & 3 & | & 10 \\ 0 & 5 & -7 & | & -2 \\ 3 & -2 & 1 & | & 26 \end{bmatrix}$ $-3R_1+R_3 \Rightarrow R_3$

$\begin{bmatrix} 1 & -2 & 3 & | & 10 \\ 0 & 5 & -7 & | & -2 \\ 0 & 4 & -8 & | & -4 \end{bmatrix}$ $\dfrac{1}{5}R_2 \Rightarrow R_2$

$\begin{bmatrix} 1 & -2 & 3 & | & 10 \\ 0 & 1 & -\dfrac{7}{5} & | & -\dfrac{2}{5} \\ 0 & 4 & -8 & | & -4 \end{bmatrix}$ $-4R_2+R_3 \Rightarrow R_3$

$\begin{bmatrix} 1 & -2 & 3 & | & 10 \\ 0 & 1 & -\dfrac{7}{5} & | & -\dfrac{2}{5} \\ 0 & 0 & -\dfrac{12}{5} & | & -\dfrac{12}{5} \end{bmatrix}$ $-\dfrac{5}{12}R_3 \Rightarrow R_3$

$\begin{bmatrix} 1 & -2 & 3 & | & 10 \\ 0 & 1 & -\dfrac{7}{5} & | & -\dfrac{2}{5} \\ 0 & 0 & 1 & | & 1 \end{bmatrix}$

$z=1$;

$y-\dfrac{7}{5}z = -\dfrac{2}{5}$

$y-\dfrac{7}{5}(1) = -\dfrac{2}{5}$

$y = -\dfrac{2}{5} + \dfrac{7}{5}$

$y = \dfrac{5}{5}$

$y = 1$;

$x-2y+3z=10$

$x-2(1)+3(1)=10$

$x+1=10$

$x=9$;

$(9,1,1)$

73. $\dfrac{x^2-1}{x} \geq 0$

$\dfrac{(x+1)(x-1)}{x} \geq 0$

$x \in [-1,0) \cup [1, \infty)$

8.7 Exercises

1. One

3. $(a + (-2b))^5$

5. Answers will vary.

7. $(x + y)^5$
$$= x^5 + 5x^4 y + 10x^3 y^2 + 10x^2 y^3 + 5xy^4 + y^5$$

9. $(2x + 3)^4$
$$= (2x)^4 + 4(2x)^3 (3) + 6(2x)^2 (3)^2$$
$$+ 4(2x)(3)^3 + 3^4$$
$$= 16x^4 + 4(24x^3) + 6(36x^2) + 4(54x) + 81$$
$$= 16x^4 + 96x^3 + 216x^2 + 216x + 81$$

11. $(1 - 2i)^5$
$$= 1^5 + 5(1)^4 (-2i) + 10(1)^3 (-2i)^2$$
$$+ 10(1)^2 (-2i)^3 + 5(1)(-2i)^4 + (-2i)^5$$
$$= 1 - 10i - 40 + 80i + 80 - 32i$$
$$= 41 + 38i$$

13. $\binom{n}{r} = \dfrac{n!}{r!(n-r)!}$
$$\binom{7}{4} = \dfrac{7!}{4!(7-4)!} = \dfrac{7 \cdot 6 \cdot 5 \cdot 4!}{4! \cdot 3 \cdot 2 \cdot 1} = \dfrac{210}{6} = 35$$

15. $\binom{n}{r} = \dfrac{n!}{r!(n-r)!}$
$$\binom{5}{3} = \dfrac{5!}{3!(5-3)!} = \dfrac{5 \cdot 4 \cdot 3!}{3! 2 \cdot 1} = \dfrac{20}{2} = 10$$

17. $\binom{n}{r} = \dfrac{n!}{r!(n-r)!}$
$$\binom{20}{17} = \dfrac{20!}{17!(20-17)!} = \dfrac{20 \cdot 19 \cdot 18 \cdot 17!}{17! 3 \cdot 2 \cdot 1}$$
$$= \dfrac{6840}{6} = 1140$$

19. $\binom{n}{r} = \dfrac{n!}{r!(n-r)!}$
$$\binom{40}{3} = \dfrac{40!}{3!(40-3)!} = \dfrac{40 \cdot 39 \cdot 38 \cdot 37!}{3 \cdot 2 \cdot 1 \cdot 37!}$$
$$= \dfrac{59280}{6} = 9880$$

21. $\binom{n}{r} = \dfrac{n!}{r!(n-r)!}$
$$\binom{6}{0} = \dfrac{6!}{0!(6-0)!} = \dfrac{6!}{6!} = 1$$

23. $\binom{n}{r} = \dfrac{n!}{r!(n-r)!}$
$$\binom{15}{15} = \dfrac{15!}{15!(15-15)!} = \dfrac{15!}{15!(0)!} = 1$$

25. $(c + d)^5$; $a = c$; $b = d$; $n = 5$
$$(c + d)^5$$
$$= \binom{5}{0} c^5 d^0 + \binom{5}{1} c^4 d + \binom{5}{2} c^3 d^2 + \binom{5}{3} c^2 d^3$$
$$+ \binom{5}{4} cd^4 + \binom{5}{5} c^0 d^5$$
$$= 1c^5 + 5c^4 d + \dfrac{5!}{2! 3!} c^3 d^2 + \dfrac{5!}{3! 2!} c^2 d^3$$
$$+ \dfrac{5!}{4! 1!} cd^4 + 1d^5$$

$$= c^5 + 5c^4 d + 10c^3 d^2 + 10c^2 d^3 + 5cd^4 + d^5$$

8.7 Exercises

27. $(a-b)^6$; $a=a$; $b=b$; $n=6$

$(a-b)^6$

$= \binom{6}{0}a^6 b^0 - \binom{6}{1}a^5 b^1 + \binom{6}{2}a^4 b^2 - \binom{6}{3}a^3 b^3$

$\quad + \binom{6}{4}a^2 b^4 - \binom{6}{5}ab^5 + \binom{6}{6}a^0 b^6$

$= 1a^6 b^0 - 6a^5 b + \dfrac{6!}{2!4!}a^4 b^2 - \dfrac{6!}{3!3!}a^3 b^3$

$\quad + \dfrac{6!}{2!4!}a^2 b^4 - \dfrac{6!}{1!5!}ab^5 + 1a^0 b^6$

$= a^6 - 6a^5 b + 15a^4 b^2 - 20a^3 b^3$

$\quad + 15a^2 b^4 - 6ab^5 + b^6$

29. $(2x-3)^4$; $a=2x$; $b=3$; $n=4$

$= \binom{4}{0}(2x)^4 (3)^0 - \binom{4}{1}(2x)^3 (3)^1 + \binom{4}{2}(2x)^2 (3)^2$

$\quad - \binom{4}{3}(2x)^1 (3)^3 + \binom{4}{4}(2x)^0 (3)^4$

$= 1\left(16x^4\right) - 4\left(8x^3\right)(3) + \dfrac{4!}{2!2!}\left(4x^2\right)(9)$

$\quad - \dfrac{4!}{3!1!}(2x)(27) + 1(1)(81)$

$= 16x^4 - 96x^3 + 6\left(36x^2\right) - 4(54x) + 81$

$= 16x^4 - 96x^3 + 216x^2 - 216x + 81$

31. $(1-2i)^3$; $a=1$; $b=2i$; $n=3$

$(1-2i)^3$

$= \binom{3}{0}1^3 (2i)^0 - \binom{3}{1}1^2 (2i) + \binom{3}{2}1(2i)^2 - \binom{3}{3}1^0 (2i)^3$

$= 1(1) - 3(1)(2i) + \dfrac{3!}{2!1!}\left(4i^2\right) - \left(8i^3\right)$

$= 1 - 6i + 3(-4) + 8i$

$= 1 - 6i - 12 + 8i$

$= -11 + 2i$

33. $(x+2y)^9$; $a=x$; $b=2y$; $n=9$

$= \binom{9}{0}x^9 (2y)^0 + \binom{9}{1}x^8 (2y)^1 + \binom{9}{2}x^7 (2y)^2 + \dots$

$= 1\left(x^9\right)(1) + 9x^8 (2y) + \dfrac{9!}{2!7!}x^7 \left(4y^2\right)$
$\quad\quad\quad\quad\quad\quad\quad\quad\quad\quad\quad + \dots$

$= x^9 + 18x^8 y + 36x^7 \left(4y\right)^2 + \dots$

$= x^9 + 18x^8 y + 144x^7 y^2 + \dots$

35. $\left(v^2 - \dfrac{1}{2}w\right)^{12}$; $a=v^2$; $b=\dfrac{1}{2}w$; $n=12$

$= \binom{12}{0}\left(v^2\right)^{12}\left(\dfrac{1}{2}w\right)^0 - \binom{12}{1}\left(v^2\right)^{11}\left(\dfrac{1}{2}w\right)^1$

$\quad + \binom{12}{2}\left(v^2\right)^{10}\left(\dfrac{1}{2}w\right)^2 + \dots$

$= v^{24} - 12\left(v^{22}\right)\left(\dfrac{1}{2}w\right) + \dfrac{12!}{2!10!}v^{20}\left(\dfrac{1}{4}w^2\right) + \dots$

$= v^{24} - 6v^{22}w + 66v^{20}\left(\dfrac{1}{4}w^2\right) + \dots$

$= v^{24} - 6v^{22}w + \dfrac{33}{2}v^{20}w^2 + \dots$

37. $(x+y)^7$, 4th term

$a=x$; $b=y$; $n=7$; $r=3$

$\binom{n}{r}a^{n-r}b^r$

$\binom{7}{3}(x)^{7-3}y^3 = \dfrac{7!}{3!4!}x^4 y^3 = 35x^4 y^3$

39. $(p-2)^8$; 7th term

$a=p$; $b=2$; $n=8$; $r=6$

$\binom{n}{r}a^{n-r}b^r$

$\binom{8}{6}p^{8-6}(2)^6 = \dfrac{8!}{6!2!}p^2 (64)$

$= 28p^2 (64) = 1792p^2$

41. $(2x+y)^{12}$; 11th term

$a = 2x;\quad b = y;\quad n = 12;\quad r = 10$

$\dbinom{n}{r}a^{n-r}b^r$

$\dbinom{12}{10}(2x)^{12-10}y^{10} = \dfrac{12!}{10!2!}(2x)^2 y^{10}$

$= 66\left(4x^2\right)y^{10} = 264x^2 y^{10}$

43. $P(k) = \dbinom{n}{k}\left(\dfrac{1}{2}\right)^k \left(\dfrac{1}{2}\right)^{n-k}$

$P(5) = \dbinom{10}{5}\left(\dfrac{1}{2}\right)^5 \left(\dfrac{1}{2}\right)^{10-5}$

$= \dfrac{10!}{5!5!}\left(\dfrac{1}{32}\right)\left(\dfrac{1}{2}\right)^5$

$= 252\left(\dfrac{1}{32}\right)\left(\dfrac{1}{32}\right)$

$= \dfrac{252}{1024}$

≈ 0.25

45. a. $P(\text{exactly }3)$

$= \dbinom{5}{3}(1-0.347)^{5-3}(0.347)^3$

$= 10(0.653)^2 (0.347)^3 \approx 0.178\,;\ 17.8\%$

b. $P(\text{at least }3) = P(3)+P(4)+P(5)$

$\dbinom{5}{3}(0.653)^2 (0.347)^3 + \dbinom{5}{4}(0.653)^1 (0.347)^4$

$+ \dbinom{5}{5}(0.653)^0 (0.347)^5$

$= 0.1782 + 0.0473 + 0.0050 \approx 0.2305\,;$
$\approx 23.0\%$

47. a. $P(\text{exactly }5)$

$= \dbinom{8}{5}(1-0.94)^{8-5}(0.94)^5$

$= 56(0.06)^3 (0.94)^5 \approx 0.0088;\, 0.88\%$

b. $P(\text{exactly }6)$

$\dbinom{8}{6}(1-0.94)^{8-5}(0.94)^6$

$= 28(0.06)^3 (0.94)^6 \approx 6.9\%$

c. $P(\text{at least }6) = P(6)+P(7)+P(8)$

$= \dbinom{8}{6}(0.06)^2 (0.94)^6 + \dbinom{8}{7}(0.06)(0.94)^7$

$+ \dbinom{8}{8}(0.06)^0 (0.94)^8$

$= 0.0695 + 0.3113 + 0.6096$
$= 0.9904 \approx 0.99;\, 99\%$

d. $P(\text{none}) = P(\text{all on time})$

$\dbinom{8}{8}(1-0.94)^{8-8}(0.94)^8$

$= (0.06)^0 (0.94)^8 \approx 0.6096;\, 61.0\%$

49. (a) $\%\text{error} = \dfrac{\text{approximate value}}{\text{actual value}}$

$= \dfrac{1.476}{(1.02)^{20}} = 0.9933 = 99.33\%$

(b) $\%\text{error} = \dfrac{\text{approximate value}}{\text{actual value}}$

$\dfrac{1.4}{(1.02)^{20}} = 0.94216 = 94.22\%$

8.7 Exercises

51. $\dbinom{n}{k} = \dbinom{n}{n-k}$

$\dbinom{6}{6} = \dbinom{6}{0}$

$\dfrac{6!}{6!0!} = \dfrac{6!}{0!6!}$

$1 = 1;$

$\dbinom{6}{4} = \dbinom{6}{2}$

$\dfrac{6!}{4!2!} = \dfrac{6!}{2!4!}$

$15 = 15;$

$\dbinom{6}{5} = \dbinom{6}{1}$

$\dfrac{6!}{5!1!} = \dfrac{6!}{1!5!}$

$6 = 6$

$\dbinom{6}{3} = \dfrac{6!}{3!3!} = 20$

53. $f(x) = \begin{cases} x+2 & x \le 2 \\ (x-4)^2 & x > 2 \end{cases}$

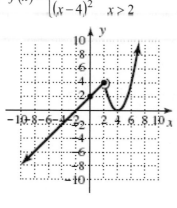

$f(3) = (3-4)^2 = (-1)^2 = 1$

55. $g(x) = x^3 - x^2 - 6x$

$x^3 - x^2 - 6x = 0$

$x(x^2 - x - 6) = 0$

$x(x-3)(x+2) = 0$

$x = 0; \quad x = 3; \quad x = -2$

x-intercepts: $(0, 0), (3, 0), (-2, 0)$

y-intercept: $(0, 0)$

$g(x)\uparrow: (-\infty, -1) \cup (2, \infty)$

$g(x)\downarrow: (-1, 2)$

$g(x) > 0 : (-2, 0) \cup (3, \infty)$

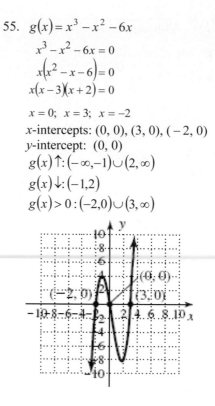

Chapter 8 Summary Exercises

1. $a_n = 5n - 4$
    $a_1 = 5(1) - 4 = 5 - 4 = 1$;
    $a_2 = 5(2) - 4 = 10 - 4 = 6$;
    $a_3 = 5(3) - 4 = 15 - 4 = 11$;
    $a_4 = 5(4) - 4 = 20 - 4 = 16$;
    $a_{10} = 5(10) - 4 = 50 - 4 = 46$
    $1, 6, 11, 16; a_{10} = 46$

3. $1, 16, 81, 256, \ldots$
    $a_n = n^4$
    $a_6 = 6^4 = 1296$

5. $\dfrac{1}{2}, \dfrac{1}{4}, \dfrac{1}{8}, \ldots$
    $a_n = \dfrac{1}{2^n}$
    $S_8 = a_1 + a_2 + a_3 + a_4 + a_5 + a_6 + a_7 + a_8$
    $S_8 = \dfrac{1}{2} + \dfrac{1}{4} + \dfrac{1}{8} + \dfrac{1}{16} + \dfrac{1}{32} + \dfrac{1}{64} + \dfrac{1}{128} + \dfrac{1}{256}$
    $S_8 = \dfrac{255}{256}$

7. $\displaystyle\sum_{n=1}^{7} n^2$
    $= 1^2 + 2^2 + 3^2 + 4^2 + 5^2 + 6^2 + 7^2$
    $= 1 + 4 + 9 + 16 + 25 + 36 + 49$
    $= 140$

9. $a_n = \dfrac{n!}{(n-2)!}$
    $a_1 = \dfrac{1!}{(1-2)!}$ is not defined;
    $a_2 = \dfrac{2!}{(2-2)!} = \dfrac{2!}{0!} = \dfrac{2 \cdot 1}{1} = 2$;
    $a_3 = \dfrac{3!}{(3-2)!} = \dfrac{3!}{1!} = \dfrac{3 \cdot 2}{1} = 6$;
    $a_4 = \dfrac{4!}{(4-2)!} = \dfrac{4 \cdot 3 \cdot 2!}{2!} = 12$;
    $a_5 = \dfrac{5!}{(5-2)!} = \dfrac{5 \cdot 4 \cdot 3!}{3!} = 20$;
    $u_6 = \dfrac{6!}{(6-2)!} = \dfrac{6 \cdot 5 \cdot 4!}{4!} = 30$
    not defined, $2, 6, 12, 20, 30$

11. $\displaystyle\sum_{n=1}^{7} n^2 + \sum_{n=1}^{7} (3n - 2)$
    $\displaystyle\sum_{n=1}^{7} \left(n^2 + 3n - 2\right)$
    $= \left(1^2 + 3(1) - 2\right) + \left(2^2 + 3(2) - 2\right)$
    $\quad + \left(3^2 + 3(3) - 2\right) + \left(4^2 + 3(4) - 2\right)$
    $\quad + \left(5^2 + 3(5) - 2\right) + \left(6^2 + 3(6) - 2\right)$
    $\quad + \left(7^2 + 3(7) - 2\right)$
    $= (1 + 3 - 2) + (4 + 6 - 2) + (9 + 9 - 2)$
    $\quad + (16 + 12 - 2) + (25 + 15 - 2) + (36 + 18 - 2)$
    $\quad + (49 + 21 - 2)$
    $= 2 + 8 + 16 + 26 + 38 + 52 + 68$
    $= 210$

13. $3, 1, -1, -1 \ldots;$ find a_{35}
    $a_1 = 3; \quad d = -2$
    $a_n = a_1 + (n-1)d$
    $a_n = 3 + (n-1)(-2)$
    $a_{35} = 3 + (35-1)(-2) = 3 + 34(-2)$
    $= 3 - 68 = -65$

15. $1+4+7+10+...+88$
$a_n = a_1 + (n-1)d$
$88 = 1 + (n-1)(3)$
$88 = 1 + 3n - 3$
$88 = -2 + 3n$
$90 = 3n$
$30 = n;$
$a_1 = 1; \quad d = 3; \quad a_n = 88; \quad n = 30;$
$S_n = \dfrac{n(a_1 + a_n)}{2}$
$S_{30} = \dfrac{30(1+88)}{2} = \dfrac{30(89)}{2} = 1335$

17. $1 + \dfrac{3}{4} + \dfrac{1}{2} + \dfrac{1}{4} + ...; S_{15}$
$a_n = a_1 + (n-1)d$
$a_{15} = 1 + (15-1)\left(\dfrac{-1}{4}\right) = 1 + 14\left(\dfrac{-1}{4}\right) = \dfrac{-5}{2};$
$a_1 = 1; \quad d = \dfrac{-1}{4}; \quad a_{15} = \dfrac{-5}{2}; \quad n = 15$
$S_n = \dfrac{n(a_1 + a_n)}{2}$
$S_{15} = \dfrac{15\left(1\dfrac{5}{2}\right)}{2} = \dfrac{15\left(\dfrac{-3}{2}\right)}{2} = -11.25$

19. $\displaystyle\sum_{n=1}^{40}(4n-1)$
$a_1 = 4(1) - 1 = 4 - 1 = 3;$
$a_2 = 4(2) - 1 = 8 - 1 = 7;$
$a_{40} = 4(40) - 1 = 160 - 1 = 159;$
$a_1 = 3; \quad d = 4; \quad a_{40} = 159; \quad n = 40;$
$S_n = \dfrac{n(a_1 + a_n)}{2}$
$S_{40} = \dfrac{40(3+159)}{2} = \dfrac{40(162)}{2} = 3240$

21. $a_1 = 4, r = \sqrt{2};$ find a_7
$a_n = a_1 r^{n-1}$
$a_7 = 4(\sqrt{2})^{7-1} = 4(\sqrt{2})^6 = 4(8) = 32$

23. $16 - 8 + 4 - ...$ find S_7
$a_1 = 16; \quad r = \dfrac{-1}{2};$
$S_n = \dfrac{a_1(1-r^n)}{1-r}$
$S_7 = \dfrac{16\left(1-\left(\dfrac{-1}{2}\right)^7\right)}{1-\left(\dfrac{-1}{2}\right)} = \dfrac{16\left(1-\left(\dfrac{-1}{128}\right)\right)}{\dfrac{3}{2}}$
$= \dfrac{16\left(\dfrac{129}{128}\right)}{\dfrac{3}{2}} = 10.75$

25. $\dfrac{4}{5} + \dfrac{2}{5} + \dfrac{1}{5} + \dfrac{1}{10} + ...$ find S_{12}
$a_1 = \dfrac{4}{5}; \quad r = \dfrac{\dfrac{2}{5}}{\dfrac{4}{5}} = \dfrac{1}{2};$
$S_n = \dfrac{a_1(1-r^n)}{1-r}$
$S_{12} = \dfrac{\dfrac{4}{5}\left(1-\left(\dfrac{1}{2}\right)^{12}\right)}{1-\dfrac{1}{2}} = \dfrac{\dfrac{4}{5}\left(1-\dfrac{1}{4096}\right)}{\dfrac{1}{2}}$
$= \dfrac{\dfrac{4}{5}\left(\dfrac{4095}{4096}\right)}{\dfrac{1}{2}} = \dfrac{819}{512}$

27. $5 + 0.5 + 0.05 + 0.005 + \ldots$

$a_1 = 5; \quad r = \dfrac{0.5}{5} = \dfrac{1}{10};$

$S_\infty = \dfrac{a_1}{1-r}$

$S_\infty = \dfrac{5}{1 - \dfrac{1}{10}} = \dfrac{5}{\dfrac{9}{10}} = \dfrac{50}{9}$

29. $\displaystyle\sum_{n=1}^{8} 5\left(\dfrac{2}{3}\right)^n$

$a_1 = 5\left(\dfrac{2}{3}\right) = \dfrac{10}{3};$

$a_2 = 5\left(\dfrac{2}{3}\right)^2 = 5\left(\dfrac{4}{9}\right) = \dfrac{20}{9};$

$r = \dfrac{\dfrac{20}{9}}{\dfrac{10}{3}} = \dfrac{2}{3};$

$a_1 = \dfrac{10}{3}; \quad r = \dfrac{2}{3};$

$S_n = \dfrac{a_1\left(1 - r^n\right)}{1 - r}$

$S_8 = \dfrac{\dfrac{10}{3}\left(1 - \left(\dfrac{2}{3}\right)^8\right)}{1 - \dfrac{2}{3}} = \dfrac{\dfrac{10}{3}\left(1 - \dfrac{256}{6561}\right)}{\dfrac{1}{3}}$

$= \dfrac{\dfrac{10}{3}\left(\dfrac{6305}{6561}\right)}{\dfrac{1}{3}} = \dfrac{63050}{6561}$

31. $\displaystyle\sum_{n=1}^{\infty} 5\left(\dfrac{1}{2}\right)^n$

$a_1 = 5\left(\dfrac{1}{2}\right) = \dfrac{5}{2};$

$a_2 = 5\left(\dfrac{1}{2}\right)^2 = 5\left(\dfrac{1}{4}\right) = \dfrac{5}{4};$

$r = \dfrac{\dfrac{5}{4}}{\dfrac{5}{2}} = \dfrac{1}{2};$

$a_1 = \dfrac{5}{2}; \quad r = \dfrac{1}{2};$

$S_\infty = \dfrac{a_1}{1-r}$

$S_\infty = \dfrac{\dfrac{5}{2}}{1 - \dfrac{1}{2}} = \dfrac{\dfrac{5}{2}}{\dfrac{1}{2}} = 5$

33. $a_1 = 121500; \quad r = \dfrac{2}{3}$

$a_n = a_1 r^n$

$a_7 = 121500\left(\dfrac{2}{3}\right)^7$

$121500\left(\dfrac{128}{2187}\right) \approx 7111.1 \text{ ft}^3$

35. $1+2+3+4+5+\ldots+n$;

The needed components are:

$a_n = n$; $a_k = k$; $a_{k+1} = k+1$;

$S_n = \dfrac{n(n+1)}{2}$; $S_k = \dfrac{k(k+1)}{2}$;

$S_{k+1} = \dfrac{(k+1)(k+2)}{2}$

1. Show S_n is true for $n=1$.

$S_1 = \dfrac{1(1+1)}{2} = \dfrac{2}{2} = 1$

Verified

2. Assume S_k is true:

$1+2+3+\ldots+k = \dfrac{k(k+1)}{2}$

and use it to show the truth of S_{k+1} follows. That is:

$1+2+3+\ldots+k+(k+1)$

$= \dfrac{(k+1)(k+1+1)}{2}$

$S_k + a_{k+1} = S_{k+1}$

Working with the left–hand side:

$1+2+3+\ldots+k+(k+1)$

$= \dfrac{k(k+1)}{2} + k+1$

$= \dfrac{k(k+1)+2(k+1)}{2}$

$= \dfrac{(k+1)(k+2)}{2}$

$= S_{k+1}$

Since the truth of S_{k+1} follows from S_k, the formula is true for all n.

37. $S_n : 4^n \geq 3n+1$

$S_k : 4^k \geq 3k+1$

$S_{k+1} : 4^{k+1} \geq 3(k+1)+1$

$S_{k+1} : 4^{k+1} \geq 3k+4$

1. Show S_n is true for $n=1$.

$S_1 :$

$3^1 \geq 2(1)+1$

$3 \geq 3$

2. Assume $S_k : 4^k \geq 3k+1$ is true and use it to show the truth of S_{k+1} follows. That is: $4^{k+1} \geq 3k+4$.

Working with the left–hand side:

$4^{k+1} = 4\left(4^k\right)$

$\geq 4(3k+1)$

$= 12k+4$

$\geq 3k+4$

Since k is a positive integer,

$12k+4 \geq 3k+4$ showing $4^{k+1} \geq 3k+4$.

Verified.

39. $3^n - 1$ is divisible by 2

1. Show S_n is true for $n=1$.

$S_n : 3^n - 1 = 2m$

$S_1 :$

$3^1 - 1 = 2m$

$2 = 2m$

2. Assume $S_k : 3^k - 1 = 2m$ for $m \in Z$.

and use it to show the truth of S_{k+1} follows.

That is: $S_{k+1} : 3^{k+1} - 1 = 2p$ for $p \in Z$.

Working with the left–hand side:

$= 3^{k+1} - 1$

$= 3\left(3^k\right) - 1$

$= 3(2m)$

$= 2(3m)$

$3^{k+1} - 1$ is divisible by 2,

which is $= S_{k+1}$.

Verified

41. (a) $10 \cdot 9 \cdot 8 = 720$
    (b) $10 \cdot 10 \cdot 10 = 1000$

43. $_{12}C_3 = \dfrac{_{12}P_3}{3!} = \dfrac{1320}{6} = 220$

45. a. $7! = 7 \cdot 6 \cdot 5 \cdot 4 \cdot 3 \cdot 2 \cdot 1 = 5040$

    b. $_7P_4 = \dfrac{7!}{(7-4)!} = \dfrac{7 \cdot 6 \cdot 5 \cdot 4 \cdot 3!}{3!} = 840$

    c. $_7C_4 = \dfrac{_7P_4}{4!} = \dfrac{840}{4 \cdot 3 \cdot 2 \cdot 1} = \dfrac{840}{24} = 35$

47. $\dfrac{_8P_8}{3!2} = \dfrac{8!}{3!2!} = \dfrac{8 \cdot 7 \cdot 6 \cdot 5 \cdot 4 \cdot 3!}{3!2 \cdot 1} = \dfrac{6720}{2} = 3360$

49. $P(\text{ten or face card}) = \dfrac{4}{52} + \dfrac{12}{52} = \dfrac{16}{52} = \dfrac{4}{13}$

51. $P(\sim 3) = 1 - P(3) = 1 - \dfrac{1}{6} = \dfrac{5}{6}$

53. $n(E) = _7C_4 \cdot _5C_3;\ n(S) = _{12}C_7$

    $P(E) = \dfrac{n(E)}{n(S)} = \dfrac{_7C_4 \cdot _5C_3}{_{12}C_7} = \dfrac{350}{792} = \dfrac{175}{396}$

55. a. $\dbinom{7}{5} = \dfrac{7!}{5!2!} = \dfrac{7 \cdot 6 \cdot 5!}{5! \cdot 2 \cdot 1} = \dfrac{42}{2} = 21$

    b. $\dbinom{8}{3} = \dfrac{8!}{3!5!} = \dfrac{8 \cdot 7 \cdot 6 \cdot 5!}{3 \cdot 2 \cdot 1 \cdot 5!} = \dfrac{336}{6} = 56$

57. a. $\left(a + \sqrt{3}\right)^8$

    $\dbinom{8}{0}a^8 + \dbinom{8}{1}a^7\left(\sqrt{3}\right)$
    $+ \dbinom{8}{2}a^6\left(\sqrt{3}\right)^2 + \dbinom{8}{3}a^5\left(\sqrt{3}\right)^3$

    $= a^8 + 8a^7\sqrt{3} + \dfrac{8!}{2!6!}a^6(3)$
    $+ \dfrac{8!}{3!5!}a^5\left(3\sqrt{3}\right)$

    $= a^8 + 8\sqrt{3}a^7 + 28a^6(3) + 56a^5\left(3\sqrt{3}\right)$

    $= a^8 + 8\sqrt{3}a^7 + 84a^6 + 168\sqrt{3}a^5$

    b. $(5a + 2b)^7$

    $\dbinom{7}{0}(5a)^7 + \dbinom{7}{1}(5a)^6(2b)$
    $+ \dbinom{7}{2}(5a)^5(2b)^2 + \dbinom{7}{3}(5a)^4(2b)^3$

    $= 78125a^7 + 7\left(15625a^6\right)(2b)$
    $+ \dfrac{7!}{2!5!}\left(3125a^5\right)\left(4b^2\right) + \dfrac{7!}{3!4!}\left(625a^4\right)\left(8b^3\right)$

    $= 78125a^7 + 218750a^6b$
    $+ 21\left(12500a^5b^2\right) + 35\left(5000a^4b^3\right)$

    $= 78125a^7 + 218750a^6b$
    $+ 262500a^5b^2 + 175000a^4b^3$

Chapter 8 Mixed Review

1. a. $120,163,206,249,...$
       Arithmetic
    b. $4,4,4,4,4,4,....$
       $a_n = 4$
    c. $1,2,6,24,120,720,5040,...$
       $a_n = n!$
    d. $2.00,1.95,1.90,1.85,...$
       Arithmetic
    e. $\dfrac{5}{8},\dfrac{5}{64},\dfrac{5}{512},\dfrac{5}{4096},....$
       Geometric
    f. $-5,5,6.05,-6.655,7.3205,....$
       Geometric
    g. $0.\overline{1},0.\overline{2},0.\overline{3},0.\overline{4}$
       Arithmetic
    h. $525,551.25,578.8125,...$
       Geometric
    i. $\dfrac{1}{2},\dfrac{1}{4},\dfrac{1}{6},\dfrac{1}{8},....$
       $a_n = \dfrac{1}{2n}$

3. $2 \cdot 25 \cdot 24 \cdot 23 = 27600$

5. $a_1 = 0.1, r = 5$
    $a_2 = 5(0.1) = 0.5$;
    $a_3 = 5(0.5) = 2.5$;
    $a_4 = 5(2.5) = 12.5$;
    $a_5 = 5(12.5) = 62.5$;
    $a_n = a_1 r^{n-1}$
    $a_{15} = 0.1(5)^{14} = 610,351,562.5$
    $0.1,0.5,2.5,12.5,62.5; a_{20} = 1907348632812$

7. $P(\sim \text{doubles}) = 1 - P(D) = 1 - \dfrac{6}{36} = \dfrac{30}{36} = \dfrac{5}{6}$

9. a. $\displaystyle\sum_{n=1}^{\infty} \left(\dfrac{2}{3}\right)^n$

    $a_1 = \dfrac{2}{3}; \quad a_2 = \left(\dfrac{2}{3}\right)^2 = \dfrac{4}{9}; \quad r = \dfrac{\frac{4}{9}}{\frac{2}{3}} = \dfrac{2}{3}$

    $S_\infty = \dfrac{a_1}{1-r}$

    $S_\infty = \dfrac{\frac{2}{3}}{1 - \frac{2}{3}} = \dfrac{\frac{2}{3}}{\frac{1}{3}} = 2$

    b. $\displaystyle\sum_{n=1}^{10} (9 + 2n)$

    $a_1 = 9 + 2(1) = 9 + 2 = 11$;
    $a_{10} = 9 + 2(10) = 9 + 20 = 29$;
    $S_n = \dfrac{n(a_1 + a_n)}{2}$
    $S_{10} = \dfrac{10(11+29)}{2} = \dfrac{10(40)}{2} = 200$

    c. $\displaystyle\sum_{n=1}^{5} 12n + \sum_{n=1}^{5}(-5) + \sum_{n=1}^{5} n^2$

    $\displaystyle\sum_{n=1}^{5} 12n$

    $a_1 = 12; \quad a_5 = 12(5) = 60$;
    $S_5 = \dfrac{5(12+60)}{2} = \dfrac{5(72)}{2} = 180$;

    $\displaystyle\sum_{n=1}^{5}(-5)$

    $a_1 = -5; \quad a_5 = -5$;
    $S_5 = \dfrac{5(-5-5)}{2} = \dfrac{5(-10)}{2} = -25$;

    $\displaystyle\sum_{n=1}^{5} n^2 = 1 + 4 + 9 + 16 + 25 = 55$;

    $\displaystyle\sum_{n=1}^{5} 12n + \sum_{n=1}^{5}(-5) + \sum_{n=1}^{5} n^2$
    $= 180 - 25 + 55 = 210$

11. $(a+b)^n$

   a. first 3 terms for $n = 20$

$$\binom{20}{0}a^{20} + \binom{20}{1}a^{19}b + \binom{20}{2}a^{18}b^2$$

$$= a^{20} + 20a^{19}b + \frac{20!}{2!\,18!}a^{18}b^2$$

$$= a^{20} + 20a^{19}b + 190a^{18}b^2$$

   b. last 3 terms for $n = 20$

$$\binom{20}{18}a^2b^{18} + \binom{20}{19}ab^{19} + \binom{20}{20}b^{20}$$

$$= \frac{20!}{18!\,2!}a^2b^{18} + 20ab^{19} + b^{20}$$

$$= 190a^2b^{18} + 20ab^{19} + b^{20}$$

   c. fifth term where $n = 35$

$$a = a;\quad b = b;\quad n = 35;\quad r = 4$$

$$\binom{n}{r}a^{n-r}b^r$$

$$\binom{35}{4}a^{35-4}b^4 = \frac{35!}{4!\,31!}a^{31}b^4$$

$$= 52360a^{31}b^4$$

   d. fifth term where $n = 35, a = 0.2, b = 0.8$

$$\binom{n}{r}a^{n-r}b^r$$

$$\binom{35}{4}(0.2)^{35-4}(0.8)^4$$

$$= \frac{35!}{4!\,31!}(0.2)^{31}(0.8)^4$$

$$= 52360(0.2)^{31}(0.8)^4$$

$$= 4.6\text{x}10^{-18}$$

13. $3 + 6 + 9 + \ldots + 3n = \dfrac{3n(n+1)}{2}$

The needed components are:

$$a_n = 3n;\; a_k = 3k;\; a_{k+1} = 3(k+1);$$

$$S_n = \frac{3n(n+1)}{2};\; S_k = \frac{3k(k+1)}{2};$$

$$S_{k+1} = \frac{3(k+1)(k+2)}{2}$$

1. Show S_n is true for $n = 1$.

$$S_1 = \frac{3(1)(1+1)}{2} = \frac{3(2)}{2} = 3$$

Verified

2. Assume S_k is true:

$$3 + 6 + 9 + \ldots + 3k = \frac{3k(k+1)}{2}$$

and use it to show the truth of S_{k+1} follows. That is:

$$3 + 6 + 9 + \ldots + 3k + 3(k+1)$$

$$= \frac{3(k+1)(k+1+1)}{2}$$

$$S_k + a_{k+1} = S_{k+1}$$

Working with the left–hand side:

$$3 + 6 + 9 + \ldots + 3k + 3(k+1)$$

$$= \frac{3k(k+1)}{2} + 3(k+1)$$

$$= \frac{3k(k+1) + 6(k+1)}{2}$$

$$= \frac{(k+1)(3k+6)}{2}$$

$$= \frac{3(k+1)(k+2)}{2}$$

$$= S_{k+1}$$

Since the truth of S_{k+1} follows from S_k, the formula is true for all n.

15. $P(E_1 \text{ or } E_2) = \dfrac{15}{2000} + \dfrac{5}{550} \approx 0.01659$

17. $0.36 + 0.0036 + 0.000036 + 0.00000036 + \ldots$

$a_1 = 0.36; \quad r = 0.01$

$S_\infty = \dfrac{a_1}{1-r}$

$S_\infty = \dfrac{0.36}{1-0.01} = \dfrac{0.36}{0.99} = \dfrac{4}{11}$

19. $\begin{cases} a_1 = 10 \\ a_{n+1} = a_n\left(\dfrac{1}{5}\right) \end{cases}$

$a_{1+1} = a_1\left(\dfrac{1}{5}\right)$

$a_2 = 10\left(\dfrac{1}{5}\right) = 2 \; ;$

$a_{2+1} = a_2\left(\dfrac{1}{5}\right)$

$a_3 = 2\left(\dfrac{1}{5}\right) = \dfrac{2}{5} \; ;$

$a_{3+1} = a_3\left(\dfrac{1}{5}\right)$

$a_4 = \dfrac{2}{5}\left(\dfrac{1}{5}\right) = \dfrac{2}{25} \; ;$

$a_{4+1} = a_4\left(\dfrac{1}{5}\right)$

$a_5 = \dfrac{2}{25}\left(\dfrac{1}{5}\right) = \dfrac{2}{125} \; ;$

$10, 2, \dfrac{2}{5}, \dfrac{2}{25}, \dfrac{2}{125}$

Chapter 8 Practice Test

1. a. $a_n = \dfrac{2n}{n+3}$

$a_1 = \dfrac{2(1)}{1+3} = \dfrac{2}{4} = \dfrac{1}{2}$;

$a_2 = \dfrac{2(2)}{2+3} = \dfrac{4}{5}$;

$a_3 = \dfrac{2(3)}{3+3} = \dfrac{6}{6} = 1$;

$a_4 = \dfrac{2(4)}{4+3} = \dfrac{8}{7}$;

$a_8 = \dfrac{2(8)}{8+3} = \dfrac{16}{11}$;

$a_{12} = \dfrac{2(12)}{12+3} = \dfrac{24}{15} = \dfrac{8}{5}$

$\dfrac{1}{2}, \dfrac{4}{5}, 1, \dfrac{8}{7}$; $a_8 = \dfrac{16}{11}, a_{12} = \dfrac{8}{5}$

b. $a_n = \dfrac{(n+2)!}{n!}$

$a_1 = \dfrac{(1+2)!}{1!} = \dfrac{3!}{1!} = 3 \cdot 2 \cdot 1 = 6$;

$a_2 = \dfrac{(2+2)!}{2!} = \dfrac{4!}{2!} = \dfrac{4 \cdot 3 \cdot 2!}{2!} = 12$;

$a_3 = \dfrac{(3+2)!}{3!} = \dfrac{5!}{3!} = \dfrac{5 \cdot 4 \cdot 3!}{3!} = 20$;

$a_4 = \dfrac{(4+2)!}{4!} = \dfrac{6!}{4!} = \dfrac{6 \cdot 5 \cdot 4!}{4!} = 30$;

$a_8 = \dfrac{(8+2)!}{8!} = \dfrac{10!}{8!} = \dfrac{10 \cdot 9 \cdot 8!}{8!} = 90$;

$a_{12} = \dfrac{(12+2)!}{12!} = \dfrac{14!}{12!} = \dfrac{14 \cdot 13 \cdot 12!}{12!} = 182$;

$6, 12, 20, 30$; $a_8 = 90, a_{12} = 182$

c. $a_n = \begin{cases} a_1 = 3 \\ a_{n+1} = \sqrt{(a_n)^2 - 1} \end{cases}$

$a_1 = 3$;

$a_{1+1} = \sqrt{(a_1)^2 - 1}$

$a_2 = \sqrt{(3)^2 - 1} = \sqrt{9-1} = \sqrt{8} = 2\sqrt{2}$;

$a_{2+1} = \sqrt{(a_2)^2 - 1}$

$a_3 = \sqrt{(2\sqrt{2})^2 - 1} = \sqrt{8-1} = \sqrt{7}$;

$a_{3+1} = \sqrt{(a_3)^2 - 1}$

$a_4 = \sqrt{(\sqrt{7})^2 - 1} = \sqrt{7-1} = \sqrt{6}$;

$a_{7+1} = \sqrt{(a_7)^2 - 1}$

$a_8 = \sqrt{(\sqrt{3})^2 - 1} = \sqrt{2}$;

$a_{11+1} = \sqrt{(a_{11})^2 - 1}$

$a_{12} = \sqrt{i^2 - 1} = \sqrt{-1-1} = \sqrt{-2} = i\sqrt{2}$;

$3, 2\sqrt{2}, \sqrt{7}, \sqrt{6}$; $a_8 = \sqrt{2}, a_{12} = i\sqrt{2}$

3. a. $7, 4, 1, -2, \ldots$

$a_1 = 7, d = -3, a_n = 10 - 3n$

b. $-8, -6, -4, -2, \ldots$

$a_1 = -8, d = 2, a_n = 2n - 10$

c. $4, -8, 16, -32, \ldots$

$a_1 = 4, r = -2, a_n = 4(-2)^{n-1}$

d. $10, 4, \dfrac{8}{5}, \dfrac{16}{25}, \ldots$

$a_1 = 10, r = \dfrac{2}{5}, a_n = 10\left(\dfrac{2}{5}\right)^{n-1}$

5. a. $7 + 10 + 13 + \ldots + 100$

$a_1 = 7; a_n = 100; d = 3$

$a_n = a_1 + (n-1)d$

$100 = 7 + (n-1)(3)$

$100 = 7 + 3n - 3$

$100 = 4 + 3n$

$96 = 3n$

$32 = n$;

$S_n = \dfrac{n(a_1 + a_n)}{2}$

$S_{32} = \dfrac{32(7+100)}{2} = \dfrac{32(107)}{2} = 1712$

b. $\displaystyle\sum_{k=1}^{37}(3k+2)$

$a_1 = 3(1)+2 = 3+2 = 5$;

$a_2 = 3(2)+2 = 6+2 = 8$;

$a_{37} = 3(37)+2 = 111+2 = 113$;

$a_1 = 5$; $a_{37} = 113$; $d = 3$; $n = 37$;

$S_n = \dfrac{n(a_1+a_n)}{2}$

$S_{37} = \dfrac{37(5+113)}{2} = \dfrac{37(118)}{2} = 2183$

c. $4-12+36-108+\ldots$ Find S_7

$a_1 = 4$; $r = \dfrac{-12}{4} = -3$;

$S_n = \dfrac{a_1\left(1-r^n\right)}{1-r}$

$S_7 = \dfrac{4\left(1-(-3)^7\right)}{1-(-3)} = \dfrac{4(1-(-2187))}{4}$

$= 2188$

d. $6+3+\dfrac{3}{2}+\dfrac{3}{4}+\ldots$

$a_1 = 6$; $r = \dfrac{1}{2}$

$S_\infty = \dfrac{a_1}{1-r}$

$S_\infty = \dfrac{6}{1-\dfrac{1}{2}} = \dfrac{6}{\dfrac{1}{2}} = 12$

7. $a_1 = 3000$; $r = 1.07$; $n = 12$

$a_n = a_1 r^n$

$a_{12} = 3000(1.07)^{12} = 6756.57$

$\$6756.57$

9. $a_n = 5n-3$; $a_k = 5k-3$;

The needed components are:

$a_{k+1} = 5(k+1)-3 = 5k+2$;

$S_n = \dfrac{5n^2-n}{2}$; $S_k = \dfrac{5k^2-k}{2}$;

$S_{k+1} = \dfrac{5(k+1)^2-(k+1)}{2}$

1. Show S_n is true for $n = 1$.

$a_1 = 5(1)-3 = 5-3 = 2$

$S_1 = \dfrac{5(1)^2-1}{2} = \dfrac{5-1}{2} = \dfrac{4}{2} = 2$

Verified

2. Assume S_k is true:

$2+7+12+\ldots+5k-3 = \dfrac{5k^2-k}{2}$

and use it to show the truth of S_{k+1} follows. That is:

$2+7+12+\ldots+5k-3+(5(k+1)-3)$

$= \dfrac{5(k+1)^2-(k+1)}{2}$

$S_k + a_{k+1} = S_{k+1}$

Working with the left–hand side:

$2+7+12+\ldots+5k-3+(5k+2)$

$= \dfrac{5k^2-k}{2}+5k+2$

$= \dfrac{5k^2-k+2(5k+2)}{2}$

$= \dfrac{5k^2+10k-k+4}{2}$

$= \dfrac{5\left(k^2+2k\right)-k+4}{2}$

$= \dfrac{5(k+1)^2-k+4-5}{2}$

$= \dfrac{5(k+1)^2-k-1}{2}$

$= \dfrac{5(k+1)^2-(k+1)}{2}$

Since the truth of S_{k+1} follows from S_k, the formula is true for all n.

Chapter 8: Additional Topics in Algebra

11. a.

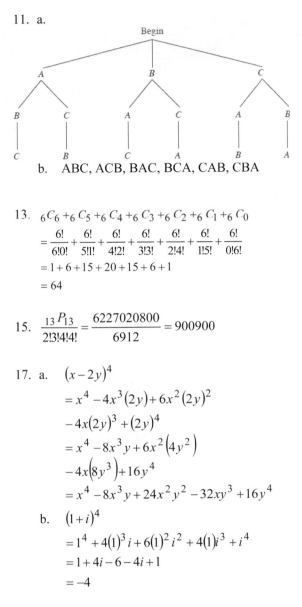

b. ABC, ACB, BAC, BCA, CAB, CBA

13. $_6C_6 + _6C_5 + _6C_4 + _6C_3 + _6C_2 + _6C_1 + _6C_0$

$$= \frac{6!}{6!0!} + \frac{6!}{5!1!} + \frac{6!}{4!2!} + \frac{6!}{3!3!} + \frac{6!}{2!4!} + \frac{6!}{1!5!} + \frac{6!}{0!6!}$$

$$= 1 + 6 + 15 + 20 + 15 + 6 + 1$$

$$= 64$$

15. $\dfrac{_{13}P_{13}}{2!3!4!4!} = \dfrac{6227020800}{6912} = 900900$

17. a. $(x - 2y)^4$

$$= x^4 - 4x^3(2y) + 6x^2(2y)^2$$

$$- 4x(2y)^3 + (2y)^4$$

$$= x^4 - 8x^3 y + 6x^2\left(4y^2\right)$$

$$- 4x\left(8y^3\right) + 16y^4$$

$$= x^4 - 8x^3 y + 24x^2 y^2 - 32xy^3 + 16y^4$$

b. $(1 + i)^4$

$$= 1^4 + 4(1)^3 i + 6(1)^2 i^2 + 4(1)i^3 + i^4$$

$$= 1 + 4i - 6 - 4i + 1$$

$$= -4$$

19. $1 - 0.011 = 0.989$

21. a. $0.05 + 0.03 = 0.08$

b. $0.02 + 0.30 + 0.60 = 0.92$

c. $0.02 + 0.30 + 0.60 + 0.05 + 0.03 = 1$

d. 0

e. $0.02 + 0.30 + 0.60 + 0.03 = 0.95$

f. 0.03

23. a. $P(\text{woman or craftsman})$

$$= \frac{50}{100} + \frac{18}{100} - \frac{9}{100} = \frac{59}{100}$$

b. $P(\text{man or contractor})$

$$= \frac{50}{100} + \frac{7}{100} - \frac{4}{100} = \frac{53}{100}$$

c. $P(\text{man and technician}) = \dfrac{13}{100}$

d. $P(\text{journeyman or apprentice})$

$$= \frac{13}{100} + \frac{34}{100} = \frac{47}{100}$$

25. $1 + 2 + 3 + \ldots + n = \dfrac{n(n+1)}{2}$

$a_n = n$; $a_k = k$; $a_{k+1} = k + 1$

$S_n = \dfrac{n(n+1)}{2}$; $S_k = \dfrac{k(k+1)}{2}$;

$S_{k+1} = \dfrac{(k+1)(k+2)}{2}$

1. Show S_n is true for $n = 1$.

$$S_1 = \frac{1(1+1)}{2} = \frac{2}{2} = 1$$

Verified

2. Assume S_k is true, show

$S_k + a_{k+1} = S_{k+1}$.

$1 + 2 + 3 + \ldots + k + (k+1)$

$$= \frac{k(k+1)}{2} + k + 1$$

$$= \frac{k(k+1) + 2(k+1)}{2}$$

$$= \frac{(k+1)(k+2)}{2}$$

$$= S_{k+1}$$

Verified

Chapters 1-8 Cumulative Review

Calculator Exploration and Discovery

1. $a_1 = \dfrac{1}{3}$ and $r = \dfrac{1}{3}$

   Using a grapher: $\dfrac{1}{2}$

3. $a_n = \dfrac{1}{(n-1)!}$

   Using a grapher: e

Strengthening Core Skills

1. $\dfrac{_4C_1 \cdot \, _{13}C_5 - 40}{_{52}C_5} \approx 0.001970$

3. $\dfrac{4 \cdot \, _{13}C_4 \cdot \, _{39}C_1}{_{52}C_5} \approx 0.042917$

Chapters 1-8 Cumulative Review

1. a. $(9, 52)\ (11, 98)$

   $m = \dfrac{98-52}{11-9} = \dfrac{46}{2} = 23$

   23 cards are assembled each hour

   b. $y - y_1 = m(x - x_1)$

   $y - 52 = 23(x - 9)$

   $y - 52 = 23x - 207$

   $y = 23x - 155$

   c. $23(8) = 184$

   184 cards

   d. $\dfrac{52}{23} \approx 2.26$

   Approximately 2.25 hours before 9am
   $\approx 6:45$am

3. $3x^2 + 5x - 7 = 0$

   $x = \dfrac{-5 \pm \sqrt{(5)^2 - 4(3)(-7)}}{2(3)}$

   $x = \dfrac{-5 \pm \sqrt{25 + 84}}{6}$

   $x = \dfrac{-5 \pm \sqrt{109}}{6}$

   $x \approx 0.91;\ x \approx -2.57$

5. $Y = \dfrac{kVW}{X}$

   $10 = \dfrac{k(5)(12)}{9}$

   $90 = 60k$

   $\dfrac{3}{2} = k;$

   $y = \dfrac{3VW}{2X}$

7. $f(x) = x^3 - 5;\quad g(x) = \sqrt[3]{x+5}$

   $y = x^3 - 5$

   $x = y^3 - 5$

   $x + 5 = y^3$

   $\sqrt[3]{x+5} = y$

   Verified

   The graphs are reflections across $y = x$.

9. a. $f(x) = 2x^2 - 3x$

   $\dfrac{f(x+h) - f(x)}{h}$

   $= \dfrac{\left(2(x+h)^2 - 3(x+h)\right) - \left(2x^2 - 3x\right)}{h}$

   $= \dfrac{\left(2\left(x^2 + 2xh + h^2\right) - 3x - 3h\right) - \left(2x^2 - 3x\right)}{h}$

   $= \dfrac{\left(2x^2 + 4xh + 2h^2 - 3x - 3h\right) - \left(2x^2 - 3x\right)}{h}$

   $= \dfrac{2x^2 + 4xh + 2h^2 - 3x - 3h - 2x^2 + 3x}{h}$

   $= \dfrac{4xh + 2h^2 - 3h}{h}$

   $= \dfrac{h(4x + 2h - 3)}{h}$

   $= 4x + 2h - 3$

b. $h(x) = \dfrac{1}{x-2}$

$\dfrac{f(x+h) - f(x)}{h}$

$= \dfrac{\dfrac{1}{x+h-2} - \dfrac{1}{x-2}}{h}$

$= \dfrac{\dfrac{x-2-(x+h-2)}{(x-2)(x+h-2)}}{h}$

$= \dfrac{\dfrac{x-2-x-h+2}{(x-2)(x+h-2)}}{h}$

$= \dfrac{\dfrac{-h}{(x-2)(x+h-2)}}{h}$

$= \dfrac{-1}{(x+h-2)(x-2)}$

11. $h(x) = \dfrac{2x^2 - 8}{x^2 - 1}$

$0 = 2x^2 - 8$

$0 = 2(x^2 - 4)$

$x^2 - 4 = 0$

$x^2 = 4$

$x = \pm 2$

x-intercepts: $(2, 0), (-2, 0)$

$h(0) = \dfrac{2(0)^2 - 8}{0^2 - 1} = 8$

y-intercept: $(0, 8)$

$0 = x^2 - 1$

$x = \pm 1$

Vertical asymptotes: $x = \pm 1$

Horizontal asymptote: $y = 2$

(deg num = deg den)

13. a. $3 = \log_x(125)$

$x^3 = 125$

b. $\ln(2x - 1) = 5$

$e^5 = 2x - 1$

15. a. $e^{2x-1} = 217$

$(2x - 1)\ln e = \ln 217$

$2x - 1 = \ln 217$

$2x = \ln 217 + 1$

$x = \dfrac{\ln 217 + 1}{2}$

$x \approx 3.19$

b. $\log(3x - 2) + 1 = 4$

$\log(3x - 2) = 3$

$10^3 = 3x - 2$

$1000 = 3x - 2$

$1002 = 3x$

$334 = x$

$x = 334$

17. $\begin{cases} 0.7x + 1.2y - 3.2z = -32.5 \\ 1.5x - 2.7y + 0.8z = -7.5 \\ 2.8x + 1.9y - 2.1z = 1.5 \end{cases}$

$A = \begin{bmatrix} 0.7 & 1.2 & -3.2 \\ 1.5 & -2.7 & 0.8 \\ 2.8 & 1.9 & -2.1 \end{bmatrix} \quad B = \begin{bmatrix} -32.5 \\ -7.5 \\ 1.5 \end{bmatrix}$

$X = A^{-1}B$

$X = \begin{bmatrix} 5 \\ 10 \\ 15 \end{bmatrix}$

$(5, 10, 15)$

19. $x^2 + 4y^2 - 24y + 6x + 29 = 0$

$(x^2 + 6x) + 4(y^2 - 6y) = -29$

$(x+3)^2 + 4(y-3)^2 = -29 + 9 + 36$

$(x+3)^2 + 4(y-3)^2 = 16$

$\dfrac{(x+3)^2}{16} + \dfrac{(y-3)^2}{4} = 1$

Center: $(-3, 3)$

Vertices: $(-7, 3), (1, 3)$

$a = 4, \quad b = 2$

$c^2 = 16 - 4 = 12$

$c = 2\sqrt{3}$

Foci: $\left(-3 - 2\sqrt{3}, 3\right)\left(-3 + 2\sqrt{3}, 3\right)$

21. $a_1 = 52; \quad a_n = 10; \quad d = -1$

$a_n = a_1 + (n-1)d$

$10 = 52 + (n-1)(-1)$

$10 = 52 - n + 1$

$10 = 53 - n$

$-43 = -n$

$43 = n;$

$S_n = \dfrac{n(a_1 + a_n)}{2}$

$S_{43} = \dfrac{43(52 + 10)}{2} = \dfrac{43(62)}{2} = 1333$

23. $P(\text{late}) = 0.04, P(\text{on time}) = 1 - 0.04 = 0.96$

(a) $P(10) = \dbinom{12}{10}(0.04)^2(0.96)^{10}$

$\approx 0.07 = 7\%$

(b) $P(x \geq 11) = \dbinom{12}{11}(0.04)^1(0.96)^{11}$

$+ \dbinom{12}{12}(0.04)^0(0.96)^{12} \approx 0.919 = 91.9\%$

(c) $P(x \geq 10) = \dbinom{12}{10}(0.04)^2(0.96)^{10}$

$+ \dbinom{12}{11}(0.04)^1(0.96)^{11}$

$+ \dbinom{12}{12}(0.04)^0(0.96)^{12} \approx 0.989 = 98.9\%$

(d) $P(x = 10)$

$= \dbinom{12}{0}(0.04)^{12}(0.96)^0 \approx 1.7 \times 10^{-17}$

virtually nil

25. $(3, 56), (5, 126), (9, 98)$

$\begin{cases} 56 = a(3)^2 + b(3) + c \\ 126 = a(5)^2 + b(5) + c \\ 98 = a(9)^2 + b(9) + c \end{cases}$

$\begin{cases} 56 = 9a + 3b + c \\ 126 = 25a + 5b + c \\ 98 = 81a + 9b + c \end{cases}$

Using grapher:

$y = -7x^2 + 91x - 154$

(a) mid-June

(b) $y = -7x^2 + 91x - 154$

Maximum ≈ 142

(c) $y = -7(4)^2 + 91(4) - 154 = 98$

(d) November to February

Notes

Notes

Notes

Notes

Notes

Notes

Notes

Notes